Springer Tracts in Civil Engineering

Springer Tracts in Civil Engineering (STCE) publishes the latest developments in Civil Engineering - quickly, informally and in top quality. The series scope includes monographs, professional books, graduate textbooks and edited volumes, as well as outstanding PhD theses. Its goal is to cover all the main branches of civil engineering, both theoretical and applied, including:

- Construction and Structural Mechanics
- Building Materials
- Concrete, Steel and Timber Structures
- Geotechnical Engineering
- Earthquake Engineering
- Coastal Engineering; Ocean and Offshore Engineering
- Hydraulics, Hydrology and Water Resources Engineering
- Environmental Engineering and Sustainability
- Structural Health and Monitoring
- Surveying and Geographical Information Systems
- Heating, Ventilation and Air Conditioning (HVAC)
- Transportation and Traffic
- Risk Analysis
- Safety and Security

Indexed by Scopus

To submit a proposal or request further information, please contact: Pierpaolo Riva at Pierpaolo.Riva@springer.com, or Li Shen at Li.Shen@springer.com

More information about this series at http://www.springer.com/series/15088

Jianhua Tao

Numerical Simulation of Water Waves

上海交通大學出版社
SHANGHAI JIAO TONG UNIVERSITY PRESS

Springer

Jianhua Tao
Tianjin University
Tianjin, China

Translated by
Haiwen Zhang
China Institute of Water Resources
and Hydropower Research
Beijing, China

Co-translator Jianhua Tao

ISSN 2366-259X ISSN 2366-2603 (electronic)
Springer Tracts in Civil Engineering
ISBN 978-981-15-2843-9 ISBN 978-981-15-2841-5 (eBook)
https://doi.org/10.1007/978-981-15-2841-5

Jointly published with Shanghai Jiao Tong University Press
The print edition is not for sale in China. Customers from China please order the print book from: Shanghai Jiao Tong University Press.

This Springer imprint is published by the registered company Springer Nature Singapore Pte Ltd.
The registered company address is: 152 Beach Road, #21-01/04 Gateway East, Singapore 189721, Singapore

Preface

The exploitations of oceans have accompanied the civilization of humankind over thousands of years. In modern times, the needs from economic development and the means from industrialization have pushed such efforts to the levels of scale that are unimaginable in just a few decades before. Water waves, as a dominant form of water motion in ocean body, are the key dynamic loads on structures in coastal and ocean engineering. With governing scientific theories well established and vast computing power at our disposal, it is important, as well as practical, to simulate the complex waves to guide engineering practices.

The author of this book, Prof. Jianhua Tao, has devoted her entire professional life to the education and scientific research in hydraulics and computational fluid mechanics over the past 55 years at Tianjin University. As one of the pioneer scholars advancing and promoting numerical simulations of water motions, the author has successfully applied the methods to multiple waterway engineering constructions in China. In 1981, she began her advanced studies on a Dutch government scholarship at the IHE Delft Institute for Water Education where she worked with the renowned Prof. M. B. Abbott. After the completion of her studies and with the recommendation from Prof. Abbott, she went on to take a visiting scholar position at Danish Hydraulic Institute (DHI) in Demark. Twice in 1992 and 1994, invited by the United Nations Development Program (UNDP) as an Expert on Mission, she went to India for training technicians and consulting on the project "Mathematical Modeling Center for Fluvial and Ocean Hydromechanics" supported by the UNDP for India.

The contents of this book compile the research achievements of Prof. Tao and her team of Computational Fluid Dynamics (CFD) group at Tianjin University for a span of over thirty-five years, covering many focused subjects in the science community on modern wave dynamics. Chapter 1 briefs the basic concepts, methods, and engineering problems that can be approached with numerical simulations. Chapter 2 gives an introduction of water wave theories. Topics from Chaps. 3–8 cover various practiced water wave problems from hydraulic, coastal, ocean, and environmental engineering, including simulations of long waves in shallow water, shallow water waves in coastal regions, wave run-up and breaking on beaches, wave loading on structures, pollutant transport under waves and tidal currents, and coastal morphological evolution. The presentations are styled to address details of numerical simulation methods followed by case studies. Chapter 9 describes Reynold's equation for incompressible viscous fluid as a model for simulating water waves. Chapter 10 establishes virtual experimental methodologies for numerical wave flume and numerical wave basin using modern simulation methods and techniques. Chapter 11, absent in the original Chinese book, is an addition to this English edition. It includes cases of numerical wave modeling in coastal regions of China and a case of numerical prediction of thirty-year shoreline evolution in the downstream of the breakwater at the Port of Nouakchott in Mauritania. The last case is of particular interest because, at the time of this writing, the construction has elapsed twenty-four years in time, which presents data (satellite images) for confirming our numerical predictions.

This treatise puts emphasis on linking theories to applications, by presenting a large amount of domestic and international real engineering cases to demonstrate the viabilities of numerical simulation theories and methods in hydraulic, coastal, ocean, and environmental engineering. Comparisons of numerical results with experimental data or observations are included as much as possible. The aim of the author is to present this book as a reference for anyone whose profession is teacher, researcher, or practice engineer in fields involving water wave dynamics.

The first translator of this book, Dr. Haiwen Zhang, is a former member of CFD group at Tianjin University, where she received her master's degree. In 2001, she visited Hong Kong University for a short term as a research assistant. She went on to Denmark to study for her doctorate at Technical University of Denmark with Ph.D. scholarship financed by DHI Water and Environment. After receiving her Ph.D. in Coastal, Maritime, and Structural Engineering in 2006, she worked at COWI A/S in Denmark, the headquarters of COWI Group, as a coastal and hydraulic modeling specialist for seven years. In 2013, she returned to China to join Chinese Institute of Water Resources and Hydropower Research (IWHR). She is currently engaged in coastal and hydraulic modeling studies as a senior staff member.

During the laborious process of translating this book, many former and current members of CFD group at Tianjin University have provided their help. Dr. Yan Wu (in UK), Dr. Zeliang Wang (in Canada), Dr. Wen Long (in USA) have carefully proofread the English translations of Chaps. 1–4, Chaps. 5–7 and Chaps. 8–10, respectively. Professor Dekui Yuan, Dr. Changgen Liu, Dr. Hongtao Nie, and Dr. Jian Sun are involved in writing the Chinese subscript of Chap. 11. Dr. Xinyu Dou (in USA) has carefully proofread the English translation of Chap. 11. It is very appropriate to state that this English edition is a dedication from the team of CFD group at Tianjin University to its readers. We hereby express sincere gratitude for their contributions.

We also wish to extend gratefulness to IWHR for its great support. Furthermore, thanks are also due to Shanghai Jiaotong University Press and Springer for making this book available to readers beyond China. Finally, any comments from readers would be appreciated.

Beijing, China Translators

Contents

Chapter 1
Introduction

1.1 Numerical Simulation of Fluid Flow

Fluid flow is one of the most common phenomena in nature, engineering, and industry. Flow of the air, movements of flying vehicles in the air, water flow in rivers, lakes, and oceans, changes of the flow field as results of constructions of water infrastructures and hydraulic structures in coastal, marine, and environmental engineering are the most general and common fields in natural science. In the past, the main approaches to study fluid flow were mathematically analytical methods and laboratory experiments. Since the 1940s, a new approach to studying fluid flow, i.e., numerical simulation, has become available following the emergence and rapid development of computer.

This book will focus on various unsteady gravity flows with free surface in nature and engineering, i.e., water waves, and introduce the theories, methodologies, and case studies on numerical simulation.

1.1.1 What Is Numerical Simulation of Fluid Flow

Numerical simulation of fluid flow is an approach to study the phenomena and mechanics of fluid flow based on numerical models that describe fluid flow and to obtain numerical solutions using computers.

The so-called numerical solutions are approximate solutions in discrete form. The accuracy of the approximation relies on the accuracy of the numerical methods. Numerical simulations have no limitation of the scale, and it is much flexibility in adjusting initial and boundary conditions, and it is possible to simulate flow for a relatively long period. When the computer capacity is large enough and computer speed is fast enough, the model extent can be very large. The biggest advantage

J. Tao, *Numerical Simulation of Water Waves*, Springer Tracts in Civil Engineering, https://doi.org/10.1007/978-981-15-2841-5_1

of numerical simulations is its capability of comparing various states or scenarios easily, which can save time, material resources, and manpower significantly.

1.1.2 Contents of Numerical Simulation

1. Understanding and analysis of the mechanism of physical phenomena

In order to study a complex flow using numerical simulation method, analysis of the mechanism of the physical problem should be carried out first. It is not always to be able to build a complete and accurate numerical model to describe a complex flow phenomenon. But if the fundamental nature and the main physical phenomena have been captured, it is possible to establish a numerical model that can solve the physical problem. For example, it is not necessary to use the Navier–Stokes equations or the Reynolds equations for the water wave study. It is possible to establish models using various water wave theories based on different parameters and characteristics of waves. The accuracy of the numerical model has been improved as the researches on physical problems become deeper and deeper. For example, the studies on turbulence make the turbulence models more reasonable and the numerical models on turbulence more accurate.

2. Setup of numerical model to describe physical phenomenon

Numerical models comprise differential equations and well-posed initial and boundary conditions. The main equations to describe the fluid flow are various partial differential equations. The three general partial differential equations, i.e., hyperbolic equations, parabolic equations, and elliptic equations, are all used in the numerical simulation of water waves. These equations have different strict demands for the corresponding well-posed initial and boundary conditions.

The so-called well-posed means that the initial and boundary conditions make unique solutions of the partial differential equations exist. Also, continuous variation of the solution is response to the changes of initial and boundary conditions. Therefore, when a numerical model describing a fluid problem is set up, the first step is to study the fundamental nature of the physical phenomenon and establish governing equations which can describe the physical problem correctly. Next, it is to set up the well-posed initial and boundary conditions before carrying out the numerical simulations.

3. Selection of numerical method and numerical techniques

A numerical model describing the physical phenomenon can be solved by computer with numerical methods after it is established. Since the normal computers are not able to solve the differential equations directly but only carry on algebraic operations, all sorts of numerical methods are to get numerical solutions by turning the differential equations into a series of algebraic equations, mainly linear algebraic equations, and then solving them. A common feature of various numerical methods is to discretize the model domain into a certain number of cells first and then discretize the governing equations and solve the unknown values in the cells. Commonly used numerical methods for fluid flow include finite difference method, finite volume method, finite element method, boundary element method, and spectral method.

Apart from numerical methods, mesh generation techniques, boundary treatment techniques, and solution techniques of discrete equations for incompressible flow are also important to accomplish the numerical simulation on fluid flow, particularly for numerical simulation of water waves.

4. Calibration and verification of numerical model

Calibration and verification are needed when building numerical models. Calibration of the numerical model is the process that calibrates some model parameters such as local bottom resistance, wind friction coefficients, and dispersion coefficients, by comparing model results against field measurements or observations in laboratories. Verification of the numerical model is the process that checks whether the calculated solution of the model is acceptable by comparing model results against specific theoretical solutions, field measurements, or observations in laboratories that are different to the ones used in the model calibration. The model can be used for applications only after the completion of model calibration and verification.

5. Graphical display and pre- and post-processing

The numerical solutions given by numerical simulations are generally in discrete forms, such as flow speed, pressure or surface elevation at each spatial point at each time step. Using computer graphical display techniques to represent the computational results with contour plots or 3D plots is an essential part of numerical simulations which could give users an overview of the simulation results.

Pre- and post-processing are the exercises of preparing model input and analyzing model output, such as building the database, generating model bathymetries and structures, collecting and exporting data.

1.1.3 Purpose of Numerical Simulation in Engineering

1. Numerical simulation used as a study tool

Numerical simulations can be deemed as doing tests on computer. For example, in the study of wave forces on a semicircular breakwater, for a long time, the problem of submerged breakwater was solved by potential flow theory as the semicircular breakwater in wave field can be considered as local phenomenon. However, through a large number of comparisons between numerical and physical tests, it was found that calculation using potential flow theory did not match with measurements in many cases. When viscous dissipation term is included in the numerical model, the simulated and measured results match well (see Sect. 6.6). This indicates that the viscosity should be considered in some local water wave phenomena. Another example is about Zhuhai Port. After hit by a typhoon in 1998, the revetment at Pingpai Hill was damaged seriously with a great loss. From the observation, the incoming wave direction was almost parallel to the revetment. What makes the waves turn and hit straight toward the revetment? Was this accidental or inevitable? By means of a large number of numerical tests, it was found that this accident was caused by multiple wave reflections in the main navigation channel and the port (see Sect. 10.3). Therefore, using numerical simulations on computer can reveal the mechanism of some unclear flow phenomena. It is a fast, economical, and effective tool for studies.

2. Numerical simulation used as a design tool

A correct scientific design should be made by comparing various scenarios. For instance, for the design of a harbor layout, comparisons among a large number of scenarios should be carried out in order to obtain an optimal layout, including the direction of navigation channels, the direction and length of breakwater, and the location of the port. But in the early phase of the project, it is not allowed to do lots of physical tests due to the limitation of time and budget. Therefore, the best way is to undertake numerical simulations using various mathematical models for different complexity at different design phases. For example, in the primary planning phase of a project, a CAD system of a single point-source model for simple wave diffraction and its drawing software can be used to calculate the wave height distribution in the harbor quickly and adjust the direction and length of breakwaters on computer in time. In the preliminary design phase, a numerical model of refraction and diffraction can be used to calculate the wave refraction and diffraction in the harbor to determine the wave height distribution in and out of the harbor and determine the design wave height for the structures designed. In structure design and construction phase, 2D or 3D viscous flow model can be used to determine wave loads on the structures, surrounding wave field and current field, and to forecast whether local scour and significant impact on ambient conditions occur after the construction of the structures.

3. Numerical simulation used as a decision-making tool

With the development of society, for any economic development activities and engineering constructions, investigations of their impact on the ecosystem and environment are needed. Comprehensive analyses of multifactor, multilevel, and multidiscipline are required in order to establish a decision-making system or expert system. Within a decision-making system of hydraulics, coastal engineering, ocean engineering, and environmental engineering, one of the core modules is the mathematical model for predicting the change of water flow, water waves, and sediment movement caused by a development. For instance, reclamation can change the distribution of tidal current, wave, sediment and plankton distribution in coastal water. With simulations of these distribution changes, it is possible to make comprehensive evaluation of rationality and environment impact and conduct countermeasures study.

Numerical simulation of water flow is a very important part for watershed management, and coastal zone and marine information systems.

1.2 Water Waves in Engineering and Classifications of Water Waves

1.2.1 Water Waves in Engineering

1. Flood wave and tidal wave in river, estuary, and coastal area

Flood wave and tidal wave are kinds of half-period wave or long period wave. Normally, they are in the form of gradually varied flow. But in case of dam-break, massive landslides around the river, lake, and coastal water, or tidal bore in estuary, discontinuous waves occur in mutation form. Numerical simulation of water waves is an important means for river flood control and management, storm surge prediction, and prediction of tidal bore and dam-break.

2. Propagation and transformation of water waves from deep sea to inshore shallow water

Coastal engineering works are mostly in shallow water areas. Wave processes such as shoaling, refraction, diffraction, and reflection due to complex bathymetry and structures are very important for marine engineering in coastal areas. In shallow water, the wave nonlinearity is relatively high and the waveforms are complex. Therefore, it is quite difficult to simulate wave propagation and deformation in coastal shallow water using numerical model. Wave propagation and deformation in coastal shallow water is also an important content of design, construction, and management in coastal and offshore engineering.

3. Surf zone and nearshore current

Coastal regions as the intersection of land and sea are extremely dynamic, especially the coast with gentle slope. The surf zone is generated by wave breaking, in which movements of sediment and pollutants are active. Simulations of nearshore current and the pattern of sediment movement in surf zone are important to the development of coastal regions.

4. Interaction between water waves and structures

Wave load is an important design load in both coastal and offshore engineering. The effect of wave and current on various engineering structures and the influence of engineering structures on wave field and current field as well as on the ambient environment are major concerns.

5. Sediment transport and coastal morphological evolution caused by waves and current

In nature, sediment transport and coastal morphology are in equilibrium in the long-term action of wave and current. When the development of coastal area is made such as construction of ports, terminals, artificial islands, reclamation, and offshore oil production engineering, the previous equilibrium state might be broken. It is very important to predict coastal morphological evolution and the conditions for the new equilibrium state due to the projects, which are directly related to whether the coastal developments are reasonable and successful.

6. Nearshore pollutant distribution and ecological environment change caused by waves and current

Flow in coastal regions is complex, especially in surf zone, where various pollutants have specific movement patterns. Currently, offshore discharge of urban sewage is inevitable. Simulations of pollutant transport in coastal regions are very important for the design of wastewater marine disposal projects and analysis on the impact of urban sewage on the ocean and ecological environment.

1.2.2 Classification of Water Waves and Wave Theories

Although currently there are lots of high-accuracy numerical methods and powerful computational tools—high performance computers, which can perform direct numerical simulations of fluid flow in various conditions, for such complex hydrodynamic phenomena as water waves, we have to use various water wave theories derived through mathematical analysis to make the simulations more simplified and more effective. All kinds of water wave theories are based on certain assumptions and based on different parameters and characteristics of water waves. Only when the characteristics of the waves and corresponding wave theories are deeply understood, a numerical model correctly describing water waves can be established, and then effective numerical simulations can be carried out.

Figure 1.1 shows definition of wave surface and basic wave parameters in the coordinate system of (x, y, z), where H is wave height, L is wave length, and d is still water depth. These are the basic parameters to describe water waves. Wave period T is defined as the time spent from the starting position of the water surface moved back to the same position when observed at a fixed point. c is the speed of wave propagation, $c = L/T$.

Below are some general dimensionless parameters to describe water wave characteristics.

Relative water depth d/L, dimensionless number d/gT^2, relative wave height H/d, wave steepness H/L, and dimensionless number H/gT^2. For the waves with high

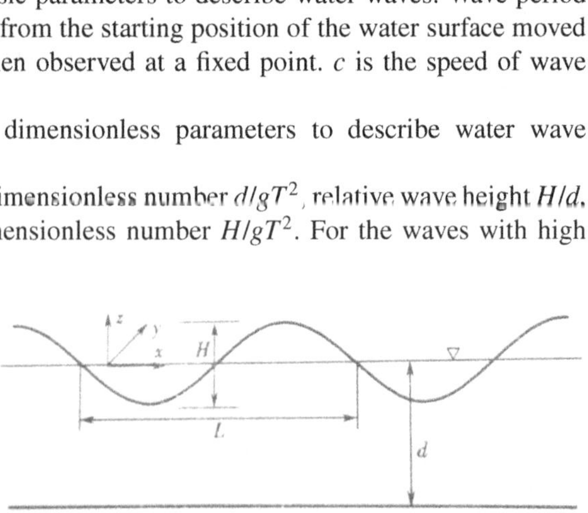

Fig. 1.1 Definition sketch of wave free surface and basic parameters

wave steepness in shallow water, Ursell number of $\mathrm{Ur} = HL^2/d^3 = H/L(L/d)^3$ is quite common to describe water wave characteristics.

Table 1.1 presents classification of water waves and corresponding wave theories.

1.3 Numerical Methods and Techniques of Water Wave Simulation

1.3.1 Finite Difference Method

1. Basic concepts

The fundamental idea of finite difference method is to form discretization of the differential equations by substituting the derivatives with finite differences.

One example is to solve the initial value problem for a unidirectional wave equation

$$\left.\begin{array}{l} \frac{\partial u}{\partial t} + c\frac{\partial u}{\partial x} = 0 \\ u(0, x) = \phi(x) \end{array}\right\} \tag{1.3.1}$$

Substituting the derivatives in Eq. (1.3.1) with finite differences at the n time step and j point,

using forward differentiation formula for $\frac{\partial u}{\partial t}$ to get

$$\frac{\partial u}{\partial t} = \frac{u_j^{n+1} - u_j^n}{\Delta t} - \frac{\Delta t}{2!}\left(\frac{\partial^2 u}{\partial t^2}\right) + \cdots = \frac{u_j^{n+1} - u_j^n}{\Delta t} + O(\Delta t) \tag{1.3.2}$$

using backward differentiation formula for $\frac{\partial u}{\partial x}$ to have

$$\frac{\partial u}{\partial x} = \frac{u_j^n - u_{j-1}^n}{\Delta x} - \frac{\Delta x}{2!}\left(\frac{\partial^2 u}{\partial t^2}\right) + \cdots = \frac{u_j^n - u_{j-1}^n}{\Delta x} + O(\Delta x) \tag{1.3.3}$$

The original differential equation can be represented as

$$\frac{u_j^{n+1} - u_j^n}{\Delta t} + c\frac{u_j^n - u_{j-1}^n}{\Delta x} = O(\Delta t, \Delta x) \tag{1.3.4}$$

The approximate difference equation is expressed as

$$\frac{u_j^{n+1} - u_j^n}{\Delta t} + c\frac{u_j^n - u_{j-1}^n}{\Delta x} = 0 \tag{1.3.5}$$

Table 1.1 Classification of water waves and corresponding wave theories

Water wave feature		Water wave parameters			Water wave characteristics				Corresponding wave theories
		Relative wave height H/d	Ursell No. $H/L\ (L/d)^3$	Relative water depth d/L	Pressure distribution in vertical	Viscosity	Particle motion	Mass transmission	
Hyperbolic wave	Linear wave	≪1	≪1	<0.05	Hydrostatic pressure	Irrotational, inviscid	Oscillation	Without	Linear theory—Airy theory
	Nonlinear wave	>1	≫1	≪0.05	Hydrostatic pressure	With bottom friction	Transmission	With	Nonlinear theory—Saint Venant theory
Dispersion wave	Linear wave	≪1	≪1	0.03–0.5	Non-hydrostatic pressure	Irrotational, inviscid	Oscillation	Without	Linear small-amplitude wave—Airy theory
	Nonlinear wave	≪1	<10	0.05–0.5	Non-hydrostatic pressure	Irrotational, inviscid	Oscillation	A small amount	Finite-amplitude wave—Stokes theory
		≪1	<10	>0.5	Non-hydrostatic pressure	Rotational, inviscid	Oscillation	Without	Deepwater finite-amplitude wave—Gerstner theory
		<1	1	<0.1	Non-hydrostatic pressure	With bottom friction	Transmission	With	Shallow water nonlinear dispersion wave–Boussinesq theory

(continued)

Table 1.1 (continued)

Water wave feature		Water wave parameters			Water wave characteristics				Corresponding wave theories
		Relative wave height H/d	Ursell No. $H/L\,(L/d)^3$	Relative water depth d/L	Pressure distribution in vertical	Viscosity	Particle motion	Mass transmission	
		<1	1	<0.1	Hydrostatic pressure (first order)	Irrotational, inviscid	Transmission	With	Unidirectional shallow water nonlinear wave—Solitary theory
		<1	1	<0.1	Hydrostatic pressure (first order)	Irrotational, inviscid	Oscillation	A small amount	Shallow water nonlinear period wave—Cnoidal theory

In which, H is wave height, L is wave length, and d is still water depth where Δt denotes time step from step n to $n+1$, Δx denotes space interval from grid point j to $j+1$. $O(\cdot)$ denotes the order of magnitude. $O(\Delta x)$ is the first order of Δx. $TE = O(\Delta t, \Delta x)$ denotes truncation error of difference equation, which is the difference between the differential equation and the difference equation. Here, TE is in the first order of Δx and Δt.

Equation (1.3.4) is in FTBS scheme, which means using forward scheme for time derivative and backward scheme for spatial derivative. This scheme is in the first order of Δx and Δt.

2. Relationship between difference equations and differential equations

In mathematics, an effective difference equation requires the following properties.

(1) Consistency

The difference equation converges to the differential equation if the truncation error of the difference equation tends to zero when both space interval and time step tend to zero. These difference equations and differential equations are consistent. Or it is said that the difference schemes have consistency.

(2) Convergence

For a consistent difference equation, if solution of the difference equation converges to solution of differential equation when both space interval and time step tend to zero, it is called that the difference schemes are convergent.

(3) Stability

When using difference equation to solve the initial value problems, if the calculation error (such as rounding error) can be effectively restrained in the calculation process so that the error disturbance occurred at a certain time step cannot increase substantially to mask the real solution, then the difference scheme is stable or stable in certain conditions. Otherwise, it is unstable.

Lax theorem indicates that for a well-posed fixed solution problem of linear differential equation, when the difference scheme has consistency, the convergence of the scheme is the necessary and sufficient condition for stability, i.e., convergence and stability are equivalent.

3. Validity of difference equations

Apart from the above mathematical requirements for a difference equation, validity in application is also needed. It means that the pseudo-physical effects are limited within a permissible range. Mainly, there are two types of pseudo-physical effects as described below.

Numerical dissipation:

Dissipation caused by the even-order derivatives in the truncation error is called numerical dissipation. For instance, the initial value problem for a unidirectional wave equation in Eq. (1.3.1) describes a unidirectional permanent wave. Physically, there is no dissipation. If using difference equation in Eq. (1.3.5) to approximate it, it is found from Eqs. (1.3.2) and (1.3.3) that the second-order derivatives are included in the truncation error. The second-order derivatives in the equations are dissipative terms. When using Eq. (1.3.5) to calculate, large dissipation could happen if the parameters are not selected properly so that the approximation for the original differential equation distorts.

Numerical dispersion:

Odd-order derivatives (above the third order) in differential equations are dispersive items. If odd-order derivatives exist in the truncation error of difference equation, numerical dispersion will occur. Equation (1.3.1) is permanent wave equation without dissipative term and dispersive item. If using the first-order scheme in Eq. (1.3.5) to compute, both numerical dissipation and numerical dispersion will occur.

Validity of a difference equation is to restrict the numerical dissipation and the numerical dispersion caused by the difference scheme within a small permissible range. Otherwise, the difference equation is not valid because of the high pseudo-physical effects even though it meets the requirements of consistency, stability, and convergence.

For detailed introduction on difference method, see Refs. [1, 2] in this chapter.

1.3.2 Finite Volume Method

Finite difference method is to establish discrete equation based on the differential equation describing fluid movement, while finite volume method is to establish discrete equation based on conservation equation in terms of integral forms.

The differential equation of 3D convection and diffusion equation in conservation form is the following,

$$\frac{\partial(\rho\Phi)}{\partial t}+\frac{\partial(\rho u\Phi)}{\partial x}+\frac{\partial(\rho v\Phi)}{\partial y}+\frac{\partial(\rho w\Phi)}{\partial z}=\frac{\partial}{\partial x}\left(k\frac{\partial\Phi}{\partial x}\right)+ \frac{\partial}{\partial y}\left(k\frac{\partial\Phi}{\partial y}\right)+\frac{\partial}{\partial z}\left(k\frac{\partial\Phi}{\partial z}\right)+S_\phi \tag{1.3.6}$$

where Φ is material function, such as temperature, concentration; k is dispersion coefficient; S_ϕ is the source item.

The above equation can be written in divergence and gradient form as the following,

$$\frac{\partial}{\partial t}(\rho\Phi)+\mathrm{div}(\rho u\Phi)=\mathrm{div}(k\,\mathrm{grad}\Phi)+S_\phi \tag{1.3.7}$$

Integrating Eq. (1.3.7) within time step Δt in control volume (CV) yields

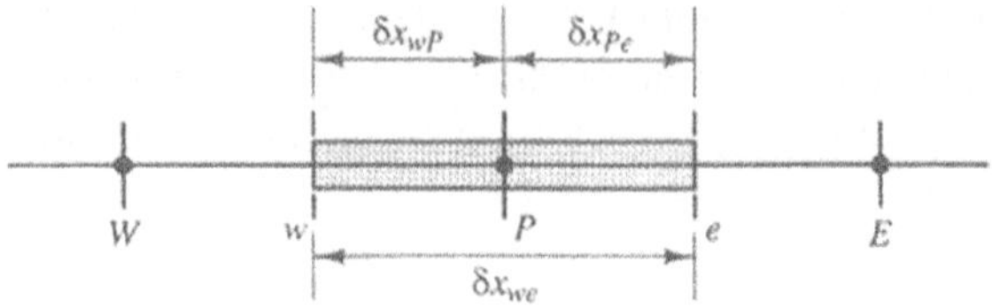

Fig. 1.2 Sketch of element of finite volume

$$\int_{C_V}\left(\int_t^{t+\Delta t}\frac{\partial}{\partial t}(\rho\phi)\mathrm{d}t\right)\mathrm{d}V+\int_t^{t+\Delta t}\left(\int_A n\cdot(\rho u\phi)\mathrm{d}A\right)\mathrm{d}t$$
$$=\int_t^{t+\Delta t}\left(\int_A n\cdot(K grad\phi)\mathrm{d}A\right)\mathrm{d}t+\int_t^{t+\Delta t}\int_{C_V}S_\phi\mathrm{d}V\mathrm{d}t \tag{1.3.8}$$

in which, divergence integral has been turned to area integral using Green formula. A is the surface area of the control volume.

The physical meaning of the equation is that the change of $\rho\Phi$ in the volume of CV in time period of Δt plus the amount of convection $\rho u\Phi$ through the surface of the control volume in Δt is equal to the amount of diffusion through the surface of the control volume in Δt plus the change of source item in the volume of CV in Δt.

For instance, 1D differential equation of unsteady thermal diffusion equation is given as

$$\rho c\frac{\partial T}{\partial t}=\frac{\partial}{\partial x}\left(k\frac{\partial T}{\partial t}\right)+\boldsymbol{S} \tag{1.3.9}$$

Figure 1.2 shows a sketch of element of finite volume.

Integrating Eq. (1.3.9) within time step Δt in Control volume (CV) yields

$$\int_t^{t+\Delta}\int_{CV}\rho c\frac{\partial T}{\partial t}\mathrm{d}V\mathrm{d}t=\int_t^{t+\Delta}\int_{CV}\frac{\partial}{\partial}\left(k\frac{\partial T}{\partial x}\right)\mathrm{d}V\mathrm{d}t+\int_t^{t+\Delta}\int_{CV}s\mathrm{d}V\mathrm{d}t \tag{1.3.10}$$

which can be written as

$$\int_w^e\left[\int_t^{t+\Delta t}\rho c\frac{\partial T}{\partial t}\mathrm{d}t\right]\mathrm{d}V=\int_t^{t+\Delta t}\left[\left(kA\frac{\partial T}{\partial x}\right)_e-\left(kA\frac{\partial T}{\partial x}\right)_w\right]\mathrm{d}t+\int_t^{t+\Delta t}\bar{S}\Delta V\mathrm{d}t \tag{1.3.11}$$

In Eq. (1.3.11), A is the area of the control volume; ΔV is the volume value; $\Delta V=A\Delta x$, Δx is the width of the control volume; $\bar{S}$ is the average source intensity in the control volume. Assuming at point P, the temperature at time of t is T_p^0, and the temperature at $t+\Delta t$ is T_P. Then Eq. (1.3.11) can be changed to

$$\rho c\left(T_P - T_P^0\right)\Delta V = \int\limits_t^{t+\Delta t}\left[k_e A\frac{T_E - T_P}{\delta x_{PE}} - k_w A\frac{T_P - T_W}{\delta x_{WP}}\right]\mathrm{d}t + \int\limits_t^{t+\Delta t}\bar{S}\Delta V\mathrm{d}t \tag{1.3.12}$$

In order to calculate the integration of T_P, T_E and T_w on the right-hand side of the above equation, a weight number $\theta \in [0, 1]$ is introduced. The integration is represented by linear relationship within t and $t + \Delta t$

$$I_T = \int\limits_t^{t+\Delta t} T_P\mathrm{d}t = \left[\theta T_P + (1-\theta)T_P^0\right]\Delta t \tag{1.3.13}$$

Then Eq. (1.3.12) can be written as

$$\rho c\left(\frac{T_P - T_P^0}{\Delta t}\right)\Delta x = \theta\left[\frac{k_e(T_E - T_P)}{\delta x_{PE}} - \frac{k_w(T_P - T_W)}{\delta x_{WP}}\right] + (1-\theta)\left[\frac{k_e(T_E^0 - T_P^0)}{\delta x_{PE}} - \frac{k_w(T_P^0 - T_W^0)}{\delta x_{WP}}\right] + \bar{S}\Delta x \tag{1.3.14}$$

The temperature T_P^0 at t time in the second item on the right side of the above equation is known. Thus this equation gives a relation among T_P, T_E and T_w at time of $t + \Delta t$. The equations for all the neighbored three nodes in the domain form a system of linear algebraic equations, from which all the unknown variables in the domain can be solved when the boundary conditions are given.

Because the basic law of fluid flow is conservation law and the discrete form of finite volume is also conservative, the method of finite volume is applied widely in calculation of fluid motions. The reference of [3] gives the details on finite volume method.

1.3.3 Finite Element Method

1. Concept of trial function method

Assuming the variable in the study of boundary value problem is $u(x, y)$, the differential equation to describe this variable normally can be written as

$$F(u) = G(u) \tag{1.3.15}$$

in which, F is differential operator, and G is a function of u. For instance, Poisson equation $\nabla^2 u = f$, here $F(u) = \nabla^2 u$, $G(u) = f$. The so-called trial function method is using a linear polynomial expression to approximate the solution of Eq. (1.3.15) u_M and using linear polynomial expression to approximate it,

$$u_M = \sum_{i=1}^{M} c_i \Phi_i \quad i = 1, 2, \ldots, M \tag{1.3.16}$$

where Φ_i are the selected linearly independent functions called basis functions which should meet boundary condition; c_i are undetermined coefficients. If M equations can be established to solve the undetermined coefficients $c_i (i = 1, 2, \ldots, M)$, a trail solution can be determined. There are two methods to establish M equations for solving the undetermined coefficients. They are calculus of variations and residual method. It is obvious that the approximation of trial solution u_M to the solution of differential equation depends on the selection of basis function and the method for determination of undetermined coefficients c_i.

The first step of calculus of variations to solve differential equation is to seek a functional $I(u)$ which corresponds to the differential equation and the boundary conditions. It can be proved that the solution of the differential equation is the function $u(x, y)$ which makes the functional $I(u)$ minimum [4]. Substituting the trial solution of (1.3.16) into the functional and finding the extreme value yields

$$\frac{\partial I(U_M)}{\partial c_i}_i = 0 \quad i = 1, 2, \ldots, M \tag{1.3.17}$$

Based on the above M equations, M coefficients c_i can be determined.

The residual method to solve differential equation is to substitute the trial solution into Eq. (1.3.15), to get the non-zero residual R of the approximation to the solution.

$$R = F(u_M) - G(u_M) \tag{1.3.18}$$

This residual is required to be minimized. The minimization criterion can be generally formulated as

$$\int_{\Omega} w_i R \mathrm{d}\Omega = 0 \quad i = 1, 2, \ldots, M \tag{1.3.19}$$

in which, the integration is applied for the computational domain; w_i are weight functions. Using M weight functions can get M integration expressions in order to find M undetermined coefficients c_i. Choosing different types of weight functions can get different methods, such as collocation method, moment method, Galerkin's method, and least square method. Galerkin's method has been commonly applied for calculation. The basis functions in the trial function are chosen for weight functions in Galerkin's method. $w_i = \phi_i, i = 1, 2, \ldots, M$.

2. Basic concept of finite element method

The difficulty of using trial function to solve the problem is the selection of series of linearly independent functions or basis functions ϕ_i. Especially when the boundary conditions are complicated, it is very difficult to make the functions meet the requirement of boundary conditions. Therefore, although variation method and residual method have been established for many years, they are not been used widely for the concrete application. The most important breakthrough of finite element method is to discretize the computational domain into a large number of small elements, to suppose trial solutions in small elements so that the boundary conditions can be easily met.

Using element vertex functions u_k to express the trial solution at any point (x, y) in each small element,

$$u_M = \sum_{k=1}^{n} N_k(x, y)u_k \quad k = 1, 2, 3, \ldots \tag{1.3.20}$$

in which, n is the node number in the element; N_k are interpolation method related (linear or nonlinear) functions, i.e., the basis functions in Eq. (1.3.16); u_k are undetermined coefficients, i.e., the unknown functions to be solved at the nodes, which are also undetermined coefficients in Eq. (1.3.16). When the element is triangular, $n = 3$, Eq. (1.3.20) can be seen as expression of trial functions, in which the basis functions N_k are linearly independent and meet the boundary conditions. Establishing three equations (using variation principle or Galerkin's method) at each triangle in the domain, the linear algebraic equations for the unknowns at all the nodes in the whole domain can be obtained. From these equations and the boundary conditions, the unknown variable u at each node in the whole domain can be solved.

For detailed introduction on finite element method based on variation principle, see Ref. [4], and finite element method based on Galerkin's method, see Ref. [2].

3. Solution procedures of finite element method

(1) Discretizing the computational domain into n elements. They are a series of line segments for one-dimensional problem, triangular or quadrangular planar region for two-dimensional problem, and blocks for three-dimensional problem. The size of elements can be different. But the elements cannot be coincident except for the points at the boundaries. Also, the union of all the elements should cover the whole computational domain.
(2) Using appropriate interpolation function to express the solution in the element by the function to be solved at each node, such as Eq. (1.3.20). The interpolation function can be linear or high-ordered. Therefore, solving the problem in the computational domain becomes solving it at the nodes in the sub-domains.
(3) Establishing M equations by means of variation principle or Galerkin formulations to determine the solution at the nodes and at any point in the element.

The characteristics of finite element method are that elements can be discretized arbitrarily, which is suitable for the case with complex boundary conditions. The elements can be small for high resolution in the area of interest. The inter-influences of elements are confined to the neighbored elements. Generally, the final obtained coefficient matrices are sparse matrices which can reduce computation time.

1.3.4 Boundary Element Method

Boundary element method has been developed from finite elementary solutions or boundary integral equation methods. It is also one type of trial function method.

For example, in the case of potential problem,

$$\nabla^2 u = 0, \quad \text{in the domain of } \Omega \tag{1.3.21}$$

$$u = \bar{u} \quad \text{on the boundary of } \Gamma_1 \tag{1.3.22}$$

$$\frac{\partial u}{\partial n} = \bar{q} \quad \text{on the boundary of } \Gamma_2 \tag{1.3.23}$$

where $\bar{u}$ and $\bar{q}$ are known functions; $\Gamma = \Gamma_1 \cup \Gamma_2$ is the boundary of the domain Ω.

The basic idea of boundary integral equation method is to satisfy the equation in the domain and not to strictly meet the boundary conditions [5]. Supposing u can be expressed by a set of complete linearly independent basis functions of $\{\Phi_i\}$ for a trial solution

$$u_M = \sum_{i=1}^{N} c_i \Phi_i \tag{1.3.24}$$

Substituting Eq. (1.3.24) into Eq. (1.3.21) to Eq. (1.3.23), in general some residuals exist

$$\varepsilon = \sum_{i=1}^{N} c_i \nabla^2 \Phi_i(x, y) \neq 0, \quad \text{in the domain} \tag{1.3.25}$$

$$\varepsilon_1 = u_M - \bar{u} = \sum_{i=1}^{N} c_i \Phi_i - \bar{u} \neq 0, \quad \text{on the boundary of } \Gamma_1 \tag{1.3.26}$$

$$\varepsilon_2 = \frac{\partial u_M}{\partial n} - \bar{q} = \sum_{i=1}^{N} c_i \frac{\partial \Phi_i}{\partial n} - \bar{q} \neq 0, \quad \text{on the boundary of } \Gamma_2 \tag{1.3.27}$$

The residual expression can be written as [2]

$$\iint \varepsilon w \mathrm{d}\Omega = \int_{\Gamma 1} -(u - \bar{u}) \frac{\partial w}{\partial n} \mathrm{d}\Gamma + \int_{\Gamma 2} (q - \bar{q}) w \mathrm{d}\Gamma \tag{1.3.28}$$

where w is weight function; ε is the residual in the domain $\varepsilon = \nabla^2 u_M$; the first item on the right side is the residual on the boundary of Γ_1, and the second item is the residual on the boundary of Γ_2.

Applying integration by parts twice on Eq. (1.3.28) yields

$$\iint \nabla^2 wu\mathrm{d}\Omega = -\int_{\Gamma 1} \frac{\partial u}{\partial n} w\mathrm{d}\Gamma + \iint_{\Gamma 2} u\frac{\partial w}{\partial n}\mathrm{d}\Gamma \tag{1.3.29}$$

If choosing the weight function w as the fundamental solution u^*, the fundamental solution of Laplace equation satisfies the following

$$\nabla^2 u^* = \delta(X - \xi) \tag{1.3.30}$$

where δ is Dirac-δ function.

Substituting the fundamental solution as weight function into Eq. (1.3.29), and using S to denote the boundary line or surface, the boundary integral equation turns into

$$c_i u_i + \int \frac{\partial u}{\partial n} u^* \mathrm{d}S = u \int \frac{\partial u^*}{\partial n}\mathrm{d}S \tag{1.3.31}$$

$$c_i = \begin{cases} 1, & \text{in the domain} \\ \frac{1}{2}, & \text{on the smooth boundary} \\ 0 & \text{outside of the domain} \end{cases}$$

Equation (1.3.31) is called boundary integral equation. For fundamental solutions of various differential equations, see Ref. [2] or mathematics handbooks.

The basic idea of boundary element is to discretize the boundary of computational domain into n elements. The boundary elements are a series of line segments for 2D computational domain and planes or surfaces for 3D computational domain. Equation (1.3.31) can be written in discrete form as

$$c_i u_i + \sum_{j=1}^{n} \int_{S_j} u\frac{\partial u^*}{\partial n}\mathrm{d}S = \sum_{j=1}^{n} \int_{S_j} u^* \frac{\partial u}{\partial n}\mathrm{d}S \tag{1.3.32}$$

The approach to solve the above equation can be found in Reference [5]. The advantage of boundary element method is to reduce the dimension of the problem by one. Green function method is also called indirect boundary element method. The basic formulation is

$$u_i = \int_S \sigma G\mathrm{d}S \tag{1.3.33}$$

$$\frac{1}{2}\sigma_i + \int_S \sigma\frac{\partial G}{\partial n}\mathrm{d}S = \left(\frac{\partial u}{\partial n}\right)_i \tag{1.3.34}$$

where σ is the source intensity; G is Green function; $\left(\frac{\partial u}{\partial n}\right)_i$ is the known natural boundary condition of the object plane. Normally, G is the fundamental solution of the equation, also called Green function. S is the whole boundary of the computational domain. The integrations of Eqs. (1.3.33) and (1.3.34) must be carried out on all boundaries. When using indirect boundary element method, Green function G is determined first. The source intensity σ in each element is solved by Eq. (1.3.34). Then the expected function u_i in each element can be solved by Eq. (1.3.33).

In order to reduce the workload of integration, Green function could be sought for partly meeting the boundary conditions, such as free surface boundary condition, radiation condition, and bottom boundary condition, so integration is only needed to be carried out on the object plane. Here, Green function is not fundamental solution but called complex Green function. Examples of derivation of complex Green function can be found in Ref. [6].

1.3.5 Numerical Techniques of Water Wave Simulation

Numerical techniques of water wave simulation include mesh generation techniques, various boundary treatment techniques, solution techniques of incompressible flow discrete equation, visualization techniques, and pre- and post-processing techniques. Most of the above numerical techniques will be introduced in this book.

1. Mesh generation techniques

The discretization of the computational domain is an important step in the simulation of fluid flow. Using simple rectangular grids cannot meet the requirement for the simulation of fluid flow with complex boundary conditions. One of the reasons is the poor approximation at the boundaries. Another reason is that, because the gradient at each position in the flow field is different, the resolution at each position is not consistent when using the uniform rectangular grids for calculation.

Mesh generation consists of structured mesh and unstructured mesh. Structured mesh is generated by using algebraic equations or differential equations. Normally, it is body-fitted mesh on the physical plane and turns the complex region into rectangular computational domain. It can also be refined locally. The structure of this mesh is fixed and the node structure is restricted. Unstructured mesh is generated with arbitrary structure by following certain rules. It is more effective in boundary fitting and local mesh refinement.

2. Boundary treatment techniques

The fixed solution of differential equation is solved by means of initial condition and boundary conditions. Particularly, the treatment of the boundary condition is important to the numerical solution. There are stringent requirements on boundary conditions for numerical simulations. First of all, whether the boundary conditions and the differential equations can compose a well-posed problem, i.e., the solution is existent and unique; secondly, whether the boundary conditions have physical meaning for an actual problem; thirdly, whether the boundary conditions with physical meaning can be expressed by mathematical formulae. The boundary conditions is accurately defined only when all the above three terms are satisfied. For numerical simulations on an actual problem, some boundary conditions can satisfy the above three terms in some cases. For instance, the current speed in normal direction being zero $u = 0$ for the impermeable solid wall condition can satisfy the above terms. But in many cases, it is difficult to make the boundary conditions meet the above requirements. Therefore, some reasonable boundary treatment techniques are needed for numerical simulations.

(1) Wave generation techniques

Wave generation is to generate given incoming waves (time series or wave spectrum) at certain positions in the computational domain. It is hard to give wave surface function or velocity distribution function due to the fluctuation interference caused by the reflection from the structures in the computational domain. Wave generation techniques mainly include source wave generation and numerical wave generator. Both of them have to deal with the reflected waves, as well as the reasonable selection of the location and the area of the wave generation.

(2) Free surface treatment

The free surface condition with physical meaning of water waves can be given by kinematic boundary condition and dynamic condition in mathematical formulae. But since the period of water waves is short in general, the computational domain is constantly changing caused by the quick change of the water surface. Therefore, how to deal with the free surface is a key issue in numerical simulations on water waves. There are several methods such as MAC, VOF, and level set for free surface treatment.

(3) Moving boundary treatment

The problem of moving boundary refers to the computational domain changes with the wave run-up and run-down on the gentle beach. How to deal with the boundary at the interface between water and land in order to make the computational domain be consistent with the real water region at certain time is the key for moving boundary treatment. Techniques of moving boundary treatment include flood and dry method, slot method, thin water-layer method.

(4) Non-reflecting boundary conditions

The computational domain is always limited to the study of interaction between water waves and structures. The boundary conditions are called non-reflecting boundary conditions, open or radiation boundary conditions if there is no reflected wave propagating toward the interior part of the domain when water waves pass through the open boundary. This is consistent with the physical concept of wave absorber set in the physical wave basin. Sommerfeld [7] defined the radiation condition as "the sources must be sources, not sinks. The energy which is radiated from the sources must scatter to infinity; no energy may be radiated from infinity into the computational domain." Sommerfeld proposed a mathematical formula for radiation condition at infinity and proved that the solution of linear diffraction wave is existent and unique by the radiation condition. This radiation condition made the analytical solution exists for the problem of linear diffraction around a cylinder. The radiation condition for nonlinear diffraction wave is much more complicated so that there is no mature mathematical expression for it so far. In practical numerical computations, it is not possible to put the open boundary of the computational domain at infinity. Therefore, the treatment techniques on non-reflecting boundary conditions or open boundary conditions are necessary. Non-reflecting boundary conditions such as sponge layer set at the open boundary of the domain can let wave propagate outward without any barrier or absorb the wave energy propagating outward completely (dissipation).

3. Solution techniques of incompressible flow equation

Water waves are kind of flow motion with free surface. The fluid density can be deemed as constant. When using flow equation of incompressible and viscous fluid to describe water waves, it is formally easier than using the compressible fluid equation. But it is hard to utilize the mature time-marching method since there is no time-term in continuity equation. The continuity equation turns into a constraint on velocity vector in various directions at each time to modify the pressure field. The modification of pressure field or iterative process is more complicated than compressible flow computation. Common computation methods used for incompressible and viscous flow motion include artificial compression method, projection method, (2D) vorticity stream function method, and iteration method.

4. Visualization, pre- and post-processing techniques

Visualization of water waves is to present the numerical simulation results of water waves using computer image display technology so that the water wave motion in a certain condition is represented or forecast on computer. In engineering design, visualization of water waves in different layouts can help engineers to directly understand the design results and the existing problems.

Pre and post processing of numerical simulations include generation of bathymetry and structures in the domain, saving and extraction in any manner on computational results.

By making the best uses of processing and visualization of computer systems, numerical simulation of water waves will be a powerful tool for research, engineering design, and decision-making.

1.4 Development of Water Wave Simulation

1.4.1 Computational Fluid Mechanics and Numerical Simulation of Water Waves

Numerical simulation of water waves has been developed with the development of the subject of computational fluid mechanics. It is an important part of computational fluid mechanics in the field of hydrodynamics.

Computational fluid mechanics is a subject to describe fluid motion using numerical methods by computer, to reveal the laws of fluid motion and to study time-space physical characteristics of fluid motion [8]. Since the first electronic computer came out in 1946, computational fluid mechanics, being an independent subject branch in 1965, has contributed to the study of fluid mechanics significantly.

Computational fluid mechanics is divided into computational aerodynamics and computational hydrodynamics according to the different fields. They have common basic theories and also their own particularities. Computational aerodynamics mainly studies aircraft motion in air and space domain. The development of computational aerodynamics plays a crucial role in aviation and aerospace industry. Computational hydrodynamics mainly focuses on the various problems of fluid flow motion in water resources, marine, coastal, environmental science, and industry.

Fluid motion in the field of hydrodynamics has diversity of forms and complexity of boundary conditions. Also, the governing equations and boundary conditions are usually different for different problems. Therefore, specific study on the problem concerned is needed in order to establish a numerical model of well-posed in mathematics and reasonable in physical meaning. Numerical methods and the treatment of boundary conditions are also different due to the difference of study objects. Numerical simulation of water waves introduced in this book focuses on unsteady hydrodynamic problems with free surface, i.e., numerical simulation of water wave problems.

1.4.2 Development of Numerical Simulation of Water Waves

Before the 1960s, physical model had been mainly used to study and solve the water wave problems in water sources, coastal, marine, and environmental science. Numerical models came into use in the late 1960s. For instance, numerical models partially replacing physical models were early applied to engineering problems by Danish Hydraulic Institute (DHI) in Denmark, Delft Hydraulics in the Netherlands, and SOGREAH in France. Along with the development of computer and computation techniques, numerical model takes up an increasing proportion compared with physical model. Appropriate combination of a numerical model and physical model becomes a quick, economic, and reliable tool for scientific researches and preliminary engineering works.

In China, numerical model has been developed rapidly since the early 1980s. It is highly effective to use numerical simulation methods for preliminary engineering works due to the complexity of dynamic factors and the large spatial extent in the engineering water wave problems. Therefore, numerical model has been developed rapidly in water resources, coastal, marine, and environmental engineering. So far, there are a number of practical models which play important roles in engineering design and construction. Regarding the development of numerical hydrodynamic models, Abbott [9], who is an international famous scholar, divided it into five phases or five generations of numerical model. The first generation is starting phase in which using computers to replace previous manual calculations to implement numerical calculations in hydrodynamics. The second generation is model group developed on the basis of the first generation models which can solve individual problems only. In this phase, general systematic software has been established. Apart from using operating system, the third generation is more precise and effective in the use of numerical methods. Modelers of the fourth generation serve the users by providing numerical model products. The development of user-oriented numerical model and pre- and post-processing techniques makes the users work with the model more conveniently and efficiently. The specific training on the model is not necessary. The design engineers can use the mathematical model software directly. In the fifth generation, with the coming of information era, mathematical models make leap to information-age models. Information theories and technology provide powerful tools for the informatization of mathematical models. The information systems established by the combination of mathematical model and artificial intelligence expert system can directly serve the engineering design, construction, and management. In the past decade, numerical hydrodynamic models have been basically developed in these five stages.

1.4.3 Relationship Among Numerical Simulation of Water Waves, Theoretical Analysis, and Experimental Study

When the mechanism and law of physical phenomena are quite clear, numerical method is relatively accurate. It has been proved in practice that numerical simulations have greater advantages than physical model tests when they reflect the essence of the problems. First of all, numerical simulations cost less human and material resources with shorter completion period so that they can be utilized to compare a number of layouts for engineering design. Next, they are useful for large-scale flow fields. With the development of computer, the computer capacity and speed are increasing rapidly. This provides a powerful condition for numerical simulations which are especially applicable to solve problems in water environment, water resource, coastal and marine engineering in large-scale fields. On the contrary, physical model tests have limitation on the scale and size of facility. In general, substantially reduced ratio is required so that there are lots of difficulties on scale effects and similarities in physical model tests. Finally, numerical models are to solve differential equations with certain boundary and initial conditions, which have good reproducibility. The conditions for numerical models are easily controlled and the models could simulate in a long period. Also, the boundary conditions, bathymetry, and wave directions can be conveniently changed in numerical models but not so in physical models. Due to the above advantages, numerical models have played important roles in solving water wave problems in engineering.

Numerical models have limitations as well. First of all, it is not possible to propose an accurate numerical model for any complex physical phenomenon. Before the mechanisms of the problems have been completely identified, it is hard to get accurate numerical models for the problems such as hydraulic jump, aeration of breaking waves, exchange of water–air interface, and dynamics of mud. Secondly, the analyses on stability and convergence are required when the established mathematical models are discretized by numerical methods. The existing analytical methods are valid for linear equations. But there is no perfect theory for nonlinear equations. Some pseudo-physical effects (such as analyses and control of numerical dispersion and dissipation) are expected to be further studied. Also, there are still some approximate numerical techniques in calculations. Therefore, numerical model results should be validated by physical model tests or prototype observations. Thirdly, numerical simulations are restricted by computer resources. For instance, as to turbulence being a key issue in fluid flow, due to the limitation of computer speed and capacity, the direct numerical simulations have not yet reached the stage of practical application. Thus, various turbulence models are needed to close the time-averaged fluid mechanics equations. Furthermore, some empirical coefficients should be determined in turbulence models. All in all, when the advantages of numerical simulations are emphasized, the limitations of them should also be seen and given their proper place. Numerical simulations, theoretical analyses, and physical model tests are mutually supportive. All of them are key tools to solve the problems.

"Hybrid model" developed recently provides a potential development direction. One type is a combination of theoretical analysis and numerical model. For a physical problem, using analytical method as much as possible reduces the numerical calculation. A good example is complex Green function method. By means of derivation, finding Green functions which meet some boundary conditions reduces the workload of boundary integration greatly. Another type of hybrid model is a combination of numerical model and physical model. This combination is using numerical model for calculation in a large area and determining the boundary conditions of local physical models by the numerical results. This approach can make some necessary physical experiments tested in a local area with a model scale not reduced too much. Therefore, this combination is an economical and efficient approach.

References

1. Anderson DA, Tannehill JC, Pletcher RH. Computational fluid mechanics and heat transfer. New York: Hemisphere Publishing Corporation; 1984.
2. Xiaokang XIN. Computational fluid dynamics (Chinese Version). Changsha: National Defense Science and Technology University Press; 1989.
3. Versteeg HK, Malalasekera W. An introduction to computational fluid dynamics: the finite, vol. method. London: Pearson Education; 1998.
4. Jianghang W, Qingshu H. Theories, methods and applications of computational fluid mechanics (Chinese Version). Beijing: Science Press; 1988.
5. Bribia CA. Boundary element method for engineers. London: Pentech Press; 1978.
6. John F. On the motion of the floating bodies II. Comm. Pure Appl. Mathm. 1950;3:45–101.
7. Sommerfeld A. Partial differential equation in physics. New York: Academic Press; 1949.
8. Dexun FU, Yanwen MA. Computational fluid dynamics (Chinese Version). Beijing: Higher Education Press; 2002.
9. Abbott MB. Hydroinformatics: information, technology and the aquatic environment. Aldershot, UK and Brookfield, USA: Ashgate; 1991.

Chapter 2
Water Wave Theories

2.1 Mathematical Model of Water Waves

The mathematical model describing various water wave phenomena can be based on general equations of fluid mechanics, and can be also based on the water wave theories in specific conditions. The mathematical model of water waves includes the governing differential equations which describe the water wave motion, and definite conditions which consist of initial condition and boundary conditions.

This chapter introduces mainly two important kinds of gravity waves. They are dispersive waves and hyperbolic waves (long waves).

2.1.1 Governing Equations for Water Waves

1. Governing equations for incompressible inviscid fluid flow

The governing equations for the incompressible inviscid fluid flow are described as follows:

Let u, v, and w be the velocity components in the x-, y-, and z-directions, respectively, at a certain point.

The continuity equation can be written as:

$$\frac{\partial u}{\partial x} + \frac{\partial v}{\partial y} + \frac{\partial w}{\partial z} = 0 \tag{2.1.1}$$

or in vector form,

$$\nabla \cdot \boldsymbol{U} = 0 \tag{2.1.2}$$

where $\boldsymbol{U} = (u, v, w)$.

J. Tao, *Numerical Simulation of Water Waves*, Springer Tracts in Civil Engineering, https://doi.org/10.1007/978-981-15-2841-5_2

The momentum equations can be written as:

$$\frac{\partial u}{\partial t} + u\frac{\partial u}{\partial x} + v\frac{\partial u}{\partial y} + w\frac{\partial u}{\partial z} = -\frac{1}{\rho}\frac{\partial p}{\partial x} + f_x \tag{2.1.3}$$

$$\frac{\partial v}{\partial t} + u\frac{\partial v}{\partial x} + v\frac{\partial v}{\partial y} + w\frac{\partial v}{\partial z} = -\frac{1}{\rho}\frac{\partial p}{\partial y} + f_y \tag{2.1.4}$$

$$\frac{\partial w}{\partial t} + u\frac{\partial w}{\partial x} + v\frac{\partial w}{\partial y} + w\frac{\partial w}{\partial z} = -\frac{1}{\rho}\frac{\partial p}{\partial z} + f_z \tag{2.1.5}$$

where p is the pressure in fluid; f_x, f_y and f_z are the unit mass forces in the x-, y-, and z-directions, respectively; ρ is the fluid density. The above equations can be written in vector form as:

$$\frac{D\boldsymbol{U}}{Dt} = -\nabla\left(\frac{p}{\rho} - \boldsymbol{f}\right) \tag{2.1.6}$$

where $\boldsymbol{f} = (f_x, f_y, f_z)$.

The above momentum equations are also called the Euler equations.

2. Governing equations for incompressible viscid fluid flow

The continuity equation can be written as:

$$\frac{\partial u}{\partial x} + \frac{\partial v}{\partial y} + \frac{\partial w}{\partial z} = 0 \tag{2.1.7}$$

or in vector form,

$$\nabla \cdot \boldsymbol{U} = 0 \tag{2.1.8}$$

The momentum equations can be written as:

$$\frac{\partial u}{\partial t} + u\frac{\partial u}{\partial x} + v\frac{\partial u}{\partial y} + w\frac{\partial u}{\partial z} = f_x - \frac{1}{\rho}\frac{\partial p}{\partial x} + \frac{\mu}{\rho}\left(\frac{\partial^2 u}{\partial x^2} + \frac{\partial^2 u}{\partial y^2} + \frac{\partial^2 u}{\partial z^2}\right) \tag{2.1.9}$$

$$\frac{\partial v}{\partial t} + u\frac{\partial v}{\partial x} + v\frac{\partial v}{\partial y} + w\frac{\partial v}{\partial z} = f_y - \frac{1}{\rho}\frac{\partial p}{\partial y} + \frac{\mu}{\rho}\left(\frac{\partial^2 v}{\partial yx^2} + \frac{\partial^2 v}{\partial y^2} + \frac{\partial^2 v}{\partial z^2}\right) \tag{2.1.10}$$

$$\frac{\partial w}{\partial t} + u\frac{\partial w}{\partial x} + v\frac{\partial w}{\partial y} + w\frac{\partial w}{\partial z} = f_z - \frac{1}{\rho}\frac{\partial p}{\partial z} + \frac{\mu}{\rho}\left(\frac{\partial^2 w}{\partial yx^2} + \frac{\partial^2 w}{\partial y^2} + \frac{\partial^2 w}{\partial z^2}\right) \tag{2.1.11}$$

where μ is the dynamic viscosity coefficient.

The momentum equations can be written in vector form as:

$$\left(\frac{\partial}{\partial t} + \boldsymbol{U}\cdot\nabla\right)\boldsymbol{U} = -\nabla\left(\frac{p}{\rho} - \boldsymbol{f}\right) + \nu\nabla^2\boldsymbol{U} \tag{2.1.12}$$

where $\nu = \mu/\rho$ is the kinematic viscosity coefficient. The above momentum equations are also called the Navier–Stokes equations or N–S equations.

With the increase of computer capacity and speed, and the improvement of calculation techniques, it becomes possible to simulate water waves based on the N-S equations. However, it is not necessary to use the N-S equations to simulate all water wave problems as some wave problems can be simplified under specific conditions.

3. Potential theory of water waves

Let Ω denotes the vortex vector in fluid motion, which is defined as,

$$\boldsymbol{\Omega} = \nabla \times \boldsymbol{U} \tag{2.1.13}$$

For the ideal, barotropic fluid under external force potentiality, it can be proved that, if there is no vortex at the initial moment ($t = 0$) in fluid motion, no vortex will occur in the flow at any time, i.e., $\boldsymbol{\Omega} = 0$ always, and it is called potential flow. Normally, wind-generated waves can be treated as potential flow. But the potential theory is not appropriate for waves in shallow water.

For the inviscid, irrotational potential flow, the velocity vector can be expressed by the gradient of a scalar function $\Phi(x, y, z, t)$,

$$\boldsymbol{U} = \nabla \Phi \tag{2.1.14}$$

$\Phi(x, y, z, t)$ is the velocity potential function.

Using the vector relation formula,

$$\boldsymbol{U} \cdot \nabla \boldsymbol{U} = \nabla \frac{\boldsymbol{U}^2}{2} - \boldsymbol{U} \times (\nabla \times \boldsymbol{U})$$

when $\boldsymbol{\Omega} = \nabla \times \boldsymbol{U} = 0$, the kinematic viscosity coefficient $\nu = 0$ and the mass force is pure gravity, Eq. (2.1.12) becomes

$$\nabla \left[\frac{\partial \Phi}{\partial t} + \frac{1}{2} |\nabla \Phi|^2 \right] = -\nabla \left(\frac{p}{\rho} + gz \right) \tag{2.1.15}$$

where g is the gravitational acceleration. Integrating the above equation yields

$$-\frac{p}{\rho} = gz + \frac{\partial \Phi}{\partial t} + \frac{1}{2} |\nabla \Phi|^2 + C(t) \tag{2.1.16}$$

The integration constant $C(t)$ is an arbitrary function related to time. Normally Φ can be redefined to make $C(t)$ equal to zero without any effect on the velocity field. Therefore, $C(t) = 0$ can be chosen.

Equation (2.1.16) is called the Bernoulli equation.

Introducing the velocity potential function Φ into the continuity equation (2.1.8) yields the Laplace equation,

$$\nabla^2\Phi = \frac{\partial^2\Phi}{\partial x^2} + \frac{\partial^2\Phi}{\partial y^2} + \frac{\partial^2\Phi}{\partial x^2} = 0 \tag{2.1.17}$$

Equations (2.1.16) and (2.1.17) are the fundamental equations for potential wave motion.

2.1.2 Boundary Conditions for Water Waves

Boundary conditions will determine the specific form of water wave motion. The free surface boundary condition and bottom boundary condition for potential waves are mainly discussed here.

1. Free surface boundary condition

The surface of water waves called free surface is exposure to the atmosphere. Since the atmosphere density is very small, it cannot provide the fluid with rigidity constraint like solid boundary. There is an additional stress from the atmosphere on the fluid surface. In general, this stress should have the tangential and normal components. The wind stress is a kind of tangential component. But for the inviscid fluid flow, the atmospheric force is only perpendicular to the surface. The force per unit area is called pressure. Generally, the air pressure p_a can be chosen as a constant. But in some special cases such as typhoon, the pressure on the surface is variable $p_a(x, y, z)$. Furthermore, the surface tension can also cause the change of pressure field. But in the study of gravity waves, the surface tension is neglectable.

(1) Kinematic condition at the free surface

Let $\eta(x, y, t)$ denote the vertical displacement of the free surface. The free surface equation reads

$$F(x, y, z, t) = z - \eta(x, y, t) = 0$$

The free surface $\eta(x, y, t)$ moves along with the fluid flow and a fluid particle located at the free surface must remain at the free surface. It means the normal velocity of the particle on the free surface must be the same as the normal velocity of the free surface. Therefore, it gives

$$\frac{DF}{Dt} = 0 \tag{2.1.18}$$

i.e.,

$$\frac{D}{Dt}(z - \eta(x, y, t)) = 0 \tag{2.1.19}$$

it yields

$$w = \frac{\partial z}{\partial t} = \frac{\partial \eta}{\partial t} + u\frac{\partial \eta}{\partial x} + v\frac{\partial \eta}{\partial y} \tag{2.1.20}$$

or,

$$\frac{\partial \Phi}{\partial z} = \frac{\partial \eta}{\partial t} + \frac{\partial \Phi}{\partial x}\frac{\partial \eta}{\partial x} + \frac{\partial \Phi}{\partial y}\frac{\partial \eta}{\partial y} \tag{2.1.21}$$

No force is considered in the boundary condition at the free surface as described in Eqs. (2.1.20) or (2.1.21). Therefore, it is called the kinematic boundary condition.

(2) Dynamic condition at the free surface

According to the fundamental equation of the potential flow Eq. (2.1.16), when the pressure on the surface is equal to the air pressure, i.e., $p = p_a$, the condition at the free surface becomes,

$$-\frac{p_a}{\rho} = g\eta + \frac{\partial \Phi}{\partial t} + \frac{1}{2}|\nabla\Phi|^2, \quad z = \eta \tag{2.1.22}$$

By eliminating η from the kinematic and the dynamic boundary conditions at the free surface in Eqs. (2.1.21) and (2.1.22), and letting p_a be a constant, the free surface condition for potential waves is obtained,

$$\frac{\partial^2 \Phi}{\partial t^2} + g\frac{\partial \Phi}{\partial z} + \frac{\partial \boldsymbol{U}^2}{\partial t} + \frac{1}{2}\boldsymbol{U} \cdot \nabla \boldsymbol{U}^2 = 0, \quad z = \eta \tag{2.1.23}$$

2. Bottom boundary condition

For the inviscid and irrotational flow of water waves, the normal velocity at the solid boundary must be zero, i.e.,

$$\frac{\partial \Phi}{\partial n} = 0 \tag{2.1.24}$$

where $\boldsymbol{n}$ is the unit normal vector to the solid wall with the direction toward the fluid.

When the impermeable sea bed is uneven without erosion and deposition, and the depth from the still water level is $d(x, y)$, the boundary condition can be written as

$$\frac{\partial \Phi}{\partial z} = \frac{\partial \Phi}{\partial x}\frac{\partial d}{\partial x} + \frac{\partial \Phi}{\partial y}\frac{\partial d}{\partial y}, \quad z = -d(x, y) \tag{2.1.25}$$

2.1.3 Initial Condition for Water Waves

The initial condition for water waves is less important than the boundary conditions. The reason is for periodic fluctuations the influence of the initial condition will disappear after several wave periods and the solution to the equations becomes periodic as well. The formulations of the initial condition for water waves in various problems will be described in the following chapters.

2.2 Dispersive Waves

2.2.1 Governing Equations and Boundary Conditions

The dispersive waves described in this chapter are mainly surface gravity waves. When using the potential flow theory to describe the incompressible, inviscid, and irrotational flow, Eqs. (2.1.17), (2.1.23), and (2.1.25) are the governing equation and boundary conditions.

$$\begin{aligned} &\nabla^2\Phi = 0, \quad \text{in the domain} \\ &\frac{\partial^2\Phi}{\partial t^2} + g\frac{\partial\Phi}{\partial z} + \frac{\partial U^2}{\partial t} + \frac{1}{2}U\cdot\nabla U^2 = 0, \quad z=\eta \\ &\frac{\partial\Phi}{\partial z} = \frac{\partial\Phi}{\partial x}\frac{\partial d}{\partial x} + \frac{\partial\Phi}{\partial y}\frac{\partial d}{\partial y}, \quad z=-d(x,y) \end{aligned}$$

2.2.2 Linear Small-Amplitude Wave Theory

To obtain an exact solution to Eqs. (2.1.17), (2.1.23), and (2.1.25) for the incompressible, inviscid, and irrotational wave problems with free surface, the most tremendous difficulty is that the free surface boundary condition described in Eq. (2.1.23) is nonlinear. Also, the shape and the position of the free surface $z = \eta$ are unknown. The simplest and basic approach to solve this problem is using the linear small-amplitude wave theory which was first proposed by Airy. Therefore, the linear small-amplitude wave theory is also called Airy wave theory. When the wave height H is much smaller than the wavelength L and the still water depth d, i.e., $H \ll L, d$, the nonlinear terms in Eq. (2.1.23) of the free surface condition can be neglected. The boundary condition is thus simplified to be linear. Also, the free surface can be simplified to $z = 0$. When considering the sea bed is approximately flat and the still water depth, d, is constant, the governing equation and boundary conditions for the velocity potential function $\Phi(x, y, z, t)$ of linear small-amplitude waves read

$$\nabla^2\Phi = 0, \quad -d < z < 0 \tag{2.2.1}$$

$$\frac{\partial \Phi}{\partial z} - \frac{\partial \eta}{\partial t} = 0, \quad z = 0 \tag{2.2.2}$$

$$\frac{\partial \Phi}{\partial t} + g\eta = 0, \quad z = 0 \tag{2.2.3}$$

$$\frac{\partial \Phi}{\partial z} = 0, \quad z = -d \tag{2.2.4}$$

Equations (2.2.2) and (2.2.3) are the linearized kinematic and dynamic free surface boundary conditions, respectively. Combing them to eliminate η yields the linear free surface condition, i.e., the linearized form of Eq. (2.1.23),

$$\frac{\partial^2 \Phi}{\partial t^2} + g\frac{\partial \Phi}{\partial z} = 0, \quad z = 0 \tag{2.2.5}$$

For the linear small-amplitude wave problems in Eqs. (2.2.1)–(2.2.4), a periodic solution to the boundary problem is found first without consideration of the initial condition, to understand the wave characteristics, and key parameters.

It is assumed that one-dimensional (1D) wave occurs in a plane of xOz as shown in Fig. 2.1. The free surface is a harmonic waveform. The angular frequency is denoted by ω. Separating time and space in variables yields

$$\eta(x, t) = \eta'(x)e^{-i\omega t} \tag{2.2.6}$$

$$\eta'(x) = ae^{ikx} \tag{2.2.7}$$

$$\Phi(x, z, t) = \phi(x, z)e^{\;i\omega t} \tag{2.2.8}$$

where a is the wave amplitude, k is the wave number.

By substituting Eqs. (2.2.6) and (2.2.8) into Eqs. (2.2.1)–(2.2.5), the governing equation and boundary conditions for ϕ and η' are written as,

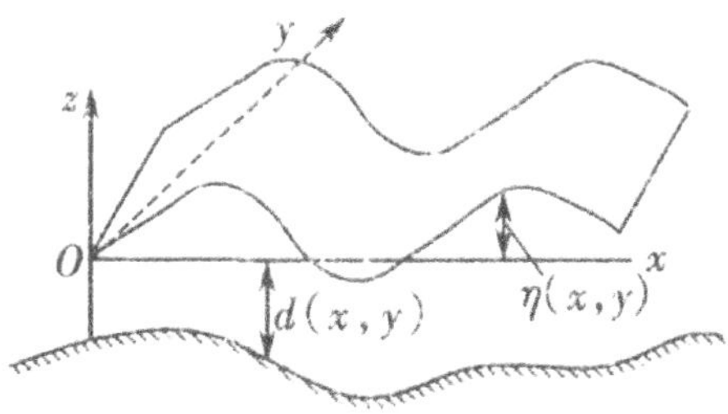

Fig. 2.1 Definition sketch

$$\nabla^2\phi = 0, \qquad -d < z < 0 \tag{2.2.9}$$

$$\frac{\partial \phi}{\partial z} = 0, \quad z = -d \tag{2.2.10}$$

$$\frac{\partial \phi}{\partial t} + i\omega\eta' = 0, \quad z = 0 \tag{2.2.11}$$

$$g\eta' - i\omega\phi = 0, \quad z = 0 \tag{2.2.12}$$

Eliminating η' from Eqs. (2.2.11) to (2.2.12) yields the free surface condition described by ϕ,

$$\frac{\partial \phi}{\partial z} - \frac{\omega^2}{g}\phi = 0 \tag{2.2.13}$$

The velocity potential function ϕ which satisfies Eqs. (2.2.9) and (2.2.10) can be expressed as

$$\phi = B\cosh k\ (z + d)e^{ikx} \tag{2.2.14}$$

where B is undetermined coefficient.

Substituting Eqs. (2.2.14) and (2.2.7) into Eq. (2.2.12) yields

$$B = -\frac{iga}{\omega}\frac{1}{\cosh kd} \tag{2.2.15}$$

The velocity potential function for small-amplitude waves becomes

$$\Phi = -\frac{iga}{\omega}\frac{\cosh k(z + d)}{\cosh kd}e^{ikx} \tag{2.2.16}$$

Substituting Eq. (2.2.16) into the free surface condition, Eq. (2.2.13) yields

$$\omega^2 = gk \tanh kd \tag{2.2.17}$$

This is called dispersion relation which gives a relation between the angular frequency ω and the wave number k (or the wavelength). It is a very important formula for dispersive waves.

Substituting wave height $H = 2a$ and Eq. (2.2.16) into Eq. (2.2.8) gives

$$\Phi = \frac{gH}{2\omega}\frac{\cosh k(z + d)}{\cosh kd}\sin\theta \tag{2.2.18}$$

$$\theta = kx - \omega t \tag{2.2.19}$$

The surface elevation is $\eta = \frac{H}{2} e^{i(kx - \omega t)}$

$$\text{or} \quad \eta = \frac{H}{2}\cos(kx - \omega t) = \frac{H}{2}\cos\theta \tag{2.2.20}$$

When x is increased by $\frac{2\pi}{k}$ or t increased by $\frac{2\pi}{\omega}$, the cosine function value is the same, i.e., η keeps unchanged. $L = \frac{2\pi}{k}$ is so-called the wavelength and $T = \frac{2\pi}{\omega}$ is the wave period. If $(kx - \omega t) = \text{const}$, i.e., $x = \text{const} + \frac{\omega}{k}t$, η will keep unchanged. It means the waveform moves forward with a speed of $c = \frac{\omega}{k} = \frac{L}{T}$, so-called the phase velocity.

The dispersion relation of Eq. (2.2.17) can also be rewritten as

$$c^2 = \frac{g}{k}\tanh kd \tag{2.2.21}$$

The phase function is defined as

$$S(x, t) = kx - \omega t \tag{2.2.22}$$

For two-dimensional (2D) free surface waves, $\eta(x, y, t) = a\cos(\boldsymbol{k}\cdot\boldsymbol{x} - \omega t) = a\cos\left(k_x x + k_y y - \omega t\right)$, the phase function reads

$$S(x, y, t) = \boldsymbol{k}\cdot\boldsymbol{x} - \omega t = k_x x + k_y y - \omega t \tag{2.2.23}$$

At a certain fixed time t, $S(x, y, t) = \text{const} = S_0$ represents a straight line. Its normal vector is

$$e_k = \left(\frac{k_x}{k}, \frac{k_y}{k}\right), k = \left(k_x^2 + k_y^2\right)^{1/2}$$

The water pressure for the linear waves can be obtained by omitting the nonlinear term in Eq. (2.1.22),

$$p = -\rho g z - \rho\frac{\partial\Phi}{\partial t} \tag{2.2.24}$$

where $-\rho g z$ is the hydrostatic pressure, $-\rho\frac{\partial\Phi}{\partial t}$ is the wave-induced pressure.

Substituting Eq. (2.2.18) into Eq. (2.2.24) yields

$$p = -\rho g z + \frac{1}{2}\rho g H\frac{\cosh k(z + d)}{\cosh kd}\cos\theta \tag{2.2.25}$$

Table 2.1 presents various wave parameters for linear small-amplitude waves. Two types of linear small-amplitude waves are discussed in the following.

Table 2.1 Wave parameter formulas for linear small-amplitude waves

Velocity potential function Φ	$\Phi = \frac{\pi H}{kT}\frac{\cosh ks}{\sinh kd}\sin\theta = \frac{gH}{2\omega}\frac{\cosh ks}{\cosh kd}\sin\theta$
Wave celerity c	$c^2 = \frac{\omega^2}{k^2} = \frac{g}{k}\tanh kd$
Wave surface elevation η	$\eta = \frac{H}{2}\cos\theta$
Horizontal particle position ξ	$\xi = -\frac{H}{2}\frac{\cosh ks}{\sinh kd}\sin\theta$
Vertical particle position ζ	$\zeta = \frac{H}{2}\frac{\sinh ks}{\sinh kd}\cos\theta$
Horizontal velocity u	$u = \frac{\pi H}{T}\frac{\cosh ks}{\sinh kd}\cos\theta$
Vertical velocity w	$w = \frac{\pi H}{T}\frac{\sinh ks}{\sinh kd}\sin\theta$
Horizontal acceleration $\partial u/\partial t$	$\frac{\partial u}{\partial t} = \frac{2\pi^2 H}{T^2}\frac{\cosh ks}{\sinh kd}\sin\theta$
Vertical acceleration $\partial w/\partial t$	$\frac{\partial w}{\partial t} = -\frac{2\pi^2 H}{T^2}\frac{\sinh ks}{\sinh kd}\cos\theta$
Pressure p	$p = -\rho gz + \frac{1}{2}\rho gH\frac{\cosh ks}{\cosh kd}\cos\theta$
Group velocity c_g	$c_g = \frac{1}{2}\left(1 + \frac{2kd}{\sinh 2kd}\right)c$
Mean energy density E	$E = \frac{1}{8}\rho gH^2$
Energy flux P	$P = Ec_g$

In the table, $\theta = kx - \omega t$, $s = z + d$

(1) Shallow water waves

When $kd < \frac{\pi}{10}$, $\frac{d}{L} < \frac{1}{20}$, i.e., the wavelength L is greater than 20 times of the water depth, it gives $\sinh kd \approx \tanh kd \approx kd$, $\cosh kd \approx 1$, and $\frac{d}{gT^2} < 0.0025$. Thus, the dispersion relation becomes

$$c^2 = gd \tag{2.2.26}$$

It indicates that when the water is very shallow the wave celerity is independent of the wavelength, so that the wave is not dispersive. This kind of long wave will be introduced in Sect. 2.4. The pressure can be express as

$$p = -\rho gz + \frac{1}{2}\rho gH\cos\theta \tag{2.2.27}$$

where the first term is the hydrostatic pressure, and the second term is the excess pressure caused by the long wave.

(2) Deepwater waves

When $kd > \pi$, $\frac{d}{L} > \frac{1}{2}$, i.e., the water depth is greater than half of the wavelength, it gives $\sinh kd \approx \cosh kd \approx \frac{1}{2}e^{kd}$, $\tanh kd \approx 1$, and $\frac{d}{gT^2} > 0.08$. Thus, the dispersion relation becomes

$$c^2 = \frac{\omega^2}{k^2} = \frac{g}{k} \tag{2.2.28}$$

and the pressure is

$$p = -\rho g z + \frac{1}{2}\rho g H e^{kz} \cos\theta \tag{2.2.29}$$

The excess pressure caused by the deepwater wave decreases exponentially with the distance away from the still water level. The particle orbital motions are circular, and the diameters also decrease with the distance away from the still water level.

Table 2.2 presents various wave parameters for the above two types of linear small-amplitude waves.

The basic starting point of the small-amplitude wave theory is linearization and the small amplitude assumption. Therefore, either $kd = O(1)$ or $kd \ll 1$, the relative wave height $\frac{H}{d} \ll 1$ is required. Its application is thus subject to strict restrictions. However, a basic understanding of the characteristics of water waves has been shown by means of the analysis of the small-amplitude wave theory.

Table 2.2 Wave parameter formulas for the two types of linear small-amplitude waves

	Shallow water waves	Deep water waves
Valid conditions	$kd < \frac{\pi}{10}$	$kd > \pi$
	$\frac{d}{L} < \frac{1}{20}$	$\frac{d}{L} > \frac{1}{2}$
	$\frac{d}{gT^2} < 0.0025$	$\frac{d}{gT^2} > 0.08$
Velocity potential function Φ	$\Phi = \frac{\pi H}{k^2 T d} \sin\theta$	$\Phi = \frac{\pi H}{kT} e^{kz} \sin\theta$
	$= \frac{gH}{2\omega} \sin\theta$	$= \frac{gH}{2\omega} e^{kz} \sin\theta$
Wave celerity c	$c^2 = \frac{\omega^2}{k^2} = gd$	$c^2 = c_0^2 = \frac{\omega^2}{k^2} = \frac{g}{k}$
Wave length L	$L = T\sqrt{gd}$	$L = L_0 = gT^2/2\pi$
Surface elevation η	$\eta = \frac{H}{2}\cos\theta$	$\eta = \frac{H}{2}\cos\theta$
Horizontal particle position ξ	$\xi = -\frac{H}{2kd}\sin\theta$	$\xi = -\frac{H}{2}e^{kz}\sin\theta$
Vertical particle position ζ	$\zeta = \frac{H}{2}\left(1 + \frac{z}{d}\right)\cos\theta$	$\zeta = \frac{H}{2}e^{kz}\cos\theta$
Horizontal velocity u	$u = \frac{\pi H}{T(kd)}\cos\theta$	$u = \frac{\pi H}{T}e^{kz}\cos\theta$
Vertical velocity w	$w = \frac{\pi H}{T}\left(1 + \frac{z}{d}\right)\sin\theta$	$w = \frac{\pi H}{T}e^{kz}\sin\theta$
Horizontal acceleration $\partial u/\partial t$	$\frac{\partial u}{\partial t} = \frac{2\pi^2 H}{T^2(kd)}\sin\theta$	$\frac{\partial u}{\partial t} = \frac{2\pi^2 H}{T^2}e^{kz}\sin\theta$
Vertical acceleration $\partial u/\partial t$	$\frac{\partial w}{\partial t} = -\frac{2\pi^2 H}{T^2}\left(1 + \frac{z}{d}\right)\cos\theta$	$\frac{\partial w}{\partial t} = -\frac{2\pi^2 H}{T^2}e^{kz}\cos\theta$
Pressure p	$p = -\rho g z + \frac{1}{2}\rho g H \cos\theta$	$p = -\rho g z + \frac{1}{2}\rho g H e^{kz}\cos\theta$
Group velocity c_g	$c_g = c$	$c_g = \frac{1}{2}c$
Mean energy density E	$E = \frac{1}{8}\rho g H^2$	$E = \frac{1}{8}\rho g H^2$
Energy flux P	$P = Ec$	$P = Ec/2$

2.2.3 Basic Concepts of Progressive Waves

1. Dispersion relation

Linear small-amplitude waves are typical dispersive waves. The phase velocity is dependent on the wavelength as presented in Eq. (2.2.21)

$$c^2 = \frac{\omega^2}{k^2} = \frac{g}{k} \tanh kd$$

which indicates that the longer the wavelength, the faster the wave celerity will be for waves in the same water depth. Any local initial disturbance can be deemed as a superposition of a number of wavelets with continuous wavelengths and periods. The longer waves travel faster than the shorter waves. When the disturbance propagates outwards, the longest and shortest waves are further and further separated and the medium waves are in between them. The waves with different frequencies and wavelengths propagate with different speeds, which lead to the phenomenon of wave dispersion.

From the above, we can see that the wave celerity of shallow water waves ($c = \sqrt{gd}$) is only dependent on the water depth (d), but not on the wavelength. Therefore, the shallow water waves are not dispersive. The following analyses will indicate that both finite-amplitude waves and nonlinear waves can be dispersive waves as long as the wave celerity is dependent on the wavelength.

2. Group velocity

Group velocity is one of the most important concepts in the dispersive wave theory. Figure 2.2 shows two small-amplitude waves propagating in the same direction with the same wave height H, but with slightly different wave numbers k and $k + \Delta k$ (or wavelengths), and slightly different wave frequencies ω and $\omega + \Delta\omega$. The surface elevation of the composite wave is a superposition of the two waves and expressed as

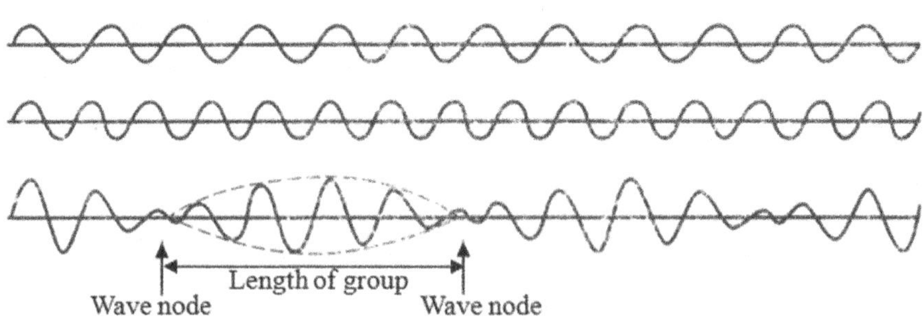

Fig. 2.2 Sketch of wave group

$$\eta_g = \frac{H}{2}\cos\left[\left(k+\frac{\Delta k}{2}\right)x-\left(\omega+\frac{\Delta\omega}{2}\right)t\right]+\frac{H}{2}\cos\left[\left(k-\frac{\Delta k}{2}\right)x-\left(\omega-\frac{\Delta\omega}{2}\right)t\right]$$
$$= H\cos\left(\frac{\Delta k}{2}x-\frac{\Delta\omega}{2}t\right)\cos(kx-\omega t) \quad (2.2.30)$$

From Eq. (2.2.30), it can be seen that the amplitude of the composite wave is $H\cos\left(\frac{\Delta k}{2}x-\frac{\Delta\omega}{2}t\right)$, the length of the wave group is $L_g = \frac{2\pi}{\Delta k/2}$, the period of the group is $T_g = \frac{2\pi}{\Delta\omega/2}$, and the propagating velocity of the wave group (called group velocity) is

$$c_g = \lim_{\substack{\Delta k\to 0\\ \Delta\omega\to 0}}\left(\frac{L_g}{T_g}\right) = \frac{d\omega}{dk} \quad (2.2.31)$$

Taking the derivative of dispersion relation in linear wave theory Eq. (2.2.17) gives

$$2\omega\frac{d\omega}{dk} = g\tanh kd + gkd\ \mathrm{sec\,h}^2\,kd$$

Thus the group velocity is

$$c_g = \frac{d\omega}{dk} = \frac{g}{2\omega}\left(\tanh kd + kd\ \mathrm{sec\,h}^2\,kd\right) = \frac{1}{2}c\left(+\frac{2kd}{\sinh 2kd}\right) \quad (2.2.32)$$

This is the formula for the group velocity. Let $n = \frac{1}{2}\left(1+\frac{2kd}{\sinh 2kd}\right)$, then we have $c_g = nc$.

In deep water, $kd \gg 1$, it gives $n = 1/2$, $c_g = c/2$. In shallow water, $kd \ll 1$, it gives $2kd/\sinh 2kd \approx 1$, $n = 1$, $c_g = c$.

In the study of wave motion, several different types of velocities should be noted. They are the particle velocity components u and w at a spatial point, the transport velocity of the individual wave c (phase velocity), and the transport velocity of the wave group c_g(group velocity).

3. Wave energy and wave energy flux

(1) Wave energy

For a wave propagating over constant water depth, the mean energy of the water volume under the free surface over a unit area and a wave period comprises of the kinetic and potential energy.

The mean kinetic energy in the water volume reads

$$E_k = \frac{\rho}{2}\overline{\int_{-d}^{\eta}(u^2+w^2)dz} \quad (2.2.33)$$

where "–" denotes the time average over a period, i.e.,

$$\overline{f} = \frac{1}{T}\int_{0}^{T} f dt \tag{2.2.34}$$

Substituting the particle velocity components u and w in linear wave theory into Eq. (2.2.33) yields

$$E_k = \frac{\rho}{4}\left(\frac{gka}{\omega}\right)^2 \frac{\sinh 2k(d+\eta)}{2k\cosh^2 kd} \tag{2.2.35}$$

When $\eta \ll d$, it gives

$$E_k = \frac{\rho g a^2}{4} \tag{2.2.36}$$

The mean potential energy in the water volume reads

$$E\rho = \overline{\int_{0}^{\eta} \rho g z dz} = \frac{1}{4}\rho g a^2 \tag{2.2.37}$$

Therefore, according to linear wave theory, we have

$$E = E_k + E_\rho = \frac{1}{2}\rho g a^2 \tag{2.2.38}$$

It is noted that here the mean kinetic energy and the mean potential energy are identical. This property is called the equipartition of energy.

(2) Energy transmission—wave energy flux

The power of dynamic pressure in the vertical section per unit width of wave front over a period reads

$$P = \rho \int_{0}^{T}\int_{-d}^{0} pudzdt \tag{2.2.39}$$

When $x = 0$, it gives

$$P = \frac{1}{2}\rho g a^2\left[\frac{1}{2}\frac{\omega}{k}\left(1 + \frac{2kd}{\sinh\ 2kd}\right)\right] = Ec_g \tag{2.2.40}$$

This is the work done by the fluid on the left side of the section upon the water volume per unit time, which is also the energy transmitted through the vertical section to the water volume. It is also called the wave energy flux.

From the above equation, it is seen that the energy transport velocity is the wave group velocity c_g which is of dynamic significance in energy transport. But the phase velocity c is a kinematic parameter which cannot be seen as a transport velocity of any physical parameter.

The kinematic significance of the wave group velocity c_g is further discussed below.

In the water body with varying depth, the wave number k and the angular frequency ω are variable. In general, it can be assumed that,

$$k = k(x, t), \ \ \omega = \omega(x, t)$$

Now the wave difference in a tiny segment $(x, x + \Delta x)$ in space is considered. The net increase of the wave entering into the segment within time of Δt is

$$\left(\frac{\omega(x, t)}{2\pi} - \frac{\omega(x + \Delta x, t)}{2\pi} \right) \Delta t$$

and the increase of the wave in the segment after Δt is

$$\left(\frac{k(x, t + \Delta t)}{2\pi} - \frac{k(x, t)}{2\pi} \right) \Delta x$$

As the wave in the segment is neither lost nor created, it gives

$$\frac{\partial k}{\partial t} + \frac{\partial \omega}{\partial x} = 0 \tag{2.2.41}$$

This represents the principle of wave number conservation. By using the dispersion relation $\omega = \omega(k)$, the above equation becomes

$$\frac{\partial k}{\partial t} + \frac{\partial \omega}{\partial k}\frac{\partial k}{\partial x} = \frac{\partial k}{\partial t} + c_g \frac{\partial k}{\partial x} = 0 \tag{2.2.42}$$

Similarly,

$$\frac{\partial k}{\partial \omega}\frac{\partial \omega}{\partial t} + \frac{\partial \omega}{\partial x} = \frac{1}{c_g}\frac{\partial \omega}{\partial t} + \frac{\partial \omega}{\partial x} = 0 \tag{2.2.43}$$

$$\frac{\partial \omega}{\partial t} + c_g \frac{\partial \omega}{\partial x} = 0 \tag{2.2.44}$$

If taking

$$\frac{dx}{dt} = c_g \tag{2.2.45}$$

Eqs. (2.2.42) and (2.2.44) turn into

$$\frac{dk}{dt} = 0 \tag{2.2.46}$$

$$\frac{d\omega}{dt} = 0 \tag{2.2.47}$$

Therefore, it can be deemed that the wave number k and the angular frequency ω remain unchanged along the curve $\frac{dx}{dt} = c_g$ in the plane of (x, t).

2.2.4 *Stokes Finite Height Waves*

First of all Stokes [8] applied the perturbation method to the nonlinear wave study, considering the finite height nonlinear waves as Fourier combination of infinite number of linear waves. A small order parameter $\varepsilon = ka$ is chosen, in which k is the wave number ($k = \frac{2\pi}{L}$); a is the wave amplitude in the lowest order, $a = H/2$. Therefore, the general valid condition of the Stokes expansion method is $H/d \ll (kd)^2$, in which kd is the relative water depth. When $kd < 1$ and $H/L \ll 1$, the potential function and the surface elevation of the finite height waves can be expressed, respectively, as

$$\Phi = \sum_{n=1}^{\infty} \varepsilon^n \Phi_n = \varepsilon \Phi_1 + \varepsilon^2 \Phi_2 + \cdots + \varepsilon^n \Phi_n + \cdots \tag{2.2.48}$$

$$\eta = \sum_{n=1}^{\infty} \varepsilon^n \eta_n = \varepsilon \eta_1 + \varepsilon^2 \eta_2 + \cdots + \varepsilon^n \eta_n + \cdots \tag{2.2.49}$$

Substituting Eqs. (2.2.48) and (2.2.49) into the 2D Laplace Eq. (2.1.17) and the bottom boundary condition Eq. (2.1.24), and then taking the term of $1, \varepsilon, \varepsilon^2, \varepsilon^3, \ldots$, respectively, yield,

$$\frac{\partial^2 \Phi_n}{\partial x^2} + \frac{\partial^2 \Phi_n}{\partial y^2} = 0, \quad \text{in the domain}, n = 1, 2, 3 \ldots \tag{2.2.50}$$

$$\frac{\partial \Phi_n}{\partial z} = 0 \quad \text{at}\, z = -d,\ n = 1, 2, 3 \ldots \tag{2.2.51}$$

which means each Φ_n in Eq. (2.2.48) satisfies the Laplace equation and the bottom boundary condition. For the free surface boundary conditions, it is better to give Φ_n and η_n in different orders, respectively, due to the nonlinear terms. The kinematic and dynamic boundary conditions are rewritten as

$$\Phi_z = \frac{\partial \eta}{\partial t} + \Phi_x \frac{\partial \eta}{\partial x} + \Phi_y \frac{\partial \eta}{\partial y}, \quad z = \eta \tag{2.2.52}$$

$$\eta = -\frac{1}{g}\left[\frac{\partial \Phi}{\partial t} + \frac{1}{2}\left(\frac{\partial \Phi}{\partial x}\right)^2 + \frac{1}{2}\left(\frac{\partial \Phi}{\partial z}\right)^2\right], \quad z = \eta \tag{2.2.53}$$

Since the free surface is always around the still water level, a Taylor expansion of Φ around the free surface $z = \eta$ can be made as

$$\Phi_{z=\eta} = \Phi|_{z=0} + \eta \left.\frac{\partial \Phi}{\partial z}\right|_{z=0} + \left.\frac{\eta^2}{2!}\frac{\partial^2 \Phi}{\partial z^2}\right|_{z=0} + \cdots + \left.\frac{\eta^n}{n!}\frac{\partial^n \Phi}{\partial z^n}\right|_{z=0} + \cdots \tag{2.2.54}$$

Substituting Eqs. (2.2.48) and (2.2.49) into Eqs. (2.2.52) and (2.2.53), using Eq. (2.2.54), and sorting out the terms according to the power of the small order parameter ε give

$$\varepsilon^1\left(\frac{\partial \Phi_1}{\partial z} - \frac{\partial \eta_1}{\partial t}\right) + \varepsilon^2\left(\frac{\partial \Phi_2}{\partial z} - \frac{\partial \eta_2}{\partial t} + \eta_1\frac{\partial^2 \Phi_1}{\partial z^2} - \frac{\partial \eta_1}{\partial x}\frac{\partial \Phi_1}{\partial x}\right) + \varepsilon^3(\cdots) + \cdots = 0 \tag{2.2.55}$$

$$\varepsilon^1\left(\frac{\partial \Phi_1}{\partial t} + g\eta_1\right) + \varepsilon^2\left\{g\eta_2 + \frac{\partial \Phi_2}{\partial t} + \eta_1\frac{\partial^2 \Phi_1}{\partial t \partial z} + \frac{1}{2}\left[\left(\frac{\partial \Phi_1}{\partial x}\right)^2 + \left(\frac{\partial \Phi_1}{\partial z}\right)^2\right]\right\} + \varepsilon^3(\cdots) + \cdots = 0 \tag{2.2.56}$$

As ε is a constant less than 1, the coefficients of ε^n should tend to be zero to make Eqs. (2.2.55) and (2.2.56) tenable. Thus, a series of partial differential equations independent of ε are obtained. For the terms of ε^1, the linear free surface boundary conditions are given as

$$\frac{\partial \Phi_1}{\partial z} - \frac{\partial \eta_1}{\partial t} = 0, \quad z = 0 \tag{2.2.57}$$

$$\eta = -\frac{1}{g}\left(\frac{\partial \Phi_1}{\partial t}\right), \quad z = 0 \tag{2.2.58}$$

For the terms of ε^2, the free surface boundary conditions of the second-order Stokes waves are given as

$$\frac{\partial \Phi_2}{\partial z} - \frac{\partial \eta_2}{\partial t} + \eta_1\frac{\partial^2 \Phi_1}{\partial z^2} - \frac{\partial \eta_1}{\partial x}\frac{\partial \Phi_1}{\partial x} = 0, \quad z = 0 \tag{2.2.59}$$

$$\eta_2 = -\frac{1}{g}\left[\frac{\partial \Phi_2}{\partial t} + \eta_1 \frac{\partial^2 \Phi_1}{\partial z \partial t} + \frac{1}{2}\left(\frac{\partial \Phi_1}{\partial x}\right)^2 + \frac{1}{2}\left(\frac{\partial \Phi_1}{\partial z}\right)^2\right], \quad z = 0 \tag{2.2.60}$$

After getting Φ_1 and η_1 in the first order, substituting the results into the Laplace equation for the second-order waves Eq. (2.2.50), and the boundary conditions Eqs. (2.2.59) and (2.2.60), then Φ_2 and η_2 can be obtained, which satisfy the Laplace equation as well as the free surface and bottom boundary conditions in the second order. Similarly, the approximate solutions of Φ_n and η_n in various order can be obtained by solving the partial differential equations from the lower to higher-order gradually.

(1) Second-order Stokes waves

The wave parameters for the second-order Stokes waves can be obtained as presented in Table 2.3 by substituting the velocity potential and the surface elevation of linear wave theory into Eqs. (2.2.59) and (2.2.60).

(2) Third-order Stokes waves

Skjelbreia [9] enhanced the accuracy of the Stokes waves up to the third-order Stokes' approximation. The main results are presented in Table 2.4.
in which,

Table 2.3 Wave parameter formulas for the second-order Stokes waves

Velocity potential function Φ	$\Phi = \frac{\pi H}{kT}\frac{\cosh ks}{\sinh kd}\sin\theta + \frac{3}{8}\frac{\pi H}{kT}\left(\frac{\pi H}{L}\right)\frac{\cosh 2ks}{\sinh^4 kd}\sin 2\theta$
Wave celerity c	$c^2 = \frac{g}{k}\tanh kd$
Surface elevation η	$\eta = \frac{H}{2}\cos\theta + \frac{H}{8}\left(\frac{\pi H}{L}\right)\frac{\cosh kd}{\sinh^3 kd}(\cosh 2kd + 2)\cos 2\theta$
Horizontal velocity u	$u = \frac{\pi H}{T}\frac{\cosh ks}{\sinh kd}\cos\theta + \frac{3}{4}\frac{\pi H}{T}\left(\frac{\pi H}{L}\right)\frac{\cosh 2ks}{\sinh^4 kd}\cos 2\theta$
Vertical velocity w	$w = \frac{\pi H}{T}\frac{\sinh ks}{\sinh kd}\sin\theta + \frac{3}{4}\frac{\pi H}{T}\left(\frac{\pi H}{L}\right)\frac{\sinh 2ks}{\sinh^4 kd}\sin 2\theta$
Horizontal acceleration $\partial u/\partial t$	$\frac{\partial u}{\partial t} = \frac{2\pi^2 H}{T^2}\frac{\cosh ks}{\sinh kd}\sin\theta + \frac{3\pi^2 H}{T^2}\left(\frac{\pi H}{L}\right)\frac{\cosh 2ks}{\sinh^4 kd}\sin 2\theta$
Vertical acceleration $\partial w/\partial t$	$\frac{\partial w}{\partial t} = -\frac{2\pi^2 H}{T^2}\frac{\sinh ks}{\sinh kd}\cos\theta - \frac{3\pi^2 H}{T^2}\left(\frac{\pi H}{L}\right)\frac{\sinh 2ks}{\sinh^4 kd}\cos 2\theta$
Pressure p	$p = -\rho g z + \frac{1}{2}\rho g H\frac{\cosh ks}{\cosh kd}\cos\theta + \frac{3}{4}\rho g H\left(\frac{\pi H}{L}\right)\frac{1}{\sinh 2kd}\left(\frac{\cosh 2ks}{\sinh^2 kd} - \frac{1}{3}\right)\cos 2\theta - \frac{1}{4}\rho g H\left(\frac{\pi H}{L}\right)\frac{1}{\sinh 2kd}(\cosh 2ks - 1)$
Mean energy density E	$E = \frac{1}{8}\rho g H^2(+O(\varepsilon^4))$
Energy flux P	$P = \frac{1}{16}\rho g H^2 c(1 + \frac{2kd}{\sinh 2kd})(+O(\varepsilon^4))$

in which, $\theta = kx - \omega t, \ s = z + d$

Table 2.4 Wave parameter formulas for the third-order stokes waves

Velocity potential function Φ	$\Phi = F_1 \cosh ks \sin\theta + F_2 \cosh 2ks \sin 2\theta + F_3 \cosh 3ks \sin 3\theta$
Wave celerity c	$c^2 = \frac{g}{k}\tanh kd\left(1 + k^2a^2\frac{8+\cosh 4kd}{8\sinh^4 kd}\right)$
Surface elevation η	$\eta = af_1\left(\frac{d}{L}\right)\cos\theta + \frac{ka^2}{4}f_2\left(\frac{d}{L}\right)\cos 2\theta + \frac{3k^2a^3}{64}f_3\left(\frac{d}{L}\right)\cos 3\theta$
Horizontal velocity u	$u = F_1 \cosh ks \cos\theta + 2F_2 \cosh 2ks \cos 2\theta + 3F_3 \cosh 3ks \cos 3\theta$
Vertical velocity w	$w = F_1 \sinh ks \sin\theta + 2F_2 \sinh 2ks \sin 2\theta + 3F_3 \sinh 3ks \sin 3\theta$
Horizontal acceleration $\partial u/\partial t$	$\frac{\partial u}{\partial t} = \frac{2\pi c}{T}F_1 \cosh ks \sin\theta + \frac{8\pi c}{T}F_2 \cosh 2ks \sin 2\theta + \frac{18\pi c}{T}F_3 \cosh 3ks \sin 3\theta$
Vertical acceleration $\partial w/\partial t$	$\frac{\partial w}{\partial t} = -\frac{2\pi c}{T}F_1 \sinh ks \sin\theta - \frac{8\pi c}{T}F_2 \sinh 2ks \sin 2\theta - \frac{18\pi c}{T}F_3 \sinh 3ks \sin 3\theta$

$$f_1\left(\frac{d}{L}\right) = 1 + \frac{k^2a^2}{16}\cdot\frac{18\cosh^4 kd - 4\cosh^2 kd + 1}{\sinh^4 kd},$$

$$f_2\left(\frac{d}{L}\right) = \frac{\cosh kd(2 + \cosh 2kd)}{\sinh^3 kd}, \quad f_3\left(\frac{d}{L}\right) = \frac{1 + 8\cosh^6 kd}{\sinh^6 kd}$$

a is a parameter related to the wave height H and kd. It can be determined by,

$$H = 2a\left(1 + \frac{k^2a^2}{64}\frac{96\cosh^6 kd - 88\cosh^4 kd + 20\cosh^2 kd - 1}{\sinh^6 kd}\right).$$

$$F_1 = \frac{ac}{\sin kd}\left(1 + \frac{k^2a^2}{16}\cdot\frac{2 + \cosh 2kd}{\sinh^4 kd}\right), \quad F_2 = \frac{3}{8}\frac{ka^2c}{\sinh^4 kd},$$

$$F_3 = k^2a^3c\cdot\frac{11 - 2\cosh 3kd}{64\sinh^7 kd}.$$

2.3 Dispersive Waves in Shallow Water

2.3.1 Nonlinear Shallow Water Wave Equations

From the above discussion, it can be seen that the linear small-amplitude wave theory is valid only when $H/L \ll 1$ and $H/d \ll 1$, and the Stokes finite height wave theory is mainly valid in deep and intermediate water depth. For the waves in shallow water such as the waves propagating to the coastal areas where the water is shallow, the wave height and nonlinearity are high. Thus, nonlinear dispersive wave theories are needed. Here, in order to analyze the shallow water waves, three important parameters are defined, which include the relative water depth $\mu = kh = \frac{2\pi}{L}h$, the relative wave height $\varepsilon = \frac{a}{h}$ and the Ursell number (U_r). Based on the general definition in Chap. 1, the Ursell number is given by

$$U_r = \frac{\varepsilon}{\mu^2} = \frac{a}{h}\frac{1}{(kh)^2} = \frac{a}{h}\left(\frac{L}{2\pi h}\right)^2 \tag{2.3.1}$$

where $h = \eta + d$.

It is noted that, in small-amplitude wave theory, the still water depth d is commonly used for the water depth as the amplitude is very small. But in nonlinear waves, h is used for the water depth.

When μ^2 and ε are in the same order, $U_r = O(1)$.

In order to analyze the nonlinear shallow water waves, the following non-dimensional parameters [3] are introduced,

$$\begin{aligned}
&(x', y') = k(x, y),\ z' = \frac{z}{h},\ t' = k(gh)^{1/2}t, \\
&\eta' = \frac{\eta}{a},\ \Phi' = \Phi\left[\frac{a}{kh}(gh)^{1/2}\right]^{-1} \\
&(u, v) = \left(\frac{\partial}{\partial x}, \frac{\partial}{\partial y}\right)\Phi = \frac{a}{h}(gh)^{1/2}\left(\frac{\partial}{\partial x'}, \frac{\partial}{\partial y'}\right)\Phi' = \frac{a}{h}(gh)^{1/2}(u', v') \\
&w = \frac{\partial \Phi}{\partial z} = \frac{1}{kh}\frac{a}{h}(gh)^{1/2}\frac{\partial \Phi'}{\partial z'} = \frac{1}{kh}\frac{a}{h}(gh)^{1/2}(w')
\end{aligned} \tag{2.3.2}$$

Non-dimensionalizing the fundamental equations for potential waves, Eqs. (2.1.17), (2.1.21), (2.1.22) and (2.1.24) yields

$$\mu^2(\Phi'_{x'x'} + \Phi'_{y'y'}) + \Phi'_{z'z'} = 0, \quad -1 < z' < \varepsilon\eta' \tag{2.3.3}$$

$$\mu^2(\eta'_{t'} + \varepsilon\Phi'_{x'}\eta'_{x'} + \varepsilon\Phi'_{y'}\eta'_{y'}) = \Phi'_{z'}, \quad z' = \varepsilon\eta' \tag{2.3.4}$$

$$\mu^2(\Phi'^2_{t'} + \eta') + \frac{1}{2}\varepsilon\left[\mu^2(\Phi'^2_{x'} + \Phi'^2_{y'}) + \Phi'^2_{z'}\right] = 0,\ z' = \varepsilon\eta' \tag{2.3.5}$$

$$\Phi'_{x'} = 0,\ z' = -1 \tag{2.3.6}$$

In Cartesian coordinate system, expanding the velocity potential function into the power series of the vertical coordinate (Omitting the apostrophe) yields

$$\Phi(x, y, z, t) = \sum_{n=0}^{\infty}(z+1)^n\Phi_n \tag{2.3.7}$$

where $\Phi_n = \Phi_n(x, y, t)$, $n = 0, 1, 2, 3\ldots$

Substituting Eqs. (2.3.7) into (2.3.3) gives

$$\mu^2\nabla^2\Phi + \frac{\partial^2\Phi}{\partial z^2} = \sum_{n=0}^{\infty}(z+1)^n\left[\mu^2\nabla^2\Phi_n + (n+2)(n+1)\Phi_{n+2}\right] = 0 \quad (2.3.8)$$

As z is an arbitrary value within $(-1, \varepsilon\eta)$, the coefficients of each power of $(z+1)$ must be zero, which gives

$$\Phi_{n+2} = \frac{-\mu^2\nabla^2\Phi_n}{(n+2)(n+1)}, \quad n = 0, 1, 2, \ldots \quad (2.3.9)$$

From the boundary condition Eq. (2.3.6), it yields $\Phi_1 \equiv 0$. From Eq. (2.3.9), it is obtained that Φ_n is zero for each odd n, i.e.,

$$\Phi_1 = \Phi_3 = \Phi_5 = \cdots = 0$$

$$\Phi_2 = \frac{-\mu^2}{2!}\nabla^2\Phi_0,\ \Phi_4 = \frac{+\mu^4}{4!}\nabla^2\nabla^2\Phi_0,\ \Phi_6 = \frac{-\mu^6}{6!}\nabla^2\nabla^2\nabla^2\Phi_0$$

From $\Phi_0 = O(\mu) = O(1)$, it yields $\Phi_2 = O\left(\mu^2\right)$ and $\Phi_4 = O\left(\mu^4\right)$. Therefore, the velocity potential with an error of $O\left(\mu^6\right)$ reads

$$\Phi = \Phi_0 - \frac{\mu^2}{2!}(z+1)^2\nabla^2\Phi_0 + \frac{\mu^4}{24}(z+1)^4\nabla^2\nabla^2\Phi_0 + O\left(\mu^6\right) \quad (2.3.10)$$

By means of deduction, the non-dimensional nonlinear wave equations with terms of $O\left(\mu^2\right)$ are obtained (the terms of $O\left(\mu^3\right)$ are zero),

$$\tilde{H}_t + \varepsilon\nabla\cdot\left(\tilde{H}U\right) = 0 \quad (2.3.11)$$

$$U_t + \varepsilon U\cdot\nabla U + \frac{\nabla\tilde{H}}{\varepsilon} + \frac{\mu^2}{6}\left(\tilde{H}^2\nabla^2 U\right) + \mu^2\nabla\left[-\frac{\varepsilon}{3}\tilde{H}^2 U\cdot\nabla^2 U + \frac{\varepsilon}{2}\tilde{H}^2(\nabla\cdot U)^2 - \frac{\tilde{H}^2}{2}\nabla\cdot U_t\right] = O\left(\mu^4\right) \quad (2.3.12)$$

$$p = (\varepsilon\eta - z) - \frac{\mu^2}{2}\left[\tilde{H}^2 - (z+1)^2\right]\left\{\nabla\cdot U_{0t} + \varepsilon\left[U_0\nabla^2 U_0 - (\nabla\cdot U_0)^2\right]\right\} = O\left(\mu^4\right) \quad (2.3.13)$$

where $\tilde{H} = 1 + \varepsilon\eta$; U is the depth-averaged velocity,

$$U = \frac{1}{h}\int_{-1}^{\varepsilon\eta}\nabla\Phi dz = U_0 - \frac{\mu^2}{6}\tilde{H}^2\nabla^2 U_0 + O\left(\mu^4\right),\ U_0 = \nabla\Phi_0$$

In Eqs. (2.3.11)–(2.3.13), when $\mu \to 0$ (the wavelength is very long) and in the condition of finite amplitude with $\varepsilon = O(1)$, the dimensional shallow water long wave equations or so-called Airy long wave equations are obtained

$$\eta_t + \nabla \cdot [(\eta + d)U] = 0 \tag{2.3.14}$$

$$U_t + U \cdot \nabla U + g\nabla \eta = 0 \tag{2.3.15}$$

$$p = \rho g(\eta - z) \tag{2.3.16}$$

It is seen that the Airy long wave is the first-order approximation of nonlinear shallow water wave described by Eqs. (2.3.11)–(2.3.13). Equations (2.3.14) and (2.3.15) are the first-order quasi-linear equations. When $\mu = \frac{2\pi h}{L}$ is very small, the long waves are hyperbolic waves, and the distribution of pressure along the water depth is hydrostatic. This will be described in the next section.

2.3.2 Boussinesq Equations

For the weakly nonlinear waves in shallow water, only terms of $O(\varepsilon)$ and $O\left(\mu^2\right)$ are retained in Eqs. (2.3.11)–(2.3.13). Taking into account $O(\varepsilon) = O\left(\mu^2\right) < 1$, then the non-dimensional Boussinesq equations are obtained,

$$\eta_t + \nabla \cdot [(\varepsilon\eta + 1)U] = 0 \tag{2.3.17}$$

$$\mathbf{U}_t + \varepsilon \mathbf{U} \cdot \nabla \mathbf{U} + \nabla \eta - \frac{\mu^2}{3}\nabla\nabla \cdot \mathbf{U}_t = 0 \tag{2.3.18}$$

$$p = \eta - z - \frac{\mu^2}{2}\left(-z^2 + 2z\right)\nabla \cdot \mathbf{U}_t \tag{2.3.19}$$

In order to analyze the properties of Boussinesq equations, transforming Eqs. (2.3.17)–(2.3.18) to dimensional form and neglecting the nonlinear terms give

$$\eta_t + dU_x = 0 \tag{2.3.20}$$

$$U_t + g\eta_x - \frac{d^2}{3}U_{xxt} = 0 \tag{2.3.21}$$

It is assumed that the waveform is 1D sinusoidal wave with,

$$\eta = ae^{i(kx - \omega t)}, \quad U = ue^{i(kx - \omega t)} \tag{2.3.22}$$

Substituting Eq. (2.3.22) into Eqs. (2.3.20) and (2.3.21) yields

$$\left.\begin{aligned} &-i\omega a + ikhu = 0 \\ &-i\omega u + ikga - \tfrac{h^2}{3}(ik)^2(-i\omega)u = 0 \end{aligned}\right\} \tag{2.3.23}$$

These two equations are homogeneous equations for a and u. The coefficient determinant must be zero to have non-trivial solution,

$$\begin{vmatrix} -i\omega & ikh \\ igk & -i\omega\left(1+\frac{k^2h^2}{3}\right) \end{vmatrix} = 0$$

Thus,

$$\omega^2 = \frac{ghk^2}{1+\frac{k^2h^2}{3}} = ghk^2\left(1-\frac{k^2h^2}{3}+\cdots\right) \tag{2.3.24}$$

The phase velocity reads

$$c \approx (gh)^{1/2}\left(1+\frac{k^2h^2}{3}\right)^{-1/2} \approx (gh)^{1/2}\left(1-\frac{k^2h^2}{3}\right)^{1/2} \tag{2.3.25}$$

From the above discussion, it is known that, for the shallow water waves in linear small-amplitude wave theory, the phase velocity (i.e., the wave celerity) $c=\sqrt{gd}$ given by Eq. (2.2.26) is independent of the wavelength so that the waves are non-dispersive. By comparing Eq. (2.3.25) with Eq. (2.2.26), it is found that the Boussinesq waves are dispersive. The phase velocity has one more term of $\frac{k^2h^2}{3}$. The wave celerity is dependent on the wave number k or the wavelength. By comparing Eq. (2.3.15) with Eq. (2.3.18), it is seen that the dispersion property of the Boussinesq equations is due to the existence of the third-order derivative terms. The dispersion of the Boussinesq equations can be understood more clearly by further comparing it with the dispersion of the first-order linear Stokes waves.

The dispersion relation of linear Stokes waves has been given in Eq. (2.2.17),

$$\omega^2 = gk\tanh kd \quad \text{or} \quad c^2 = gd\frac{\tanh kd}{kd}$$

Making Padé expansion [2] of $\frac{\tanh kd}{kd}$ yields

$$c^2 = gd\left(1-\frac{1}{3}(kd)^2+\frac{2}{15}(kd)^4-\frac{17}{315}(kd)^6+O\left(\mu^8\right)\right) \tag{2.3.26}$$

By comparing Eqs. (2.3.25) and (2.3.26), it is seen that the phase velocity of the shallow water waves described by the Boussinesq Eqs. (2.3.17) and (2.3.18) correspond to taking the terms up to the second-order from the linear Stokes dispersion relation. Therefore, the dispersion of the Boussinesq equations is weak. The waves described by the Boussinesq equations are weakly dispersive waves.

2.3.3 Unidirectional Dispersive Waves—KDV Equation

The shallow water nonlinear dispersive waves described by the Boussinesq equations contain the waves moving both to the left and to the right. A kind of unidirectional dispersive wave equation called Korteweg–deVries (KDV) equation will be derived from the Boussinesq equations as below.

The 1D non-dimensional Boussinesq Eqs. (2.3.17) and (2.3.18) read

$$\eta_t + \varepsilon(\eta U)_x + U_x = 0 \tag{2.3.27}$$

$$U_t + \varepsilon U U_x + \eta_x = \frac{\mu^2}{3} U_{xxt} \tag{2.3.28}$$

By using the coordinate transformation $\sigma = x - t$ and $\tau = \varepsilon t$, the derivatives $\frac{\partial}{\partial x}$ and $\frac{\partial}{\partial t}$ turn to $\frac{\partial}{\partial x} \to \frac{\partial}{\partial \sigma}$ and $\frac{\partial}{\partial t} \to -\frac{\partial}{\partial \sigma} + \varepsilon \frac{\partial}{\partial \tau}$. Then Eqs. (2.3.27) and (2.3.28) turn to,

$$\varepsilon \frac{\partial \eta}{\partial \tau} - \frac{\partial \eta}{\partial \sigma} + \frac{\partial U}{\partial \sigma} + \varepsilon \frac{\partial U \eta}{\partial \sigma} = 0 \tag{2.3.29}$$

$$\varepsilon \frac{\partial U}{\partial \tau} - \frac{\partial U}{\partial \sigma} + \frac{\partial \eta}{\partial \sigma} + \varepsilon U \frac{\partial U}{\partial \sigma} = \frac{-\mu^2}{3} U_{\sigma\sigma\sigma} + O\left(\varepsilon \mu^2\right) \tag{2.3.30}$$

Adding the above two equations and retaining the terms of $O(1)$ and $O\left(\mu^2/\varepsilon\right)$ leads to,

$$\frac{\partial (U + \eta)}{\partial \tau} + U \frac{\partial U}{\partial \sigma} + \frac{\partial U \eta}{\partial \sigma} + \frac{\mu^2}{3\varepsilon} U_{\sigma\sigma\sigma} = O\left(\mu^2\right) \tag{2.3.31}$$

By using the first-order approximation of the Bernoulli equation, it yields

$$U = \eta + O\left(\varepsilon, \mu^2\right)$$

Equation (2.3.31) can be rewritten as

$$\frac{\partial U}{\partial \tau} + \frac{3}{2} U \frac{\partial U}{\partial \sigma} + \frac{\mu^2}{6} \frac{1}{\varepsilon} U_{\sigma\sigma\sigma} = 0 \tag{2.3.32}$$

or

$$\frac{\partial \eta}{\partial \tau} + \frac{3}{2} \eta \frac{\partial \eta}{\partial \sigma} + \frac{\mu^2}{6} \frac{1}{\varepsilon} \eta_{\sigma\sigma\sigma} = 0 \tag{2.3.33}$$

Expressed by the dimensional variables in the stationary coordinate system, the above turns to

$$\frac{\partial \eta}{\partial t} + \sqrt{gd}\left(1 + \frac{3}{2}\frac{\eta}{d}\right)\frac{\partial \eta}{\partial x} + \frac{d^2}{6}\sqrt{gd}\frac{\partial^3 \eta}{\partial x^3} = 0 \tag{2.3.34}$$

Equation (2.3.34) is the well-known KDV equation.

2.3.4 *Permanent Wave Solutions to the KDV Equation—Solitary Waves*

Solitary waves are the unidirectional permanent wave solutions to the KDV equation. They are also the unidirectional propagating wave solutions to the Boussinesq equations. Being a traveling wave, a solitary wave has only one single wave crest. The water particles move only in the direction of wave propagation. Figure 2.3 shows a sketch of wave profile and particle paths under a solitary wave. In nature, when a ship is sailing in a channel, it is possible to generate solitary waves. Landslides in lakes and oceans may also generate solitary waves in front of them. When the waves propagating to the nearshore shallow water nearly break, the wave shape and movement properties are similar to solitary waves.

For solitary waves, when $|x| \to \infty$, the wave amplitude tends to zero.

The surface elevation of a solitary wave reads

$$\eta = H\text{sech}^2\left[\sqrt{\frac{3H}{4d^3}}(x - ct)\right] \tag{2.3.35}$$

$$z_s = d + \eta \tag{2.3.36}$$

where the origin of x is set to the position of the wave crest. H is the wave height.

The wave celerity is

$$c^2 = g(d + H) \tag{2.3.37}$$

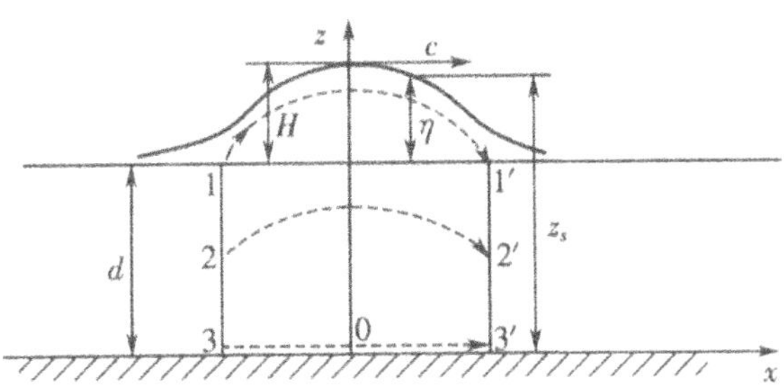

Fig. 2.3 Wave profile and particle paths under a solitary wave

The horizontal velocity is

$$u = \sqrt{gd}\frac{H}{d}\mathrm{sech}^2\left[\sqrt{\frac{3H}{4d^3}}(x - ct)\right] \tag{2.3.38}$$

The vertical velocity is

$$w = \sqrt{3gH}\frac{H}{d}\frac{z}{d}\mathrm{sech}^2\left[\sqrt{\frac{3H}{4d^3}}(x - ct)\right]\tanh\left(\sqrt{\frac{3H}{4d^3}}(x - ct)\right) \tag{2.3.39}$$

The pressure is

$$p(x, z, t) = \rho g\left\{(d - z) + H\mathrm{sech}^2\left[\sqrt{\frac{3H}{4d^3}}(x - ct)\right]\right\} \tag{2.3.40}$$

From the above, it is seen that the main factor to determine the solitary wave properties is H/d. Similar to the phenomenon of the Stokes waves: Waves break when the wave steepness is increased to a certain value. When H/d of the solitary wave goes up to a certain value, waves break at the surface elevation around the crest. Normally, the critical value of H/d is within 0.714–1.03. The commonly used value is 0.78.

Solitary waves are non-periodic since one solitary wave has only one crest. In view of a solitary wave with infinite wavelength, the particle motions are concentrated near the wave crest, and there is almost no motion far from the crest. For a periodic wave propagating to the shallow water, there is no connection between the adjacent crests when the wavelength is large enough although it is finite. Therefore, it is possible to treat the segment of the periodic wave near the crest as a segment of a solitary wave. The so-called effective wavelength L_e is defined as a criterion,

$$L_e = 2\pi/\sqrt{\frac{3H}{d}} \tag{2.3.41}$$

A periodic wave might be treated as a solitary wave when the periodic wavelength $L > L_e$.

2.3.5 *Propagating Wave Solution to the Boussinesq Equations—Cnoidal Wave*

The nonlinear dispersive long waves might be unidirectional or periodic permanent waves. The nonlinearity makes the wave crest steeper but the dispersion restrains this trend. As a result, the waves with different wavelengths are separated. Solitary waves are unidirectional permanent waves, and cnoidal waves are the periodic permanent wave solutions to the Boussinesq equations. Cnoidal waves are finite-amplitude long waves in shallow water. As the surface elevation η is expressed by the Jacobian elliptic cosine function, these waves are called cnoidal waves.

Figure 2.4 shows a sketch of cnoidal wave profile. Based on cnoidal wave theory, some key results can be presented as follows.

The free surface elevation above the bottom is

$$z_s = z_t + H cn^2\left[2K(\kappa)\left(\frac{x}{L} - \frac{t}{T}\right), \kappa\right] \tag{2.3.42}$$

where z_t is the distance from the bottom to the wave trough given by

$$z_t = d - H + \frac{16d^3}{3L^2} K(\kappa)[K(\kappa) - E(\kappa)] \tag{2.3.43}$$

in which, $cn()$ is the Jacobian elliptic cosine function; κ is the elliptic parameter $(0 \leq \kappa < 1)$; K is the complete elliptic integral of the first kind; and E is the complete elliptic integral of the second kind.

$$K(\kappa) = \int_0^{\frac{\pi}{2}} \frac{d\theta}{\sqrt{1 - \kappa^2 \sin^2\theta}} \tag{2.3.44}$$

$$E(\kappa) = \int_0^{\frac{\pi}{2}} \sqrt{1 - \kappa^2 \sin^2\theta} d\theta \tag{2.3.45}$$

The dispersion relation for cnoidal waves reads

$$\frac{16}{3}[\kappa \cdot K(\kappa)]^2 = \left(\frac{L}{d}\right)^2 \cdot \frac{H}{d} \tag{2.3.46}$$

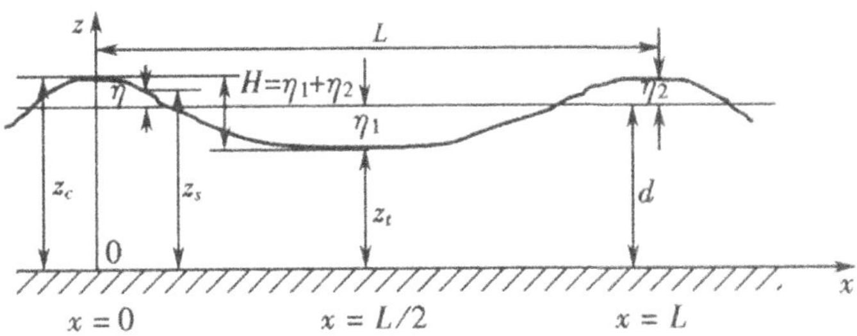

Fig. 2.4 Cnoidal wave profile

The wavelength is given by

$$L = \sqrt{\frac{16\,d^3}{3\,H}}\kappa \cdot K(\kappa) \tag{2.3.47}$$

The wave celerity c and the wave period T are determined, respectively, by

$$c^2 = gd\left[1 + \frac{H}{d}\frac{1}{\kappa^2}\left(-\kappa^2 + 2 - 3\frac{E(\kappa)}{K(\kappa)}\right)\right] \tag{2.3.48}$$

$$T = \frac{4d}{\sqrt{3gH}}\left\{\frac{\kappa \cdot K(\kappa)}{\sqrt{1 + \frac{H}{d}\frac{1}{\kappa^2}\left(-\kappa^2 + 2 - 3\frac{E(\kappa)}{K(\kappa)}\right)}}\right\} \tag{2.3.49}$$

The wave celerity c does not only depend on κ, but also H. This is an important property of nonlinear dispersive waves.

The two limiting cases of cnoidal waves are presented as follows.

(1) $\kappa \to 1$, $E(1) = 1$, $K(1) \to \infty$, $L \to \infty$, $\mathrm{cn}^2 u \to \mathrm{sec\,h}^2 u$, which gives

$$\eta = H\,\mathrm{sec\,h}^2\left[\sqrt{\frac{3H}{4d^3}}(x - ct)\right] \tag{2.3.50}$$

As a limiting case when the wavelength is infinitely long, it describes a solitary wave.

(2) $\kappa \to 0$, $H \to 0$, $c^2 \to gd$, $\mathrm{cn}\,u \to \cos u$, $K(0) = \pi/2$, which yields

$$\eta = \frac{H}{2}\cos\frac{2\pi}{L}(x - ct) \tag{2.3.51}$$

As a limiting case when the wave amplitude is very small, this describes a linear wave.

2.4 Long Waves

The dispersive waves and the dispersive waves in shallow water have been described, respectively, in the above sections. In this section, another type of waves (long waves or hyperbolic waves) will be introduced. Whitham [12] divided the wave motions into two main types, i.e., hyperbolic waves and dispersive waves. When the wavelength is relatively larger than the wave height and the water depth, the waves are called long waves such as tidal waves and flood waves. The long waves can be described by the first-order quasi-linear hyperbolic equation, i.e., the St. Venant equation. Therefore, they are also called hyperbolic waves.

In the following sections, the flood waves in open channels will be discussed in detail as examples of the hyperbolic waves.

2.4.1 Basic Equations

Based on the Euler Eqs. (2.1.1)–(2.1.5), the shallow water wave equations can be derived. They can describe the water flow in open channels, surface runoff, and the water flow in lakes, estuaries, and coastal areas.

The vertical two-dimensional (2D) Euler equations read

$$\frac{\partial u}{\partial x} + \frac{\partial w}{\partial z} = 0 \tag{2.4.1}$$

$$\frac{\partial u}{\partial t} + u\frac{\partial u}{\partial x} + w\frac{\partial u}{\partial z} = -\frac{1}{\rho}\frac{\partial p}{\partial x} \tag{2.4.2}$$

$$\frac{\partial w}{\partial t} + u\frac{\partial w}{\partial x} + w\frac{\partial w}{\partial z} = -\frac{1}{\rho}\frac{\partial p}{\partial z} - g \tag{2.4.3}$$

It is assumed that an open channel has constant water depth. The origin of the z-coordinate is set to the bottom for convenience. See Fig. 2.5 for the definitions. $z = \eta(x, t) = h(x, t)$ is the free surface.$u = u(x, z, t)$ and $w = w(x, z, t)$ are the horizontal and the vertical velocity components, respectively.

The kinematic condition at the free surface Eq. (2.1.20) gives

$$w|_{z=h} = \left(\frac{\partial \eta}{\partial t} + u\frac{\partial h}{\partial x}\right)_{z=h} \tag{2.4.4}$$

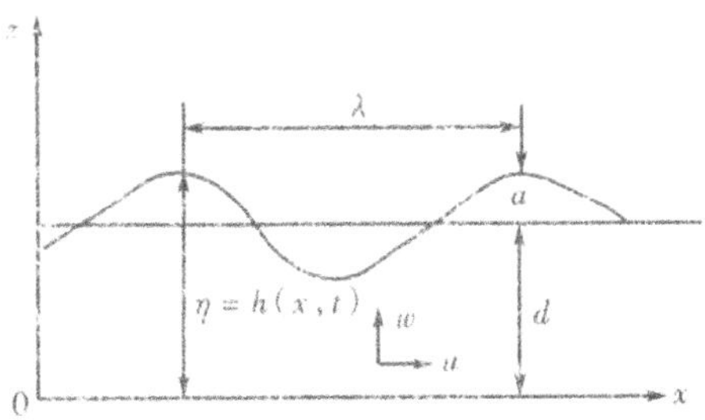

Fig. 2.5 Definitions for long waves

The depth-averaged velocity is defined as

$$U = \frac{1}{h(x,t)} \int_{0}^{h(x,t)} u(x,z,t)dz \tag{2.4.5}$$

Integrating the continuity Eq. (2.4.1) over the water depth and by means of Leibniz's law, it yields,

$$w = -\int_{0}^{h} \frac{\partial u}{\partial x} dz = -\frac{\partial}{\partial x} \int_{0}^{h} u dz + u_{\eta} \frac{\partial h}{\partial x} \tag{2.4.6}$$

where u_{η} is the surface velocity. From the kinematic condition at the free surface Eq. (2.4.4) and the continuity equation in integral form Eq. (2.4.6), it gives

$$\frac{\partial h}{\partial t} + u_{\eta} \frac{\partial h}{\partial x} = -\frac{\partial (hU)}{\partial x} + u_{\eta} \frac{\partial h}{\partial x}$$

i.e.,

$$\frac{\partial h}{\partial t} + \frac{\partial}{\partial x}(hU) = 0 \tag{2.4.7}$$

This is another form of the continuity Eq. (2.4.1) under the condition of free surface. In this equation, the variables are the water depth h and the discharge per unit width Uh.

Integrating the momentum equation in x-direction Eq. (2.4.2) from $z = 0$ to $z = h(x,t)$ yields

$$\frac{\partial}{\partial t}(hU) + \frac{\partial}{\partial x}\left[h\left(\overline{u^2} + \bar{p}\right)\right] = 0 \tag{2.4.8}$$

where $h\overline{u^2} = \int_{0}^{h} u^2 dz = \alpha' h U^2$, in which α' is a momentum correction coefficient which can be taken as $\alpha' \approx 1$; $h\bar{p} = \frac{1}{\rho} \int_{0}^{h} p dz$.

Integrating the momentum equation in the z-direction Eq. (2.4.3) yields

$$\frac{p(x,z,t)}{\rho} = -\int_{h}^{z} \left(g + \frac{Dw}{Dt}\right) dz \tag{2.4.9}$$

where $\frac{Dw}{Dt} = \frac{\partial w}{\partial t} + u\frac{\partial w}{\partial x} + v\frac{\partial w}{\partial z}$,

$$h\bar{p}(x,t) = \int_0^h \frac{p(x,z,t)}{\rho} dz = \int_0^h \left[\int_z^h \left(g + \frac{Dw}{Dt} \right) dz \right] dz$$
$$= \frac{gh^2}{2} + \int_0^h \left(z\frac{Dw}{Dt} \right) dz \tag{2.4.10}$$

Substituting Eqs. (2.4.10) into (2.4.8) yields the following depth-integrated momentum equation,

$$\frac{\partial}{\partial t}(Uh) + \frac{\partial}{\partial x}\left(\alpha' U^2 h + \frac{1}{2} g h^2\right) + \frac{\partial}{\partial x}\left[\int_0^h z\frac{Dw}{Dt} dz\right] = 0 \tag{2.4.11}$$

Equations (2.4.7) and (2.4.11) are the governing equations for 1D shallow water waves. The last term on the left side of Eq. (2.4.11) is dependent on the vertical velocity component w.

The shallow water waves with different Ursell numbers (U_r) have different properties. When $U_r \ll 1$, they are called shallow water long waves. When $U_r = O(1)$, they are shallow water short waves, i.e., the dispersive waves in shallow water as described in Sect. 2.3.

When $U_r \ll 1$, and a/h is an arbitrary value, then $(h/L)^2 \ll 1$. In other words, the wavelength L is much larger than the water depth h. The streamline curvature is very small, and the vertical acceleration can be ignored. Thus, the last term on the left side of Eq. (2.4.11) can be omitted. In general, the flood waves and tidal waves in open channels belong to this kind. Therefore, the governing equations for shallow water long waves read

$$\frac{\partial h}{\partial t} + \frac{\partial}{\partial x}(Uh) = 0 \tag{2.4.7}$$

$$\frac{\partial}{\partial t}(Uh) + \frac{\partial}{\partial x}\left(\alpha' U^2 h + \frac{1}{2} g h^2\right) = 0 \tag{2.4.12}$$

Equations (2.4.7) and (2.4.12) are the 1D first-order quasi-linear hyperbolic equations, which are the basic equations for unsteady flow in open channels. They are the first-term approximation of nonlinear shallow water wave Eqs. (2.3.14) and (2.3.25) in dimensional conservation form. They are also called the St. Venant equations.

Rewriting Eqs. (2.4.7) and (2.4.12) in vector form gives

$$\frac{\partial \boldsymbol{f}}{\partial t} + \frac{\partial \boldsymbol{g}(\boldsymbol{f})}{\partial x} = 0 \tag{2.4.13}$$

$$\boldsymbol{f} = \begin{bmatrix} h \\ Uh \end{bmatrix}, \quad \boldsymbol{g} = \boldsymbol{g}(\boldsymbol{f}) = \begin{bmatrix} Uh \\ \alpha' U^2 h + \frac{gh^2}{2} \end{bmatrix} \tag{2.4.14}$$

These equations are suitable for the calculation of 1D unsteady flow in horizontal open channels having rectangular cross section without erosion.

It is supposed that a lateral inflow with a discharge per unit width $q(\mathrm{m}^2/\mathrm{s})$ enters to a straight horizontal channel. The channel width is denoted by b. The lateral inflow enters to the channel with a velocity component u_i in the positive direction of the channel as shown in Fig. 2.6. Then the basic equations read

$$\frac{\partial h}{\partial t} + \frac{\partial (Uh)}{\partial x} - \frac{q}{b} = 0 \tag{2.4.15}$$

$$\frac{\partial}{\partial t}(Uh) + \frac{\partial}{\partial x}\left(\alpha' U^2 h + \frac{gh^2}{2}\right) - \frac{qu_i}{b} = 0 \tag{2.4.16}$$

Considering the bottom friction, wind stress over the surface and the bottom slopes, the equations can be rewritten as

$$\frac{\partial h}{\partial t} + \frac{\partial}{\partial x}(Uh) = 0 \tag{2.4.17}$$

$$\frac{\partial}{\partial t}(Uh) + \frac{\partial}{\partial x}\left(\alpha' U^2 h + \frac{gh^2}{2}\right) + s_f - \frac{\tau_w}{\rho} - ghs_0 = 0 \tag{2.4.18}$$

where $s_f = \frac{\bar{u}|\bar{u}|}{C^2 h}$; C is the Chezy number; τ_w is the wind stress; and s_0 is the bottom slope.

Naturally, the bottom elevation and the shape of the cross section vary along the channel. Thus, it is convenient to use the water level of the free surface η and the flux Q to replace the variables of h and U in the above equations. Here $Q = U \cdot A$, in which A is the submerged cross-sectional area; U is the averaged velocity on the cross section, i.e., $U = \frac{1}{A}\int u dA$, in which u is the particle velocity. Thus, $Q = \int_A u dA$ is obtained.

Figure 2.7 shows a definition sketch of a cross section in natural rivers. $\eta = z_0 + h$ is the water level, and $z_0 = z_0(x)$ is the bottom elevation. $A = A(h)$ is the cross-sectional area, and $R = A/\chi$ is the hydraulic radius, in which χ is the total boundary

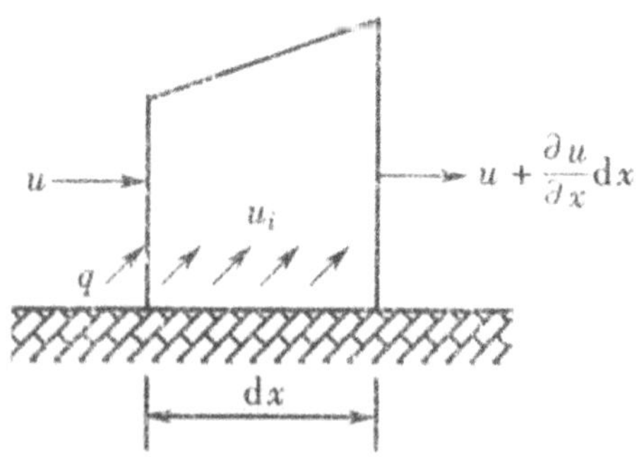

Fig. 2.6 Incoming flow from side of the channel

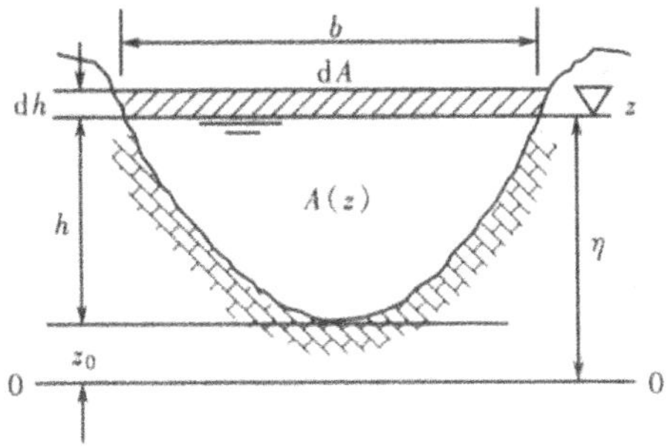

Fig. 2.7 Sketch of the cross-section

perimeter of the submerged cross section; $b = b(\eta, x) = \frac{\partial A}{\partial \eta}$ is the surface width. The basic equations become

$$\frac{\partial \eta}{\partial t} + \frac{1}{b}\frac{\partial Q}{\partial x} = 0 \tag{2.4.19}$$

$$\frac{\partial Q}{\partial t} + \frac{\partial}{\partial x}\left(\frac{\alpha' Q^2}{A}\right) + gA\frac{\partial \eta}{\partial x} + g\frac{|Q|Q}{C^2 RA} = 0 \tag{2.4.20}$$

These above equations can be used to calculate 1D unsteady flow in natural rivers such as flood waves and tidal waves.

In the case of continuous lateral inflow of $q(\mathrm{m}^2/\mathrm{s})$ entering with a direction of Φ from the axis of the channel, and u_i is the averaged velocity, the equations turn to,

$$\frac{\partial \eta}{\partial t} + \frac{1}{b}\frac{\partial Q}{\partial x} - \frac{q}{b} = 0 \tag{2.4.21}$$

$$\frac{\partial Q}{\partial t} + \frac{\partial}{\partial x}\left(\frac{\alpha' Q^2}{A}\right) + gA\frac{\partial \eta}{\partial x} + g\frac{|Q|Q}{C^2 RA} - u_i \cos\Phi q = 0 \tag{2.4.22}$$

For the shallow water flow in large areas, such as estuary and coast areas, 1D equations are not suitable for use. If the water depth and the vertical velocity components w are much smaller than the spatial scales in the- x and y-directions and the horizontal velocities, respectively, the flow can be thought as 2D horizontal flow. U and V are defined as the depth-averaged velocity components in the x- and y-directions, respectively. The continuity and momentum equations read

$$\frac{\partial \eta}{\partial t} + \frac{\partial}{\partial x}(Uh) + \frac{\partial}{\partial y}(Vh) = 0 \tag{2.4.23}$$

$$\frac{\partial}{\partial t}(Uh) + \frac{\partial}{\partial x}\left(U^2 h\right) + \frac{\partial}{\partial y}(UVh) + gh\frac{\partial \eta}{\partial x} + \frac{gU\left(U^2 + V^2\right)^{\frac{1}{2}}}{C^2} - \gamma h V = 0 \tag{2.4.24}$$

$$\frac{\partial}{\partial t}(Vh) + \frac{\partial}{\partial x}(UVh) + \frac{\partial}{\partial y}\left(V^2h\right) + gh\frac{\partial \eta}{\partial y} + \frac{gV\left(U^2+V^2\right)^{\frac{1}{2}}}{C^2} + \gamma hU = 0 \tag{2.4.25}$$

where γ is the Coriolis parameter, $\gamma = 2\omega_e \sin \Phi_{lat}$, in which ω_e is the angular rate of revolution and Φ_{lat} is the geographic latitude. The Coriolis force should be considered for the calculation in large areas. The detailed derivation of Eqs. (2.4.24) and (2.4.25) will be described in Sect. 3.3.1.

2.4.2 *Theory of Characteristics for 1D Long Waves in Channel*

Equations (2.4.7) and (2.4.12) are the first-order quasi-linear hyperbolic partial differential equations. There are two characteristic lines and two characteristic relations in the plane of (x, t). It is possible to use four ordinary differential equations to solve the problem instead of using the two partial differential Eqs. (2.4.7) and (2.4.12).

The method of characteristics is not only an accurate method to solve hyperbolic partial differential equations, but also an important theoretical basis to establish the difference schemes in finite difference for this type of partial differential equations.

Equations (2.4.7) and (2.4.12) can be rewritten in Euler form as

$$\frac{\partial h}{\partial t} + U\frac{\partial h}{\partial x} + h\frac{\partial U}{\partial x} = 0 \tag{2.4.26}$$

$$\frac{\partial U}{\partial t} + U\frac{\partial U}{\partial x} + g\frac{\partial h}{\partial x} = 0 \tag{2.4.27}$$

A new variable $c = \sqrt{gh}$ is introduced, which is the propagating velocity of the long wave in still water. Differentiating c^2 yields

$$2c\frac{\partial c}{\partial x} = g\frac{\partial h}{\partial x} \tag{2.4.28a}$$

$$2c\frac{\partial c}{\partial t} = g\frac{\partial h}{\partial t} \tag{2.4.28b}$$

Substituting Eqs. (2.4.28a) and (2.4.28b) into Eqs. (2.4.26) and (2.4.27), the differential equations with variables of c and U are obtained

$$2\frac{\partial c}{\partial t} + 2U\frac{\partial c}{\partial x} + c\frac{\partial U}{\partial x} = 0 \tag{2.4.29}$$

$$\frac{\partial U}{\partial t} + 2c\frac{\partial c}{\partial x} + U\frac{\partial U}{\partial x} = 0 \tag{2.4.30}$$

Adding and subtracting Eqs. (2.4.29) and (2.4.30), respectively, give

$$\frac{\partial}{\partial t}(U+2c)+(U+c)\frac{\partial}{\partial x}(U+2c)=0 \tag{2.4.31}$$

$$\frac{\partial}{\partial t}(U-2c)+(U-c)\frac{\partial}{\partial x}(U-2c)=0 \tag{2.4.32}$$

In Eq. (2.4.31), if

$$\frac{dx}{dt}=U+c \tag{2.4.33}$$

then it becomes a complete differential equation

$$\frac{d(U+2c)}{dt}=0 \quad \text{or} \quad U+2c=\text{const} \tag{2.4.34}$$

In Eq. (2.4.32), if

$$\frac{dx}{dt}=U-c \tag{2.4.35}$$

then it becomes a complete differential equation,

$$\frac{d(U-2c)}{dt}=0 \quad \text{or} \quad U-2c=\text{const} \tag{2.4.36}$$

Two groups of characteristic lines are defined by Eqs. (2.4.33) and (2.4.35) in the plane of (x, t). The directions of the lines are called the positive and negative characteristic line directions, respectively. Equations (2.4.34) and (2.4.36) show that there are two constants of $u+2c$ and $u-2c$ along the two groups of characteristic lines. They are called the positive Riemann invariant and the negative Riemann invariant, respectively. Equations (2.4.33)–(2.4.36) consist of four ordinary equations. Solving the two partial differential Eqs. (2.4.26) and (2.4.27) can be replaced by solving these four ordinary equations.

It is assumed that, in a prismatic channel the bottom friction $s_f \neq 0$, the bottom slope $s_0 \neq 0$, and the initial conditions of $U_{x_1}, c_{x_1}, U_{x_2}, c_{x_2}, \ldots$ at $t=0$ are known. The characteristic lines through x_2 and x_3 defined by Eqs. (2.4.33) and (2.4.35) are intersected at the point M with the coordinate of (x_m, t_m). U_m, c_m, x_m, and t_m are unknown. These four unknowns can be obtained by solving the four Eqs. (2.4.33)–(2.4.36). The discrete equations can be written as

$$x_m = x_2 + \int_{t_{x_2}}^{t_m} (U + c)dt = x_2 + (\bar{U} + \bar{c})(t_m - t_{x_2}) \quad (2.4.37)$$

$$U_m + 2c_m = U_{x_2} + 2c_{x_2} + \int_{t_{x_2}}^{t_m} g(s_0 - s_f)dt = U_{x_2} + 2c_{x_2} + g(s_0 - \bar{s}_f)(t_m - t_{x_2}) \quad (2.4.38)$$

$$x_m = x_3 + \int_{t_{x_3}}^{t_m} (U - c)dt = x_3 + (\bar{U} - \bar{c})(t_m - t_{x_3}) \quad (2.4.39)$$

$$U_m - 2c_m = U_{x_3} - 2c_{x_3} + \int_{t_{x_3}}^{t_m} g(s_0 - s_f)dt = U_{x_3} - 2c_{x_3} + g(s_0 - \bar{s}_f)(t_m - t_{x_3}) \quad (2.4.40)$$

where $\bar{U}, \bar{c}, \bar{s}_f$ are the averages of the values at the two points. The discrete points $x_1, x_2, x_3, \ldots$ are selected along the line at $t = 0.x_m, t_m, U_m, h_m$ at the point M can be obtained by solving the above four equations by means of iteration. Similarly, the solutions at the points 1, 2, and P in the domain can be obtained (see Fig. 2.8 for the sketch).

When using the method of characteristics to solve a problem of 1D long waves in an open channel, the distance between the discrete points $x_1, x_2, x_3, \ldots$ should be small enough. At the beginning, the slope at the initial point can be used to replace the average with that at the unknown point M for the slope of characteristic line. Then the iteration process is carried out. The computing workload of this method is large. At present, the method of characteristics is rarely used directly to solve the long wave problems, and the difference methods based on the theory of characteristics are generally used.

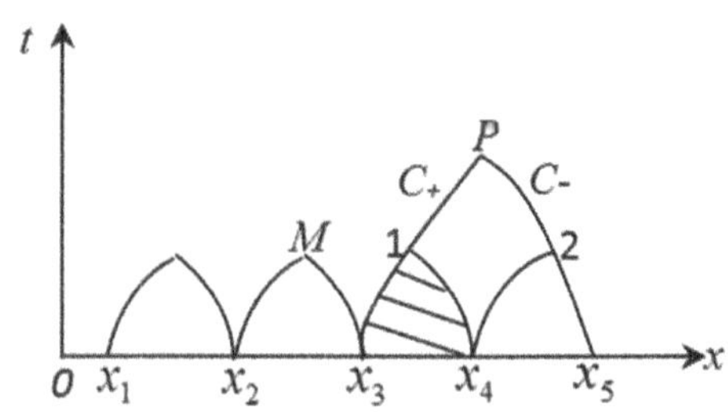

Fig. 2.8 Sketch of characteristics

2.4.3 Influenced Zone of the Solution to Hyperbolic Equations, Requirements of Boundary, and Initial Conditions

As the governing equations for 1D unsteady flow in channels are the first-order quasi-linear hyperbolic partial differential equations, there are two groups of characteristic lines in the plane of (x, t). From Fig. 2.8, it can be seen that the influenced zone of the solution with the initial conditions of x_3 and x_4 is the shaded area x_3x_41 enclosed by the two characteristic lines x_3 1 and x_4 1, and it has no influence on the other points outside of the shaded area. In other words, the solution at the point 1 only depends on the initial conditions in the interval between x_3 and x_4, and other initial conditions have no effect on this point. Therefore, the interval between x_3 and x_4 is the dependent area of the solution at the point 1. Similarly, the initial conditions in the interval between x_3 and x_5 can only affect the area enclosed by the two lines x_31P and x_52P. Therefore, this area is the affected zone of x_3, x_4, and x_5.

The above property of the hyperbolic equations reflects the process of wave propagation. For example, it involves a wave propagation process for flood waves or tidal waves in a channel to reach a certain location. It takes a long time when the solution in the domain is affected by both the upstream and downstream boundaries and the initial conditions. The initial and boundary conditions for the unsteady flow in a channel are discussed below.

In Fig. 2.9 for the case of slow-flow with $Fr < 1$, the domain in the plane of (x, t) is defined by $x = x_0$, $x = x_1$ and $t \geq 0$. It can be seen that the characteristic line C_- of the point. A intersects with the boundary $x = x_0$ at the point L. At this moment, U and h at the point A have been given by the initial conditions. But U and h at the point L cannot be given independently, and they must be compatible with the characteristic line C_- that goes through the point A. As there is a relation between U and h at the two points on C_- determined by the requirement of the negative Riemann invariant being constant, either one quantity of U and h or the relation between U and h at the point L on the boundary might be given. In fact, this principle is applicable to all the points on the straight line $x = x_0$. Similarly, the positive characteristic line C_+ gives a condition of the positive Riemann invariant for the point R on the straight

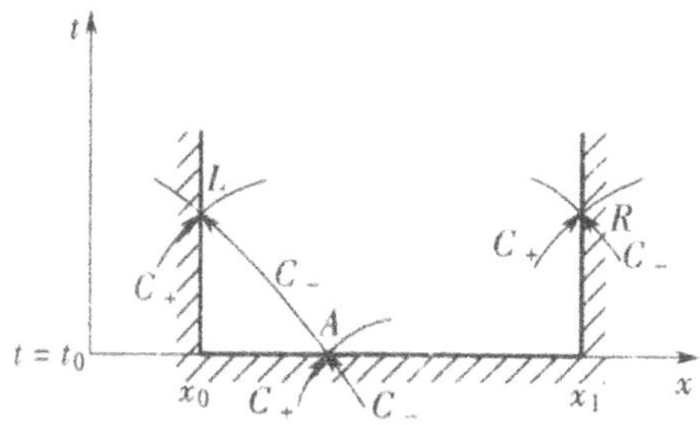

Fig. 2.9 Construction of characteristics at the boundaries for slow-flow ($Fr < 1$)

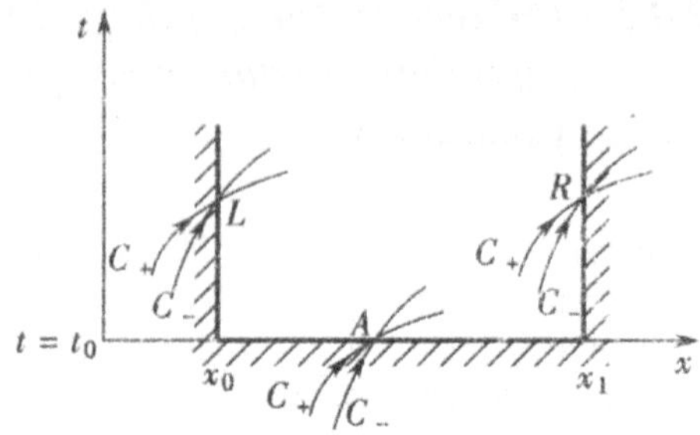

Fig. 2.10 Construction of characteristics at the boundaries for rapid-flow ($Fr > 1$)

line $x = x_1$. Therefore, there is also only one boundary condition at the boundary $x = x_1$.

Now it can be said that, for the slow-flow in the domain defined by $x = x_0$, $x = x_1$ and $t \geq 0$, two initial conditions are needed along $t = 0$ within $x_0 \leq x \leq x_1$, *and* on the boundaries at $x = x_0$ and $x = x_1$ for $t \geq 0$, one condition for each boundary is necessary. The two initial conditions together with the two boundary conditions can make the slow-flow problem well-posed in this domain. Figure 2.10 shows a sketch of the characteristic lines for rapid-flow with $Fr > 1$. To sum up, in a bounded domain in the plane of (x, t), one boundary or initial condition must be given when a characteristic line enters into the domain through the boundary. When $Fr = \frac{U}{\sqrt{gh}} < 1$, it is seen from Fig. 2.9 that there are two characteristic lines entering into the domain at $t = t_0$. Thus, two initial conditions should be given. Furthermore, each of the two boundaries $x = x_0$ and $x = x_1$ has one characteristic line to enter. Thus, a boundary condition should be given for each of the upstream and downstream boundaries. But for the case of rapid-flow with $Fr = \frac{U}{\sqrt{gh}} > 1$ as shown in Fig. 2.10, both of the two groups of characteristic lines are positive. There are two characteristic lines entering into the domain at $t = t_0$. Thus, two initial conditions should be given. There are also two characteristic lines entering at the upstream boundary $x = x_0$, but there is no characteristic line entering at the downstream boundary $x = x_1$. Thus, two boundary conditions at the upstream boundary should be given. It is not necessary to give any boundary condition at the downstream boundary. The physical explanation is that the flow velocity is greater than the wave celerity, i.e., $U > \sqrt{gh}$, so that the disturbance at the downstream does not affect the upstream. This is the idea for giving well-posed initial and boundary conditions to solve long wave equations in a channel.

2.4.4 Discontinuous Waves and Weak Solutions

Equations (2.4.26) and (2.4.27) are nonlinear equations. The slope of the characteristic line at each point in the domain of (x, t) is dependent on the variables of U and h. Even if the initial values are smooth, the characteristic lines in the same group may intersect as shown in Fig. 2.11. Then the two groups of U and h solutions at

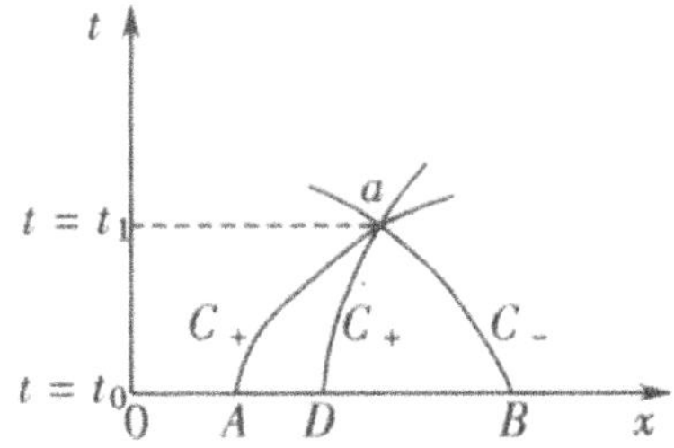

Fig. 2.11 Sketch of intersecting characteristic lines

the point a might be obtained. One group is given from Aa and Ba, another is from Da and Ba. Then there are multi-solutions, and after $t = t_1$, they are not dependent on the initial conditions on the straight line $t = t_0$. This situation corresponds to discontinuous waves. Physically, this phenomenon is called shock or discontinuity such as dam-break waves in channels and tidal bore.

Obviously, the St. Venant equations are not applicable for the discontinuous waves. At the discontinuity, the surface curvature is very large, the vertical acceleration cannot be ignored, and also the hypothesis of static pressure distribution is no longer valid. A generalized solution must be sought.

1. Weak solutions

For conservative Eqs. (2.4.7) and (2.4.12), they can be written as

$$\frac{\partial f}{\partial t} + \frac{\partial g(f)}{\partial x} = 0 \tag{2.4.13}$$

where

$$f = \begin{bmatrix} h \\ Uh \end{bmatrix}, \; g(f) = \begin{bmatrix} Uh \\ U^2 h + \frac{gh^2}{2} \end{bmatrix} \tag{2.4.14}$$

Physically, the weak solution is a kind of generalized solution, with which the differential equations are satisfied on both sides of the discontinuity and the integral equations or algebraic equations obtained by the differential equations are satisfied at the location of discontinuity. The well-known Rankine–Hugoniot relations are the algebraic equations at the discontinuity, which can be obtained by the following method.

The continuity equation in Eq. (2.4.13) reads

$$\frac{\partial h}{\partial t} + \frac{\partial}{\partial x}(Uh) = 0 \tag{2.4.41}$$

As shown in Fig. 2.12 for the computational domain, U is the unknown solution to Eq. (2.4.41) within the domain D. Integrating the equation over D yields

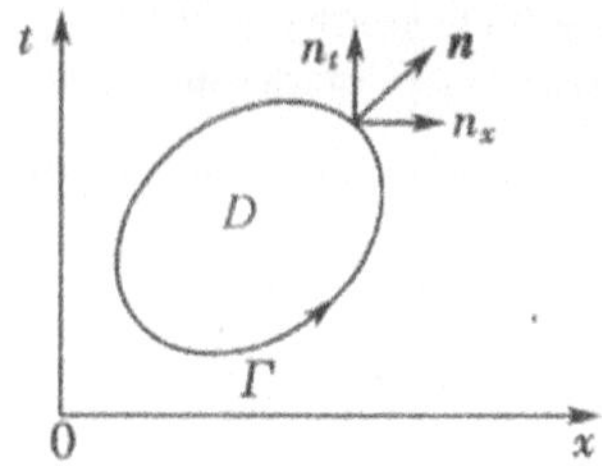

Fig. 2.12 Sketch of computational domain

$$\iint_D [h_t + (Uh)_x] dxdt = 0 \tag{2.4.42}$$

According to Gauss's law, U satisfies

$$\oint_\Gamma [hn_t + (Uh)n_x] d\Gamma = 0 \tag{2.4.43}$$

where Γ is the boundary of D domain, n_t and n_x are the components of the unit vector on Γ in the x- and t-directions, respectively.

As shown in Fig. 2.13 for a rectangular domain with discontinuity L, Eq. (2.4.43) is valid in the region $A_1B_1C_1D_1$ which covers the discontinuity. But the differential equations in Eq. (2.4.13) are not applicable. A weak solution can be obtained by solving Eq. (2.4.43) as described below.

Let the velocities on the two sides of the discontinuity be $U_1(t)$ and $U_2(t)$, respectively, and $d_1 = d_2 \to 0$. Equation (2.4.43) becomes

$$\Delta s(h_1 n_{1t} + U_1 h_1 n_{1x} + h_2 n_{2t} + U_2 h_2 n_{2x}) = 0$$

where $n_1 = -n_2$. Thus,

$$n_{1t}(h_1 - h_2) + n_{1x}(U_1 h_1 - U_2 h_2) = 0$$

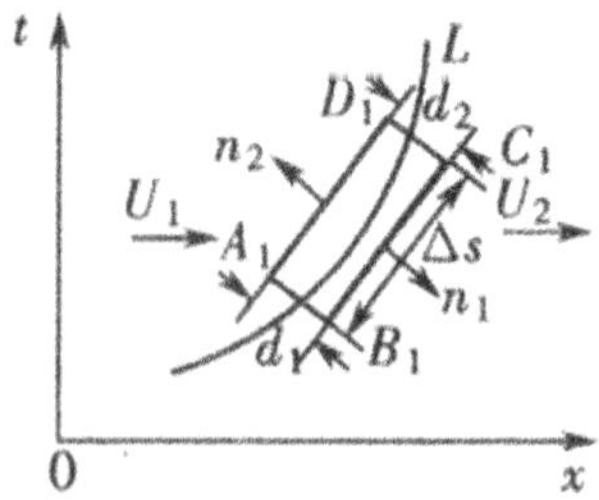

Fig. 2.13 Sketch of rectangular domain with discontinuity

The tangent of L is the moving velocity of the discontinuity,

$$S = \frac{dx}{dt} = -\frac{n_{1t}}{n_{2x}} = \frac{U_1 h_1 - U_2 h_2}{h_1 - h_2}$$

or

$$S[h] + [Uh] = 0 \tag{2.4.44}$$

where [] denotes the difference of the quantities on the right and left of the discontinuity. Equation (2.4.44) is called a Rankine–Hugoniot relation, abbreviated as R–H relation. This is the R–H relation for the continuity equation.

The R–H relation for the momentum Eq. (2.4.13) is written as

$$S = \frac{dx}{dt} = \frac{\left(U_1^2 h_1 + \frac{gh_1^2}{2}\right) - \left(U_2^2 h_2 + \frac{gh_2^2}{2}\right)}{U_1 h_1 - U_2 h_2} \tag{2.4.45}$$

2. Uniqueness of weak solutions

Weak solutions are not unique. For example, the weak solution to Eq. (2.4.13) depends on the vector form of Eq. (2.4.13).

For shallow water equations,

when the variables in Eq. (2.4.13) are in the following form

$$f = \begin{bmatrix} h \\ Uh \end{bmatrix}, \ g(f) = \begin{bmatrix} Uh \\ U^2 h + \frac{gh^2}{2} \end{bmatrix} \tag{2.4.14}$$

then the weak solution reads

$$S\begin{bmatrix} U \\ Uh \end{bmatrix}_1^2 + \begin{bmatrix} Uh \\ U^2 h + \frac{gh^2}{2} \end{bmatrix}_1^2 = 0 \tag{2.4.46}$$

when the variables are

$$f = \begin{bmatrix} h \\ Uh \end{bmatrix}, \ g(f) = \begin{bmatrix} Uh \\ \frac{U^2}{2} + gh \end{bmatrix} \tag{2.4.47}$$

then the weak solution reads

$$S\begin{bmatrix} h \\ U \end{bmatrix}_1^2 + \begin{bmatrix} Uh \\ \frac{U^2}{2} + gh \end{bmatrix}_1^2 = 0 \tag{2.4.48}$$

The moving velocity of the discontinuity S in Eqs. (2.4.46) and (2.4.48) is completely different. The former equations are conservative equations and the latter are

Eulerian equations. Thus, the different forms of the same equations make weak solutions different. For the uniqueness of the weak solution, additional condition such as so-called entropy condition should be introduced. The entropy condition and the uniqueness of the weak solution had been discussed by Li et al. [15]. The analysis showed that the weak solution in Eq. (2.4.46) obtained by the conservative form of shallow water equations satisfies the entropy condition and is rational. See Ref. [1] for the detailed analysis. Therefore, to calculate the discontinuity of shallow water waves, the conservative equations of Eq. (2.4.14) instead of Eulerian equations of Eq. (2.4.47) should be used.

2.5 Waves in Current

Various waves with different properties have been described in the above sections. In this section, waves in current will be discussed. Coexistence of waves and currents is common in practical engineering. For instance, when waves propagate from the deep water to the coastal or estuarine area, the wave elements and the vertical structure can be changed due to the effect of current field. When the wave direction and current direction are inverse, the wave height is increased so that the safety of navigation and engineering structures is influenced. When the waves and the oblique shear flow are encountered, wave reflection occurs. A brief introduction to the influence of current on the wave elements and the vertical structure of waves will be given in this section.

2.5.1 Waves in Steady Current

1. Waves in steady uniform current

In order to study the influence of current on the waves, one of the most basic phenomena, i.e., the interaction between stable regular small-amplitude waves and the steady uniform current is studied first. Based on this study, some basic features of the current influence on the waves can be obtained.

Let α denotes the angle between the wave direction and the steady uniform current with a speed of U. c is the wave celerity relative to the current, and ω is the angular frequency relative to the current. Then the wave celerity and the angular frequency in the stationary coordinate system are

$$c_c = c + U\cos\alpha \tag{2.5.1}$$

$$\omega_c = \omega + kU\cos\alpha \tag{2.5.2}$$

which means that c_c and ω_c are the wave celerity and the angular frequency with existence of the current c and ω are the wave celerity and the angular frequency without current or when the coordinate system moves together with the current. It can be assumed that the wave period is unchanged whether the coordinate system is stationary or moves with the current.

With existence of the current, the dispersion relation reads

$$\omega = \omega_c - kU\cos\alpha = \pm[gk\tanh(kd)]^{1/2} \tag{2.5.3}$$

$$\omega_c - kU\cos\alpha = \pm[gk\tanh(kd)]^{1/2} \tag{2.5.4}$$

The current influence on the waves can be analyzed from Eq. (2.5.4). By taking the wave number k as an independent variable and drawing the curves for the functions on the two sides of Eq. (2.5.4), the intersections of the two groups of the curves are the solutions of k under the wave–current interaction.

Let the function on the left side of Eq. (2.5.4) be

$$y_1 = \omega_c - kU\cos\alpha \tag{2.5.5}$$

and on the right side be

$$y_2 = \pm[gk\tanh(kd)]^{1/2} = \pm\omega \tag{2.5.6}$$

Figure 2.14 shows the curves for Eqs. (2.5.5) and (2.5.6). It can be seen that y_1 can be express by three straight lines for the three cases of the slope of y_1:$U\cos\alpha > 0$, $U\cos\alpha = 0$, and $U\cos\alpha < 0$, while two curves are plotted for y_2 when it takes the positive and the negative value, respectively.

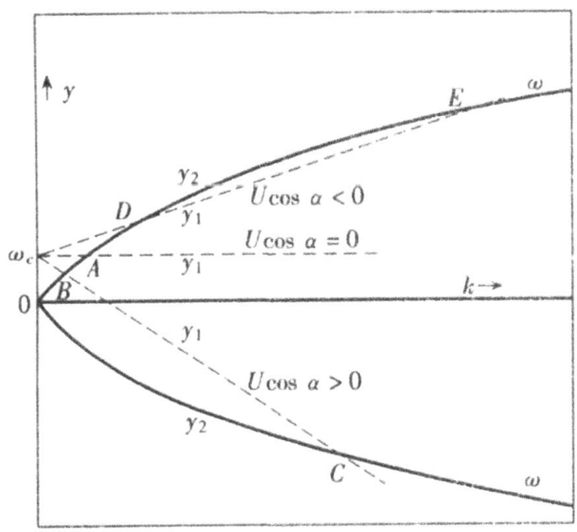

Fig. 2.14 Curves of Eqs. (2.5.5) and (2.5.6)

(1) When $U\cos\alpha = 0$, $y_1 = \omega_c$ corresponds to $U = 0$ without current, or the current direction is perpendicular to the wave direction with $\alpha = \pm\pi/2$. The straight line $y_1 = \omega_c$ and the upper curve of y_2 intersect at the point A. The wave number k_A at A is the original wave number unaffected by the current.
(2) When $U\cos\alpha > 0$, i.e., when the wave and the current have the same direction, there may be two solutions at the points B and C. B is the intersection of y_1 and y_2 with the positive value. It is found that $k_B < k_A$, which means the wave number decreases and the wavelength increases when the wave and the current have the same direction. C is the intersection of y_1 and y_2 with the negative value. $k_C > k_A$ shows that some wave with very short wavelength might occur over the current.
(3) When $U\cos\alpha < 0$, i.e., when the wave and the current have the opposite directions, there may be two solutions at the points D and E. Both D and E are the intersections of y_1 and y_2 with the positive value. It can be seen that $k_D > k_A$, i.e., the wave number increases and the wavelength decreases when the wave and the current have the opposite directions. The wave number at E is even larger, which corresponds to some wave with shorter wavelength.

When the counter current speed is relatively large, D and E might coincide. It means the line y_1 with $U\cos\alpha < 0$ becomes the tangent line of $+y_2$. Thus,

$$-U\cos\alpha = \frac{d\omega}{dk} = c_g$$

i.e.,

$$c_g + U\cos\alpha = 0 \tag{2.5.7}$$

Equation (2.5.7) indicates that when the opposing current speed reaches the wave group velocity, the wave energy is not able to propagate in the opposing current. As the group velocity is the propagating velocity of the wave energy, the current makes the wave "stop" when the current speed is up to the wave group velocity.

In deep water, the wave celerity is c_0 and $c_g = \frac{1}{2}c_0$. In the case of $\alpha = 0$, the wave energy cannot propagate in the opposing current when $U > \frac{1}{2}c_0$.

In shallow water, $c_g = c = \sqrt{gd}$. Thus, the wave energy cannot propagate in the opposing current when $U \geq \sqrt{gd}$. This is why the small wave disturbance cannot propagate in the opposing current when $Fr = \frac{U}{\sqrt{gd}} \geq 1$.

2. Current influence on wave elements

When the bathymetry is flat and the current is parallel to the wave direction, the ratio of the wavelength in the current L_c to the wavelength in still water L can be calculated by [16],

$$\frac{L_c}{L} = \frac{c_c}{c} = \left(1 - \frac{U}{c_c}\right)^{-2} \frac{\tanh k_c d}{\tanh kd} \tag{2.5.8}$$

where k_c is the wave number in the current.

The ratio of the wave height in the current H_c to the wave height in still water H can be calculated by,

$$\frac{H_c}{H} = \left(1 - \frac{U}{c_c}\right)^{1/2} \left(\frac{L}{L_c}\right)^{1/2} \left(\frac{N}{N_c}\right)^{1/2} \left(1 + \left(\frac{U}{c_c}\right)\frac{2 - N_c}{N_c}\right)^{1/2} \tag{2.5.9}$$

where

$$N_c = 1 + \frac{2k_c d}{\sinh 2k_c d} \tag{2.5.10}$$

$$N = 1 + \frac{2kd}{\sinh 2kd} \tag{2.5.11}$$

The wave period T in the coordinate system moving with the current speed U can be obtained by,

$$T = \sqrt{\frac{2\pi L}{g \tanh kd}} \tag{2.5.12}$$

In the case of an opposing current, when the wave steepness after deformation exceeds the limit, the wave height after deformation should be determined by the wavelength after deformation and the limit of the wave steepness.

See Ref. [16] for the calculation of wave amplitude and wave height after deformation in the other cases of the current.

2.5.2 Vertical Structure Under the Wave–Current Interaction

Wave–current interaction not only causes the changes of wave elements, but also affects the vertical distribution of the horizontal velocity components. The vertical distribution of the horizontal velocity under the wave–current interaction, especially the flow pattern in the boundary layer near the bottom, has significant influence on the sediment transport and the pollutant dispersion in the coastal area. Kemp and Simons [4, 5] carried out experiments on the waves propagating in the conditions of same and opposite directions of current, for both smooth bed and rough bed. Klopman [6] carried out experiments on the interaction between the non-breaking waves and steady current on the bed with large roughness. These experiments have indicated that waves have great influence on the vertical profile of the horizontal current velocity. As shown in Fig. 2.15 [2], when waves propagate in the direction of the current, the upper current velocity decreases and the lower current velocity increases due to the additional shear stress by the waves. But when the waves propagate in the opposite

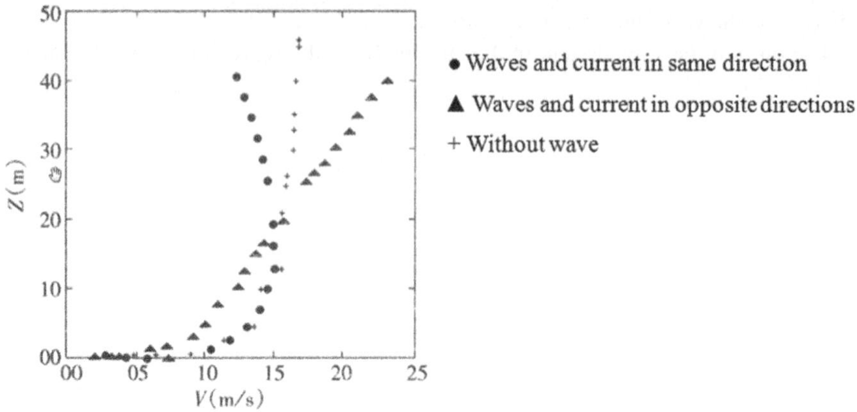

Fig. 2.15 Wave influence on current velocity profile [6]

direction of the current, the upper current velocity increases and the lower current velocity decreases.

Theoretical models of the current velocity distribution under the wave–current interaction have been studied by Thomas [10], [11], You [13, 14], Sun et al. [17], and some other researchers.

2.6 Introduction of Random Wave Theory

The water wave theories introduced in the above sections are all based on the fundamental equations of fluid mechanics, to study how to describe the wave motion, and the mathematical and physical characteristics of the water waves under certain conditions. The above methods are deterministic methods. However, the waves in the sea change rapidly in nature. They seem to be chaotic at first. However,by the statistical analyses on a large number of observation data, it is found that as random phenomena the waves have statistical characteristics which could be studied by using probability theory. The theory to study water waves by using probability theory is called random wave theory, which is a stochastic method

2.6.1 Statistical Characteristics of Random Functions

1. Mathematical expectation (mean) and mean square value

The mathematical expectation denoted by μ_x or $M[x]$ is the population mean of a random function $x(t)$. If the probability density of the random function $x(t)$ is $p(t)$, the mathematical expectation is defined as

$$\mu_x = M[x] = \int_{-\infty}^{\infty} xp(x)dx \quad (2.6.1)$$

Mean square value denoted by $M[x^2]$ is the population mean of the square of the random function $x(t)$,

$$M[x^2] = \int_{-\infty}^{\infty} x^2 p(x)dx \quad (2.6.2)$$

2. Variance and root-mean-square deviation

Let the deviation between the random function $x(t)$ and the corresponding mathematical expectation μ_x be $x - \mu_x$. The mathematical expectation of the squared deviation can be used to measure how far the random function $x(t)$ is spread out from the corresponding mathematical expectation μ_x. The expectation of the squared deviation of the random function is called variance and denoted by $D[x]$,

$$D[x] = M[(x - \mu_x)^2] = \int_{-\infty}^{\infty} (x - \mu_x)^2 p(x)dx = M[x^2] - \mu_x^2 \quad (2.6.3)$$

The square root of the variance $D[x]$ of the random function $x(t)$ (with positive value) is called root-mean-square deviation and denoted by σ_x,

$$\sigma_x = \sqrt{D[x]} \quad (2.6.4)$$

The mean and the variance are two important parameters to describe the characteristics of a random variable.

3. Autocorrelation function and covariance function

The autocorrelation function of the random function $x(t)$ describes the correlation of $x(t)$ with $x(t + \tau)$ at different time. When the time difference τ is very small, the random values $x(t)$ and $x(t + \tau)$ are closely correlated When τ becomes larger, the correlation of $x(t)$ with $x(t + \tau)$ decreases until it becomes zero. The autocorrelation function of the random function $x(t)$ is defined as

$$R(t, t + \tau) = M[x(t) \cdot x(t + \tau)] \quad (2.6.5)$$

The covariance function is defined as

$$C_{0v}(t, t + \tau) = M\{[x(t) - M(x(t))] \cdot [x(t + \tau) - M(x(t + \tau))]\} \quad (2.6.6)$$

When the population mean of a stochastic process does not vary with time, it is called a stationary random process. If the time-averaged mean, the autocorrelation function and other time-averaged statistical characteristic values are equal to the corresponding population means of characteristic values, then the stationary random process is said to be ergodic. For a stationary random process, a sufficient and necessary condition of ergodicity is

$$\lim R(\tau) = 0 \tag{2.6.7}$$

i.e., when $\tau \rightarrow \infty$ the autocorrelation function is zero, which means the correlation is very weak. Therefore, for a stationary random process with ergodicity, it is possible to select a single sample of the process to represent the entire process, and the statistical properties are independent of the starting point in time. Consequently, the statistical characteristic values read

the mean value:

$$M[x(t)] = \overline{x(t)} = \lim_{T \to \infty} \frac{1}{T} \int_0^T x(t)dt \tag{2.6.8}$$

the mean square value:

$$M[x^2(t)] = \overline{x^2(t)} = \lim_{T \to \infty} \frac{1}{T} \int_0^T x^2(t)dt \tag{2.6.9}$$

the variance:

$$D[x(t)] = \sigma_x^2 = \overline{[x(t) - \overline{x(t)}]^2} = \lim_{T \to \infty} \frac{1}{T} \int_0^T [x(t) - \overline{x(t)}]^2 dt \tag{2.6.10}$$

the autocorrelation function:

$$R(\tau) = \overline{x(t) \cdot x(t+\tau)} = \lim_{T \to \infty} \frac{1}{T} \int_0^T x(t) \cdot x(t+\tau)dt \tag{2.6.11}$$

the covariance function:

$$C_{Ov}(\tau) = R(\tau) - \overline{[x(t)]^2} \tag{2.6.12}$$

the autocorrelation coefficient:

$$\rho(\tau) = \frac{R(\tau)}{\overline{x^2(t)}} = \frac{\overline{x(t) \cdot x(t+\tau)}}{\overline{x^2(t)}} \quad (2.6.13)$$

When $\tau \to 0$, there is $\rho(\tau) = 1$, and when, $\tau \to \infty$ $\rho(\tau) = 0$.

2.6.2 Description of Random Waves and the Concept of Spectrum

A surface elevation $\eta(t)$ recorded at a fixed station gives a stochastic shape with irregular ups and downs as shown in Fig. 2.16. A criterion is required to separate an individual wave from a recording. Presently, zero-crossing counting method is commonly used. The method to define an individual wave by starting from a zero-upcrossing point (at mean sea level) to the next zero-upcrossing point is called zero-upcrossing method. Wave period of the individual wave denoted by T is defined by the time difference between the two successive zero-upcrossing points. Wave height is defined by the difference of the wave crest and the trough within the two successive zero-upcrossing points. Zero-downcrossing method by starting from a zero-downcrossing point to the next zero-downcrossing point can also be used to define an individual wave from a time series of surface elevation. In general, zero-upcrossing method is used if it is not specified. Due to the stochasticity of the surface elevation recordings, wave height H and wave period T are also stochastic parameters, and the characteristic wave height and wave period in random wave theory are defined as below.

$\bar{H}$ and $\bar{T}$ denote the mean wave height and the mean wave period of all the individual waves, respectively. They are the arithmetic mean values of the wave heights and periods, namely mathematical expectations. H_{max} and $T_{H_{max}}$ denote the maximum wave height and the corresponding period in the wave recordings. $H_{1/10}$ and $T_{H_{1/10}}$ denote the averages of the wave heights and periods of the one-tenth highest waves. $H_{1/3}$ and $T_{H_{1/3}}$ denote the averages of the wave heights and periods

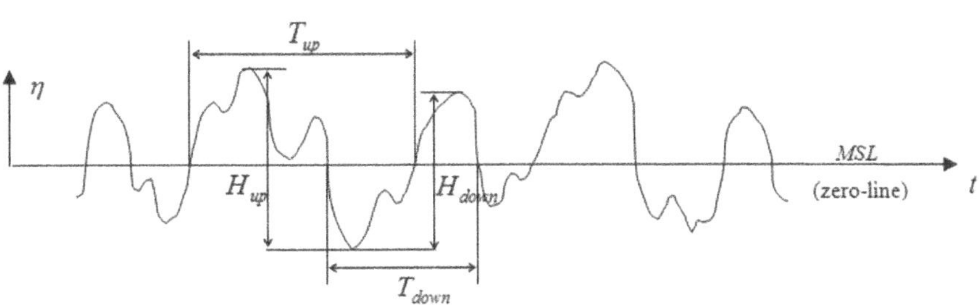

Fig. 2.16 Definition of zero-upcrossing and zero-downcrossing waves

of the one-third highest waves. $H_{1/3}$ is also called the significant wave height. The average wave height of a proportion of large waves reflects a significant portion of the sea waves, which are concerned in the design of marine navigation and marine engineering.

The waves in stable sea state are considered as stationary random processes with ergodicity. Longuet-Higgins [7] proposed such a wave model that waves could be regarded as a superposition of infinite number of simple cosine waves with different amplitudes, different frequencies, different initial phases propagating in different directions with different angles of θ from x in the plane of (x, y). The amplitude of each wavelet is infinitely small. Then the wave surface elevation $\eta(x, y, t)$ could be expressed as

$$\eta(x, y, t) = \sum_{n=1}^{\infty} a_n \cos(k_n x \cos\theta_n + k_n y \sin\theta_n - \omega_n t - \varepsilon_n) \tag{2.6.14}$$

where a_n, ω_n, and k_n are the amplitude, the angular frequency and the wave number of each wave component, respectively; θ_n is the wave direction of each wave component, $0 < \theta_n \leq 2\pi$; ε_n, as a random quantity with uniform distribution, is the initial phase of each wave component.

According to the small-amplitude wave theory, the average wave energy per unit area in the vertical column of the wave component is

$$E_n = \frac{1}{2}\rho g a_n^2 \tag{2.6.15}$$

If a random wave is composed of infinite number of component waves with different amplitudes and different frequencies, and the wave frequencies are distributed continuously in between $0 \sim \infty$, then the energy of each wave component in the frequency band $\omega \sim \omega + d\omega$ is $\sum_{\omega}^{\omega+d\omega} \frac{1}{2}a_n^2$. The energy spectral density $S_\eta(\omega)$ is defined as

$$S_\eta(\omega)d\omega = \sum_{\omega}^{\omega+d\omega} \frac{1}{2}a_n^2 \tag{2.6.16}$$

$$S_\eta(\omega) = \frac{1}{\Delta\omega} \sum_{\omega}^{\omega+d\omega} \frac{1}{2}a_n^2 \tag{2.6.17}$$

Obviously, $S_\eta(\omega)$ is the average energy in the frequency band $\omega \sim \omega + d\omega$. When $\Delta\omega \to 0$,$S_\eta(\omega)$ gives a distribution of the energy density with respect to the frequency in the random wave, which constitutes the frequency domain characteristics of the random sea wave process.

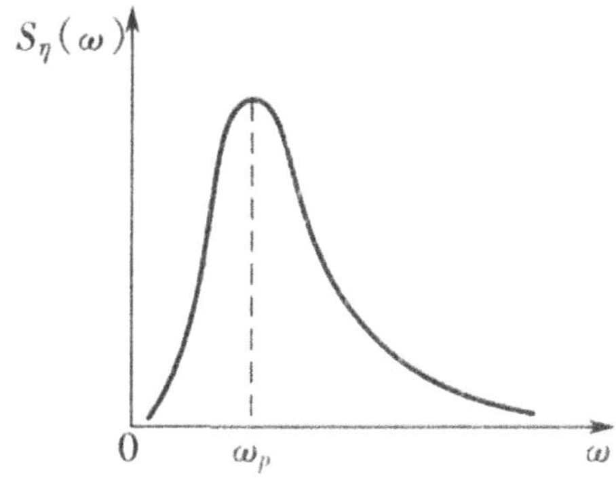

Fig. 2.17 Sketch of sea wave spectrum

Figure 2.17 shows a sketch of spectral density distribution. It can be seen that, $S_\eta(\omega)$ is very small around $\omega = 0$. With the increase of ω, $S_\eta(\omega)$ increases dramatically to the maximum. At this point with the maximum value, the corresponding frequency ω_p is called the peak frequency. With further increase of ω, $S_\eta(\omega)$ decreases. Lastly, when $\omega \to \infty$, $S_\eta(\omega) \to 0$. The area under the wave spectrum curve, namely the integral of the spectral density function $S_\eta(\omega)$ in the whole frequency range ($\omega = 0 \sim \infty$) denotes the total energy of all wave components expressed as

$$E = \int_0^\infty S_\eta(\omega)d\omega = \sum_{n=1}^{\infty} \frac{1}{2} a_n^2 \tag{2.6.18}$$

$S_\eta(\omega)$ is called the energy spectral density function of the random wave $\eta(t)$, spectral density function or spectral density for short.

The spectral moments can be used to characterize the distribution of spectral density and defined as below.

The 0th moment, m_0:

$$m_0 = \int_0^\infty S_\eta(\omega)d\omega \tag{2.6.19}$$

The second-order moment, m_2:

$$m_2 = \int_0^\infty \omega^2 S_\eta(\omega)d\omega \tag{2.6.20}$$

The nth-order moment, m_n:

$$m_n = \int_0^\infty \omega^n S_\eta(\omega)d\omega \tag{2.6.21}$$

According to the definition of variance, the variance of surface elevation reads

$$\sigma_\eta^2 = \lim \frac{1}{T} \int_0^T [\eta(t) - \bar{\eta}(t)]^2 dt \tag{2.6.22}$$

Considering the mean value of the surface elevation $\overline{\eta(t)} = 0$, it gives

$$\sigma_\eta^2 = \int_0^\infty S_\eta(\omega) d\omega = m_0 \tag{2.6.23}$$

2.6.3 *Statistical Distribution of Random Wave Elements*

The sea waves are regarded as stochastic processes and proved to meet the requirements of stability and ergodicity in physics. Therefore, it is possible to find out the relevant statistical characteristic values to represent the overall statistical characteristics of the sea waves, by statistically analyzing wave recordings observed at several stations over an appropriate time period.

1. Distribution of surface elevation

The surface elevation at a certain fixed location ($x = 0$, $y = 0$) is

$$\eta(t) = \sum_{n=1}^{\infty} a_n \cos(\omega_n t + \varepsilon_n) \tag{2.6.24}$$

The dispersion relation for each linear small-amplitude wave component is

$$\omega_n^2 = g k_n \tanh k_n h \tag{2.6.25}$$

Based on a large number of field data analyses, many researchers have found that the distribution of surface elevation match the Gaussian normal distribution. The probability density is expressed by

$$p(\eta) = \frac{1}{\sqrt{2\pi}\sigma_\eta} \exp\left(-\frac{\eta^2}{2\sigma_\eta^2}\right) \tag{2.6.26}$$

where $\sigma_\eta^2\left(= \overline{\eta^2(t)}\right) = m_0$ denotes the variance of the surface elevation.

2. Distribution of wave surface maxima

The extremes of wave surface refer to the points at the wave surface with the first derivative $\eta'(t) = 0$. The maxima refers to the point with $\eta'(t) = 0$ and the second derivative $\eta''(t) < 0$, and the minima refers to the point with $\eta'(t) = 0$ and $\eta''(t) > 0$.

The wave surface elevation $\eta(t)$ is a random function denoted by η_1. The first derivative of the surface elevation with respect to time $\eta'(t)$ and the second derivative $\eta''(t)$ are also random functions and denoted by η_2 and η_3, respectively. Utilizing Eq. (2.6.24) for the surface elevation yields

$$\begin{cases} \eta_1 = \eta(t) = \sum\limits_{n=1}^{\infty} a_n \cos(\omega_n t + \varepsilon_n) \\ \eta_2 = \eta'(t) = -\sum\limits_{n=1}^{\infty} a_n \omega_n \sin(\omega_n t + \varepsilon_n) \\ \eta_3 = \eta''(t) = -\sum\limits_{n=1}^{\infty} a_n \omega_n^2 \cos(\omega_n t + \varepsilon_n) \end{cases} \tag{2.6.27}$$

The joint probability distribution density of η_1, η_2, and η_3 reads

$$p(\eta_1, \eta_2, \eta_3) = \frac{1}{(2\pi)^{3/2}(D \cdot m_2)^{1/2}} \exp\left[-\frac{1}{2}\left(\frac{\eta_2^2}{m_2} + \frac{m_4\eta_1^2 + 2m_2\eta_1\eta_3 + m_0\eta_3^2}{D}\right)\right] \tag{2.6.28}$$

where $D = m_0 m_4 - m_2^2$, m_0, m_2, and m_4 are the zero moment, the second-order moment, and the forth-order moment of the wave spectrum $S_\eta(\omega)$, respectively.

3. Distribution of wave height

The wave height of the individual wave is taken as a random variable, which fits to a certain distribution. Through analyses of a large number of wave records, if the wave energy is confined to a narrow frequency range the wave height distribution can be expressed by the Rayleigh distribution,

$$p(H) = 2\alpha^2 \frac{H}{H_*^2} \exp\left(-\alpha^2 \frac{H^2}{H_*^2}\right) \tag{2.6.29}$$

where $p(H)$ is the probability density function of the wave height;

$$\alpha = \begin{cases} \frac{\pi}{4}, & (H_* = \bar{H}) \text{ for the mean wave height} \\ 1, & (H_* = H_{rms}) \text{ for the root--mean square wave height} \\ \frac{1}{8}, & (H_* = \sigma_\eta) \text{ for the root--mean square of surface elevation} \end{cases} \tag{2.6.30}$$

4. Relation between wave spectrum and wave elements

When the Rayleigh distribution is used as the wave height distribution, the various wave height expectations can be expressed by the zero moment of wave spectrum as follows.

Mean wave height:

$$\bar{H} = 2.057\sqrt{m_0} \tag{2.6.31}$$

The square of the root-mean-square wave height:

$$H_{rms}^2 = 8m_0 \tag{2.6.32}$$

The characteristic wave heights:

$$H_{1/10} = 5.090\eta_{rms} = 2.031\bar{H} = 1.800H_{rms} = 5.091\sqrt{m_0} \tag{2.6.33}$$

$$H_{1/3} = 4.004\eta_{rms} = 1.597\bar{H} = 1.416H_{rms} = 4.005\sqrt{m_0} \tag{2.6.34}$$

$$\bar{H} = \sqrt{2\pi}\,\eta_{rms} = (\sqrt{\pi}/2)H_{rms} \tag{2.6.35}$$

where H_{rms} is the root-mean-square wave height.

According to the definition of the wave period T by zero-upcrossing method from wave records, the mean value of the wave periods is called the mean period $\bar{T}$ defined by zero-upcrossing method, which is expressed as

$$\bar{T} = T_{0.2} = 2\pi\left(\frac{m_0}{m_2}\right)^{1/2} \tag{2.6.36}$$

2.6.4 *Frequency Spectrum and Directional Spectrum*

In the above sections, sea waves are treated as random processes and the concept of spectrum has been introduced. Utilizing wave spectrum to describe sea waves has become an important approach. Therefore, determining the form of wave spectrum is an important part of studying random waves. The wave spectrum is divided into frequency spectrum and directional spectrum. The frequency spectrum $S(\omega)$ only takes into account the distribution of the wave energy in the frequency domain. The directional spectrum $S(\omega, \theta)$ takes into account the distribution of the wave energy in both frequency and direction. The forms of the frequency and directional spectra come from the statistics of a large amount of measurement data.

1. Frequency spectrum

Theoretically, $S_\eta(\omega)$ is distributed in the whole frequency range, $\omega = 0 \sim \infty$, but its significant part is confined to a narrow frequency band. In the wave components of sea waves, the components with very small and very large frequencies provide little wave energy. The wave energy mainly comes from the wave components in a narrow frequency band.

Converting the spectrum $S_\eta(\omega)$ into $S_\eta(f)$ expressed by the frequency f gives

$$S_\eta(f) = 2\pi S_\eta(\omega) \tag{2.6.37}$$

Here are several commonly used frequency spectra.

(1) PM spectrum

PM spectrum is developed from measurements in the North Atlantic and used for a fully developed sea.

$$S_\eta(\omega) = 0.78\omega^{-5} \exp\left[-\frac{3.11}{\omega^4 H_{1/3}^2}\right] \tag{2.6.38}$$

(2) JONSWAP spectrum

JONSWAP spectrum was drawn out from the North Sea data during the Joint North Sea Wave Project undertaken by the researchers from the UK, the Netherlands, the USA, and Germany.

$$S_\eta(\omega) = ag^2 \frac{1}{\omega^5} \exp\left[-\frac{5}{4}\left(\frac{\omega_m}{\omega}\right)^4\right] r^{\exp[-(\omega-\omega_m)^2/(2\sigma^2\omega_m^2)]} \tag{2.6.39}$$

$$\sigma = \begin{cases} \sigma_a \ (\omega \le \omega_{\mathrm{m}}), \\ \sigma_b \ (\omega > \omega_m), \end{cases} \quad \gamma = 3.3$$

$\sigma_a = 0.07$; $\sigma_b = 0.09$; $a = 0.076(\bar{X})^{-0.22}$; $\bar{X} = 10^{-1} \sim 10^5$, $\omega_{\mathrm{m}} = 22(g/U)(\bar{X})^{-0.33}$, $\bar{X} = gx/U^2$, x is the wind fetch, U is the wind speed at 10 m above the sea level (m/s).

(3) Bretschneider spectrum

$$S_\eta(\omega) = 400.5\left(\frac{H_{1/3}}{T_{H_{1/3}}^2}\right)^2 \frac{1}{\omega^5} \exp\left[-1605\frac{1}{(T_{H_{1/3}}\omega)^4}\right] \tag{2.6.40}$$

which is suitable for the wind wave spectrum when the wind fetch is limited.

(4) Wen's spectrum [18] for coastal wave spectrum in China

It was put forward by Wen et al. based on the analysis of the coastal measurement data in China.

$$\begin{aligned} S_\eta(f) = {} & 0.0687 H_{1/3}^2 P \exp\left\{-95 \ln\left[\frac{P(5.813-5.137H^*)}{(6.77-1.088P+0.013P^2)(1.307-1.426H^*)}\right]\right. \\ & \left. \cdot (1.1T_{1/3}f - 1)^{12/5}\right\}, \quad \text{for} \quad 0 \le f \le 1.05/T_{1/3} \end{aligned} \tag{2.6.41}$$

$$S_\eta(f) = 0.0687 H_{1/3}^2 T_{1/3}^{\frac{(6.77-1.088P+0.013P^2)(1.307-1.426H^*)}{5.813-5.137H^*}} \cdot \left[\frac{1.05}{T_{1/3} f}\right]^{(4-2H^*)},$$
$$\text{for } f > 1.05/T_{1/3} \tag{2.6.42}$$

where $H^* = 0.626 H_{1/3}/d$, d is still water depth; $P = 95.3 H_{1/3}^{1.35}/T_{1/3}^{2.7}$, $T_{1/3}$ is the significant wave period

2. Directional spectrum

The wave surface elevation at a fixed location is related to both the frequency of the wave components and the propagation directions. Thus, the surface elevation reads

$$\eta(x, y, t) = \sum_{i=1}^{\infty} \sum_{j=1}^{\infty} a_{ij} \cos(k_i x \cos(\theta_p + \theta_i) + k_j y \sin(\theta_p + \theta_i) - \omega_i t - \varepsilon_{ij}) \tag{2.6.43}$$

The wave energy is distributed over the frequency ω and the propagation direction θ of the wave components. In Eq. (2.6.43), a_{ij} is the amplitude of the wave component at the ith frequency in the jth direction, θ_p is the dominant wave direction, θ_i is the deflected angle of the ith direction from the dominant direction, ω_i is the representative frequency in the ith divided frequency sub-area, ε_{ij} is the initial phase of the wave component at the ith frequency in the jth direction. The wave spectrum for the wave energy distributed over both frequency and direction is called the directional spectrum $S_\eta(\omega, \theta)$.

$$S_\eta(\omega, \theta) d\omega d\theta = \sum_{\omega}^{\omega+d\omega} \sum_{\theta}^{\theta+d\theta} \frac{1}{2} a_n^2 \tag{2.6.44}$$

Figure 2.18 shows a sketch of sea wave directional spectrum, which gives a distribution of the wave energy with respect to frequency in different directions.

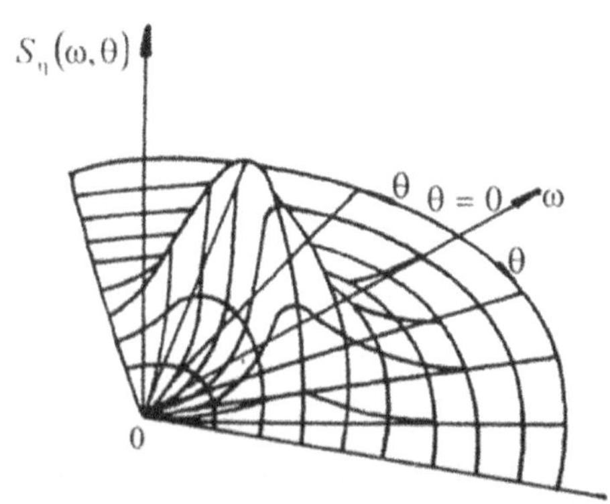

Fig. 2.18 Sketch of sea wave directional spectrum

The wave energy per unit area in the vertical water column at a fixed location reads

$$E = \int_0^{\infty}\int_{-\pi}^{\pi} S_\eta(\omega, \theta) d\omega d\theta = \int_0^{\infty} S_\eta(\omega) d\omega \tag{2.6.45}$$

Thus,

$$S_\eta(\omega) = \int_{-\pi}^{\pi} S_\eta(\omega, \theta) d\theta \tag{2.6.46}$$

The variance of the wave surface elevation is

$$\sigma_\eta^2 = \int_0^{\infty} S_\eta(\omega) d\omega = m_0 \tag{2.6.47}$$

The directional spectrum is rewritten in general form as

$$S_\eta(\omega, \theta) = S_\eta(\omega) \cdot G(\omega, \theta) \tag{2.6.48}$$

where $G(\omega, \theta)$ is called the directional spreading function which satisfies

$$\int_{-\pi}^{\pi} G(\omega, \theta) d\theta = 1 \tag{2.6.49}$$

Here are several commonly used directional spreading functions.

(1) Mitsuyasu spectrum

$$G(f, \theta) = G_0(S)\left|\cos\frac{\theta}{2}\right|^{2S} \tag{2.6.50}$$

where $G_0(S) = \frac{1}{\pi}2^{2S-1} \cdot \frac{\Gamma^2(S+1)}{\Gamma(2S+1)}$, S is a direction distribution parameter.

(2) SWOP spectrum

$$G(\omega, \theta) = \frac{1}{\pi}\left\{1 + \left(0.5 + 0.82\exp\left[-\frac{1}{2}\left(\frac{\omega U}{g}\right)^4\right]\right)\cos 2\theta + 0.32\exp\left[-\frac{1}{2}\left(\frac{\omega U}{g}\right)\right]^4 \cos 4\theta\right\} \tag{2.6.51}$$

where $|\theta| \leq \frac{\pi}{2}$.

(3) Simple empirical formula

$$G(f,\theta) = G(\theta) = \frac{1}{\sqrt{\pi}} \frac{\Gamma(S+1)}{\Gamma(S+2)} \cos^{2S}(\theta - \theta_0) \tag{2.6.52}$$

where Γ is the gamma function, $|\theta - \theta_0| \leq \frac{\pi}{2}$, θ_0 is the dominant direction of incoming waves, S is a direction distribution parameter.

References

1. Abbott MB, Marshall G, Ohno T. On Weak solution of the equations of nearly-horizontal flow, Rep. Series No. 3. The Netherlands: IHE; 1970.
2. Amilcare P. Applications of padé approximation theory in fluid dynamics. World Scientific; 1993.
3. Mei CC. The applied dynamics of ocean surface waves. Wiley; 1983.
4. Kemp PH, Simons RR. The interaction between waves and a turbulent current: waves progating with the current. J Fluid Mech. 1982;116:227–50.
5. Kemp PH, Simons RR. The interaction between waves and a turbulent current: waves progating against the current. J Fluid Mech. 1983;130:73–89.
6. Klopman G. Vertical structure of the flow due to waves and current: Laser-Dopple flow measurements for wave following or opposing a current. Delft Hydraulics, Progress Report H840.30. Part II; 1994.
7. Longuet-Higgins MS. The statistical analysis of a random, moving surface. Phil Trans Roy Ser A. 1957;249:321–87.
8. Stokes GG. On the theory of oscillatory waves. Trans Camb Phil Soc. 1847;8:441–55.
9. Skjelbreia L. Gravity wave, stokes. Third order approximation: tables of functions. Berkely, California: The Engineering Foundation Council on Wave Research; 1958.
10. Thomas GP, Wave-current interactions: an experimental and numerical study. Part 1. Linear Waves J Fluid Mech. 1981;110:475–74.
11. Thomas GP, Wave-current interactions: an experimental and numerical study. Part 2. Nonlinear Waves J Fluid Mech. 1990;216:505–36.
12. Whitham GB. Linear and nonlinear waves. New York: Wiley Interscience; 1974.
13. You ZJ. A simple model for current velocity profiles in combined wave current flows. Coast Eng. 1994;23:289–304.
14. You ZJ. Eddy viscosities and velocities in combined wave-current flows. Ocean Eng. 1994;21(1):81–97.
15. Deyuan Li, et al. Numerical methods for two dimensional unsteady fluid dynamics (in Chinese). Beijing: Science Press; 1987.
16. Li Y, Chen S, Yu Y, et al. Translation of handbook of coastal and marine engineering (in Chinese),, vol. 1. Dalian: Dalian University of Technology Press; 1992.
17. Sun H, Han G, Tao J. The flow structure along the vertical line under the interaction of current and wave. J Hydraul Eng. 2001;32(7):63–8.
18. Chinese Industry Standard. Code of hydrology for sea harbour, (JTJ21398). Beijing: China Communications Press; 1998.

Chapter 3
Numerical Simulation of Long Waves in Shallow Water

3.1 Introduction

Long waves in shallow water will be described in this chapter. Shallow water long wave theory can be used to study unsteady flow problems in hydrodynamics such as flood waves in channels, river networks and lake systems, tidal waves in estuaries and coastal waters, and the wave phenomena at the downstream channels caused by the daily regulation of hydropower stations.

Natural shallow water long wave motion is essentially a three-dimensional (3D) problem, i.e., all the velocity components in x, y and z directions in the Cartesian coordinate system change. In order to simplify the numerical simulation and reduce the computational workload, shallow water long waves can be studied as one-dimensional (1D), two-dimensional (2D) or three-dimensional (3D) problems respectively, according to the geometric scales.

The flood waves in channels or tidal waves in tidal rivers can be deemed as 1D problems, because the width of water surface and the water depth are much less than the length of channels or rivers. The averaged velocity (or flux) over the cross-section and water level are used as variables. Only the variation in the flow direction is considered, without describing the change of the velocity and water depth within the cross-section directly. In the shallow lakes, estuaries and coastal waters, the velocity can be averaged over the depth and the variation in the two horizontal directions can be considered in the simulation of long waves as the horizontal length scale is much larger than the water depth. Thus they can be considered as 2D problems. For the lakes, estuaries and coastal waters with great variation in bathymetry, the changes of horizontal velocity in the vertical direction and the vertical velocity cannot be ignored. Therefore, the water flow should be studied by using 3D models. The 3D models of long waves can be divided into vertically stratified quasi-3D models and full 3D models. The vertically stratified quasi-3D models can be used for shallow water long waves in large areas when the depth scale is much smaller than the horizontal scale. The full 3D models can be used for the local areas with large velocities and large

J. Tao, *Numerical Simulation of Water Waves*, Springer Tracts in Civil Engineering, https://doi.org/10.1007/978-981-15-2841-5_3

velocity changes in horizontal and vertical directions, such as the local areas near the estuaries or outfalls. The numerical hydrodynamic models (including 1D, 2D and 3D) established on the basis of shallow water nonlinear long wave equations are applicable to the simulation of unsteady flow such as flood waves and tidal waves in rivers, lakes, estuaries and coastal areas. They also provide hydrodynamic conditions for the studies on sediment transport, morphology, advection-dispersion of pollutants and ecological processes. Therefore, the numerical hydrodynamic models are very important basic models.

3.2 Mathematical Model of 1D Long Waves

3.2.1 Governing Equations

The equations for 1D shallow water long waves in a channel with rectangular cross-section have been given by Eqs. (2.4.15) and (2.4.16) in Chap. 2.

Continuity equation:

$$\frac{\partial h}{\partial t} + \frac{\partial (Uh)}{\partial x} - \frac{q}{b} = 0 \tag{2.4.15}$$

Momentum equation:

$$\frac{\partial (Uh)}{\partial t} + \frac{\partial}{\partial x}\left(U^2 h + gh^2/2\right) - \frac{qu_i}{b} = 0 \tag{2.4.16}$$

where, h is the water depth; U is the averaged velocity over the cross section defined by $U = \frac{1}{A}\int\limits_A u\mathrm{d}A$, in which A is the submerged cross-sectional area with $A = bh$ for a rectangular cross section; $q(\mathrm{m}^2/\mathrm{s})$ is the lateral inflow per unit width; b is the channel width; u_i is the lateral inflow velocity component; t is time; x is the distance along the axis of the channel.

For natural channels, the governing equations are given by Eqs. (2.4.19) and (2.4.20),

Continuity equation:

$$\frac{\partial \eta}{\partial t} + \frac{1}{b}\frac{\partial Q}{\partial x} = 0 \tag{2.4.19}$$

Momentum equation:

$$\frac{\partial Q}{\partial t} + \frac{\partial}{\partial x}\left(\alpha \frac{Q^2}{A}\right) + gA\frac{\partial \eta}{\partial x} + g\frac{|Q|Q}{C^2 AR} = 0 \tag{2.4.20}$$

where η is the surface elevation; $A = A(\eta.x)$ is the submerged cross-sectional area; $Q = UA$ is the flux; $b = b(\eta, x)$ is the surface width; $h = \eta - z_b$ is the water depth; z_b is the bottom elevation; C is the Chezy coefficient, $C = \frac{1}{n}R^{1/6}$, in which n is the bottom roughness; $R = A/\chi$ is the hydraulic radius, in which χ is the total boundary perimeter of the submerged cross section; α is the momentum correction coefficient defined by $\alpha = \frac{A}{Q^2}\int_A u^2 \mathrm{d}A$.

3.2.2 Difference Equations

Difference equations by using finite difference method are established to solve the 1D long wave equations as follows.

The computational domain is discretized first when solving Eqs. (2.4.15) and (2.4.16) or Eqs. (2.4.19) and (2.4.20) with the given initial and boundary conditions. The channel is divided into a certain number of cells (channel segments) with a requirement of the submerged cross-sectional area (the channel width and the water depth) in each individual cell being similar. Then, the differential equations are discretized to establish the difference equations.

A weighted implicit scheme can be adopted for the difference scheme, in which the weighted average method with second-order accuracy in both time and space is used.

Let f denote a variable of the equations, j denote the serial number of cell, and n denote the serial number of time. The difference scheme is expressed as

$$\frac{\partial f}{\partial x} = \theta \frac{f_{j+1}^{n+1} - f_j^{n+1}}{\Delta x_j} + (1-\theta)\frac{f_{j+1}^n - f_j^n}{\Delta x_j} \quad 0.5 \le \theta \le 1 \tag{3.2.1}$$

$$\frac{\partial f}{\partial t} = \varphi \frac{f_{j+1}^{n+1} - f_{j+1}^n}{\Delta t} + (1-\varphi)\frac{f_j^{n+1} - f_j^n}{\Delta t} \quad 0.5 \le \varphi \le 1 \tag{3.2.2}$$

$$f = \frac{1}{2}\left(f_j^{n+\frac{1}{2}} + f_{j+1}^{n+\frac{1}{2}}\right) \tag{3.2.3}$$

where θ and φ are adjustable parameters in between 0.5 and 1.0. Δx_j is the spatial step along the channel, and Δt is the time step.

The following method can be used for the calculation of the nonlinear terms in Eqs. (2.4.19) and (2.4.20)

$$\frac{\partial}{\partial x}\left(\frac{Q^2}{A}\right) = \frac{\theta}{\Delta x}\left(\frac{Q_{j+1}^n Q_{j+1}^{n+1}}{A_{j+1}^{n+1/2}} - \frac{Q_j^n Q_j^{n+1}}{A_j^{n+1/2}}\right) + (1-\theta)\frac{1}{\Delta x}\left(\frac{\left(Q_{j+1}^n\right)^2}{A_{j+1}^{n+1/2}} - \frac{\left(Q_j^n\right)^2}{A_j^{n+1/2}}\right) \tag{3.2.4}$$

$$\frac{|Q|Q}{K^2} = \frac{1}{2}\left\{\frac{\left|Q_j^n\right|Q_j^{n+1}}{\left(K_j^{n+1/2}\right)^2} + \frac{\left|Q_{j+1}^n\right|Q_{j+1}^{n+1}}{\left(K_{j+1}^{n+1/2}\right)^2}\right\}$$

$$= \frac{1}{2}\left\{\frac{n_j^2\left|Q_j^n\right|Q_j^{n+1}}{\left(A_j R_j^{4/3}\right)^{n+1/2}} + \frac{n_{j+1}^2\left|Q_{j+1}^n\right|Q_{j+1}^{n+1}}{\left(A_{j+1} R_{j+1}^{4/3}\right)^{n+1/2}}\right\} \tag{3.2.5}$$

where

$$K = C^2 AR \tag{3.2.6}$$

The function values at $n + \frac{1}{2}$ time step can be determined as the mean values at $n\Delta t$ and $(n + 1)\Delta t$ time steps.

The derivatives with respect to time and space in each term of the equations can be discretized by using the above difference scheme. Then, the following linear algebraic equations are obtained:

$$A_{1j}\, Q_j^{n+1} + B_{1j}\, \eta_j^{n+1} + C_{1j}\, Q_{j+1}^{n+1} + D_{1j}\, \eta_{j+1}^{n+1} = S_{1j}^n \tag{3.2.7}$$

$$A_{2j}\, Q_j^{n+1} + B_{2j}\, \eta_j^{n+1} + C_{2j}\, Q_{j+1}^{n+1} + D_{2j}\, \eta_{j+1}^{n+1} = S_{2j}^n \tag{3.2.8}$$

where A_1, B_1, C_1, D_1 and A_2, B_2, C_2, D_2 are the known coefficients of the unknown quantities at $n + 1$ time step. All the variable values at n time step are put in S_{1j}^n and S_{2j}^n which are also known. Together with the initial and boundary conditions, the above equations can be solved by various methods for solving linear algebraic equations.

The spatial step Δx_j is normally selected according to the change of the cross section in the calculation. Although an implicit scheme is adopted, for the consideration of accuracy, Δt cannot be too large. In general, the Courant number $C_r = c\frac{\Delta t}{\Delta x} \le 1$, where $c \approx \sqrt{gh}$ is the wave speed in still water.

3.2.3 Solving Method

A double-sweep method is described below, which is a convenient memory-saving method.

For the case of slow-flow with a Froude number $F_r = U^2/gh < 1$, a boundary condition should be given for each of the upstream and downstream boundaries. For instance, water-level condition $\eta = \eta(t)$ is given at the downstream boundary, and the flux condition $Q = Q(t)$ or the water-level condition $\eta = \eta(t)$ is given at the upstream boundary.

Equations (3.2.7) and (3.2.8) are transformed into triangular systems of equations as follows:

Using $\frac{D_{1j}}{D_{2j}}\times$ Eqs. (3.2.8)–(3.2.7) gives

$$P_{1j}\,Q_j^{n+1} + N_{1j}\,\eta_j^{n+1} + R_{1j}\,Q_{j+1}^{n+1} = T_{1j} \tag{3.2.9}$$

where

$$P_{1j} = \frac{D_{1j}}{D_{2j}}A_{2j} - A_{1j},\quad N_{1j} = \frac{D_{1j}}{D_{2j}}B_{2j} - B_{1j}$$
$$R_{1j} = \frac{D_{1j}}{D_{2j}}C_{2j} - C_{1j},\quad T_{1j} = \frac{D_{1j}}{D_{2j}}S_{2j} - S_{1j}$$

Using $\frac{A_{2j}}{A_{1j}}\times$ Eqs. (3.2.7)–(3.2.8) gives

$$P_{2j}\,\eta_j^{n+1} + N_{2j}\,Q_{j+1}^{n+1} + R_{2j}\,\eta_{j+1}^{n+1} = T_{2j} \tag{3.2.10}$$

where

$$P_{2j} = \frac{A_{2j}}{A_{1j}}B_{1j} - B_{2j},\quad N_{2j} = \frac{A_{2j}}{A_{1j}}C_{1j} - C_{2j}$$
$$R_{2j} = \frac{A_{2j}}{A_{1j}}D_{1j} - D_{2j},\quad T_{2j} = \frac{A_{2j}}{A_{1j}}S_{1j} - S_{2j}$$

Let

$$Q_j^{n+1} = E_{1j}\,\eta_j^{n+1} + F_{1j} \tag{3.2.11}$$

$$\eta_j^{n+1} = E_{2j}\,Q_{j+1}^{n+1} + F_{2j} \tag{3.2.12}$$

According to Eq. (3.2.9), it yields

$$\eta_j^{n+1} = \frac{-R_{1j}}{P_{1j}\,E_{1j} + N_{1j}}Q_{j+1}^{n+1} + \frac{T_{1j} - P_{1j}\,F_{1j}}{P_{1j}\,E_{1j} + N_{1j}} \tag{3.2.13}$$

$$Q_{j+1}^{n+1} = \frac{-R_{2j}}{P_{2j}\,E_{2j} + N_{2j}}\eta_{j+1}^{n+1} + \frac{T_{2j} - P_{2j}\,F_{2j}}{P_{2j}\,E_{2j} + N_{2j}} \tag{3.2.14}$$

Comparing Eq. (3.2.13) with Eqs. (3.2.12) and (3.2.14) with Eq. (3.2.11) gives

$$E_{2j} = \frac{-R_{1j}}{P_{1j}\,E_{1j} + N_{1j}} \tag{3.2.15a}$$

$$F_{2j} = \frac{T_{1j} - P_{1j} F_{1j}}{P_{1j} E_{1j} + N_{1j}} \tag{3.2.15b}$$

$$E_{1j+1} = \frac{-R_{2J}}{P_{2j} E_{2j} + N_{2j}} \tag{3.2.16a}$$

$$F_{1j+1} = \frac{T_{2j} - P_{2j} F_{2j}}{P_{2j} E_{2j} + N_{2j}} \tag{3.2.16b}$$

When the flux is given as $Q_1 = Q(t)$ at the upstream boundary $j = 1$, by comparing Eq. (3.2.9) with Eq. (3.2.11), it is found that $E_{11} = 0$, $F_{11} = Q_1$.

When E_{11} and F_{11} are known, E_{12} and F_{12} can be solved by Eq. (3.2.16a). Thereby, E_{1j}, F_{1j}, E_{2j}, and F_{2j} at each node can be solved from the upstream to the downstream.

When the water level is given as $\eta_N = \eta(t)$ at the downstream boundary $j = N$, Q_N can be solved by Eq. (3.2.11) and η_{N-1} solved by Eq. (3.2.12). The water level η_j and the flux Q_j in the whole domain can be solved by alternately using Eqs. (3.2.11) and (3.2.12).

When the water level is given as $\eta_1 = \eta(t)$ at the upstream boundary $j = 1$, according to Eq. (3.2.11), it yields

$$\eta_i = \frac{Q_1}{E_{11}} - \frac{F_1}{E_{11}} \tag{3.2.17}$$

As η_1 is given by the boundary condition which is independent of Q_1, $E_{11} = \alpha$ and $F_{11} = -\alpha\eta_1$ can be used. Here, α is a number with a high order of magnitude, normally selected as $\alpha = 10^4 \sim 10^6$. After E_{11} and F_{11} are known, E_{1j}, F_{1j}, E_{2j} and F_{2j} can be solved from the upstream to the downstream. Then, the water level η_j and the flux Q_j at each node can be solved from the downstream to the upstream by using the water-level condition at the downstream boundary.

For the case of having $Q(t)$ at the downstream boundary and the case of having the relation $Q(\eta)$ at both upstream and downstream boundaries, the coefficients E and F and the solution formulas can be deduced according to the above principle.

3.2.4 Boundary Conditions

As the conditions for the fixed solution to the differential equations, the boundary conditions are given conditions or rules that are not influenced by the flow in the computational domain. The boundary conditions for 1D channel are divided into external boundaries and internal boundaries. The external boundaries refer to the upstream and downstream boundaries of the computational domain. The internal boundaries refer to the nodes with inflow or outflow in the domain, or where there are given constraint conditions of the flow caused by the structures such as weirs, bridge piers, and dams.

1. External boundary conditions

The upstream and downstream boundary conditions of the computational domain may have three forms: time series of water level $\eta = \eta(t)$, time series of flux $Q = Q(t)$, and the relation between the water level and the flux $Q = Q(\eta)$. When the Froude number $F_r = \frac{U^2}{gh} < 1$, the flow is slow-flow (subcritical flow). According to the characteristics of the first-order quasi-linear hyperbolic equation, one boundary condition is needed for each of upstream and downstream boundaries. When $F_r = \frac{U^2}{gh} > 1$, the flow is rapid-flow (supercritical flow), and both of the two boundary conditions should be at the upstream boundary.

When determining the locations of the upstream and downstream boundaries, i.e., where the boundary conditions are given, the selected sections should have stable time series of water level, flux, or relations between the water level and the flux, which are not influenced by flood or tide. For example, the downstream boundary of the tidal river should be at the location having stable water-level time series $\eta = \eta(t)$ which could be the downstream boundary condition. In case of having a large lake or reservoir in the lower reaches of the river, when the water level of the lake/reservoir is unchanged or the relation between the water level and the flux is stable, the location where the river enters into the lake/reservoir can be selected as the downstream boundary with the boundary condition of the water level or the relation between the water level and the flux of the lake/reservoir. The upstream boundary could be selected at the cross section having a stable relation between the water level and the flux, or the section with reservoir control.

2. Internal boundary conditions

Internal boundary conditions refer to the conditions given by the influence of the weirs, gates, and dams on the water flow and at the outflow or inflow sections in the channel.

(1) Tributary inflow and outflow

Let the flux of the lateral inflow be q. Two sections j and $j + 1$ are set on both sides of the intake which form an imaginary segment with distance close to zero as shown in Fig. 3.1. In this segment, the original differential equations are no longer applicable. The equations under certain conditions will be used instead.

The water level at the two imaginary upstream and downstream sections can be assumed to be equal. The flux at the section $j + 1$ is the sum of the flux at the

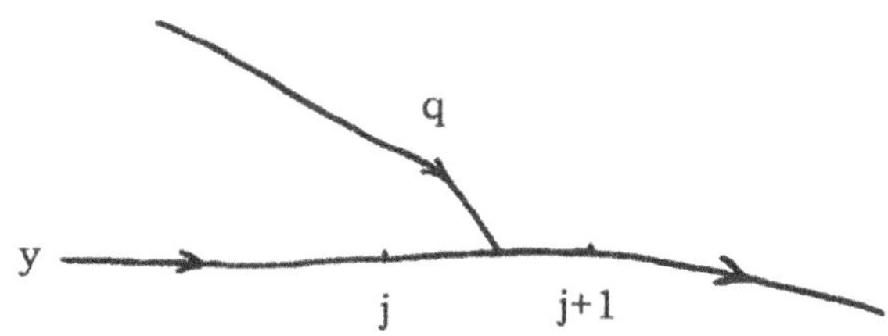

Fig. 3.1 Sketch of an imaginary segment with lateral inflow

section j and the flux of the lateral inflow q. Therefore, the continuity equation for this segment is

$$Q_j + q = Q_{j+1} \tag{3.2.18}$$

where q is positive for inflow and negative for outflow.

The condition of equal water level at the upstream and downstream sections can be used to replace the momentum equation

$$\eta_j = \eta_{j+1} \tag{3.2.19}$$

In the double-sweep process, the coefficients in Eqs. (3.2.9) and (3.2.10) for this imaginary segment read

$$\left.\begin{aligned} &-P_{1j} = R_{1j} = 1 \\ &N_{1j} = 0 \\ &T_{1j} = q \end{aligned}\right\} \tag{3.2.20}$$

$$\left.\begin{aligned} &P_{2j} = -R_{2j} = 1 \\ &N_{2j} = 0 \\ &T_{2j} = 0 \end{aligned}\right\} \tag{3.2.21}$$

Therefore, in case of having inflow or outflow in the computational channel reach, a cross section can be set on both sides of the intake/outfall. Then, the coefficients in the linear algebraic Eqs. (3.2.9) and (3.2.10) can be calculated by Eqs. (3.2.20) and (3.2.21). The problem can be solved by the method described in Sect. 3.2.3.

(2) Water-blocking structures such as weirs and piers

Two sections j and $j + 1$ are set on both sides of the structure as shown in Fig. 3.2. The fluxes at the upstream and the downstream sections are equal according to the continuity equation

$$Q_j^{n+1} = Q_{j+1}^{n+1} \tag{3.2.22}$$

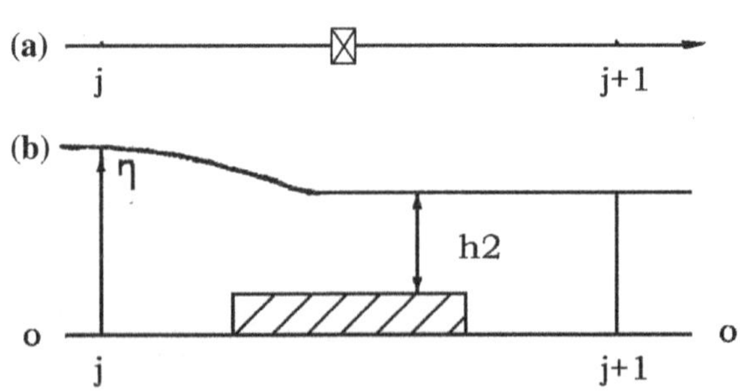

Fig. 3.2 Sketch of an imaginary segment with a weir or gate. **a** Plan view, **b** elevation view

The momentum equation can be replaced by the Weir formula

$$Q = \mu h_2 b\sqrt{2g(\eta_j - \eta_{j+1})} \tag{3.2.23}$$

where μ is the flow coefficient of the Weir, h_2 is the water depth at the crest of the weir, and b is the width of the weir.

Equation (3.2.23) is written in the discrete form as

$$\left|Q_{j+1}^n\right|Q_{j+1}^{n+1} = 2g\mu^2 b^2\left(h_{j+1}^n\right)^2\left(\eta_j^{n+1} - \eta_{j+1}^{n+1}\right) \tag{3.2.24}$$

The coefficients in Eqs. (3.2.9) and (3.2.10) read

$$\left.\begin{aligned} P_{1j} &= 1 \\ N_{1j} &= T_{1j} = 0 \\ R_{1j} &= -1 \end{aligned}\right\} \tag{3.2.25}$$

$$\left.\begin{aligned} P_{2j} &= 2g\mu^2 b^2\left(h_{j+1}^n\right)^2 \\ N_{2j} &= -\left|Q_{j+1}^n\right| \\ R_{2j} &= -P_{2j} \\ T_{2j} &= 0 \end{aligned}\right\} \tag{3.2.26}$$

In case of having structures such as gates and dams in the computational channel reach, the calculation principle is the same as above except for using the regulated flow at the gates or dams to replace the momentum equation.

3.2.5 *Flood Waves in Tidal Rivers*

The drainage of Yongding New River is taken as an example to observe the movement of flood waves and tidal waves in a river [1]. Yongding River is located in northern China, which is the main flood source in Haihe River Basin. Historically, flood disasters occurred frequently in this region. In 1970, an artificial channel called Yongding New River was built (in northern Tianjin) to discharge the water from Yongding River into Bohai Bay directly. Here, only the 1D computational results are described. Figure 3.3 shows a plan view of segmented Yongding New River. Thirty-two sections had been selected from the upstream to the downstream for Yongding New River which is bifurcated into north and south two channels at Qujiadian, and then the two channels merge into one again.

Time series of flood discharge for different return periods were given at the upstream boundary. The flood discharge lasted about 16 days in which two peaks occurred as shown in Fig. 3.4a. Time series of water level were given for the downstream boundary condition as shown in Fig. 3.4d. The internal boundary conditions

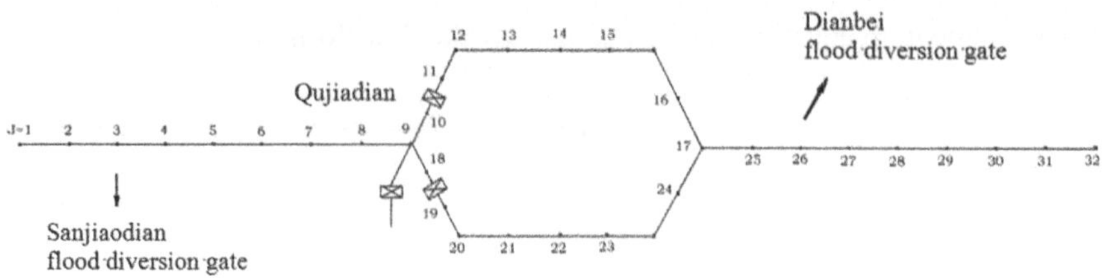

Fig. 3.3 Plan view of segmented Yongding new river

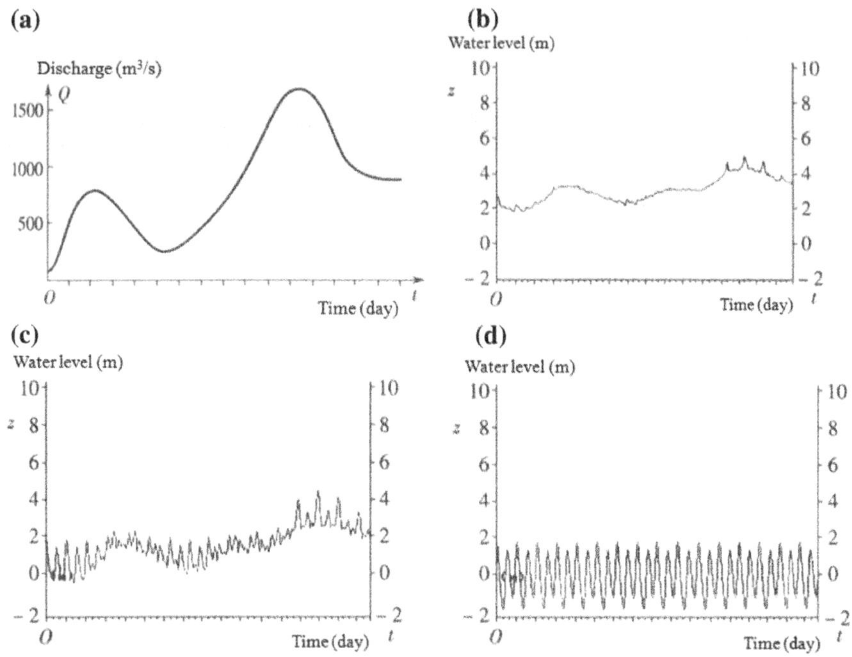

Fig. 3.4 Time series of discharge and water level at different sections in Yongding new river. **a** Flood discharge at the section of No. 1 (the upstream boundary condition), **b** water level at the section of No. 29, **c** water level at the section of No. 31, **d** water level at the section of No. 32 (the downstream boundary condition)

included branches at Qujiandian. The two branches were controlled by the two gates, respectively.

The channel roughness was set to $n = 0.0225$, and the bank roughness was $n = 0.1$.

In the case of no flood diversion, the 1D computational results are shown in Fig. 3.4, which is a typical 1D flow under the action of flood and tidal waves. As mentioned above, Fig. 3.4a shows the time series of flood discharge with two peaks for the upstream boundary condition. Figure 3.4b shows the time series of water level at the section of No. 29 which was affected by the flood with two peaks.

Figure 3.4c shows the time series of water level at the section of No. 1 influenced by the tidal waves and the flood. Figure 3.4d shows the water level at the downstream boundary which was not affected by the flood in the channel. It is found that the river flow in the channel was mainly affected by the upstream and downstream boundary conditions. When the catastrophic flood from the upstream hits the high tide from the downstream, serious disaster might occur if no measure is taken.

3.3 Mathematical Model of 2D Long Waves

In estuaries, lakes, bays, shallow reservoirs, and other water bodies with wide surface, the horizontal length scale is much larger than the vertical scale. The magnitude and variation of flow velocity in the vertical direction are much smaller than those in the horizontal direction, so that the velocity in the vertical direction can be represented by the depth-averaged velocity. Thus, the 2D depth-integrated long wave equations are derived, which can be used for the establishment of a 2D hydrodynamic model.

3.3.1 2D Depth-Integrated Long Wave Equations

With assumptions that the vertical acceleration is much smaller than the gravitational acceleration and can be ignored, and also the water density is constant, the 3D Reynolds equations are simplified to

$$\frac{\partial u}{\partial x} + \frac{\partial v}{\partial y} + \frac{\partial w}{\partial z} = 0 \tag{3.3.1}$$

$$\frac{\partial u}{\partial t} + \frac{\partial u^2}{\partial x} + \frac{\partial (uv)}{\partial y} + \frac{\partial (uw)}{\partial z} = -\frac{1}{\rho}\frac{\partial p}{\partial x} + f_x + \nu\left(\frac{\partial^2 u}{\partial x^2} + \frac{\partial^2 u}{\partial y^2} + \frac{\partial^2 u}{\partial z^2}\right) - \left(\frac{\partial \overline{u}'^2}{\partial x} + \frac{\partial \left(\overline{u'v'}\right)}{\partial y} + \frac{\partial \left(\overline{u'w'}\right)}{\partial z}\right) \tag{3.3.2}$$

$$\frac{\partial v}{\partial t} + \frac{\partial (uv)}{\partial x} + \frac{\partial v^2}{\partial y} + \frac{\partial (vw)}{\partial z} = -\frac{1}{\rho}\frac{\partial p}{\partial y} + f_y + \nu\left(\frac{\partial^2 v}{\partial x^2} + \frac{\partial^2 v}{\partial y^2} + \frac{\partial^2 v}{\partial z^2}\right) - \left(\frac{\partial \left(\overline{u'v'}\right)}{\partial x} + \frac{\partial \left(\overline{v'^2}\right)}{\partial y} + \frac{\partial \left(\overline{v'w'}\right)}{\partial z}\right) \tag{3.3.3}$$

$$f_z = \frac{1}{\rho}\frac{\partial p}{\partial z} \tag{3.3.4}$$

where u, v, w are the time-averaged velocity components in the x-, y-, and z-directions; u', v' w' are the fluctuating velocity components in the x-, y-, and z-directions;

Fig. 3.5 Sketch of coordinate system

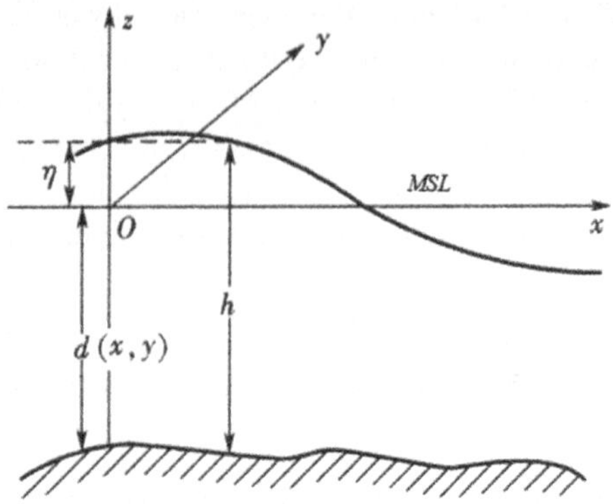

f_x, f_y, f_z are the mass forces per unit mass in the x-, y- and z-directions. If the Coriolis force is only considered for the horizontal mass force and the gravity is only for the vertical mass force, $f_x = \gamma v$ and $f_y = -\gamma u$, in which $\gamma = 2\omega \sin\varphi$, ω is the angular rate of revolution and φ is the geographic latitude; $f_z = g$; ρ is the density of fluid; p is the pressure; ν is the kinematic viscosity coefficient; $\rho\overline{u'^2}$, $\rho\overline{u'v'}$ and $\rho\overline{u'w'}$ are the Reynolds stress components (Fig. 3.5).

The average value of the integral over the water depth is defined as

$$\bar{F}(x.y) = \frac{1}{h}\int_{-d}^{\eta} f(x.y.z)\mathrm{d}z \tag{3.3.5}$$

Then the depth-averaged horizontal velocity components read,

$$U = \frac{1}{h}\int_{-d}^{\eta} u\mathrm{d}z \quad V = \frac{1}{h}\int_{-d}^{\eta} v\mathrm{d}z \tag{3.3.6}$$

When $f_z = g$, the pressure p can be obtained by integrating Eq. (3.3.4),

$$p(z) = \rho g(\eta - z) + P_a(x, y) \tag{3.3.7}$$

where, η is the free surface elevation, P_a is the atmospheric pressure. Then it gives,

$$\frac{\partial p}{\partial x} = \rho g\frac{\partial \eta}{\partial x} + \frac{\partial P_a}{\partial x} \tag{3.3.8}$$

$$\frac{\partial p}{\partial y} = \rho g\frac{\partial \eta}{\partial y} + \frac{\partial P_a}{\partial y} \tag{3.3.9}$$

Integrating the continuity Eq. (3.3.4), considering the kinematic condition at the free surface Eq. (2.4.4) and omitting the nonlinear terms yield,

$$\frac{\partial}{\partial x}\int_{-d}^{\eta} u\mathrm{d}z + \frac{\partial}{\partial y}\int_{-d}^{\eta} v\mathrm{d}z + \int_{-d}^{\eta}\frac{\partial w}{\partial z}\mathrm{d}z = 0$$

$$\frac{\partial \eta}{\partial t} + \frac{\partial (Uh)}{\partial x} + \frac{\partial (Vh)}{\partial y} = 0 \tag{3.3.10}$$

The momentum equation for the x component is integrated over the water depth, and given below as an example.

The momentum equation for the x component reads,

$$\underset{(1)}{\frac{\partial u}{\partial t}} + \underset{(2)}{\frac{\partial (u^2)}{\partial x}} + \underset{(3)}{\frac{\partial (uv)}{\partial y}} + \underset{(4)}{\frac{\partial (vw)}{\partial z}} = \underset{(5)}{-\frac{1}{\rho}\frac{\partial p}{\partial x}} + \underset{(6)}{f_x} + \underset{(7)}{\nu\left(\frac{\partial^2 u}{\partial x^2} + \frac{\partial^2 u}{\partial y^2} + \frac{\partial^2 u}{\partial z^2}\right)} - \underset{(8)}{\left(\frac{\partial \overline{u'^2}}{\partial x} + \frac{\partial\left(\overline{u'v'}\right)}{\partial y} + \frac{\partial\left(\overline{v'w'}\right)}{\partial z}\right)} \tag{3.3.11}$$

According to the Leibniz's law, such as

$$\int_{-d}^{\eta}\frac{\partial u}{\partial t}\mathrm{d}z = \frac{\partial}{\partial t}\int_{-d}^{\eta} u\mathrm{d}z - u_{\eta}\frac{\partial \eta}{\partial t} + u_{-d}\frac{\partial(-d)}{\partial t}|_{-d} \tag{3.3.12}$$

and by defining the depth-averaged velocity components U, V and the total water depth h as,

$$u = U + U' \quad v = V + V' \quad h = \eta + d$$

where, U' and V' are the difference between the time averaged velocity components (u, v) and the depth-averaged velocity components (U, V) respectively, it yields the integral of all the above terms,

(1) $\int_{-d}^{\eta}\frac{\partial u}{\partial t}\mathrm{d}z = \frac{\partial}{\partial t}(Uh) - u_{\eta}\frac{\partial \eta}{\partial t}$

(2) $$\int_{-d}^{\eta}\frac{\partial u^2}{\partial x}\mathrm{d}z = \frac{\partial}{\partial x}\int_{-d}^{\eta} u^2\mathrm{d}z - u_{\eta}^2\frac{\partial \eta}{\partial x}|_{\eta} + u_{-d}^2\frac{\partial(-d)}{\partial x}|_{-d}$$

$$= \frac{\partial}{\partial x}\int_{-d}^{\eta}(U^2 + 2UU' + U'^2)\mathrm{d}z - u_{\eta}^2\frac{\partial \eta}{\partial x}|_{\eta} + u_{-d}^2\frac{\partial(-d)}{\partial x}|_{-d}$$

$$= \frac{\partial}{\partial x}(U^2 h) + \frac{\partial}{\partial x}\int_{-d}^{\eta} U'^2\mathrm{d}z - u_{\eta}^2\frac{\partial \eta}{\partial x}|_{\eta} + u_{-d}^2\frac{\partial(-d)}{\partial x}|_{-d}$$

(3) $\int_{-d}^{\eta}\frac{\partial (uv)}{\partial y}\mathrm{d}z = \frac{\partial}{\partial y}(UhV) + \frac{\partial}{\partial y}\int_{-d}^{\eta} U'V'\mathrm{d}z - (uv)\Big|\eta\frac{\partial \eta}{\partial y}\Big|_{\eta} + (uv)\Big|_{-d}\frac{\partial(-d)}{\partial y}\Big|_{-d}$

(4) $\int_{-d}^{\eta} \frac{\partial(uw)}{\partial z}\mathrm{d}z = (uw)\big|_{\eta} - (uw)|_{-d}$

(5) $-\frac{1}{\rho}\int_{-d}^{\eta} \frac{\partial p}{\partial x}\mathrm{d}z = -\frac{1}{\rho}\int_{-d}^{\eta}\left(\rho g\frac{\partial \eta}{\partial x} + \frac{\partial P_a}{\partial x}\right)\mathrm{d}z = -gh\frac{\partial \eta}{\partial x} - \frac{h}{\rho}\frac{\partial P_a}{\partial x}$

(6) $\int_{-d}^{\eta} f_x \mathrm{d}z = \int_{-d}^{\eta}(\gamma \cdot v)\mathrm{d}z = \gamma \cdot Vh$

(7) $\int_{-d}^{\eta} \nu\left(\frac{\partial^2 u}{\partial x^2} + \frac{\partial^2 u}{\partial y^2} + \frac{\partial^2 u}{\partial z^2}\right)\mathrm{d}z = \int_{-d}^{\eta} \nu\left(\frac{\partial^2 u}{\partial x^2} + \frac{\partial^2 u}{\partial y^2}\right)\mathrm{d}z + \int_{-d}^{\eta}\frac{\partial^2 u}{\partial z^2}\mathrm{d}z$

(8)
$$\int_{-d}^{\eta}\left(\frac{\partial \overline{u'}^2}{\partial x} + \frac{\partial\left(\overline{u'v'}\right)}{\partial y} + \frac{\partial\left(\overline{u'w'}\right)}{\partial z}\right)\mathrm{d}z = \int_{-d}^{\eta}\left(\frac{\partial \overline{u'}^2}{\partial x} + \frac{\partial\left(\overline{u'v'}\right)}{\partial y}\right)\mathrm{d}z + \int_{-d}^{\eta}\frac{\partial\left(\overline{u'w'}\right)}{\partial z}\mathrm{d}z$$
$$= \int_{-d}^{\eta}\left(\frac{\partial \overline{u'}^2}{\partial x} + \frac{\partial\left(\overline{u'v'}\right)}{\partial y}\right)\mathrm{d}z + \frac{\tau_{xz}}{\rho}\bigg|_{\eta} - \frac{\tau_{xz}}{\rho}\bigg|_{-d}$$
$$= \int_{-d}^{\eta}\left(\frac{\partial \overline{u'}^2}{\partial x} + \frac{\partial\left(\overline{u'v'}\right)}{\partial y}\right)\mathrm{d}z + \frac{\tau_{wx}}{\rho} - \frac{\tau_{bx}}{\rho}$$

where, τ_{wx} and τ_{bx} are the x components of the surface wind stress and bottom stress, respectively; u_η, v_η, w_η are the velocity components at the surface.

By substituting the following bottom condition and the kinematic condition at the free surface,

$$u|_{-d} = \upsilon|_{-d} = w|_{-d} = 0$$

$$\frac{d\eta}{dt} = \left(\frac{\partial \eta}{\partial t} + u\frac{\partial \eta}{\partial x} + \upsilon\frac{\partial \eta}{\partial y}\right) = w_\eta$$

the depth-integrated momentum equation for the x component is obtained,

$$\frac{\partial(Uh)}{\partial t} + \frac{\partial(U^2h)}{\partial x} + \frac{\partial(UVh)}{\partial y} + gh\frac{\partial \eta}{\partial x} + \frac{h}{\rho}\frac{\partial P_a}{\partial x}$$
$$-\gamma \cdot Vh + \frac{\tau_{bx}}{\rho} - \frac{\tau_{wx}}{\rho} - \int_{-d}^{\eta}\nu\left(\frac{\partial^2 u}{\partial x^2} + \frac{\partial^2 u}{\partial y^2} + \frac{\partial^2 u}{\partial z^2}\right)\mathrm{d}z$$
$$+\int_{-d}^{\eta}\left(\frac{\partial \overline{u'}^2}{\partial x} + \frac{\partial\left(\overline{u'v'}\right)}{\partial y}\right)\mathrm{d}z + \frac{\partial}{\partial x}\int_{-d}^{\eta}\left(U'\right)^2\mathrm{d}z + \frac{\partial}{\partial y}\int_{-d}^{\eta}(U'V')\mathrm{d}z = 0 \quad (3.3.13)$$

Similarly, the depth-integrated momentum equation for the y component can be obtained as well.

Let $P = Uh$, $Q = Vh$, and the last four terms on the left side of Eq. (3.3.13) including the term of viscous stress, the term of lateral turbulent stress and the (last two) terms of integral stress caused by the non-uniform velocity distribution over the depth, be combined into a function of the depth-averaged velocity $E_e\left(\frac{\partial^2 P}{\partial x^2} + \frac{\partial^2 P}{\partial y^2}\right)$, where E_e is the comprehensive viscosity coefficient which consists of the kinematic viscosity coefficient ν, the turbulent diffusion coefficient ν_e and the integral viscosity coefficient ν_i,

$$E_e = \nu + \nu_e + \nu_i \tag{3.3.14}$$

Then, the 2D depth-integrated long wave equations become,

$$\frac{\partial P}{\partial x} + \frac{\partial Q}{\partial y} + \frac{\partial \eta}{\partial t} = 0 \tag{3.3.15}$$

$$\begin{aligned}\frac{\partial P}{\partial t} + \frac{\partial}{\partial x}\left(\frac{P^2}{h}\right) + \frac{\partial}{\partial y}\left(\frac{PQ}{h}\right) &= \gamma Q - gh\frac{\partial \eta}{\partial x} + \frac{h}{\rho}\frac{\partial P_a}{\partial \chi} \\ &+ \rho\frac{\tau_{wx}}{\rho} - \frac{\tau_{bx}}{\rho} + E_e\left(\frac{\partial^2 P}{\partial x^2} + \frac{\partial^2 P}{\partial y^2}\right)\end{aligned} \tag{3.3.16}$$

$$\begin{aligned}\frac{\partial Q}{\partial t} + \frac{\partial}{\partial x}\left(\frac{PQ}{h}\right) + \frac{\partial}{\partial y}\left(\frac{Q^2}{h}\right) &= -\gamma P - gh\frac{\partial \eta}{\partial y} + \frac{h}{\rho}\frac{\partial P_a}{\partial y} \\ &+ \frac{\tau_{wy}}{\rho} - \frac{\tau_{yb}}{\rho} + E_e\left(\frac{\partial^2 Q}{\partial x^2} + \frac{\partial^2 Q}{\partial y^2}\right)\end{aligned} \tag{3.3.17}$$

where, $P_a(x, y)$ is the distribution of atmospheric pressure at the free surface, τ_{bx} and τ_{by} are the x and y components of the bottom stress which can be determined by,

$$\frac{\tau_{bx}}{\rho} = g\frac{\mathrm{P}\sqrt{\mathrm{P}^2 + Q^2}}{C^2 h} \tag{3.3.18}$$

$$\frac{\tau_{by}}{\rho} = g\frac{Q\sqrt{\mathrm{P}^2 + Q^2}}{C^2 h} \tag{3.3.19}$$

where, C is the Chezy number $C = \frac{1}{n}h^{1/6}$, in which n is the Manning coefficient.

τ_{wx} and τ_{wy} are the x and y components of the wind stress given by,

$$\frac{\tau_{wx}}{\rho} = C_w w_x\sqrt{w_x^2 + w_y^2} \tag{3.3.20}$$

where, w_x and w_y are the x and y components of the wind velocity (generally 10 m above the water surface); C_w is the drag coefficient of air which can be determined by empirical formulas.

Based on the actual flow situation, the kinematic viscosity coefficient ν, the turbulent eddy viscosity coefficient ν_e and the integral viscosity coefficient ν_i in the comprehensive viscosity coefficient E_e can be determined or ignored due to the order of magnitude. In general, compared with ν_e and ν_i, the kinematic viscosity coefficient ν is much smaller and can be ignored. The integral viscosity coefficient ν_i is much higher than the eddy viscosity coefficient ν_e, even tens or hundreds of times higher in some cases. Therefore, the determination of the integral viscosity coefficient ν_i is more important in addition to the correct calculation of the eddy viscosity coefficient ν_e when using the depth-integrated 2D equations to calculate the flow field [2].

3.3.2 *Definite Conditions and Boundary Treatments*

Boundary conditions:

On the land boundary of the coast or island, the normal velocity is zero, i.e. $\boldsymbol{V_n} = 0$ in which, $\boldsymbol{n}$ is the unit vector in the direction perpendicular to the land boundary. On the inflow boundary, the water level condition $\eta = \eta(t)$ or the flux conditions $P = P(t)$ and $Q = Q(t)$ are given.

Initial condition:

Normally the water level variation at the open boundary of the sea or estuary is periodic. $\eta(x, y) = 0$, $P(x, y) = 0$ and $Q(x, y) = 0$ might be used for the initial condition at the time $t = t_0$.

In a bay with wide intertidal zone, the slope is relatively gentle. The location of shoreline boundary varies with the tide level and time. Meanwhile the area and the shape of the water surface change as well. Therefore, it is needed to treat the moving boundaries. There are many techniques to deal with the moving boundaries, such as 'wetting and drying' method by which the wetting and drying boundary is determined at each time step and then the boundary location is changed explicitly. 'Thin water layer' method is a quite simple approach with an assumption that a very thin water layer exists over the entire dry region at all times artificially. But the calculation results might be interfered due to the gravity term in the equations since the thin-layer water on the slope could flow into the computational domain. In the slot method, it is assumed that a very narrow slot exists in unit width on the beach. This makes the water in the slot connect with the water in the computational domain. The fixed boundary can thus be located in the narrow slot on the beach to make the moving boundary be a fixed boundary. See Chap. 5 for the detail of the slot method.

3.3.3 Difference Equations

An alternating direction implicit (ADI) scheme on staggered grids is described below.

Staggered grids are set in the computational domain as shown in Fig. 3.6. Equations (3.3.15)–(3.3.17) are discretized using an ADI scheme. Ignoring the variation of atmospheric pressure, the difference momentum equation for the x component is presented below as an example,

In the half time step $n\Delta t \to \left(n+\frac{1}{2}\right)\Delta t$:

$$\frac{\eta_{i,j}^{n+\frac{1}{2}} - \eta_{i,j}^{n}}{\frac{\Delta t}{2}} + \frac{P_{i+\frac{1}{2},j}^{n+\frac{1}{2}} - P_{i-\frac{1}{2},j}^{n+\frac{1}{2}}}{\Delta x} + \frac{Q_{i,j+\frac{1}{2}}^{n} - Q_{i,j-\frac{1}{2}}^{n}}{\Delta y} = 0 \tag{3.3.21}$$

$$\begin{aligned}
&\frac{P_{i+\frac{1}{2},j}^{n+\frac{1}{2}} - P_{i+\frac{1}{2},j}^{n-\frac{1}{2}}}{\Delta t} + \left[\frac{(\hat{P}\hat{U})_{i+\frac{3}{2},j}^{n} - (\hat{P}\hat{U})_{i-\frac{3}{2},j}^{n}}{2\Delta x} + \frac{(\hat{P}\bar{V})_{i+\frac{1}{2},j+\frac{1}{2}}^{n} - (\hat{P}\hat{V})_{i+\frac{1}{2},j-\frac{1}{2}}^{n}}{\Delta y}\right] \\
&= \gamma\bar{Q}_{i+\frac{1}{2},j}^{n} - \frac{gh_{i+\frac{1}{2},j}^{n}}{2\Delta x}\left(\eta_{i+1,j}^{n+\frac{1}{2}} + \eta_{i+1,j}^{n-\frac{1}{2}} - \eta_{i,j}^{n+\frac{1}{2}} - \eta_{i,j}^{n-\frac{1}{2}}\right) + C_w w_x\sqrt{w_x^2 + w_y^2} \\
&\quad - \frac{g\left(P_{i+\frac{1}{2},j}^{n+\frac{1}{2}} + P_{i+\frac{1}{2},j}^{n-\frac{1}{2}}\right)\sqrt{\left(\hat{P}_{i+\frac{1}{2},j}^{n}\right)^2 + \left(\hat{Q}_{i+\frac{1}{2},j}^{n}\right)^2}}{2\left(h_{i+\frac{1}{2},j}^{n} C_{i+\frac{1}{2},j}^{n}\right)^2} \\
&\quad + \frac{E_e}{\Delta x^2}\left(\hat{P}_{i+\frac{3}{2},j}^{n} + \hat{P}_{i-\frac{1}{2},j}^{n} - 2\hat{P}_{i+\frac{1}{2},j+1}^{n}\right) \\
&\quad + \frac{E_e}{\Delta y^2}\left(\hat{P}_{i+\frac{1}{2},j+1}^{n} + \hat{P}_{i+\frac{1}{2},j-1}^{n} - 2\hat{P}_{i+\frac{1}{2},j}^{n}\right)
\end{aligned} \tag{3.3.22}$$

where, $\widehat{U}$ is the iterative correction value, $\widehat{V}$ is the average value of the surrounding grid points; $\widehat{P}$ is the value using an upwind scheme. They are expressed by,

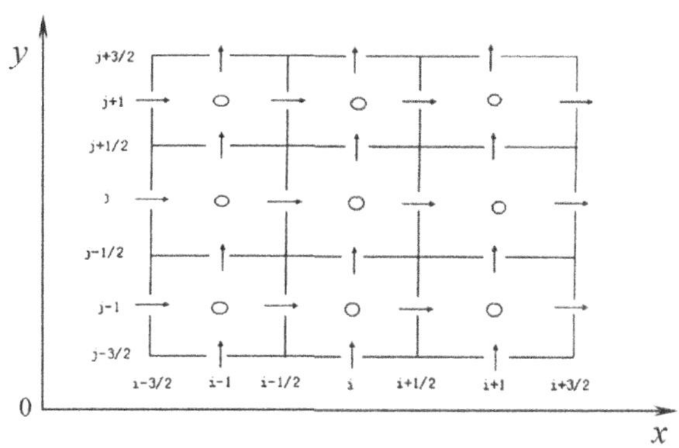

Fig. 3.6 Grid notation (○: η, → : P, ↑: Q)

$$\hat{U}^n = \begin{cases} U^{n-\frac{1}{2}} & \text{the first iteration} \\ \frac{1}{2}\left(U^{n+\frac{1}{2}} + U^{n-\frac{1}{2}}\right) & \text{the subsequent iterations} \end{cases} \tag{3.3.23}$$

$$\overline{V}^n_{i+\frac{1}{2},j+\frac{1}{2}} = \frac{1}{2}\left(V^n_{i,j+\frac{1}{2}} + V^n_{i+1,j+\frac{1}{2}}\right) \tag{3.3.24}$$

$$\widehat{P}^n_{i+\frac{1}{2},j} = \begin{cases} p^n_{i+\frac{1}{2},j-1} & V^n_{i+\frac{1}{2},j} > 0 \\ p^n_{i+\frac{1}{2},j+1} & V^n_{i+\frac{1}{2},j} < 0 \end{cases} \tag{3.3.25}$$

Equations (3.3.21) and (3.3.22) are sorted into,

$$-A_i\eta_i^{n+\frac{1}{2}} + B_i p_{i+\frac{1}{2}}^{n+\frac{1}{2}} + C_i\eta_{i+\frac{1}{2}}^{n+\frac{1}{2}} = T_i^n \tag{3.3.26}$$

$$-D_i P_{i-\frac{1}{2}}^{n+\frac{1}{2}} + E_i\eta_i^{n+\frac{1}{2}} + F_i P_{i+\frac{1}{2}}^{n+\frac{1}{2}} = S_i^n \tag{3.3.27}$$

$\eta_i^{n+\frac{1}{2}}$ and $P_{i\pm\frac{1}{2}}^{n+\frac{1}{2}}$ can be obtained by solving the tridiagonal system of the above equations using chasing method.

In the half time step $(n+\frac{1}{2})\Delta t \to (n+1)\Delta t$:

$$\frac{\eta^{n+1}_{i,j} - \eta^{n+\frac{1}{2}}_{i,j}}{\frac{\Delta t}{2}} + \frac{P^{n+\frac{1}{2}}_{i+\frac{1}{2},j} - P^{n+\frac{1}{2}}_{i-\frac{1}{2},j}}{\Delta x} + \frac{Q^{n+1}_{i,j+\frac{1}{2}} - Q^{n+1}_{i,j-\frac{1}{2}}}{\Delta y} = 0 \tag{3.3.28}$$

$$\begin{aligned}
&\frac{Q^{n+\frac{1}{2}}_{i,j+\frac{1}{2}} - Q^n_{i,j+\frac{1}{2}}}{\Delta t} + \left[\frac{(\hat{Q}\hat{V})^{n+\frac{1}{2}}_{i,j+\frac{3}{2}} - (\hat{Q}\hat{V})^{n+\frac{1}{2}}_{i,j-\frac{1}{2}}}{2\Delta x} + \frac{(\hat{Q}\bar{U})^{n+\frac{1}{2}}_{i+\frac{1}{2},j+\frac{1}{2}} - (\hat{Q}\bar{U})^{n+\frac{1}{2}}_{i+\frac{1}{2},j-\frac{1}{2}}}{\Delta y}\right] \\
&= -\gamma\bar{P}^{n+\frac{1}{2}}_{i,j+\frac{1}{2}} - \frac{gh^{n+\frac{1}{2}}_{i,j+\frac{3}{2}}}{2\Delta y}\left(\eta^{n+1}_{i,j+1} + \eta^n_{i,j-1} - \eta^{n+1}_{i,j} - \eta^n_{i,j}\right) + C_w w_y\sqrt{w_x^2 + w_y^2} \\
&\quad - \frac{g\left(Q^{n+1}_{i,j+\frac{1}{2}} + Q^n_{i,j+\frac{1}{2}}\right)\sqrt{\left(\hat{P}^{n+\frac{1}{2}}_{i,j+\frac{1}{2}}\right)^2 + \left(\hat{Q}^{n+\frac{1}{2}}_{i,j+\frac{1}{2}}\right)^2}}{2\left(h^{n+\frac{1}{2}}_{i,j+\frac{1}{2}}C^{n+\frac{1}{2}}_{i,j+\frac{1}{2}}\right)^2} \\
&\quad + \frac{E_e}{\Delta y^2}\left(Q^{n+\frac{1}{2}}_{i,j+\frac{3}{2}} + Q^{n+\frac{1}{2}}_{i,j-\frac{1}{2}} - 2Q^{n+\frac{1}{2}}_{i,j+\frac{1}{2}}\right) \\
&\quad + \frac{E_e}{\Delta x^2}\left(\hat{Q}^{n+\frac{1}{2}}_{i+1,j+\frac{1}{2}} + \hat{Q}^{n+\frac{1}{2}}_{i-1,j+\frac{1}{2}} - 2\hat{Q}^{n+\frac{1}{2}}_{i,j+\frac{1}{2}}\right)
\end{aligned} \tag{3.3.29}$$

where, $\widehat{V}$, $\widehat{U}$ and $\widehat{Q}$ have the same meanings as above and their expressions are similar to Eqs. (3.3.23)–(3.3.25).

Equations (3.3.28) and (3.3.29) can also be sorted into,

$$-a_j\eta_j^{n+1} + b_j q_{j+\frac{1}{2}}^{n+1} + c_j\eta_{j+1}^{n+1} = A_j^{n+\frac{1}{2}} \tag{3.3.30}$$

$$-d_j q_{j-\frac{1}{2}}^{n+1} + e_j \eta_j^{n+1} + f_j q_{j+\frac{1}{2}}^{n+1} = B_j^{n+\frac{1}{2}} \tag{3.3.31}$$

η_j^{n+1} and $Q_{j+\frac{1}{2}}^{n+1}$ can be obtained by solving the tridiagonal system of the above equations using chasing method.

The above scheme which has second order accuracy in time and in space are unconditionally stable. But in order to achieve the necessary calculation precision, the time step must be limited. The maximum Courant number is,

$$C_f = 2\Delta t \sqrt{gH\left(\frac{1}{\Delta x^2} + \frac{1}{\Delta y^2}\right)} \leq 4\sqrt{2} \tag{3.3.32}$$

3.3.4 Numerical Simulation of Circulation

The numerical simulation of circulation is of great significance in engineering practice. For instance, the circulation occurred near the breakwaters or downstream of the hydraulic structures may cause local scour and deposition. The circulation flow field should be determined in the study of pollutant dispersion and thermal dispersion. In estuaries and coastal waters, the existence of circulation has great influence on sediment transport, pollutant dispersion, coastal morphology and navigation. Therefore, it is necessary to forecast the formation and development of circulation.

1. Generation of circulation

For a sudden-expansion flow of fully developed turbulence, there might be a large scale vortex (circulation) in the area of sudden-expansion. Appearance and generation of the circulation had been analyzed for the expansion flow of steady turbulence in Ref. [3] (1985). Steady flow in a regular open channel tends to be uniform under the condition of a certain friction resistance. If there is a sudden expansion in the channel, the uniform flow is observed before the sudden expansion and far downstream of the sudden expansion. In the sudden expansion, due to the flow inertia, the main stream separates from the original boundary to turn smoothly and extends to the downstream wall afterwards. The surface elevation in the main-stream zone is higher than that in the non-main-stream zone at the first turning due to the existence of centripetal acceleration. But with the spreading of the main-stream, the corresponding water depth is increased gradually. When the water in the non-main-stream zone cannot flow downstream, the radial flow occurs and results in circulation. It indicates that the inertial force and resistance play leading roles in the generation of circulation. Therefore, it can be considered that the convection term and the resistance term in Eqs. (3.3.16) and (3.3.17) are the key factors in the simulation of circulation. It can be seen from the derivation of the 2D depth-integrated equations that the viscosity term is also important. The viscosity term mainly depends on the viscous stress, lateral turbulent stress and the uniformity of the velocity distribution over the water depth. As the velocity distribution over the depth is relatively uniform and the turbulence intensity is not high in the shallow water circulation, the magnitude of the viscosity term is not large. But if the flow is in a state of strong turbulence with surface roller or breaking, the viscosity term will play a significant role due to the high turbulence intensity and non-uniformity of the velocity distribution over the depth.

The convection term, the bottom resistance term and the viscosity term are all important for the simulation of circulation. But among these three terms, the convection term is more important and more sensitive. Therefore, the accuracy of modelling the convection term should be improved for better simulation of the circulation.

2. Circulation in a channel with sudden-expansion

Intensive experiments on unsteady flow in an open channel with sudden-expansion had been carried out in the laboratory of fluid mechanics at Department of civil engineering, Delft University of technology, the Netherlands (1983) [3]. In the open channel as shown in Fig. 3.7, a time series of sinusoidal water discharge was given at the 0–0 section. The time series of surface elevation was measured at the G section. The velocity and turbulence characteristics at each layer over the depth were measured by laser Doppler velocimeter. Analyzing the experimental data using the system average method, the depth-averaged velocity vectors at different time had been plotted and the development of circulation over time had been described in detail. In order to study the method of numerical simulation on circulation, a numerical model had been carried out under the same experimental conditions in a computational domain as shown in Fig. 3.8. The model grid number was 66 × 17. The grid spacing was $\Delta x = 10\,\text{cm}$ and $\Delta y = 5\,\text{cm}$ According to the experimental conditions, the water depth was 10.9 cm, and the inflow boundary condition was $Q = 9 \sin \frac{\pi}{T} t$ (l/s), where $T = 60$ s. The measured water level at the G section was used as the outflow

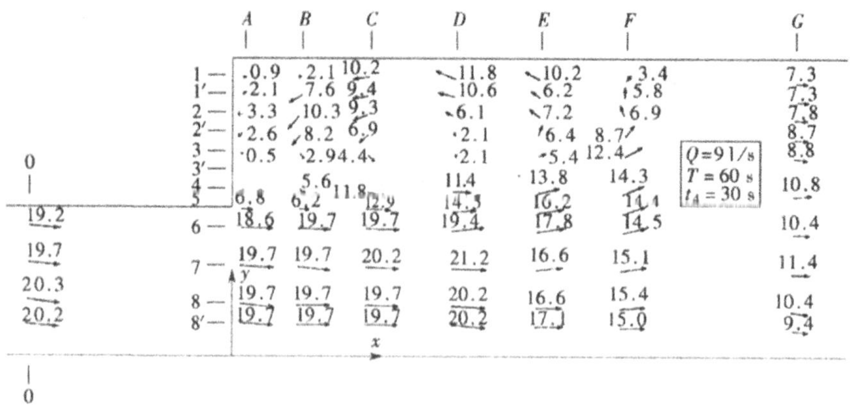

Fig. 3.7 Measured depth-averaged velocity $Q_{\max} = 9$ l/s, $T = 60$ s, $t = 30$ s

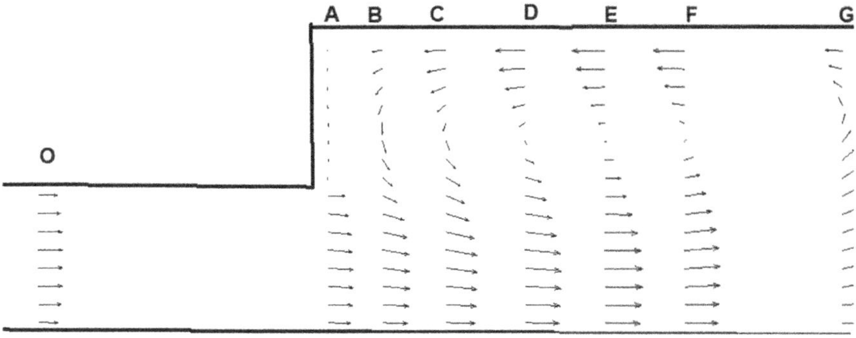

Fig. 3.8 Simulated depth-averaged velocity $Q_{\max} = 9$ l/s, $T = 60$ s, $t = 30$ s

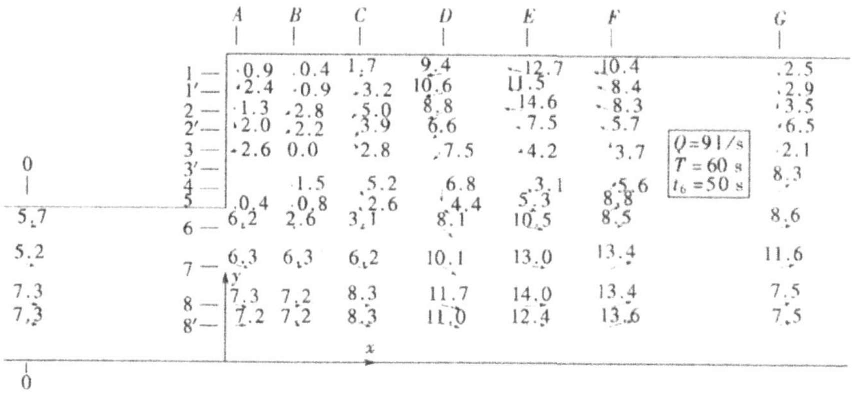

Fig. 3.9 Measured depth-averaged velocity $Q_{\max} = 9$ l/s, $T = 60$ s, $t = 50$ s

boundary condition. Figures 3.7 and 3.9 show the experimental results. Figures 3.8 and 3.10 show the numerical results by using a 2D long wave model, which matched the experimental results well. It had been proved in the numerical modelling that the convection term and the bottom friction term (roughness) are the most sensitive to the results.

3.3.5 Tidal Current Field in Bohai Sea and Its Validation

Bohai Sea is the only semi-enclosed sea in China. The total area is approximately 78,000 km^2 and the averaged water depth is about 18 m. As the water depth is small compared with the horizontal scale, a 2D long wave model can be used to study the hydrodynamic characteristics and the water quality of the Bohai Sea.

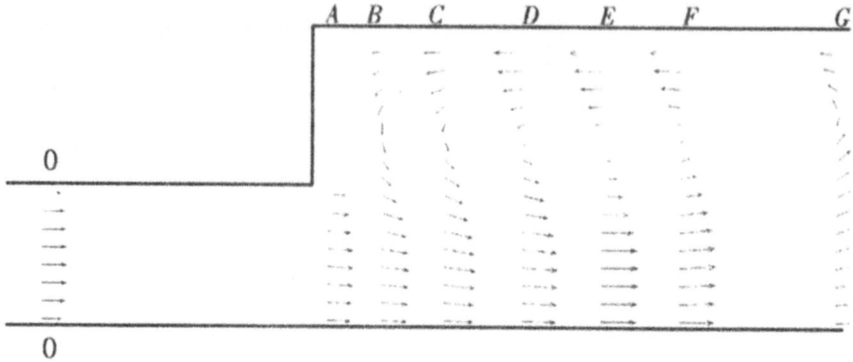

Fig. 3.10 Simulated depth-averaged velocity $Q_{max} = 9$ l/s, $T = 60$ s, $t = 50$ s

In the numerical model, rectangular grids with size of 2500 m × 2500 m had been used. The roughness was set to $n = 0.012$ and the viscosity coefficient was set to $E = 40$. The measured time series of tidal elevation at the Bohai strait had been used as the open boundary condition for the calculation of flow field in the Bohai Sea over one year period. In order to display the variation of the flow field, the positions of the particles set on the initial sea surface had been obtained at different time using marked point method [4, 5]. Figure 3.11 shows the particle distribution at the beginning and after 3 months, 6 months and 12 months in a typical year.

Figure 3.12 shows the simulated surface elevation compared with the measurement at Tanggu station (117° 41′ E, 39° 00′ N) for a period of 120 h from 31/07/1993 to 03/08/1993.

Figure 3.13 shows a comparison of current speed between the simulation and the measurement at F1 station (117° 58′ E, 39° 00′ N) for 1 day in 01/08/1993.

3.4 Mathematical Model of 3D Shallow Water Long Waves

For the study of flood and tidal current in estuaries and coast waters with great bathymetry changes, a 2D depth-integrated hydrodynamic model is not valid because the vertical flow velocity and the variation are relatively large. Then a 3D hydrodynamic model should be used. If the horizontal length scale is still much larger than the vertical scale, a quasi 3D model can be used to simulate the shallow water long waves. In general, both the grid number and grid size in the vertical direction are much smaller than in the two horizontal directions in a 3D shallow water long wave model.

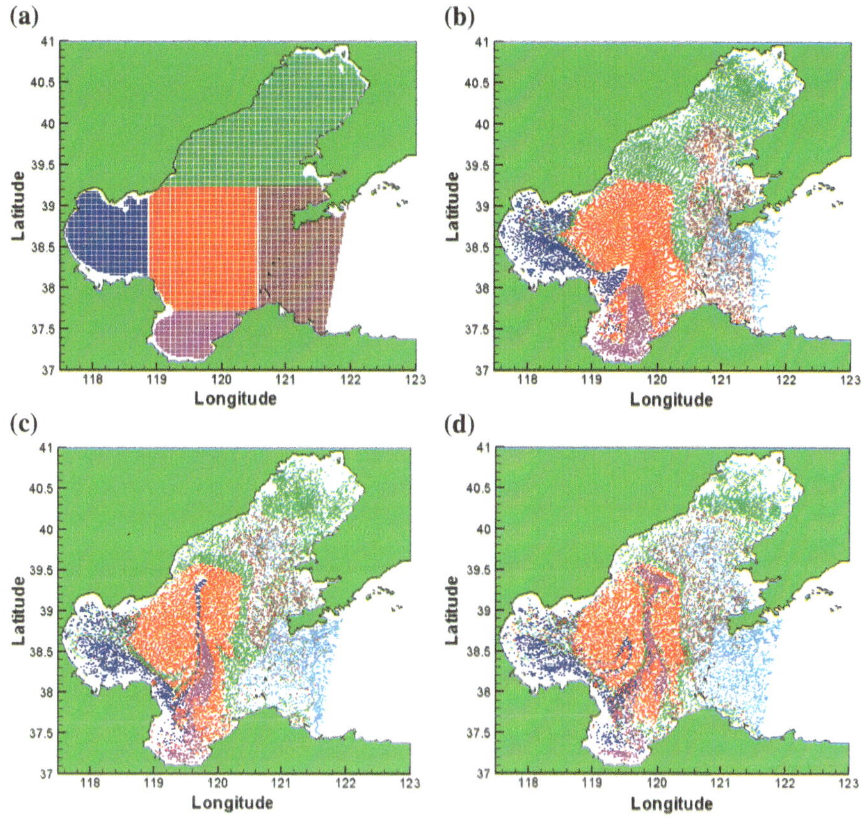

Fig. 3.11 Particle movements in Bohai Sea. **a** At the beginning, **b** after 3 months, **c** after 6 months, **e** after 12 months

3.4.1 Governing Equations

The Reynolds equations for incompressible viscous flow read,

$$\frac{\partial u}{\partial x}+\frac{\partial v}{\partial y}+\frac{\partial w}{\partial z}=0 \tag{3.4.1}$$

$$\frac{\partial u}{\partial t}+\frac{\partial (u^2)}{\partial x}+\frac{\partial (uv)}{\partial y}+\frac{\partial (uw)}{\partial z}=\gamma v-\frac{1}{\rho}\frac{\partial p}{\partial x}+\frac{1}{\rho}\left[\frac{\partial \tau_{xx}}{\partial x}+\frac{\partial \tau_{xy}}{\partial y}+\frac{\partial \tau_{xz}}{\partial z}\right] \tag{3.4.2}$$

$$\frac{\partial v}{\partial t}+\frac{\partial (vu)}{\partial x}+\frac{\partial (v^2)}{\partial y}+\frac{\partial (vw)}{\partial z}=-\gamma u-\frac{1}{\rho}\frac{\partial p}{\partial y}+\frac{1}{\rho}\left[\frac{\partial \tau_{yx}}{\partial x}+\frac{\partial \tau_{yy}}{\partial y}+\frac{\partial \tau_{yz}}{\partial z}\right] \tag{3.4.3}$$

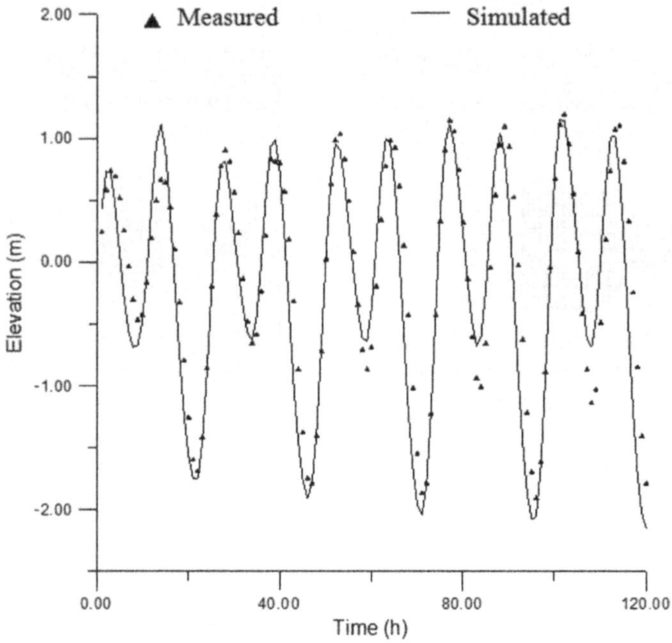

Fig. 3.12 Simulated and measured surface elevation at Tanggu station

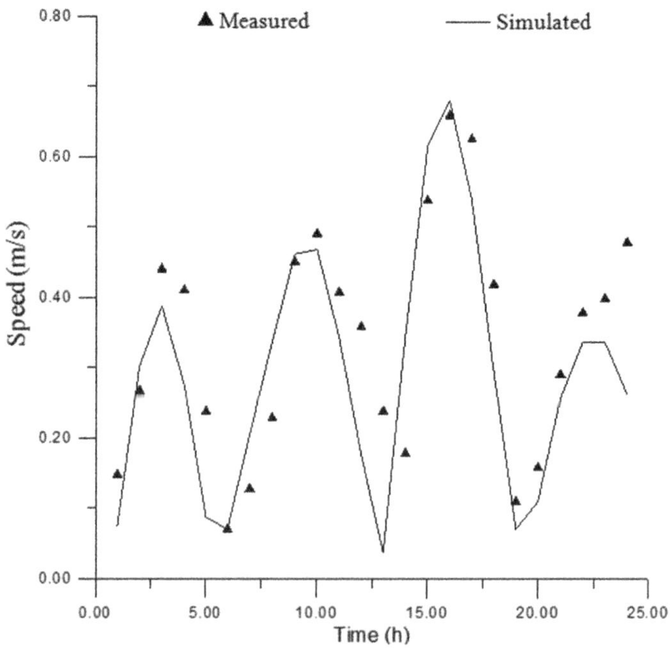

Fig. 3.13 Simulated and measured current speed at F1 station

$$\frac{\partial w}{\partial t}+\frac{\partial(uw)}{\partial x}+\frac{\partial(vw)}{\partial y}+\frac{\partial(w^2)}{\partial z}=-\frac{1}{\rho}\frac{\partial p}{\partial y}+\frac{1}{\rho}\left[\frac{\partial\tau_{zx}}{\partial x}+\frac{\partial\tau_{zy}}{\partial y}+\frac{\partial\tau_{zz}}{\partial z}\right] \tag{3.4.4}$$

where, t is time; x, y and z are the Cartesian coordinates; u, v and w are the velocity components in the x, y and z directions; p is the pressure; ρ is the fluid density; γ is the Coriolis parameter; g is the gravitational acceleration; $\tau_{xx}, \tau_{xy}, \tau_{xz}, \tau_{yx}, \tau_{yy}, \tau_{yz}, \tau_{zx}, \tau_{zy}, \tau_{zz}$ are the components of the Reynolds stress tensor.

There are two main difficulties in the establishment of a 3D long wave model based on the Reynolds equations. First, the computational domain is generally large so that the workload is heavy. Secondly, it is difficult to capture the free surface which varies with time. Therefore, it is necessary and also possible to do some simplified treatment according to the characteristics of the shallow water long waves. A commonly used method is assuming that the vertical acceleration of water particle can be neglected and simplifying the momentum equation in the z direction to the hydrostatic pressure formula. The model by using this method is called a quasi-3D long wave model. In this kind of model, stratified integration in the vertical direction can be made [6], or the computational domain with free surface variation can be transformed into a fixed computational domain by using a vertical sigma transformation. Considering the error caused by the assumption of vertical hydrostatic pressure, Stelling et al. proposed a quasi-3D long wave model with correction of non-hydrostatic pressure in [7, 8].

In full 3D models of shallow water long waves, the sigma transformation method is commonly used which transforms the free surface varying with time into a fixed position in the sigma coordinate system. 3D long wave models had been established in the sigma coordinate system by Lin and Tao [9], Xie [10] and Lin and Li [11].

A full 3D long wave model can also be established in the curvilinear coordinate system. Therefore, in addition to the coordinate transformation in the vertical direction, the horizontal coordinate transformation is also carried out.

3.4.2 *Quasi-3D Long Wave Model by Stratified Integral*

As the long wave surface curvature is small, the vertical acceleration of water particles is much less than the gravitational acceleration. Then it can be assumed that the gravitational acceleration is the only acceleration in the vertical direction, and the pressure is hydrostatic. Thus the momentum equation in the vertical direction can be simplified, and the 3D incompressible Reynolds equations in a conservation form turn to,

$$\frac{\partial u}{\partial x}+\frac{\partial v}{\partial y}+\frac{\partial w}{\partial z}=0 \tag{3.4.5}$$

$$\frac{\partial u}{\partial t}+\frac{\partial (u^2)}{\partial x}+\frac{\partial (uv)}{\partial y}+\frac{\partial (uw)}{\partial z}=\gamma v-\frac{1}{\rho}\frac{\partial p}{\partial x}+\frac{1}{\rho}\left[\frac{\partial \tau_{xx}}{\partial x}+\frac{\partial \tau_{xy}}{\partial y}+\frac{\partial \tau_{xz}}{\partial z}\right] \tag{3.4.6}$$

$$\frac{\partial v}{\partial t}+\frac{\partial (vu)}{\partial x}+\frac{\partial (v^2)}{\partial y}+\frac{\partial (vw)}{\partial z}=-\gamma u-\frac{1}{\rho}\frac{\partial p}{\partial y}+\frac{1}{\rho}\left[\frac{\partial \tau_{yx}}{\partial x}+\frac{\partial \tau_{yy}}{\partial y}+\frac{\partial \tau_{yz}}{\partial z}\right] \tag{3.4.7}$$

$$\frac{\partial p}{\partial z}+\rho g=0 \tag{3.4.8}$$

There are normally two kinds of coordinate systems in the process of discretization of Eqs. (3.4.5)–(3.4.8) and the computational domain. One is the Cartesian coordinate system, in which the computational domain is discretized directly and the water column is divided into several layers in the z direction. The other is the so-called the sigma coordinate system, in which the coordinate transformation method is used to change the computational domain into a fixed domain with unchanged x and y coordinates and the sigma coordinate transformed from the z coordinate. Quasi-3D long wave models in both the Cartesian coordinate system and the sigma coordinate system are widely used in estuaries and coastal waters. Both coordinate systems have their advantages and disadvantages which had been discussed in detail by Stelling in [7, 8].

A quasi-3D long wave model using stratified integral is described below. The computational domain can be divided into structured rectangular grids in the horizontal directions. In the vertical direction, water column can be divided into several layers with equidistant or non-equidistant intervals. The number of vertical layers can vary with x and y in order to fit the changes of surface and bathymetry. The adjacent vertical layers are interacted through shear stress. Figure 3.14 shows a sketch of grids in a vertical plane of (x, z) [6].

1. Stratified integral equations

Similar to the integral method for the 2D depth-integrated equations, integration over each vertical layer has been carried out here. Assuming the total number of layers is K, let the free surface be at the first layer, i.e., $k=1$, and the bottom at the Kth layer, i.e., $k=K$. $k \pm \frac{1}{2}$ denotes the interface between $k+1$ and k, and the interface between k and $k-1$, respectively. The integrated continuity equation at the kth layer reads,

$$\int_{k+\frac{1}{2}}^{k-\frac{1}{2}}\left(\frac{\partial u}{\partial x}+\frac{\partial v}{\partial y}+\frac{\partial w}{\partial z}\right)\mathrm{d}z=0 \tag{3.4.9}$$

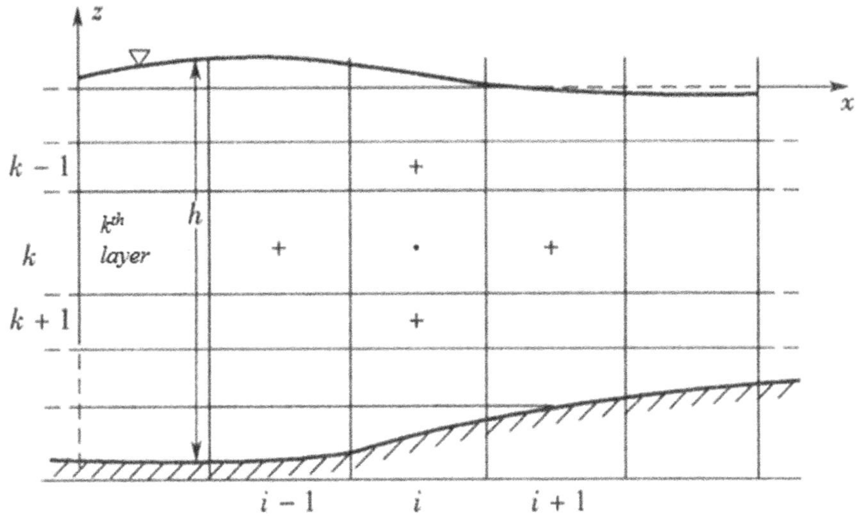

Fig. 3.14 Sketch of grids in a vertical plane

By using the Leibniz's law and the boundary condition, the vertical velocity component at the interface $k-\frac{1}{2}$ can be derived as,

$$w_{k-\frac{1}{2}} = w_{k+\frac{1}{2}} + \int\limits_{k+\frac{1}{2}}^{k-\frac{1}{2}} \left(\frac{\partial u}{\partial x} + \frac{\partial v}{\partial y}\right)\mathrm{d}z$$

$$w_{k-\frac{1}{2}} = -\sum_{k=k}^{K}\left\{\frac{\partial(hu)}{\partial x} + \frac{\partial(hv)}{\partial y}\right\} \tag{3.4.10}$$

Substituting the free surface condition $w_\eta = \partial\eta/\partial t$ into the above yields,

$$\frac{\partial\eta}{\partial t} + \sum_{k=1}^{K}\left\{\frac{\partial(hu)}{\partial x} + \frac{\partial(hv)}{\partial y}\right\} = 0 \tag{3.4.11}$$

The averaged velocity at the kth layer is defined as,

$$\bar{u}_k = \frac{1}{\Delta Z_k}\int\limits_{k+\frac{1}{2}}^{k-\frac{1}{2}} u(x, y, z, t)\mathrm{d}z \tag{3.4.12}$$

where, ΔZ_k is the interval of the kth layer.

The assumption of Boussinesq is adopted,

$$\left.\begin{aligned} \tau_{xx} &= 2\rho\nu_h \frac{\partial \bar{u}}{\partial x} \\ \tau_{xy} &= \rho\nu_h\left[\frac{\partial \bar{u}}{\partial y} + \frac{\partial \bar{v}}{\partial x}\right] \\ \tau_{xz} &= \rho\nu_v\left[\frac{\partial \bar{u}}{\partial z} + \frac{\partial \bar{w}}{\partial x}\right] \\ \tau_{yz} &= \rho\nu_v\left[\frac{\partial \bar{v}}{\partial z} + \frac{\partial \bar{w}}{\partial y}\right] \end{aligned}\right\} \tag{3.4.13}$$

where, ν_h and ν_v are the horizontal and vertical eddy viscosity coefficients respectively.

The vertical eddy viscosity coefficient is usually larger than the horizontal one in large estuaries. The horizontal eddy viscosity coefficient can be assumed to be constant over the water depth. The vertical eddy viscosity coefficient can be obtained by the following expression [12],

$$\nu_v = l^2\left[\left(\frac{\partial u}{\partial z}\right)^2 + \left(\frac{\partial v}{\partial z}\right)^2\right]^{\frac{1}{2}} \tag{3.4.14}$$

where,

$l = kz$ for $kz < 0.1h$;
$l = 0.1h$ for $kz > 0.1h$.

in which, the Karman's constant k is set to $k = 0.41$.

As we have,

$$\frac{\partial p}{\partial x} \approx \rho g \frac{\partial \eta}{\partial x}, \quad \frac{\partial p}{\partial y} \approx \rho g \frac{\partial \eta}{\partial y} \tag{3.4.15}$$

with definitions of $P_k = \bar{u}_k \Delta z_k$ and $Q_k = \bar{v}_k \Delta z_k$, Eqs. (3.4.2) and (3.4.3) in integral form at the kth layer become,

$$\begin{aligned} \frac{\partial P_k}{\partial t} + \frac{\partial(\bar{u}_k P_k)}{\partial x} + \frac{\partial(\bar{v}_k P_k)}{\partial y} &= \gamma Q_k - g\Delta z_k \frac{\partial \eta}{\partial x} + (w\bar{u})_{k+\frac{1}{2}} - (w\bar{u})_{k-\frac{1}{2}} \\ &\quad + \frac{(\tau_{xz})_{k-\frac{1}{2}} - (\tau_{xz})_{k+\frac{1}{2}}}{\rho} + \nu_h\left(\frac{\partial^2 P_k}{\partial x^2} + \frac{\partial^2 P_k}{\partial y^2}\right) \end{aligned} \tag{3.4.16}$$

$$\begin{aligned} \frac{\partial Q_k}{\partial t} + \frac{\partial(\bar{u}_k Q_k)}{\partial x} + \frac{\partial(\bar{v}_k Q_k)}{\partial y} &= \gamma P_k - g\Delta z_k \frac{\partial \eta}{\partial y} + (w\bar{v})_{k+\frac{1}{2}} - (w\bar{v})_{k-\frac{1}{2}} \\ &\quad + \frac{(\tau_{yz})_{k-\frac{1}{2}} - (\tau_{yz})_{k+\frac{1}{2}}}{\rho} + \nu_h\left(\frac{\partial^2 Q_k}{\partial x^2} + \frac{\partial^2 Q_k}{\partial y^2}\right) \end{aligned} \tag{3.4.17}$$

The terms of $(w\bar{u})_{k-\frac{1}{2}}$ and $(w\bar{v})_{k-\frac{1}{2}}$ at the free surface ($k = 1$) can be eliminated by using the free surface condition and the Leibniz's law. They are zero at the bottom according to the non-slip boundary condition.

When the layer number is 1, Eqs. (3.4.11), (3.4.16) and (3.4.17) become the 2D depth-integrated governing equations.

$$\frac{\partial \eta}{\partial t} + \frac{\partial P}{\partial x} + \frac{\partial Q}{\partial y} = 0 \tag{3.4.18}$$

$$\begin{aligned}\frac{\partial P}{\partial t} + \frac{\partial}{\partial x}\left(\frac{\alpha_1' P^2}{h}\right) + \frac{\partial}{\partial y}\left(\frac{\alpha_2' PQ}{h}\right) &= \gamma Q - gh\frac{\partial \eta}{\partial x} + \frac{\tau_{xz}|_\eta - \tau_{xz}|_{-d}}{\rho} \\ &\quad + \nu_h\left(\frac{\partial^2 P}{\partial x^2} + \frac{\partial^2 P}{\partial y^2}\right)\end{aligned} \tag{3.4.19}$$

$$\begin{aligned}\frac{\partial Q}{\partial t} + \frac{\partial}{\partial x}\left(\frac{\alpha_2' PQ}{h}\right) + \frac{\partial}{\partial y}\frac{\left(\alpha_3' Q^2\right)}{h} &= \gamma P - gh\frac{\partial \eta}{\partial x} + \frac{\tau_{yz}|_\eta - \tau_{yz}|_{-d}}{\rho} \\ &\quad + \nu_h\left(\frac{\partial^2 Q}{\partial x^2} + \frac{\partial^2 Q}{\partial y^2}\right)\end{aligned} \tag{3.4.20}$$

where, α_1', α_2' and α_3' are the momentum correction coefficients,

$$\alpha_1' = \int_{-d}^{\eta} u^2 \mathrm{d}z / \left(\frac{P^2}{h}\right) \tag{3.4.21}$$

$$\alpha_2' = \int_{-d}^{\eta} uv \mathrm{d}z / \left(\frac{PQ}{h}\right) \tag{3.4.22}$$

$$\alpha_3' = \int_{-d}^{\eta} v^2 \mathrm{d}z / \left(\frac{Q^2}{h}\right) \tag{3.4.23}$$

$\tau_{xz}|_\eta$ and $\tau_{yz}|_\eta$ are the wind stress components; $\tau_{xz}|_{-d}$ and $\tau_{yz}|_{-d}$ are the bottom stress components; h is the water depth, $h = \eta + d$; P and Q are the depth-integrated flux per unit width in the x and y directions respectively, $P = Uh$ and $Q = Vh$; U and V are the depth-averaged horizontal velocity components in the x and y directions respectively.

Equations (3.4.19) and (3.4.20) are different depth-integrated equations to Eqs. (3.3.16) and (3.3.17). In Eqs. (3.3.16) and (3.3.17), the integral viscosity caused by the non-uniform velocity distribution over the depth is included in the viscosity term. But in Eqs. (3.4.19) and (3.4.20), the momentum correction coefficients α_1', α_2' and α_3' are introduced, and the coefficient of the viscosity term is called the eddy viscosity coefficient.

2. Boundary conditions

Wind stress is the shear stress at the free surface given by,

$$\tau_{xz}|_{\eta} = \rho_a C_w w w_x \tag{3.4.24}$$

$$\tau_{yz}|_{\eta} = \rho_a C_w w w_y \tag{3.4.25}$$

$$w = \sqrt{w_x^2 + w_y^2} \tag{3.4.26}$$

where, C_w is the air-fluid drag coefficient; ρ_a is the density of air; w_x and w_y are the wind velocity components in the x and y directions.

At the bottom,

$$\frac{\tau_b}{\rho} = (u_b^*)^2 \tag{3.4.27}$$

where, τ_b is the bottom stress; u_b^* is the friction velocity. By assuming a logarithmic velocity profile near the bottom boundary, the bottom stress is given by

$$\frac{\tau_b}{\rho} = \left|\frac{\bar{u}^{n+\frac{1}{2}} + \bar{u}^{n-\frac{1}{2}}}{2}\right| \frac{(\bar{u}^{n+\frac{1}{2}} + \bar{u}^{n-\frac{1}{2}})}{2} \left[2.5 \ln\left(\frac{30 d_K}{2.72 k_s}\right)\right]^{-2} \tag{3.4.28}$$

where, d_K is the bottom layer thickness; k_s is the roughness height.

3. Procedure of solving a quasi-3D long wave model by stratified integral

(1) Solving the 2D depth-integrated Eqs. (3.4.18), (3.4.19) and (3.4.20) to obtain the surface elevation and its gradient. Here $\alpha_1' = \alpha_2' = \alpha_3' = 1$ can be chosen.
(2) When the layer number is larger than 1, solving the stratified integral Eqs. (3.4.16) and (3.4.17) to get the averaged horizontal velocity at each layer.
(3) Integrating the averaged horizontal velocity at each layer over the water depth to get the momentum correction coefficients α_1', α_2' and α_3',, and the stress terms.
(4) Substituting the momentum correction coefficients and the stress terms into the 2D model, and repeating the steps (1) and (2).
(5) Determining the vertical velocity at each layer by Eq. (3.4.7).
(6) Judging the wet and dry condition, repeatedly modifying the surface layer thickness corresponding to the new surface elevation.

Quasi-3D long wave models by stratified integral have been widely used. The thickness of the vertical layers can be constant or variable, and the grids are orthogonal. The disadvantage of this method is the staircase-like bottom boundary. Especially when the bathymetry variation is large, the staircase-like grids at the bottom are more obviously, which may affect the bottom stress and the accuracy of the convection term.

3.4.3 Full 3D Long Wave Model in the Sigma Coordinate System

Full 3D long wave models solve the 3D Reynolds equations directly instead of assuming a hydrostatic pressure distribution in the vertical direction.

The coordinate transformation is applied only in the vertical direction into the sigma coordinate system. The free surface varied with time is transformed into $\sigma = 1$, and the bottom into $\sigma = 0$ by using an algebraic formula. Then the calculation is carried out in a fixed computational domain. Figure 3.15 shows configurations for the coordinate transformation in a vertical plane. The sigma transformation is an effective method to capture the free surface. In the sigma coordinate system which is boy-fitted, the number of grids (layers) in the vertical direction is the same in the entire computational domain.

The advantage of this approach is much convenience in the calculation. The computational domain in the plane of (σ, x) becomes rectangular in a boy-fitted coordinate system. The number of vertical layers in the whole computational domain is constant, which means the grids are fine in the shallow area and coarse in the deep area. The disadvantage of this approach is that the grids are not orthogonal which may cause the horizontal gradient error. The sigma transformation accuracy is poor in the case of large and rapid variations in bathymetry and free surface.

The 3D long wave equations in the sigma coordinate system can be quasi-3D with the assumption of hydrostatic pressure distribution in the vertical direction. They can also be full 3D.

A full 3D long wave model in the sigma coordinate system is described here.

The Reynolds Eqs. (3.4.1)–(3.4.4) can be written in a conservation form as,

$$\frac{\partial M}{\partial t} + \frac{\partial E}{\partial x} + \frac{\partial F}{\partial y} + \frac{\partial G}{\partial z} = H \quad (3.4.29)$$

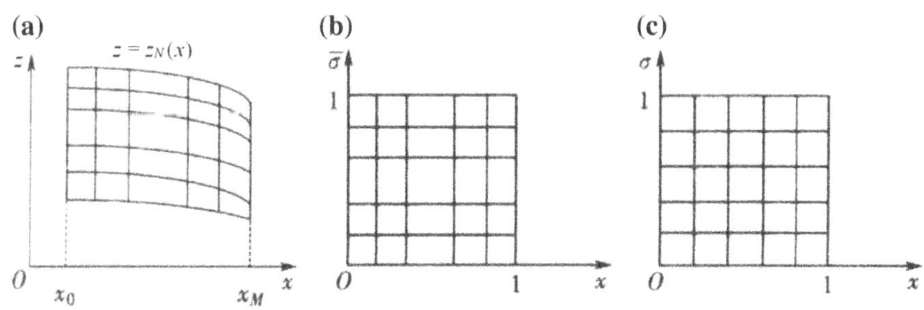

Fig. 3.15 Configurations for the sigma coordinate transformation in a vertical plane. **a** Grids in physical plane; **b** grids in transitional plane; **c** grids in computational plane

where,

$$M = \begin{bmatrix} 0 \\ u \\ v \\ w \end{bmatrix}, \quad H = \begin{bmatrix} 0 \\ f_x \\ f_y \\ f_z \end{bmatrix} \tag{3.4.30}$$

$$E = \begin{bmatrix} u \\ u^2 + \frac{p-\tau_{xx}}{\rho} \\ uv - \frac{\tau_{xy}}{\rho} \\ uw - \frac{\tau_{xz}}{\rho} \end{bmatrix}, \quad F = \begin{bmatrix} v \\ vu - \frac{\tau_{yx}}{\rho} \\ v^2 + \frac{p-\tau_{yy}}{\rho} \\ vw - \frac{\tau_{yz}}{\rho} \end{bmatrix}, \quad G = \begin{bmatrix} w \\ wu - \frac{\tau_{zx}}{\rho} \\ wv - \frac{\tau_{zy}}{\rho} \\ w^2 + \frac{p-\tau_{zz}}{\rho} \end{bmatrix} \tag{3.4.31}$$

where,

$$\tau_{ij} = \tilde{\tau}_{ij} - \rho\overline{u'_i u'_j} \tag{3.4.32}$$

here, $\tilde{\tau}_{ij} = \mu\left(\frac{\partial u_j}{\partial x_i} + \frac{\partial u_i}{\partial x_j}\right)$ denotes the shear stress in laminar flow, $\rho\overline{u'_i u'_j}$ denotes the Reynolds stress.

Coordinate transformation: the vertical coordinate is transformed using an algebraic formula first, to change the free surface that varies with time into a fixed position. This coordinate transformation carried out only in the vertical direction is called sigma transformation.

$$t_1 = t$$

$$\xi = x$$

$$\eta = y$$

$$\sigma = \frac{z - B_i(x, y)}{h(x, y, t) - B_i(x, y)}$$

The transformation relations are rewritten in a matrix form as,

$$\begin{bmatrix} t_1 \\ \xi \\ \eta \\ \sigma \end{bmatrix} = \begin{bmatrix} 1 & 0 & 0 & 0 \\ 0 & 1 & 0 & 0 \\ 0 & 0 & 1 & 0 \\ l_1 & l_2 & l_3 & l_4 \end{bmatrix} \cdot \begin{bmatrix} t \\ x \\ y \\ z \end{bmatrix} \tag{3.4.33}$$

where, $(l_1, l_2, l_3, l_4)^{\mathrm{T}} = L$ is a nonlinear operator which satisfies

$$(L, X) = \frac{z - B_t(x, y)}{h_f(x, y, t) - B_t(x, y)} \tag{3.4.34}$$

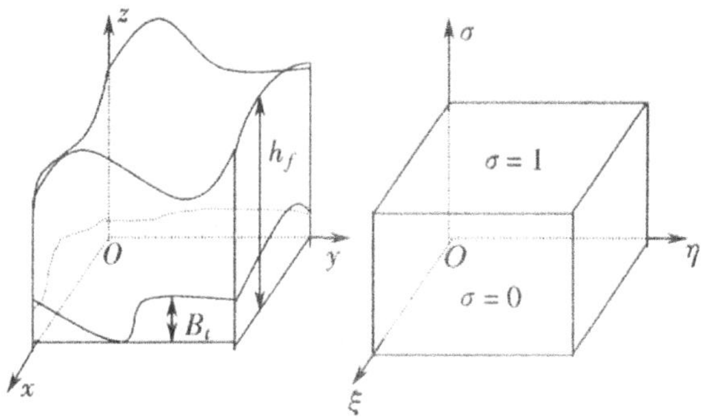

Fig. 3.16 Sketch of coordinate transformation

where, $X = (t, x, y, z)^T$; $(L, X) = L \cdot X$ is the inner product between the vectors L and X.

Figure 3.16 shows a sketch of coordinate transformation. $B_t(x, y)$ is the height function of the bottom; $h_f(x, y, t)$ is the height function of the free surface. It is assumed that both $B_t(x, y)$ and $h_f(x, y, t)$ are single valued functions of x and y. Then it leads to $\sigma \in [0, 1]$ where $\sigma = 0$ corresponds to the bottom and $\sigma = 1$ corresponds to the free surface. Therefore, the free surface is in the plane of $\sigma = 1$ during the whole calculation process, and the computational domain does not change over time.

In order to reflect the large gradient of physical quantities near the bottom, a transformation with local mesh refinement has been used. In this approach, equidistant fine grids are adopted near both the bottom and the free surface. A logarithmic distribution of grid thickness is adopted in the middle layers for the connection between the bottom and the surface, i.e.,

$$\tau = t_1$$

$$s = \xi$$

$$\varphi = \eta$$

$$n = f(\sigma)$$

where,

$$f(\sigma) = \begin{cases} A(\sigma - 1) + 1, & \delta \le \sigma \le 1 \\ B \ln(1 + b\sigma) + C, & \delta_0 \le \sigma \le \delta \\ D\sigma, & 0 \le \sigma \le \delta_0 \end{cases} \tag{3.4.35}$$

Rewriting them in a matrix form yields

$$\begin{bmatrix} \tau \\ s \\ \varphi \\ n \end{bmatrix} = \begin{bmatrix} 1 & 0 & 0 & 0 \\ 0 & 1 & 0 & 0 \\ 0 & 0 & 1 & 0 \\ 0 & 0 & 0 & f(\cdot) \end{bmatrix} \cdot \begin{bmatrix} t_1 \\ \xi \\ \eta \\ \sigma \end{bmatrix} \tag{3.4.36}$$

The parameters A, B, C and D are determined by the continuity of n and the continuity of its first derivative; b is an adjustable encryption parameter; δ_0 and δ are the parameters for the range of equidistant grids near the bottom and the surface respectively. Here δ_0 is a constant slightly larger than zero, and δ is a constant slightly less than 1.

Let

$$\boldsymbol{P} = \begin{bmatrix} 1 & 0 & 0 & 0 \\ 0 & 1 & 0 & 0 \\ 0 & 0 & 1 & 0 \\ 0 & 0 & 0 & f(\cdot) \end{bmatrix} \cdot \begin{bmatrix} 1 & 0 & 0 & 0 \\ 0 & 1 & 0 & 0 \\ 0 & 0 & 1 & 0 \\ l_1 & l_2 & l_3 & l_4 \end{bmatrix} \tag{3.4.37}$$

which yields,

$$\begin{bmatrix} \tau \\ s \\ \varphi \\ n \end{bmatrix} = \boldsymbol{P} \cdot \begin{bmatrix} t \\ x \\ y \\ z \end{bmatrix} \tag{3.4.38}$$

In this transformation, the relations between the first derivatives can be written as

$$\begin{bmatrix} \frac{\partial}{\partial t} \\ \frac{\partial}{\partial x} \\ \frac{\partial}{\partial y} \\ \frac{\partial}{\partial z} \end{bmatrix} = \begin{bmatrix} 1 & 0 & 0 & n_t \\ 0 & 1 & 0 & n_x \\ 0 & 0 & 1 & n_y \\ 0 & 0 & 0 & n_z \end{bmatrix} \cdot \begin{bmatrix} \frac{\partial}{\partial \tau} \\ \frac{\partial}{\partial s} \\ \frac{\partial}{\partial \varphi} \\ \frac{\partial}{\partial n} \end{bmatrix} \tag{3.4.39}$$

In the transformed coordinate system, Eq. (3.4.29) is transformed into,

$$\left(\frac{M}{J}\right)_\tau + \left(\frac{E}{J}\right)_s + \left(\frac{F}{J}\right)_\varphi + \left(\frac{n_t M + n_x E + n_y F + n_z G}{J}\right)_n = \frac{H}{J} \tag{3.4.40}$$

where,

$$\mathrm{J} = \frac{\partial(\tau, s, \varphi, n)}{\partial(t, x, y, z)} = \mathrm{n}_z = \det \begin{bmatrix} 1 & 0 & 0 & n_t \\ 0 & 1 & 0 & n_x \\ 0 & 0 & 1 & n_y \\ 0 & 0 & 0 & n_z \end{bmatrix} \tag{3.4.41}$$

More information on the applications of the turbulence models and Reynolds stress in the above model, the determination of the initial and boundary conditions, the numerical methods and procedures of this model are available in the Refs. [9, 10].

3.5 Mathematical Model of Dam-Break Waves in Channel

In the design and management of hydraulic engineering, it is important to predict the catastrophic flow phenomenon of dam-break. As occurrence of dam-break is always sudden, the negative and positive waves propagate quickly in the upper and lower reaches. The sudden drop of upstream water level may cause embankment collapse in the reservoir area. The fast downstream moving steep waves are very destructive. Therefore, dam-break problems have been the concerns of hydraulic engineers and researchers for many years. In this section, a level set method and a high-resolution TVD method for calculation of dam-break waves are described.

3.5.1 Calculation of Dam-Break Waves by a Level Set Method

1. Level set method

The level set method proposed by Osher and Sethain [13] in 1988 is a method for capturing the interface of two-phase flow. The basic idea of level set method for solving free surface problems in hydrodynamics is that the free surface is regarded as the interface between air and water. The solution of the problem with free surface can be obtained by solving the governing equations in the whole domain including air and water, and using a level set function to implicitly track the free surface.

The level set function is also called the distance function denoted by $\phi(x, y, t)$, which is defined as the symbolic vertical distance of each point in the computational domain to the interface between the two fluids. At time t, the distance function $\phi(x, y, t)$ is positive for the points above the interface, and negative for the points below the interface. The interface is formed by the points with $\phi(x, y, t) = 0$. It can be proved that the distance function satisfies the following differential equation.

$$\phi_t + (\boldsymbol{u} \cdot \nabla)\phi = 0 \tag{3.5.1}$$

It is difficult to determine the distance function at each point according to the definition. By solving the biharmonic equation [10, 14], the distance function in Eq. (3.5.1) can be determined conveniently,

$$\begin{cases} \Delta^2\phi = 0 \\ \frac{\partial\phi}{\partial n}|_\Gamma = 1 \\ \phi|_\Gamma = 0 \end{cases} \tag{3.5.2}$$

$\phi(x, y)$ satisfying Eq. (3.5.2) clearly satisfies the definition of the distance function. The boundary integral method can be used to solve Eq. (3.5.2). The intermediate variables φ and q at the boundary are introduced,

$$\begin{cases} \varphi = [\Delta\phi] = \Delta\phi|_{\Gamma\text{in}} - \Delta\phi|_{\Gamma\text{out}} \\ q = \left[\frac{\partial\Delta\phi}{\partial n}\right] = \frac{\partial\Delta\phi}{\partial n}|_{\Gamma\text{in}} - \frac{\partial\Delta\phi}{\partial n}||_{\Gamma\text{out}} \end{cases} \tag{3.5.3}$$

Using the boundary integral method to determine ϕ at each point in the computational domain has advantages of quick convergence and high precision, and it solves the problem encountered when the level set function was first defined.

Due to the great difference between the densities of water and air, the strong discontinuity in density at the interface may result in the instability of solution if using the densities of water and air directly. The density near the interface can be smoothed by

$$\rho = \begin{cases} \rho_a & \phi > \alpha \\ \rho_a + \frac{\rho_w - \rho_a}{\rho_w}(\alpha - \phi) & -\alpha \le \phi \le \alpha \\ \rho_w & \phi < -\alpha \end{cases} \tag{3.5.4}$$

which ensures the stability of the difference scheme. Here α is a very small distance which can be chosen as 1.5–2.5 times of the grid width.

2. Solution of dam-break wave and its verification

The following N–S equations are used to solve the flow of two fluids, i.e. water and air after dam-break,

$$\begin{cases} \boldsymbol{u}_t + (\boldsymbol{u} \cdot \nabla)\boldsymbol{u} = -\frac{1}{\rho}\nabla P + \nu\nabla^2\boldsymbol{u} \\ \nabla \cdot \boldsymbol{u} = 0 \end{cases} \tag{3.5.5}$$

where, $\boldsymbol{u} = (u, v)$, ρ is determined by Eq. (3.5.4).

The above N–S equations are solved by using the 2D projection method. A detailed description of the projection method will be presented in Sect. 9.4.

Experiments on dam-break wave were studied by Lauber and Hager in [15, 16]. The rectangular flume used for the experiments was 14 m long, 0.5 m wide, 1.2 m high in the upstream part and 0.7 m high in the downstream part. The bottom slope of the flume was adjustable between 0 and 0.5 m (the inclined angle of 0°–26.5°). The error caused by the facility was ±1 mm. The bottom and one side wall of the flume were made of 3 mm thick black PVC material. The other side wall was made of glass. The maximum surface roughness was 5×10^{-6} m. The upstream part was closed by a wall. The downstream part was an open channel, and could be extended to any position. A shock absorber was utilized in the experiments to reduce the force on the flume when it was turned on or off.

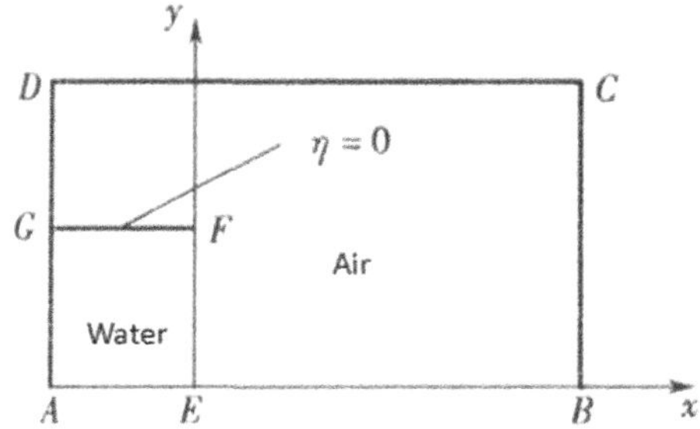

Fig. 3.17 Initial state of the flume with flat bottom before dam-break occurred

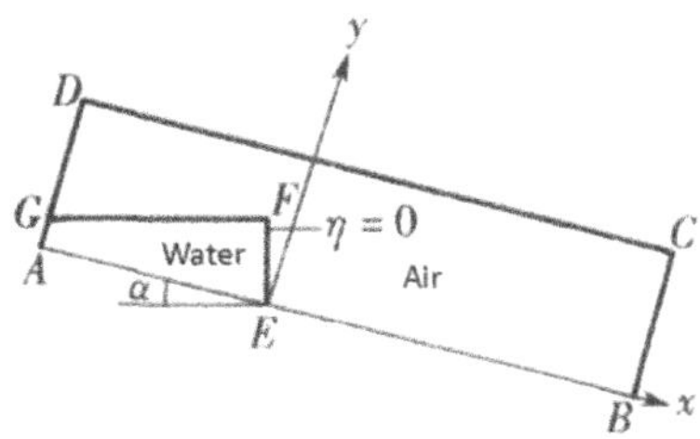

Fig. 3.18 Initial state of the flume with bottom slope before dam-break occurred

The dam-break wave tested in the experiments was simulated by the level set method to verify the effectiveness of the method in the calculation of dam-break waves [17]. Figures 3.17 and 3.18 show the initial states of the flume with and without bottom slope, respectively, before the dam-break occurred. In the figures, GF denotes the water surface (horizontal) in the upstream part; α is the inclined angle of the flume bottom (x axis) from the horizontal, i.e., the bottom slope is $I = \tan\alpha$; EF denotes the vertical section; AB, CD, AD are solid walls; BC is the open boundary. The dimensions are $AB = CD = 40$ m, $AD = BC = 2$ m, $AE = 11.05$ m, and $EF = 1$ m. The water and air are still at the initial moment.

Boundary conditions:
The solid wall boundary condition at AD and AB: $\boldsymbol{u} = 0$;

The slip boundary conditions at the surface CD:

$$\begin{cases} \boldsymbol{n} \cdot \boldsymbol{u} = 0 \\ \boldsymbol{n} \cdot \nabla \boldsymbol{u} = 0 \end{cases}$$

The derivative boundary condition at BC: $\frac{\partial v}{\partial x}|_{BC} = 0$.

Numerical simulations were carried out for three different bottom slopes of $I = 0$, 0.1 and 0.5. Figure 3.19 shows the experimental results undertaken by Lauber and Hager [15, 16]. The dam-break wave profiles at $t = 0.914n$ s ($n = 1, 2, \ldots 7$) for $I = 0$, $t = 0.73n$ s ($n = 1, 2, \ldots 7$) for $I = 0.1$ and $t = 0.438n$ s ($n = 1, 2, \ldots 7$) for $I = 0.5$ are shown in (a), (b) and (c), respectively, in Fig. 3.19.

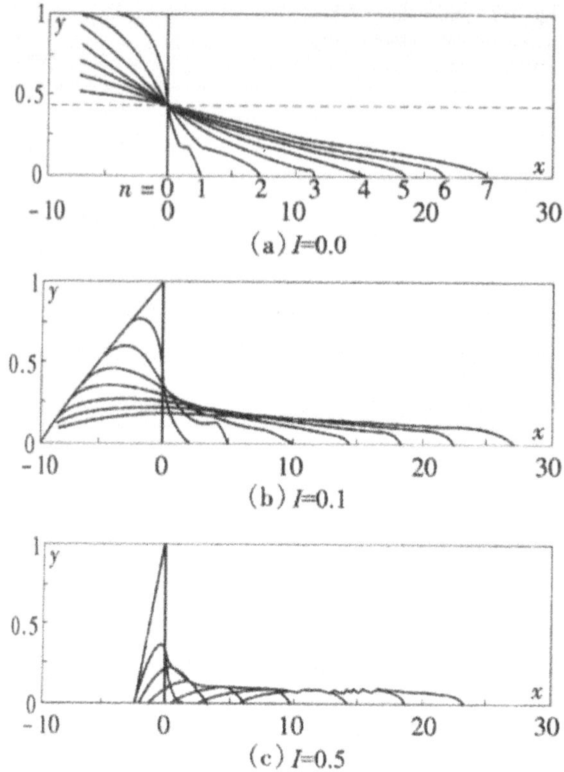

Fig. 3.19 2D dam-break wave profiles varying with time (experimental results)

Figure 3.20 shows the simulated dam-break wave profiles by using the level set method which corresponds to the three scenarios shown in Fig. 3.19. The numerical results agreed well with the experimental results. It indicates that the level set method is effective in capturing the instantaneous change of free surface such as dam-break problems.

The flow fields of air and water at the time of $n = 4$ for different bottom slopes are shown in Fig. 3.21. The coordinates in this figure correspond with Fig. 3.17 for the case of flat bottom and Fig. 3.18 for the cases of inclined bottom, respectively. The E point being the coordinate origin is the location of the water boundary at the dam bottom at the initial moment. It is found that with the propagation of dam-break wave, a strong air vortex is generated near the wave front, which reflects the movement of air and water in the wave propagation process of dam-break clearly.

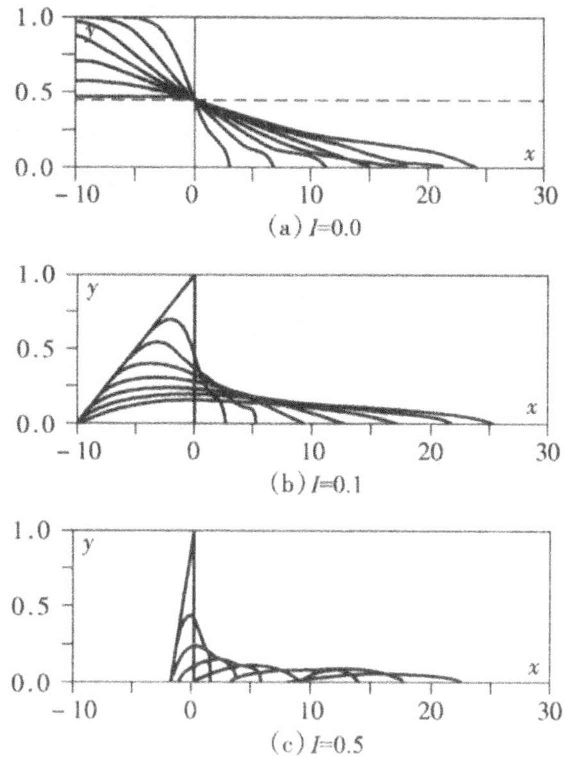

Fig. 3.20 2D dam-break wave profiles varying with time (simulated results)

3.5.2 Calculation of Dam-Break Waves by a TVD Method

The discontinuous waves (such as dam-break waves) in open channels and shock waves in gas dynamics can be described by the conservative hyperbolic equation and corresponding weak solutions. The use of a high resolution scheme utilized in gas dynamics to deal with shock waves, namely total variation non-increasing (TVIN) scheme, in solving the dam-break wave problem in hydrodynamics is presented below.

There are generally two classes of methods to calculate shock waves in gas dynamics. One is a class of shock-fitting methods for accurately positioning discontinuity. The other is a class of shock-capturing methods for calculating through discontinuity. In shock-capturing methods the location of discontinuity is not considered, and the entire domain including discontinuity is calculated using the same difference scheme. A low order accurate scheme may cause a large dissipation in the calculation. But a higher order accurate scheme such as second-order scheme could make oscillation. In 1983 Harten [18] introduced a concept of total variation non-increasing and established a class of TVD (Total Variation Diminishing) scheme or called TVNI (total variation non-increasing) scheme. This scheme has second-order accuracy and

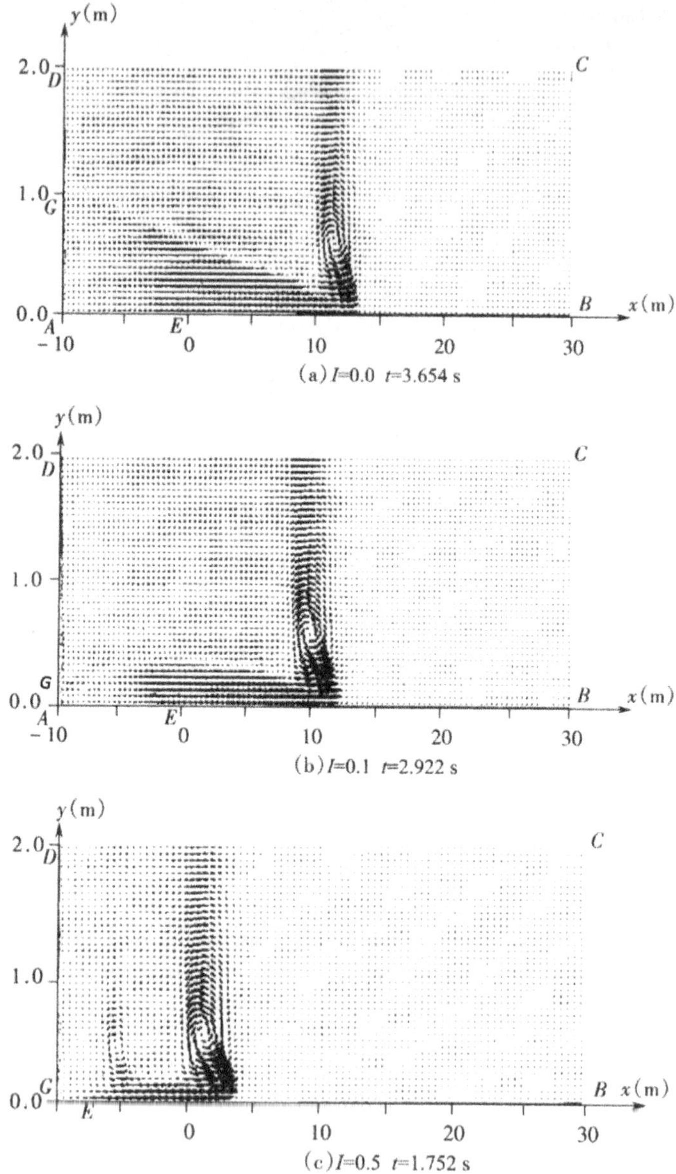

Fig. 3.21 Flow fields for different bottom slopes ($n = 4$)

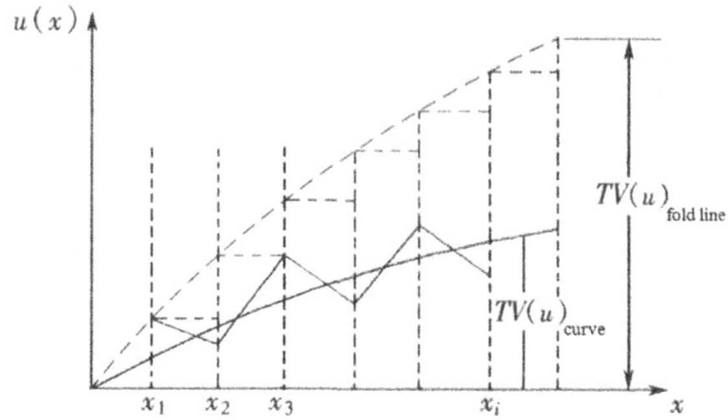

Fig. 3.22 Sketch of total variation

eliminates the oscillation, which is a widely accepted high resolution approach for the simulation of shock waves.

1. Total variation non-increasing scheme

For a single independent variable function $u(x)$ defined within an interval (x_0, x_J), the interval (x_0, x_J) is divided into several sub-intervals (x_0, x_1), (x_1, x_2),..., (x_{J-1}, x_J) and $\sum \left|u(x_{j-1}) - u(x_j)\right|$ is calculated as shown in Fig. 3.22. It is seen that the sums of the absolute difference between the two adjacent points on the smooth curve and on the fold line around the curve are different. The sum values may vary with different sub-intervals.

The supremum is taken and defined as the total variation,

$$\mathrm{TV}(u) = \mathrm{Sup} \sum_{j=0}^{J-1} \left|u_{j+1} - u_j\right| \tag{3.5.6}$$

The total variation reflects the magnitude of the function oscillation. In Fig. 3.22, the total variation by the fold line $\mathrm{TV}(u)_{\text{fold line}}$ is much larger than the total variation by the curve $\mathrm{TV}(u)_{\text{curve}}$. The larger oscillation of the fold line, the larger $\mathrm{TV}(u)$. If the total variation in the calculation does not increase with time, it indicates that the oscillation of the solution does not increase with time, which is called the total variation non-increasing denoted by TVNI. The TVNI scheme can achieve high resolution and reduce the oscillation error. Here the idea of TVNI scheme is illustrated by an initial value problem for a single hyperbolic equation.

$$u_t + f(u)_x = 0 \tag{3.5.7}$$

$$u(x, 0) = \phi(x), \quad -\infty < x < \infty \tag{3.5.8}$$

By using $(2\,k + 1)$-point explicit scheme, the difference equation can be written as,

$$u_j^{n+1} = u_j^n - \lambda\left[\bar{f}\left(u_{j-k+1}^n, \ldots, u_{j+k}^n\right) - \bar{f}\left(u_{j-k}^n, \ldots, u_{j+k-1}^n\right)\right] \tag{3.5.9}$$

where, $\lambda = \frac{\Delta t}{\Delta x}$, $\bar{f}$ is a numerical flux function.

If the difference equation is consistent with the differential equation, then

$$\bar{f}(u, u, \ldots, u) = f(u) \tag{3.5.10}$$

The numerical flux function by the Lax-Wendroff scheme and the Lax-Friendrichs scheme read,

$$\bar{f}_{\text{L-W}} = \frac{1}{2}\left[f_{j+1} + f_j - \lambda\left(\alpha_{j+\frac{1}{2}}\right)^2\left(u_{j+1} - u_j\right)\right] \tag{3.5.11}$$

$$\bar{f}_{\text{L-F}} = \frac{1}{2}\left[f_{j+1} + f_j - \frac{1}{\lambda}\left(u_{j+1} - u_j\right)\right] \tag{3.5.12}$$

Both of the above two schemes satisfy the consistency condition Eq. (3.5.10), thus the numerical flux function $\bar{f}$ can be written as

$$\bar{f} = \frac{1}{2}\left[f_{j+1} + f_j\right] - \frac{1}{2\lambda}Q\left(\lambda\alpha_{j+\frac{1}{2}}\right)\Delta_{j+\frac{1}{2}}u \tag{3.5.13}$$

where,

$$\alpha = \frac{\partial f}{\partial u} \tag{3.5.14}$$

$$\alpha_{j+\frac{1}{2}} = \begin{cases} \frac{f(u_{j+1}) - f(u_j)}{u_{j+1} - u_j} & \text{for } u_{j+1|} - u_j \neq 0 \\ a(u_j) & \text{for } u_{j+1} - u_j = 0 \end{cases} \tag{3.5.15}$$

$$\Delta_{j+\frac{1}{2}}u = u_{j+1} - u_j \tag{3.5.16}$$

Here $Q(x)$ is some function, which is often referred to as the coefficient of numerical viscosity.

By using the above two schemes, $Q(x)$ respectively read,

$$\left.\begin{aligned} Q_{\text{L-W}} &= \left(\lambda\alpha_{j+\frac{1}{2}}\right)^2 \\ Q_{\text{L-F}} &= 1 \end{aligned}\right\} \tag{3.5.17}$$

The above two schemes are not TVNI schemes although the numerical flux function satisfies the consistency condition. Constructing a TVNI scheme is to construct $Q(x)$ in a special form. It had been proved by Harten that, $Q(x)$ should satisfy the following condition for a TVNI scheme,

$$|x| \leq Q(x) \leq 1 \quad \text{for} \quad 0 \leq |x| \leq \mu \leq 1 \tag{3.5.18}$$

where, μ is a constant.

Harten had constructed a second-order 5-point TVNI scheme satisfying the uniqueness of weak solutions.

For the initial value problem,

$$u_t + f(u)_x = 0, \quad t > 0 \tag{3.5.19}$$

$$u(x, 0) = \phi(x), \quad -\infty < x < \infty \tag{3.5.20}$$

The difference equation reads

$$u_j^{n+1} = u_j^n - \lambda\left(\bar{f}_{j+\frac{1}{2}}^n - \bar{f}_{j-\frac{1}{2}}^n\right) \tag{3.5.21}$$

where, $\lambda = \Delta t/\Delta x$ and the numerical flux function $\bar{f}_{j+\frac{1}{2}}^n = \bar{f}\left(u_{j-k+1}^n, \ldots, u_{j+k}^n\right)$ satisfies $\bar{f}(u, u, \cdots u) = f(u)$. When $k = 2$, Eq. (3.5.21) becomes a 5-point scheme. Let f_j^M be a modified numerical flux function with a modification g,

$$f_j^M = f(u_j) + \frac{1}{\lambda} g_j,$$

where $g_j = g(u_{j-1}, u_j, u_{j+1})$

$$u_j^{n+1} = u_j^n - \lambda\left(\bar{f}_{j+\frac{1}{2}}^M - \bar{f}_{j-\frac{1}{2}}^M\right)$$

$$\bar{f}_{j+\frac{1}{2}}^M = \frac{1}{2}\left[f\left(u_j^n\right) + f\left(u_{j+1}^n\right)\right] + \frac{1}{2\lambda}\left[g_j + g_{j+1} - Q\left(\lambda\tilde{\alpha}_{j+\frac{1}{2}} + r_{j+\frac{1}{2}}\right)\Delta_{j+\frac{1}{2}} u\right] \tag{3.5.22}$$

where,

$$r_{j+\frac{1}{2}} = \begin{cases} \frac{g_{j+1} - g_j}{\Delta_{j+\frac{1}{2}} u}, & \Delta_{j+\frac{1}{2}} u \neq 0 \\ 0, & \Delta_{j+\frac{1}{2}} u = 0 \end{cases}$$

Constructing the modification g in the follow way,

$$g_j = S_{j+\frac{1}{2}} \max\left[0, \min\left(\left|\bar{g}_{j+\frac{1}{2}}\right|, \bar{g}_{j-\frac{1}{2}} \cdot S_{j+\frac{1}{2}}\right)\right] \tag{3.5.23}$$

where, $\bar{g}_{j+\frac{1}{2}} = \frac{1}{2}\left[Q\left(\lambda\bar{\alpha}_{j+\frac{1}{2}}\right) - \left(\lambda\bar{\alpha}_{j+\frac{1}{2}}\right)^2\right]\Delta_{j+\frac{1}{2}} u$, $\quad S_{j+\frac{1}{2}} = \text{sign}\left(\bar{g}_{j+\frac{1}{2}}\right)$.

$$Q(x) = \begin{cases} x^2/4\varepsilon, & x < 2\varepsilon \\ |x|, & x \geq 2\varepsilon \end{cases} \tag{3.5.24}$$

The difference scheme constructed by Eqs. (3.5.21)–(3.5.24) is a second order 5-point explicit TVNI scheme.

2. Calculation of 1D dam-break wave using a TVD method

The governing equation for 1D dam-break problem in vector form reads

$$\frac{\partial \boldsymbol{U}}{\partial t}+\frac{\partial \boldsymbol{F}}{\partial x}=\boldsymbol{S} \tag{3.5.25}$$

where, $\boldsymbol{U}=\begin{bmatrix} h \\ P \end{bmatrix}$, $\boldsymbol{F}=\begin{bmatrix} P \\ \frac{P^2}{h}+\frac{gh^2}{2} \end{bmatrix}$, $\boldsymbol{S}=\begin{bmatrix} 0 \\ ghS_B-ghS_o \end{bmatrix}$;

h is the water depth; $P=hu$, in which u is the average velocity in the section; g is the gravitational acceleration; S_B is the bottom slope with $S_B=\frac{\partial Z_B}{\partial x}$, in which Z_B is the bottom elevation; S_0 is the friction slope with $S_0=n^2P|P|/h^{\frac{10}{3}}$, in which n is Manning roughness coefficient.

A TVD scheme for Eq. (3.5.25) reads,

$$U_j^{n+1}=U_j^n-\lambda\left(\bar{f}^*_{j+\frac{1}{2}}-\bar{f}^*_{j-\frac{1}{2}}\right)+S_j^n\Delta t+\mathrm{SW}_j^nS_j^n\frac{\Delta t^2}{2} \tag{3.5.26}$$

where,

$$\bar{f}^*_{j+\frac{1}{2}}=\bar{f}_{j+\frac{1}{2}}+\frac{1}{2}\left(A^n_{j+\frac{1}{2}}S^n_{j+\frac{1}{2}}\Delta t+\mathrm{SW}^n_{j+\frac{1}{2}}F^n_{j+\frac{1}{2}}\Delta t\right) \tag{3.5.27}$$

$$\mathbf{A}=\frac{\partial F}{\partial U}=\begin{bmatrix} 0 & 1 \\ gh-u^2 & 2u \end{bmatrix},\quad \mathbf{SW}=\begin{bmatrix} 0 & 0 \\ gS_b+\frac{7}{3}gS_0 & \frac{-2gn^2P}{h^{7/3}} \end{bmatrix},$$

where, $\boldsymbol{A}$ is the Jacobian matrix; $\bar{f}$ is the numerical flux function of homogeneous Eq. (3.5.25).

$$\begin{aligned}\bar{f}_{j+\frac{1}{2}}&=\frac{1}{2}[F(U_j)+F(U_{j+1})]\\&\quad+\frac{1}{2\lambda}\sum_{k=1}^{m}R^k_{j+\frac{1}{2}}\left[g_j^k+g_j^{k+1}-Q\left(\gamma^k_{j+\frac{1}{2}}+r^k_{j+\frac{1}{2}}\right)\alpha^k_{j+\frac{1}{2}}\right]\end{aligned} \tag{3.5.28}$$

where, m is the dimension of Jacobian matrix $\boldsymbol{A}$; $\boldsymbol{R}$ and $\boldsymbol{L}$ are the left and right eigenvector matrices of $\boldsymbol{A}$ having $\boldsymbol{L}\cdot\boldsymbol{R}=\mathrm{I}$.

$\boldsymbol{R}=\begin{bmatrix} 1 & 1 \\ u-c & u+c \end{bmatrix}$, $\boldsymbol{L}=\begin{bmatrix} \frac{u+c}{2c} & -\frac{1}{2c} \\ -\frac{u-c}{2c} & \frac{1}{2c} \end{bmatrix}$, $c=\sqrt{gh}$, $\boldsymbol{I}$ is the identity matrix.

$$\alpha^h_{j+\frac{1}{2}} = L^h_{j+\frac{1}{2}} \Delta_{j+\frac{1}{2}} V, \quad \gamma^k_{j+\frac{1}{2}} = \lambda \alpha^h \left(U_{j+\frac{1}{2}} \right),$$

$$g^k_j = S^k_{j+\frac{1}{2}} \cdot \max\left[0, \min\left(\sigma \left|\bar{g}^k_{j+\frac{1}{2}}\right|, \left|\bar{g}^k_j\right|, \sigma \bar{g}^k_{j-\frac{1}{2}} \cdot S^k_{j+\frac{1}{2}}\right)\right],$$

$$\sigma = 2, \quad S^k_{j+\frac{1}{2}} = \mathrm{sign}(\bar{g}^k_{j+\frac{1}{2}}),$$

$$r^k_{j+\frac{1}{2}} = \begin{cases} (g^k_{j+1} - g^k_j)/\alpha^k_{j+\frac{1}{2}} & \text{for } \alpha^k_{j+\frac{1}{2}} \neq 0 \\ 0 & \text{for } \alpha^k_{j+\frac{1}{2}} = 0 \end{cases},$$

$$\bar{g}^k_{j+\frac{1}{2}} = \frac{1}{2}\left[Q^k\left(\gamma^k_{j+\frac{1}{2}}\right) - \left(r^k_{j+\frac{1}{2}}\right)^2\right]\alpha^k_{j+\frac{1}{2}},$$

$$\bar{g}^k_j = \frac{1}{2}\left(\bar{g}^k_{j-\frac{1}{2}} + \bar{g}^k_{j+\frac{1}{2}}\right),$$

$$Q(x) = \begin{cases} x^2/4\varepsilon + \varepsilon & \text{for } x < 2\varepsilon \\ |x| & \text{for } x \geq 2\varepsilon \end{cases},$$

$$\lambda = \Delta t / \Delta x,$$

$$0.01 < \varepsilon < 0.02.$$

3. Analysis and comparison of TVD scheme and Lax-wendroff scheme

It is assumed that a dam is located in the middle of an 2000 m long open channel with rectangular cross-section. The water depth upstream of dam is 10 m. The downstream water depth is 2 m for the case with water and 0.1 m for the case without water, respectively. The dam is broken at $t = 0$. The dam-break wave at $t = 60$ s has been calculated. Figures 3.23 and 3.24 show the simulated results using a TVNI scheme compared with the analytic solutions of Stokes. The bottom roughness is set to $n = 0.0$. It is found that the TVNI results match the Stokes solutions very well with high resolution at wave front.

Figure 3.25 shows a comparison of simulated results between the TVNI method and the Lax-Wendroff (L-W) method that has a second order accuracy. The Q functions in the two methods had been modified as follows.

$$\text{TVNI}: \quad Q(x) = \begin{cases} \frac{x^2}{4\varepsilon} + \varepsilon, & |x| < 2\varepsilon \\ |x|, & |x| \geq 2\varepsilon \end{cases}$$

$$\text{L-W}: \quad Q(x) = \begin{cases} \frac{x^2}{4\varepsilon} + \varepsilon, & |x| < 2\varepsilon \\ x^2, & |x| \geq 2\varepsilon \end{cases}$$

It was indicated from the calculation that, when ε is relatively large in the L-W method, the discontinuity is smoothed and the resolution becomes low. But the results using the TVNI method are almost unaffected by ε. Here $\varepsilon = 0.2$ was chosen in the

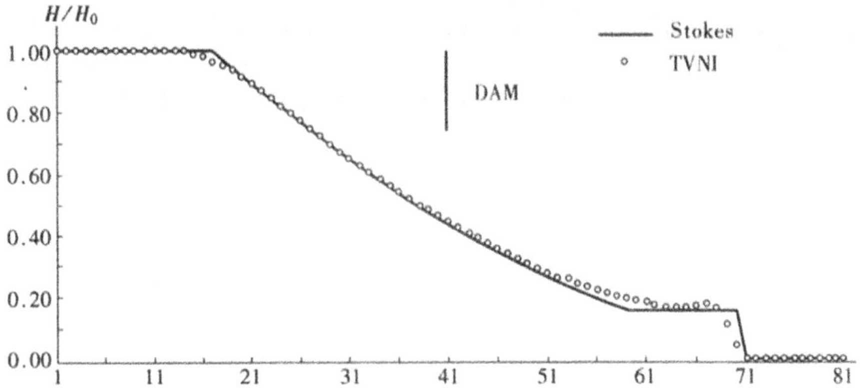

Fig. 3.23 Simulated result by TVNI method compared with the analytic solution of stokes for the case without water in the downstream

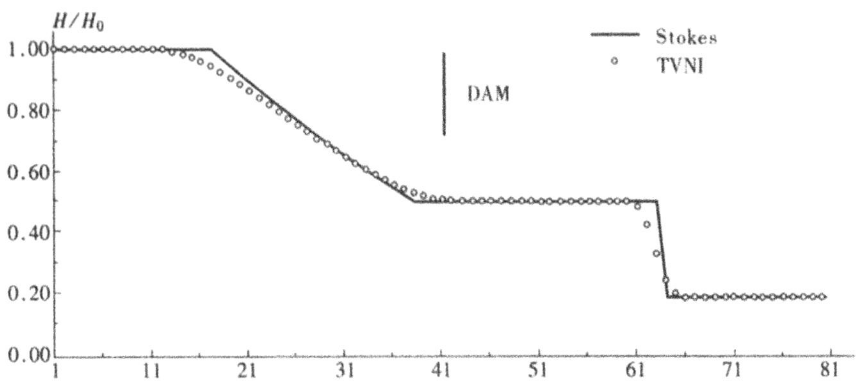

Fig. 3.24 Simulated result by TVNI method compared with the analytic solution of Stokes for the case with water in the downstream

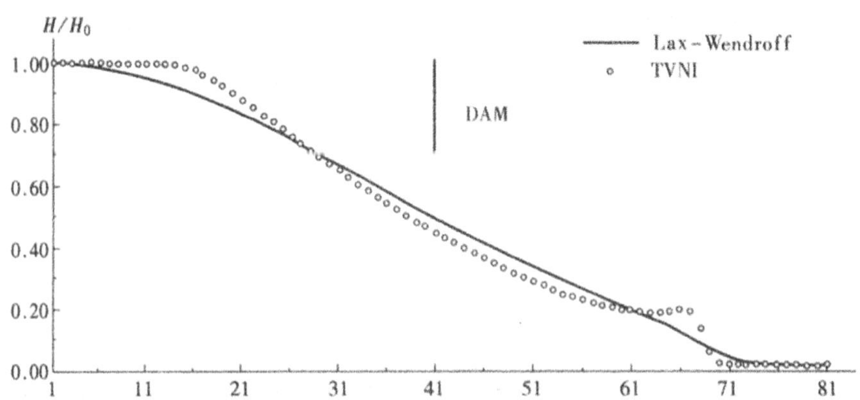

Fig. 3.25 Comparison of simulated results between the TVNI method and the Lax-Wendroff method

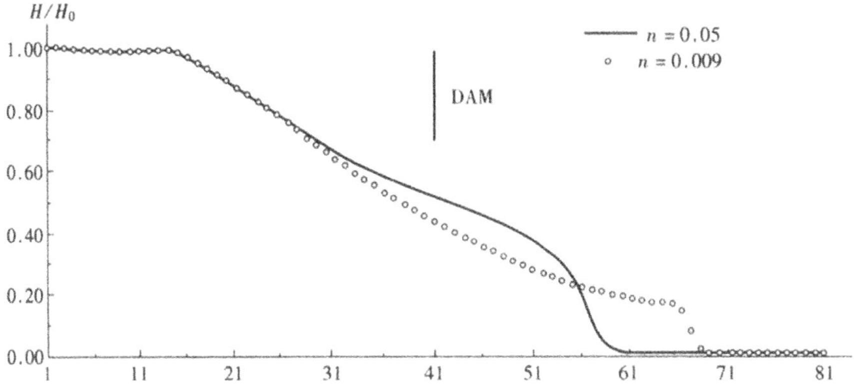

Fig. 3.26 Comparison of simulated results by the TVNI method with different bottom roughness

calculations. It is found clearly from Fig. 3.25 that the TVNI method has a higher resolution in this case.

Figure 3.26 shows a comparison of simulated results by the TVNI method with different bottom roughness $n = 0.05$ and $n = 0.009$. The bottom friction has a great effect on the propagation of the positive wave front. The smoother the bottom, the faster the propagation of the wave front. But the bottom friction has little effect on the propagation of the negative wave. Thus it is very important to estimate the bottom friction coefficient n for the calculation of dam-break wave. However it is difficult to determine this coefficient in the open channel.

References

1. Jianhua T et al. Report on the study of the structural measures of flood control for Yongding new river. Department of Mechanics, Tianjin University; 1989.
2. Jianhua T. Mathematical simulation of circulation and its verification. Acta Mech Sin. 1985;17(1):9–14.
3. Koppel T. Experiments on unsteady separating flow in an open channel report no. 3–81. In: Laboratory of Fluid Mechanics, Department of Civil Engineering Delft University of Technology; 1981.
4. Yuan D et al. Report on the study of tidal current, residual current and convection-diffusion numerical model and calculation method of water exchange. Department of Mechanics, Tianjin University; 2003.
5. Yuan D et al. Report on the study of pre and post processing of numerical model and image display. Department of Mechanics, Tianjin University; 2003.
6. Namin MM, Lin B, Falcone RA. An implicit numerical algorithm for solving non-hydrostatic free-surface flow problems. Int J Numer Meth Fluids. 2001;35:341–56.
7. Stelling GS. Numerical modeling of free surface flows in 1, 2, and 3 dimensions. In: Proceedings of the 1st international workshop on coastal eutrophication. Hong Kong; 2001. p. 2.
8. Stelling GS. Developments in free surface flow models. In: Proceedings of the 2nd international workshop on coastal eutrophication Tianjin; 2002. p. 10.

9. Lin M, Tao J. A study on numerical simulation of 3D free surface boundary problem. J Hydrodyn Ser A. 1994;9(4):477–86.
10. Xie W. Numerical modelling study of viscous hydrodynamic problems with free surface. In: Ph.D. dissertation. Department of Mechanics, Tianjin University; 1999.
11. Lin P, Li CW. A σ-coordinate three-dimensional numerical model for free surface wave propagation. Int J Numer Meth Fluids. 2002;38:1045–68.
12. Rodi W. Examples of turbulence model applications, Simulation of turbulence models and their application, vol. 2. Paris, France: Eyrolles; 1984.
13. Osher SJ, Sethain JA. Fronts propagating with curvature-dependent speed: algorithms based on Hamilton-Jacobi formulations. J Comp Phys. 1988;79.
14. Chen Y, Bi Y. Spatial spline and body-fitted coordinates. In: Proceedings of the 1995 national conference on hydrodynamics; 1995.
15. Lauber G, Hager WH. Experiments to dam-break wave: horizontal channel. J Hydraulic Res. 1998;36.
16. Lauber G, Hager WH. Experiments to dam-break wave: sloping channel. J Hydraulic Res. 1998;36.
17. Jianhua T, Weisong X. The level set method for calculation of dam-break wave. J Hydraulic Eng. 1999;10:17–22.
18. Harten A. High Resolution 60 V hyperbolic conservation caw. J Comput Phys. 1983;49:357–93.

Chapter 4
Numerical Simulation of Shallow Water Waves in Coastal Regions

4.1 Introduction

When the waves propagate in coastal waters, some wave phenomena such as shoaling, refraction, diffraction, reflection, and wave breaking may occur due to the effect of the bathymetry and the coastal structures, so that the characteristic parameters of waves change. Wave breaking is the main cause for the generation of nearshore currents in coastal regions, and it is also an important dynamic factor for sediment transport, coastline deformation, and pollutant transport in coastal regions. Moreover, wave forces are important for the design of coastal structures. Therefore, it is of great significance to correctly simulate the wave propagation and deformation for the study of the dynamic process in the coastal regions in coastal engineering. After a long period of research and practice, a series of numerical models for the simulation of wave motion in coastal regions have been developed. An appropriate model can be chosen and applied according to the different requirements.

In the early stage of numerical wave propagation models, a ray-tracing method based on the linear small-amplitude wave theory was commonly used to calculate the variation of wave parameters. In this method, it is assumed that the wave energy propagates along the wave ray (perpendicular to the wave crest lines) according to the laws of geometrical optics, and the influence of the wave height variation along the crest lines on the wave propagation is ignored. This simple method is mainly used to calculate the wave shoaling and refraction due to the change of bathymetry with uniform slope, but without considering wave diffraction. In the condition of complex natural bathymetry, this method may lead to the so-called caustic or focus phenomena, so that the calculation results are distorted or even not able to be solved. Although some complicated ray theories have been developed to compensate for the shortcomings, they are not applicable to calculate the wave field in complex bathymetries since the wave diffraction is not taken into account.

J. Tao, *Numerical Simulation of Water Waves*, Springer Tracts in Civil Engineering, https://doi.org/10.1007/978-981-15-2841-5_4

In 1972, Berkhoff [1] developed a second-order elliptic partial differential equation called the mild slope equation, which describes the combined effects of diffraction and refraction for water waves, under an assumption of mild slope by using the perturbation method. Since then, many studies on the mild slope equation had been done including the improvements of the equation, the establishments of numerical methods, and the engineering applications. The mild slope equation derived by Berkhoff is an elliptic equation. To solve this equation in the entire domain, all the boundary conditions should be specified. Due to the restriction of open boundary conditions, hybrid element methods were commonly used for solving the equation. But the application had been limited because of the high requirements on the speed and memory of the computer. Another form of mild slope equation is a time-dependent mild slope equation. In 1967, Luke [2] derived first-order hyperbolic mild slope equations when using a variational principle to study the motion of water waves, which can be used for the simulation of wave diffraction and refraction with much less computational time compared to the elliptic equation. In order to improve the efficiency of solving the mild slope equation, Radder [3] derived a parabolic mild slope equation by using a method of splitting the transmitted and reflected waves and neglecting the wave reflection under an assumption of waves propagating in a principal direction. Subsequently, further studies and improvements on the parabolic mild slope equations have been carried out in the aspects of the equations and numerical methods. At present, the parabolic mild slope equation has become an effective approach to calculate wave field in a large area with a gentle slope.

When the wave reflection must be taken into account due to the complex bathymetry or existence of the artificial islands, breakwaters, and other structures, the parabolic mild slope equation is no longer applicable. Boussinesq models can simulate combined effects of wave shoaling, refraction, diffraction, and reflection, especially the wave field near the structure. The Boussinesq equations were initially derived by Boussinesq [4] in 1872 when he studied the dispersion of nonlinear shallow water equations for the waves propagating over constant water depth. In 1967, Peregrine [5] developed equations for the waves propagating over varying water depth, so that the so-called classical Boussinesq equations were derived. Since then, the Boussinesq-type equations have been gradually formed through the improvements on the dispersion and nonlinearity. They have become effective methods for simulating wave propagation in coastal regions.

Boussinesq equations describe the combined effects of important wave phenomena including shoaling, refraction, diffraction, reflection, and the interaction between waves and structures. It can be applied to the calculation of wave field in complex bathymetry. In engineering practice, a combination of a parabolic mild slope equation and Boussinesq equations can be used by using the parabolic mild slope equation for large-scale wave field and using Boussinesq equations for the local wave field in the vicinity of the structures.

4.2 Mathematical Models of Wave Shoaling and Wave Refraction

In the process of wave propagation, wave parameters such as wave height and wavelength change as the water depth becomes shallow. This process is called shoaling. Similar to the refraction phenomenon in optics, the wave direction may also change. In the coastal regions closely related to the production and activities of human beings, wave shoaling and refraction can be found everywhere. They are important factors which must be taken into account in the study of shoreline deformation and the coastal engineering construction.

4.2.1 Wave Shoaling

When the incoming waves propagate on uniformly sloping seabed in the direction perpendicular to the shoreline as shown in Fig. 4.1, the wave deformation is called shoaling. It can be seen as a vertical two-dimensional (2D) problem. In the figure, the x-axis is located on the still water surface; the z-axis is perpendicular to the still water surface and positive in the upward direction; the origin "o" is located at the shoreline; d is the still water depth; c is the wave celerity (phase velocity); $\tan\alpha$ is the slope of seabed. The direction of wave propagation is opposite to the x-axis.

The wave shoaling on the sloping seabed can be solved by an analytical method [6]. The energy method based on the conservation of wave energy is commonly used in engineering.

The assumptions of the energy method include (1) the energy loss due to the bottom friction and permeability can be ignored, and the wave energy is conserved in the wave propagation; (2) the wave period is unchanged in the wave propagation; (3) seabed changes gently, so that the linear wave theory in a constant water depth can be applied in each local zone.

According to the assumptions (1) and (2), the conservation of wave energy flux per unit width reads

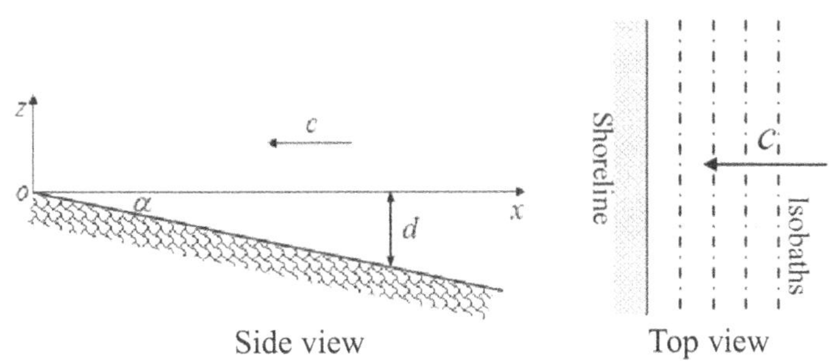

Fig. 4.1 Sketch of uniformly sloping seabed

$$P = \mathrm{Ecn} = (\mathrm{Ecn})_\infty = \mathrm{const} \tag{4.2.1}$$

$$n = \frac{1}{2}\left[1 + \frac{2kd}{\sinh(2kd)}\right] \tag{4.2.2}$$

where P is the wave energy flux; E is the wave energy density; c is the phase velocity; n is the ratio of the group velocity to the phase velocity; $(\mathrm{Ecn})_\infty$ is the deepwater wave energy flux (the subscript "∞" for the wave parameters in deep water far from the shoreline); k is the wave number.

According to the assumptions (2) and (3), the dispersion relation of linear waves reads

$$\omega^2 = gk\tanh(kd) = gk_\infty = \omega_\infty^2 \tag{4.2.3}$$

which gives,

$$\frac{L}{L_\infty} = \frac{c}{c_\infty} = \frac{k_\infty}{k} = \tanh\left(\frac{2\pi d}{L}\right) = \tanh\left(\frac{2\pi d/L_\infty}{L/L_\infty}\right) \tag{4.2.4}$$

where L is the wavelength.

The wave energy density given by the linear wave theory is

$$E = \frac{1}{8}\rho g H^2 \tag{4.2.5}$$

where ρ is the fluid density, g is the gravitational acceleration, and H is the wave height.

According to Eqs. (4.2.1) and (4.2.5), the relation between the wave height H in any water depth and the deepwater wave height can be obtained as

$$K_s = \frac{H}{H_\infty} = \left(\frac{1}{2n}\frac{c_\infty}{c}\right)^{1/2} = \left(\frac{2\cosh^2 hd}{2kd + \sinh 2kd}\right)^{1/2} \tag{4.2.6}$$

where K_s is the wave shoaling coefficient.

According to Eqs. (4.2.4) and (4.2.6), the wave parameters in any water depth can be easily calculated if the deepwater wave parameters are known. The wave parameters can be obtained by a computer program, or by checking a wave diagram.

It is noted that wave breaking occurs when the ratio of the wave height to the water depth reaches a critical value. If wave breaking occurs, the energy method is invalid and a corresponding wave breaking model must be used for the study of wave motion.

4.2.2 *Wave Refraction*

1. Ray theory of wave refraction

When the waves obliquely enter shallow water from deep water, the phase velocities on the same wave crest line are different because of different water depths. The phase velocity is faster in deep water than in shallow water (see Fig. 4.2). The wave crest lines tend to be parallel to the shoreline. The direction of wave propagation changes, which is similar to the refraction phenomenon in optics and called wave refraction. The effect of wave refraction on the gentle slope seabed can be studied preliminarily by the ray theory of linear small-amplitude waves.

The so-called ray line is that the tangential direction of this line at each point is identical to the direction of the local wave number vector $\boldsymbol{k}$ and perpendicular to the wave crest line which has a constant phase function $S = S_i = \text{Const.}$

According to the assumptions of (1) and (2), Eq. (4.2.1) can be rewritten as

$$\nabla \cdot \left(E c_g\right) = 0 \tag{4.2.7}$$

where C_g is the group velocity defined as $C_g = nc$.

Different rays can be drawn from different locations in deep water. The two adjacent rays form a ray tube. As shown in Fig. 4.2, δ_{b_0} and δ_b are the widths at the two ends of the ray tube, respectively. By using Gauss's divergence theorem and considering that the wave number factor $\boldsymbol{k}$ is tangent to the ray, integrating Eq. (4.2.7) along the closed curve formed by the boundaries of the ray tube yields

$$E c_g \delta_b = \left(E c_g \delta_b\right)_0 = \text{const} \tag{4.2.8}$$

which indicates the conservation of wave energy along the ray tube. The wave height along the ray can be obtained by Eqs. (4.2.5) and (4.2.8)

$$\frac{H}{H_0} = \left(\frac{c_{g0}}{c_g} \frac{\delta_{b0}}{\delta_b}\right)^{\frac{1}{2}} \tag{4.2.9}$$

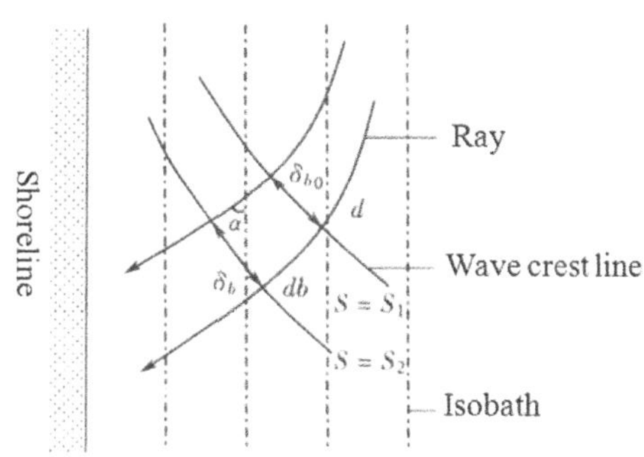

Fig. 4.2 Sketch of wave crest lines, rays, and ray tube

It is known from Eq. (4.2.9) that once the equation (or location) of the ray or the perpendicular phase line is determined, the wave height on the whole ray can be deduced from the given wave height at any point on the ray.

The derivation of the ray equation is as follows.

Under the assumption of irrotational and inviscid fluid, the free surface of the linear small-amplitude waves can be described as

$$\zeta(x, y, t) = \mathrm{Re}[Ae^{i(\boldsymbol{k}\cdot\boldsymbol{x}-\omega t)}] \tag{4.2.10}$$

where $\boldsymbol{k} = (k_x, k_y)$, $\boldsymbol{x} = (x, y)$, k_x, and k_y are the wave numbers in the x- and y-directions; ω is the angular frequency.

The phase function is

$$S(x, y, t) = k_x x + k_y y - \omega t = \boldsymbol{k}\cdot\boldsymbol{x} - \omega t \tag{4.2.11}$$

$$|\nabla S|^2 = \left(\frac{\partial S}{\partial x}\right)^2 + \left(\frac{\partial S}{\partial y}\right)^2 = k^2,$$

$$\omega = -\frac{\partial S}{\partial t} \tag{4.2.12}$$

The gradient direction of S is the direction of wave propagation. Then, the wave propagation path, i.e., the ray, can be obtained. Let $y = y(x)$ denote the ray. It leads to

$$y' = \frac{\mathrm{d}y}{\mathrm{d}x} = \frac{\partial S}{\partial y}\bigg/\frac{\partial S}{\partial x} \tag{4.2.13}$$

Equation (4.2.12) gives

$$\left.\begin{aligned} \frac{\partial S}{\partial x} &= \frac{k}{\sqrt{1+y'^2}} \\ \frac{\partial S}{\partial y} &= \frac{ky'}{\sqrt{1+y'^2}} \end{aligned}\right\}, \tag{4.2.14}$$

where, $k = |\boldsymbol{k}|$.

From the above two equations, it is obtained that,

$$\frac{d}{dx}\frac{ky'}{\sqrt{1+y'^2}} = \sqrt{1+y'^2}\frac{\partial k}{\partial y} \tag{4.2.15}$$

where $k = k(x, y(x))$. Equation (4.2.15) is a nonlinear ordinary differential equation of $y(x)$. The wave propagation path, i.e., the ray $y = y(x)$, can be obtained by solving Eq. (4.2.15).

2. Refraction of waves crossing a navigation channel

Along with the development of the shipping industry and the increasing of ship tonnage, deepwater navigation channels and deepwater harbors have been increasing at home and abroad. The effects of navigation channels, harbors, and the complex bathymetry in the ports on the local wave fields have been given big attention. In the following, the refraction phenomenon of waves traveling over typical bathymetry has been analyzed by taking waves crossing navigation channel as an example.

Let the y-axis be located at the central line of the navigation channel. The cross section of the channel is shown in Fig. 4.3a. Because the isobaths are parallel to the y-axis, the water depth and wave number are functions of x as shown in Fig. 4.3a, b, i.e., $h = h(x)$, $k = k(x)$ and $\frac{\partial k}{\partial y} = 0$. Therefore, Eq. (4.2.15) becomes

$$\frac{ky'}{\sqrt{1 + y'^2}} = K = \text{const} \tag{4.2.16}$$

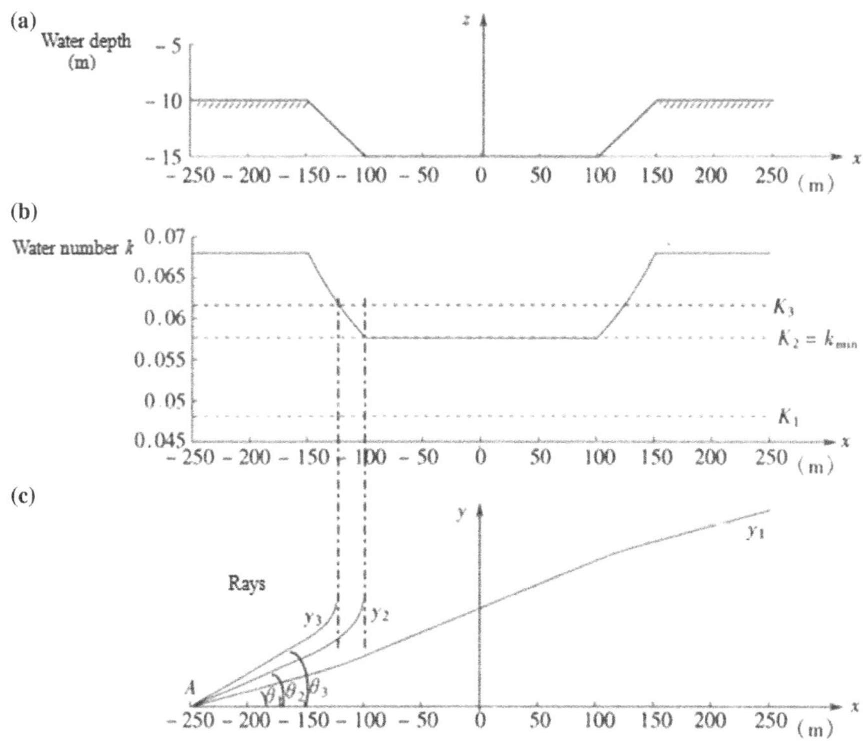

Fig. 4.3 Sketch of waves crossing a navigation channel, **a** cross section of the channel, **b** wave number along the section, **c** rays in the channel

Since $\sin\theta = dy/ds = y'/\sqrt{1+y'^2}$, where θ is the angle between the ray and the positive x-axis, it leads to

$$k\sin\theta = K = \text{const} \tag{4.2.17}$$

Equation (4.2.17) is the Snell's law. It indicates that when the waves obliquely enter over bathymetry with parallel isobaths, $k(x)\sin\theta(x)$ along the ray is constant. Here, $k(x)$ is the wave number at any point on the ray, and $\sin\theta(x)$ is the tangent of the ray at this point.

Then, it can be obtained that

$$y' = dy/dx = \pm K/\sqrt{(k^2 - K^2)} \tag{4.2.18}$$

Integrating Eq. (4.2.18) yields

$$y - y_0 = \pm\int_{x_0}^{x}\left(K/\sqrt{k^2 - K^2}\right)dx \tag{4.2.19}$$

Equation (4.2.19) is the ray equation.

Figure 4.3c shows the trajectories of the three rays starting from the A point on one side of the navigation channel but having different incident angles with the cross section (the x-axis) after they encounter the channel. Let the wave incident angle be θ_i, $K_i = k_A\sin(\theta_i)$, $i = 1, 2, 3$. The following condition should be satisfied along the x-direction in order to have a real solution of Eq. (4.2.19):

$$k(x) > K = k_A\sin(\theta) \tag{4.2.20}$$

For a navigation channel with certain water depth, the minimum wave number at the bottom is k_{min}. If $k^2_{\text{min}} \le k_A^2\sin^2\theta_i$, Eq. (4.2.19) has no real solution and the rays cannot cross the channel. It is seen from Fig. 4.3 that $K_3 > K_2 = k_{\text{min}}$, and thus y_2 and y_3 cannot cross the channel. But $K_1 < k_{\text{min}}$ so that y_1 can cross the channel. These phenomena may also be explained by the Snell's law. When the ray enters the channel, the wave number k decreases with the increase of the water depth. As the minimum wave number k_{min} is known, the possible maximum $\sin(\theta)$ is $\sin\frac{\pi}{2} = 1$. When $k_{\text{min}} < k_A\sin\theta_A$, $k\sin\theta$ cannot keep constant on the ray, and the ray is not able to cross the channel such as y_2 and y_3 in Fig. 4.3c. But the y_1 ray with smaller incident angle θ can cross the channel.

Therefore, there is a critical angle α_c defined as the angle between the ray and the channel axis, for waves crossing the navigation channel. It is defined as

$$\sin\theta_c = \sin\left(\frac{\pi}{2} - \alpha_c\right) = \frac{k_{\text{min}}}{k_A} \tag{4.2.21}$$

i.e.,

$$\alpha_c = \frac{\pi}{2} - \sin^{-1}\frac{k_{\min}}{k_A} \tag{4.2.22}$$

When the angle between the ray and the channel axis $\alpha > \alpha_c$, the ray can cross the navigation channel. When $\alpha < \alpha_c$, the waves are refracted back to the same side of the incident waves.

In practice, the bathymetries in the coastal regions or in the ports are complex. The wave ray paths also show complex variation. In most cases, the linear small-amplitude wave theory is not applicable to obtain wave heights. Then, a numerical model of nonlinear shallow water waves could be used.

3. Numerical analysis of the critical angle for waves crossing a navigation channel

Due to the limitation of the linear small-amplitude wave theory and the complexity of waves crossing a navigation channel, a Boussinesq equation model of nonlinear dispersive shallow water waves is used in the following to analyze the refraction phenomenon of waves crossing a navigation channel and the corresponding critical angle. The Boussinesq model results have also been compared with the result by the linear wave theory. The Boussinesq equation model will be described in Sect. 4.5 of this chapter.

Assuming the critical angle is a function of d, Δd, H_0, T, and g, here d is the water depth outside the navigation channel; Δd is the difference of water depth between inside and outside the channel; H_0 is the incident wave height; T is the incident wave period; g is the gravitational acceleration.

$$\alpha_c = f(d, \Delta d, H_0, T, g) \tag{4.2.23}$$

Making it dimensionless yields

$$\alpha_c = F\left(\frac{\Delta d}{d}, \frac{gT^2}{d}, \frac{H_0}{d}\right) \tag{4.2.24}$$

A series of numerical model tests by using a Boussinesq equation model had been carried out for finding the relation between α_c, $\Delta d/d$, gT^2/d, and H_0/d. In the numerical model, the critical angle α_c is defined as the angle between the incident wave direction and the channel axis when the ratio of the wave height of waves crossing the channel to the incident wave height (H/H_0) is just less than 0.1.

(1) Relation between α_c and $\Delta d/d$

Figure 4.4 shows a distribution of the relative wave height H/H_0 for the case of the critical angle obtained by the numerical model ($\Delta d/d = 0.6$, $H_0/d = 0.5$, and $\alpha_c = 31.27°$). The wave height profile at the cross section $II - II$ is shown in Fig. 4.5.

It is found that when the incident wave angle is α_c, the wave heights on the side of incident waves are large. Most of the wave energy is blocked on this side.

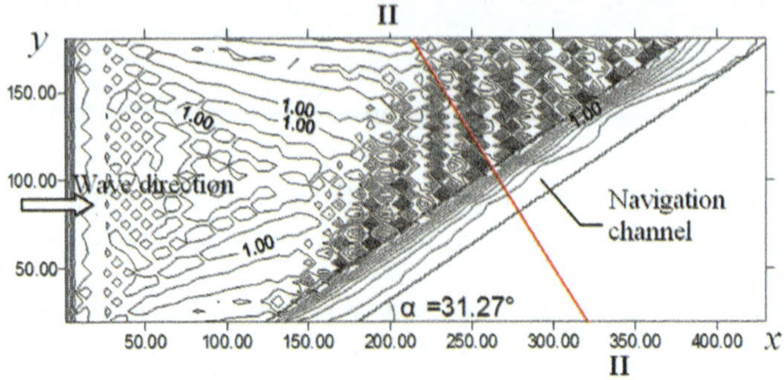

Fig. 4.4 Distribution of the relative wave height H/H_0 in the x-y plane

Fig. 4.5 Wave height profile at the cross section $II - II$

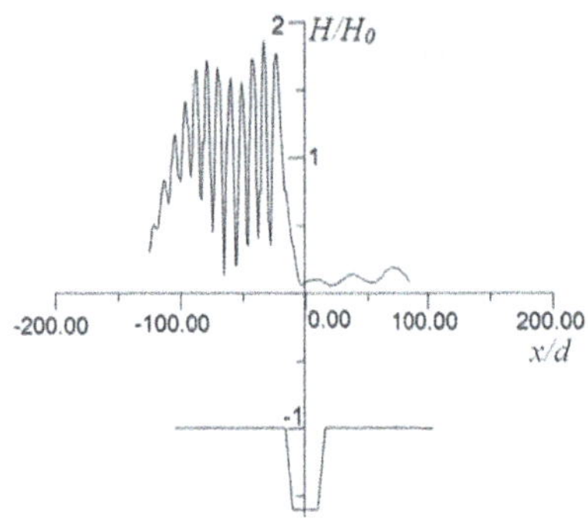

The numerical results showed that a small amount of energy still passed over the navigation channel in the critical refraction case (as shown in Fig. 4.5).

Figure 4.6 shows a comparison of critical angles α_c obtained by the numerical model with the results of the linear theory for different $\Delta d/d$. It is seen that the

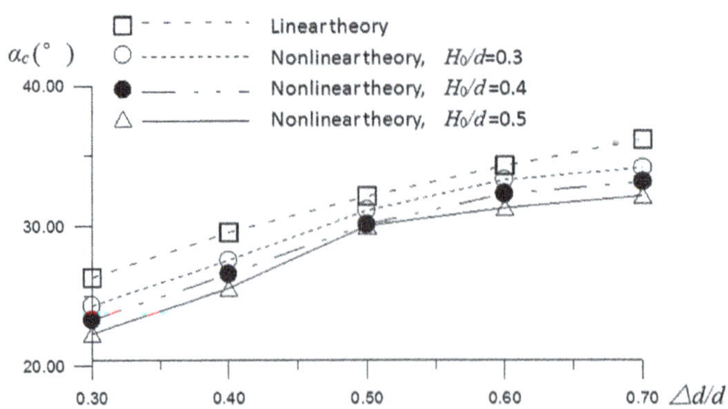

Fig. 4.6 Critical angles for different $\Delta d/d$

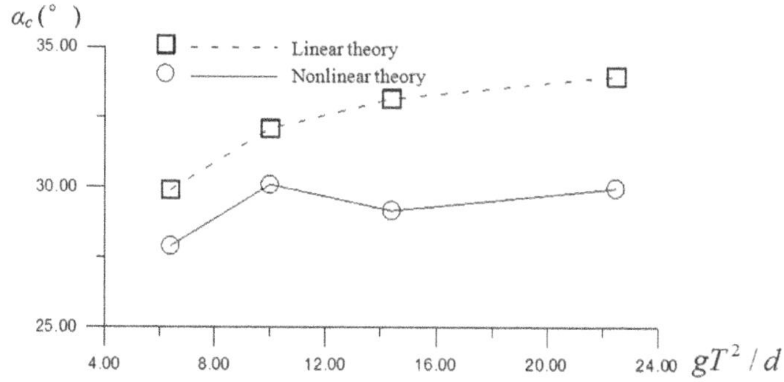

Fig. 4.7 Critical angles for different gT^2/d

results given by the nonlinear theory are smaller than the linear results. Also, α_c is related to the relative wave height. α_c decreases with increase of the wave height, which cannot be resolved by the linear theory.

The critical angle of the wave refraction is very important to the deep navigation channels and harbors which should be given enough attention in the port design. The angle of incident waves with the channel axis being less than the critical angle should be avoided as far as possible. The critical angle by the linear theory might be used for the determination of the navigation channel axis in the planning phase. The nonlinear numerical model should be used for the key layouts in the detail design [7, 8].

(2) Relation between α_c and gT^2/d

The relation between α_c and gT^2/d is shown in Fig. 4.7. Similar to the above discussion, α_c values obtained by the nonlinear theory are smaller than those by the linear theory. The difference between the nonlinear and linear results becomes larger as the wave period T goes up.

4. Wave refraction induced by current

The wave refraction caused by the bathymetry change (such as the navigation channel) has been discussed in the above section. In fact, this wave refraction phenomenon is induced by the variation of wave number k in the process of wave propagation. When waves encounter oblique shear current in the process of wave propagation, the wave number k may change due to the current effects, so that the wave refraction induced by current occurs. When waves encounter the current with an angle less than a certain value, the phenomenon that the waves cannot pass over the current may occur. This phenomenon should be given much attention in estuaries.

It is assumed that a shear current on the flat seabed has a constant velocity component in the y-direction (current direction), i.e., $\frac{\partial}{\partial y} = 0$ and various velocity component in the x-direction $u(x)$ as shown in Fig. 4.8a. It is also assumed that the waves through point A have the wave number k_A but three different directions, the angles between the three rays, and the positive axis denote θ_i, $i = 1, 2, 3$ as shown in Fig. 4.8c. According to Eq. (4.2.17), K_i is constant on each ray, $K_i = k_A \sin \theta_i$.

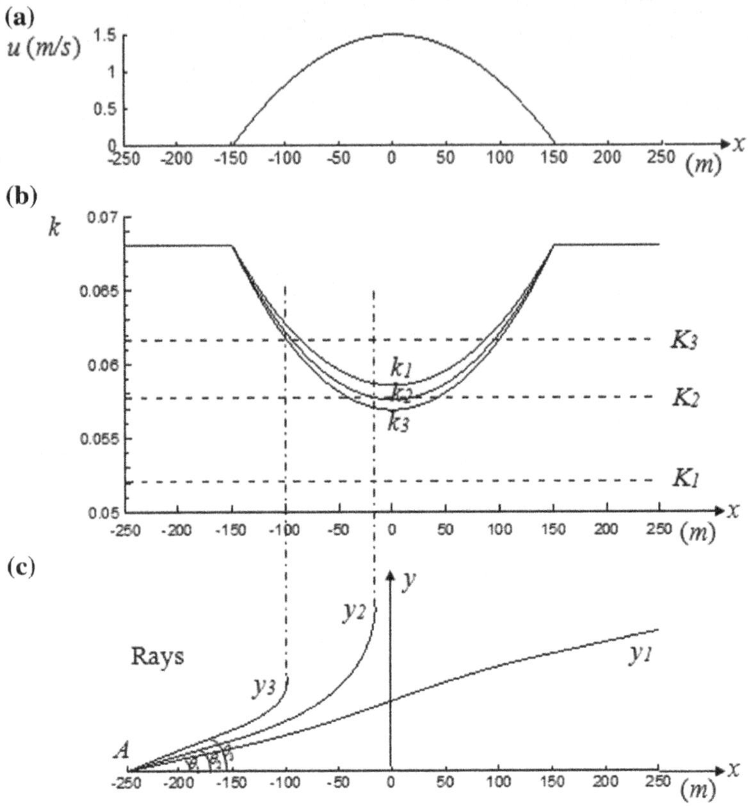

Fig. 4.8 Sketch of waves passing over shear current, **a** shear current velocity profile, **b** wave number along the velocity profile, **c** wave rays affected by the current

The wave number changes after the waves encounter the shear current as shown in Fig. 4.8b. $k(x)$ can be given by the dispersion relation as

$$\omega - u(x)K_i = [gk(x)\tanh k(x)d]^{1/2} \tag{4.2.25}$$

$$K_i = k_A \sin\theta_i = \text{const}, \quad i = 1, 2, 3.$$

Similar to Eq. (4.2.19), the ray equation reads

$$y - y_0 = \pm\int_{x_0}^{x} \frac{K}{\left[k^2(x) - K^2\right]^{1/2}}\mathrm{d}x \tag{4.2.26}$$

Figure 4.8b shows the wave numbers $k(x)$ varying along with x calculated by Eq. (4.2.25) for the waves starting from the A point with different incident angles. It is found that there is one minimum wave number k_{min} for each incident angle.

Figure 4.8c shows the trajectories of the three rays starting from point A after they encounter the shear current. When the angle between the incident wave and the current θ_1 is small, $K_1 = k_A \sin\theta_1$ and $0 < K_1 < k_{1\text{min}}$, then Eq. (4.2.5) has a real solution, and the wave can pass over the current as y_1 in the figure. When the angle between the incident wave and the current θ_2 is relatively large, $K_2 = k_A \sin\theta_2$, $K_2 > k_{2\text{min}}$, and the wave cannot pass over the flow as y_2 in the figure. When the angle is further increased to θ_3, the y_3 ray as shown in the figure cannot pass over the current either. When the waves are not able to pass over the shear current, the wave energy gathers on the side of the incident waves, and the wave height increases. The wave height distribution after wave energy gathering should be analyzed by a numerical model of nonlinear waves instead of linear ray theory.

4.3 Mathematical Model of Wave Diffraction

When waves propagate to coastal regions and encounter with coastal structures such as breakwaters and artificial islands, the phenomenon of diffraction occurs due to the effects of the structures. A part of wave energy is blocked by the structures. and the other part of energy bypasses the structures into the water behind, which is similar to the optical diffraction. Calculation of wave diffraction is needed for the effects of coastal structures on wave field, the wave force calculation, and the sheltering effect of breakwaters for ports.

4.3.1 Linear Wave Diffraction Theory

1. Basic equations

It is assumed that the incident waves are linear small-amplitude waves, the seabed is gentle enough to be approximately flat, i.e., the wave number k is constant, and only diffraction occurs without refraction in the process of wave propagation. The mathematical model can be simplified to be two-dimensional (2D) when the structures (such as breakwaters and jetties) are approximated as the columns having a uniform cross-section above the water surface.

Under the assumption of inviscid, irrotational, and incompressible fluid, the potential theory can be used. The velocity potential function at any point in the domain can be written as

$$\Phi(x, y, z, t) = \phi(x, y, z)\mathrm{e}^{-i\omega t} \tag{4.3.1}$$

The linear wave diffraction is described as

$$\nabla^2\Phi = 0, \quad -d \le z \le 0 \tag{4.3.2}$$

$$\frac{\partial^2\Phi}{\partial t^2} + g\frac{\partial\Phi}{\partial z} = 0, \quad z = 0 \tag{4.3.3}$$

$$\eta = -\frac{1}{g}\frac{\partial\Phi}{\partial t}, \quad z = 0 \tag{4.3.4}$$

$$\frac{\partial\Phi}{\partial n} = 0, \quad \text{on the body surface} \tag{4.3.5}$$

$$\frac{\partial\Phi}{\partial z} = 0, \quad z = -d \tag{4.3.6}$$

Radiation condition at infinity.

According to the assumption of linearity, the total potential function $\phi(x, y, z)$ at each point in the domain can be decomposed into the incident wave potential function ϕ_I and the scattered wave potential function ϕ_S caused by the obstacle disturbance

$$\phi(x, y, z) = \phi_I + \phi_S \tag{4.3.7}$$

The incident wave potential function ϕ_{I} can be given by Eq. (2.2.16)

$$\phi_I = \frac{gH}{2\omega}\frac{\cosh[k(z+d)]}{\cosh(kd)}\mathrm{e}^{ikx} = Z(z)\phi_I(x, y)$$

where

$$Z(z) = A\frac{\cosh[k(z+d)]}{\cosh(kd)} \tag{4.3.8}$$

in which $A = \frac{gH}{2\omega}$.

The scattered wave potential function is

$$\phi_S(x, y, z) = Z(z)\psi_S(x, y) \tag{4.3.9}$$

Substituting Eqs. (4.3.7) and (4.3.9) into Eqs. (4.3.2) and (4.3.6) yields the equation and boundary conditions for the 2D scattered wave potential function ϕ_S, which is a boundary value problem of the 2D Helmholtz equation in the plane of (x, y).

$$\nabla^2\phi_S + k^2\phi_S = 0 \quad \text{in the domain} \tag{4.3.10}$$

$$\frac{\partial\phi_S}{\partial n} = -\frac{\partial\phi_I}{\partial n} \quad \text{body surface boundary condition} \tag{4.3.11}$$

$$\lim_{r\to\infty}\sqrt{kr}\left(\frac{\partial\phi_S}{\partial r}-ik\phi_S\right)=0 \quad \text{radiation condition} \tag{4.3.12}$$

Equation (4.3.12) is the Sommerfeld radiation condition for linear small-amplitude waves. It can be proved that this radiation condition makes the solution of the diffraction problem described by Eqs. (4.3.10) to (4.3.12) exist and be unique [9].

After getting ϕ_S by solving the boundary value problem of Eqs. (4.3.10) to (4.3.12), $\phi(x, y, z)$ is solved by Eqs. (4.3.9) and (4.3.7). Then, $\phi(x, y, z, t)$ can be obtained by Eq. (4.3.1).

2. Diffraction coefficient

The diffracted wave height H_d or the diffraction coefficient K_d in the project area is concerned in engineering design. For regular waves, the diffraction coefficient K_d is defined as the ratio of the diffracted wave height H_d to the incident wave height H_I.

$$K_d=\frac{H_d}{H_I} \tag{4.3.13}$$

By using the free surface kinematic condition for linear waves in Eq. (4.3.4), the surface elevation at the still water level is obtained as

$$\eta=-\frac{1}{g}\frac{\partial\Phi}{\partial t}\bigg|_{z=0}=\frac{iH_I}{2}\phi(x,y)\mathrm{e}^{-i\omega t} \tag{4.3.14}$$

The diffracted wave height reads

$$H_d=2|\eta|_{\max}=H_I|\phi(x,y)| \tag{4.3.15}$$

Then, the diffraction coefficient can be written as

$$K_d=\frac{H_d}{H_I}=|\phi(x,y)|=\sqrt{\phi_{\mathrm{Re}}^2+\phi_{\mathrm{Im}}^2} \tag{4.3.16}$$

where ϕ_{Re} and ϕ_{Im} are the real and imaginary parts of the 2D potential function $\phi(x, y)$. Here, $\phi(x,y)=\phi_I(x,y)+\phi_S(x,y)$.

3. Numerical method

The linear wave diffraction problem described by Eqs. (4.3.10) to (4.3.12) can be solved by using a boundary element method [10]. The boundary element methods may be classified into the direct and indirect. The indirect boundary element method is also called Green's function method, source distribution method, or single point-source method. A Green's function method for solving the wave diffraction problem is described in the following. The scattered wave potential function $\phi_S(P)$ at any point P in the domain can be calculated as

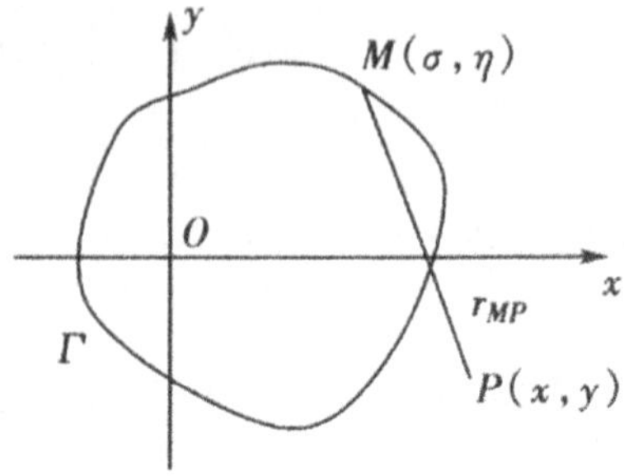

Fig. 4.9 Sketch for boundary integral

$$\phi_S(P) = \int_\Gamma \mu(M)\frac{1}{2i}H_0'(kr_{\mathrm{MP}})\mathrm{d}\Gamma \tag{4.3.17}$$

where $\mu(M)$ is the source intensity on the boundary; As shown in Fig. 4.9, the point $M(\sigma, \eta)$ is located on the boundary, and r_{MP} is the distance between the points P and M; Γ is the boundary of the domain; i is the imaginary unit; $H_0'(kr)$ is a fundamental solution of the 2D Helmholtz equation and is the zero-order Hankel function of the first kind which satisfies the Sommerfeld radiation condition at infinity. Therefore, the boundary Γ of the boundary integral Eq. (4.3.17) only includes boundaries of structures.

The exterior normal derivative of ϕ_S at the point P on the boundary Γ is

$$\left(\frac{\partial \phi_I}{\partial n}\right)_p = \mu(P) + \int_\Gamma \mu(M)\frac{\partial}{\partial n_P}\left[\frac{1}{2i}H_0'(kr_{\mathrm{MP}})\right]\mathrm{d}\Gamma \tag{4.3.18}$$

For solving the problem, the boundary of the domain is divided into several elements (line segments) first, and then Eqs. (4.3.17) and (4.3.18) are discretized. The source intensity μ is solved by Eq. (4.3.18). After that ϕ_S is obtained by substituting μ into Eq. (4.3.17).

4.3.2 *Mathematical Model of Wave Diffraction by Breakwaters*

1. Insular breakwater

For the wave diffraction by an insular breakwater, Eqs. (4.3.10) to (4.3.12) can be solved directly by using a boundary element method with sources distributed on the inner and outer sides of the breakwater. To simplify the calculation, the sources could be set only on the leeside of the breakwater which corresponds to that the incident waves are fully absorbed. Figure 4.10 shows a numerical solution of wave diffraction by an insular breakwater [11, 12].

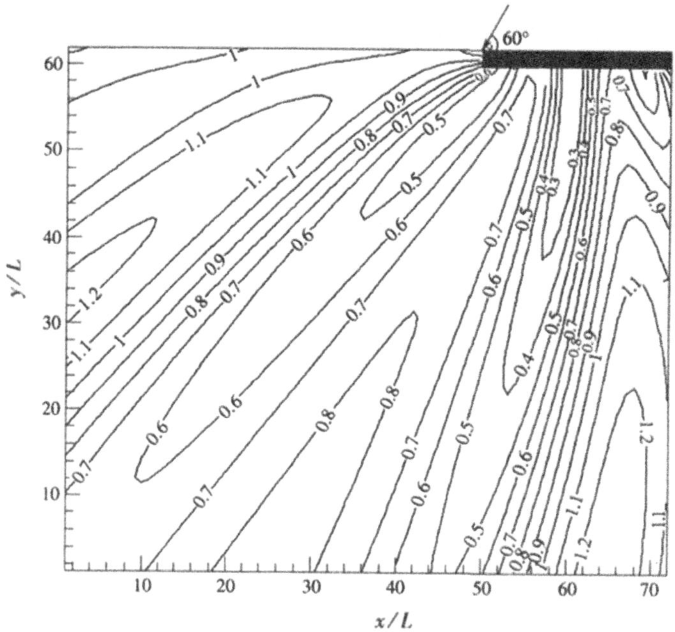

Fig. 4.10 Simulated wave diffraction coefficient by an insular breakwater

2. Semi-infinite single breakwater

For the wave diffraction by a semi-infinite single breakwater, it can be solved approximately by regarding this breakwater as an insular breakwater with a certain length.

By calculating the wave diffraction of a semi-infinite single breakwater within 10 times of the wavelength, it can be approximated by the half of the result of an insular breakwater with a length of 30–40 times of the wavelength.

3. Breakwater gap with a small gap width and a harbor of arbitrary shape

The "small" gap width refers to that the ratio of the gap width B to the incident wavelength L is not greater than 2.5, i.e., $B/L \leq 2.5$. It is assumed that the port area is divided into inside and outside the harbor by the breakwater gap. As the gap width is small, inside the harbor the incident wave potential function can be ignored and only the scattered wave potential function at the gap and the inner boundary is considered. Outside the harbor, the scattered wave potential function at the gap and the incident wave potential function are taken into account, but the scattered wave potential function at the outer side of the boundary is ignored.

Let the point P be inside the harbor and P' be outside the harbor. The following formulas can be written

Inside the harbor,

$$\phi(P) = \phi_S(P) = \int\limits_{C+\Gamma} \mu_I(M)\left[\frac{1}{2i} H_0'(kr_{MP})\right]\mathrm{d}s \tag{4.3.19}$$

$$\begin{aligned}\left(\frac{\partial\phi}{\partial n}\right)_{\mathrm{P}} = \left(\frac{\partial\phi_S}{\partial n}\right)_P &= \mu_I(P) \\ &+ \int\limits_{C+\Gamma} \mu_I(M)\frac{\partial}{\partial n_p}\left[\frac{1}{2i} H_0'(kr_{\mathrm{MP}})\right]\mathrm{d}s\end{aligned} \tag{4.3.20}$$

where μ_{I} is the source intensity on the boundary of inside harbor, C is the internal boundary of the gap, and $\boldsymbol{n}$ is the interior normal unit vector. Γ is the internal boundary of the harbor or breakwater.

Outside the harbor,

The total potential function includes the incident wave potential function and the scattered wave potential function

$$\begin{aligned}\phi(P') &= \phi_S(P') + \phi_I(P') \\ &= \int\limits_{C'} \mu_{\mathrm{II}}(M)\left[\frac{1}{2i} H_0'(kr_{\mathrm{MP'}})\right]\mathrm{d}s + \phi_I(P')\end{aligned} \tag{4.3.21}$$

$$\begin{aligned}\left(\frac{\partial\phi}{\partial n'}\right)_{P'} = \frac{\partial\phi_s(P')}{\partial n'} + \frac{\partial\phi_I(P')}{\partial n'} &= \frac{\partial\phi_I(P')}{\partial n'} + \mu_{\mathrm{II}}(P') \\ &+ \int\limits_{C'} \mu_{\mathrm{II}}(M)\frac{\partial}{\partial n'_{P'}}\left[\frac{1}{2i} H_0'(kr_{\mathrm{MP'}})\right]\mathrm{d}s\end{aligned} \tag{4.3.22}$$

where μ_{II} is the source intensity at the gap of outside harbor, C' is the external boundary of the gap, and n' is the exterior normal unit vector. ϕ_I is the incident wave potential function.

The continuity conditions of the potential function and the corresponding derivative at the gap C give the following two equations:

$$\phi(P) = \phi(P') \tag{4.3.23}$$

$$\frac{\partial\phi(P')}{\partial n'} = -\frac{\partial\phi(P)}{\partial n} \tag{4.3.24}$$

Considering at the gate C,

$$\int\limits_{C} \mu\frac{\partial}{\partial n}\left[\frac{1}{2i} H_0'(kr)\right]\mathrm{d}s = 0 \tag{4.3.25}$$

by substituting Eqs. (4.3.19) and (4.3.21) into Eq. (4.3.23), substituting Eqs. (4.3.20) and (4.3.22) into Eq. (4.3.24), the six equations of Eqs. (4.3.19) to (4.3.24) can be turned to two equations with two unknowns μ_I and μ_{II}. They can be obtained by discretizing these two equations. Then, $\phi(P)$ and $\phi(P')$ are obtained by solving Eqs. (4.3.19) and (4.3.21). Finally, the diffraction coefficient at any point inside and outside the harbor can be obtained by substituting $\phi(P)$ or $\phi(P')$ into Eq. (4.3.16).

For the case of the harbor wall with partial reflection, the boundary condition Eq. (4.3.11) can be rewritten as

$$\frac{\partial \phi}{\partial n} + R\, i\, k\phi = 0 \tag{4.3.26}$$

i.e.,

$$\frac{\partial \phi_S}{\partial n} + R\, i\, k\phi_S = -\frac{\partial \phi_I}{\partial n} - R\, i\, k\phi_I \tag{4.3.27}$$

where R is the reflection factor, $R = 0$ for full reflection and $R = 1$ for full absorption. The relation between the reflection factor R and the reflection coefficient K_R reads

$$R = \cos\frac{1 - K_R e^{ir}}{1 + K_R e^{ir}} \tag{4.3.28}$$

When the phase difference between the reflected wave and the incident wave r in the above equation is zero, Eq. (4.3.28) becomes

$$R = \cos\frac{1 - K_R}{1 + K_R} \tag{4.3.29}$$

The reflection coefficient K_R can be referred to the relevant manual.

The wave diffraction by a breakwater gap with a small gap width and a harbor of arbitrary shape can be calculated by the above model. Figure 4.11 shows a calculated result of wave diffraction by a breakwater gap [12].

4. Breakwater gap with a large gap width and a harbor of arbitrary shape

For the diffraction by a breakwater gap with a large gap width, the calculated results by using the formulas for a breakwater gap with a small gap width are quite different from the analytical solutions. The correction can be made in two cases.

(1) Considering the effects of incident waves in the harbor, it yields that inside the harbor,

$$\begin{aligned}\phi(P) &= \phi_I(P) + \phi_S(P) \\ &= \phi_I(P) + \int_{C+\Gamma} \mu_I \left[\frac{1}{2i} H_0'(kr_{\text{MP}})\right] ds\end{aligned} \tag{4.3.30}$$

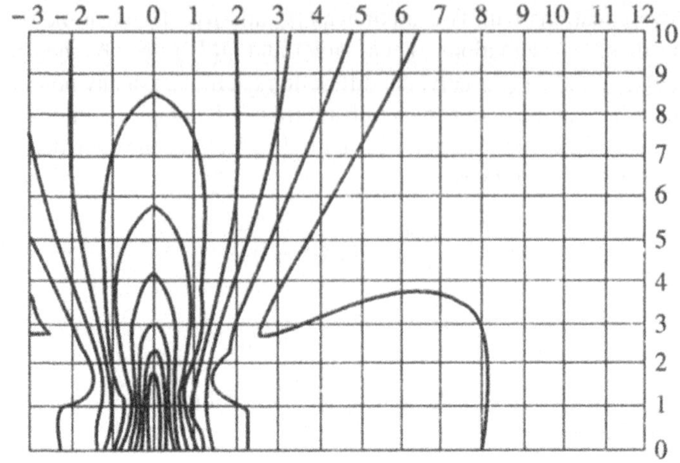

Fig. 4.11 Wave diffraction coefficient by a breakwater gap with a small gap width

$$\left(\frac{\partial\phi}{\partial n}\right)_P = \left(\frac{\partial\phi_I}{\partial n}\right)_P + \mu(P) + \int\limits_{C+\Gamma} \mu_I \frac{\partial}{\partial n_P}\left[\frac{1}{2i}H_0'(kr_{MP})\right]\mathrm{d}s \tag{4.3.31}$$

The expressions for the potential function and the corresponding derivative outside the harbor and the continuity conditions at the gate are identical to Eqs. (4.3.21) to (4.3.24).

(2) Considering the scattered wave potential function at the solid boundary outside the harbor, the sources are set at outer side of the boundary. Then, the integral boundary in Eqs. (4.3.21) and (4.3.22) turns to $C' + \Gamma'$ as the following

$$\phi(P') = \phi_I(P') + \int_{C'+\Gamma'} \mu_{II}\left[\frac{1}{2i}H_0'(kr_{MP})\right]\mathrm{d}s \tag{4.3.32}$$

$$\left(\frac{\partial\phi}{\partial n'}\right)_{P'} = \left(\frac{\partial\phi_I}{\partial n'}\right)_{P'} + \mu_{II}(P') + \int\limits_{C'+\Gamma'} \mu_{II} \frac{\partial}{\partial n'_{P'}}\left[\frac{1}{2i}H_0'(kr_{MP})\right]\mathrm{d}s \tag{4.3.33}$$

The other expressions for the potential function and the corresponding derivative inside the harbor and the continuity conditions at the gap are identical to those for the breakwater gap with a small gap width. It had been shown that the calculated results by using the above model for the breakwater gap with a big gap width were closer to the analytical solutions.

5. CAD system for the plane design of breakwaters

For pure wave diffraction problems, it is quite simple to use the Green's function method. The numerical calculation is fast so that a man–computer interactive CAD system can be formed. The system consists of parameter input, numerical calculation, and graphic display of output. It allows real-time modification of breakwater layout design on the computer screen according to the calculated results [13].

The wave diffraction coefficient or wave height in the domain can be calculated after starting the program and inputting all the parameters. The wave height distribution around the breakwater can be shown directly on the screen. If the results do not meet the requirements, the parameters such as the length, direction, and the reflection coefficient of the breakwater can be modified to achieve an optimized layout with reasonable wave conditions.

It is suitable to use the numerical model of wave diffraction to modify the breakwater layout design in the phase of preliminary planning or pre-feasibility study. As only wave diffraction is taken into account, the effects of complex bathymetry and the refraction due to the change of water depth are ignored, and the calculation is very fast. Thus, the length and direction of the breakwater can be primarily determined. In the further design phase, the model with more factors taken into account should be used for the calculation of wave field.

4.3.3 Diffraction of Random Waves

Based on the linear superposition theory, the irregular waves can be expressed by a superposition of an infinite number of regular waves with different amplitudes, different frequencies, different directions, and different phases.

$$\eta(x \cdot y \cdot t) = \sum_{i=1}^{\infty}\sum_{j=1}^{\infty} a_{ij} \cos\left\{k_i x \cos\left(\theta_p + \theta_j\right) + k_l y \sin\left(\theta_p + \theta_j\right) - 2\omega_i t - \varepsilon_{ij}\right\} \tag{4.3.34}$$

where a_{ij} is the wave amplitude of the wave components at the ith frequency in the jth direction, θ_p is the orientation angle of the principal direction of waves, θ_j is the deflected angle of the jth direction from the principal direction, ω_i is the representative frequency in the ith divided frequency sub-area, and ε_{ij} is the initial phase of the wave component at the ith frequency in the jth direction. The dispersion relation for each wave component reads

$$\omega_i^2 = gk_i \tanh k_i h \tag{4.3.35}$$

According to the definition of directional spectral density,

$$\sum_{\omega}^{\omega+\mathrm{d}\omega}\sum_{\theta}^{\theta+\mathrm{d}\theta} \frac{1}{2}a_{ij}^2 = S\left(\omega_i, \theta_j\right)\mathrm{d}\omega\mathrm{d}\theta \tag{4.3.36}$$

where $S(\omega, \theta)$ is the directional spectral density and directional spectrum for short.

Let ω_i denote the representative frequency in the interval $[\omega, \omega + \mathrm{d}\omega]$, θ_j denote the representative direction angle in the interval $[\theta, \theta + \mathrm{d}\theta]$, and the wave amplitude be

$$a_{ij} = \sqrt{2S(\omega_i, \theta_j)\mathrm{d}\omega\mathrm{d}\theta} \tag{4.3.37}$$

then, the diffracted wave height by the wave component at the ith frequency in the jth direction with amplitude of a_{ij} reads

$$(a_{ij})_\mathrm{d} = \sqrt{2S(\omega_i, \theta_j)\mathrm{d}\omega\mathrm{d}\theta} \times (K_\mathrm{d})_{ij} = \sqrt{2S_d(\omega_i, \theta_j)\mathrm{d}\omega\,\mathrm{d}\theta} \tag{4.3.38}$$

If the incident wave surface elevation $\eta(x, y, t)$ is

$$\begin{aligned}\eta(x, y, t) = &\sum_{i=1}^{m}\sum_{j=1}^{n}\sqrt{2S(\omega_i, \theta_j)\mathrm{d}\omega\mathrm{d}\theta}\\ &\times \exp\{ik_i[x\cos(\theta_p + \theta_j) + y\sin(\theta_p + \theta_j)] - 2\omega_i t + \varepsilon_{ij}\}\end{aligned} \tag{4.3.39}$$

the diffracted wave surface elevation $\eta_d(x, y, t)$ reads

$$\begin{aligned}\eta_d(x, y, t) = &\sum_{i=1}^{m}\sum_{j=1}^{n}\sqrt{2S_\mathrm{d}(\omega_i, \theta_j)\mathrm{d}\omega\mathrm{d}\theta}\\ &\times \exp\{ik_i[x\cos(\theta_p + \theta_j) + y\sin(\theta_p + \theta_j)] - 2\omega_i t + \varepsilon_{ij}\}\\ = &\sum_{i=1}^{m}\sum_{j=1}^{n}\sqrt{2S(\omega_i, \theta_j)\mathrm{d}\omega\mathrm{d}\theta} \times (K_d)_{ij}\\ &\times \exp\{ik_i[x\cos(\theta_p + \theta_j) + y\sin(\theta_p + \theta_j)] - 2\omega_i t + \varepsilon_{ij}\}\end{aligned} \tag{4.3.40}$$

The diffraction coefficient of irregular waves is defined as the ratio of the significant wave height after diffraction to the incoming wave height before diffraction,

$$(K_d)_H = \frac{(H_{1/3})_d}{(H_{1/3})_I} \tag{4.3.41}$$

The diffraction coefficient expressed by spectrum reads,

$$(K_d)_H = \frac{\sqrt{\sum_{i=1}^{n}\sum_{j=1}^{n} S(\omega_i, \theta_j)\mathrm{d}\omega_i\mathrm{d}\theta_j(K_d)_{ij}^2}}{\sqrt{\sum_{i=1}^{m}\sum_{j=1}^{n} S(\omega_i, \theta_j)\mathrm{d}\omega_i\mathrm{d}\theta_j}} \tag{4.3.42}$$

The directional spectrum S(ω,θ) can be rewritten as

$$S(\omega, \theta) = S(\omega) \times G(\theta) \tag{4.3.43}$$

In the calculation, the diffraction coefficient $(K_d)_{ij}$ for each regular wave component is calculated by the numerical model of regular wave first. Then, substituting them into Eq. (4.3.42) yields the result of irregular wave diffraction.

In the numerical calculation, the frequency spectra and directions need to be divided into finite intervals. A representative frequency ω_i and a representative direction θ_i are used for the calculation in each interval. Equal frequency interval method or equal spectrum area method can be used for the dividing method. The number of intervals can be determined by trial.

Figure 4.12 shows wave diffraction coefficients by a single breakwater calculated by a regular wave model and an irregular wave model, respectively. The difference between the regular wave diffraction and the irregular wave diffraction is significant. In the sheltered area, the irregular wave diffraction coefficients are significantly larger than the regular wave diffraction coefficients. But in the open area, the regular wave diffraction coefficients are larger. For a single breakwater, the regular wave diffraction coefficient on the geometric shadow line is about 0.5, but the corresponding value of the irregular waves goes up to about 0.7. The wave height calculated by using the regular wave model for the wave perturbation in the harbor might be underestimated, resulting in an unsafe design. Therefore, it is necessary to consider the effects of irregular waves. In the calculation of irregular wave diffraction, it is quite sensitive to the directional distribution function. Thus, the form of the directional distribution function $G(\theta)$ should be selected carefully for the calculation. The exponential forms of the cosine such as $\cos^4\theta$, $\cos^6\theta$, and $\cos^8\theta$ can be selected for $G(\theta)$. The number of power reflects the concentration of wave energy nearby the principal wave direction. In general, $\cos^4\theta$ is chosen.

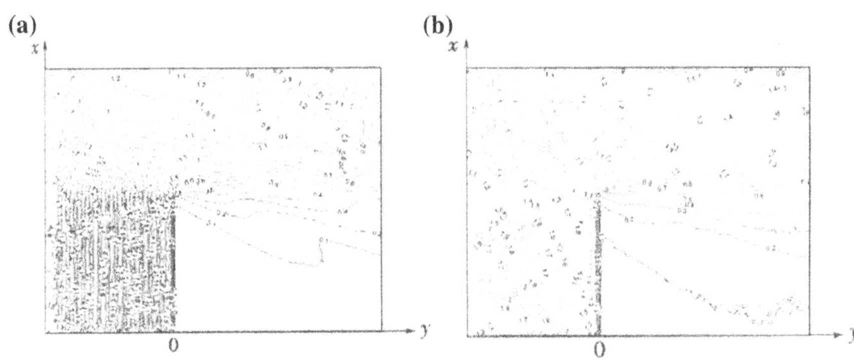

Fig. 4.12 Wave diffraction coefficients by a single breakwater, **a** regular waves, **b** unidirectional irregular waves

The comparisons of wave diffraction coefficients by a single breakwater calculated by using regular waves (*R*), irregular waves with unidirectional frequency spectrum (*F*), the directional distribution function (*G*), and the directional spectrum (*FG*) are shown in Fig. 4.13 for the profile at $y/L = 3$ behind the single breakwater and Fig. 4.14 for the profile at $y/L = 6$ behind the single breakwater [12]. The diffraction coefficients using regular waves (*R*) and using unidirectional frequency spectrum (*F*) are close, and the results using directional distribution function (*G*) and

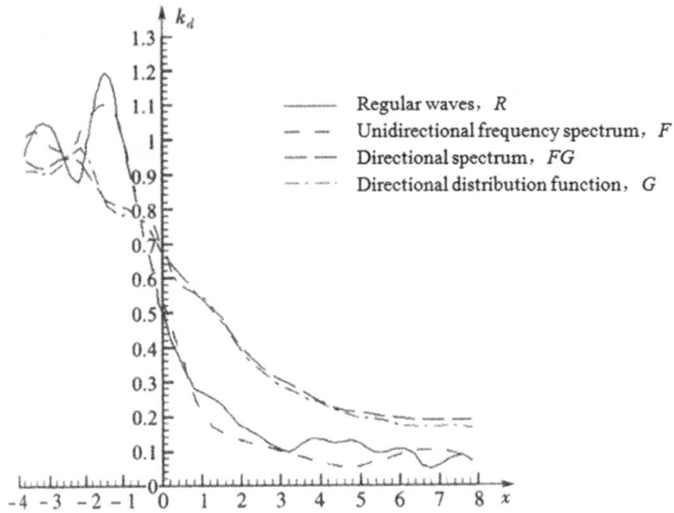

Fig. 4.13 Wave diffraction coefficients behind the single breakwater ($y/L = 3$)

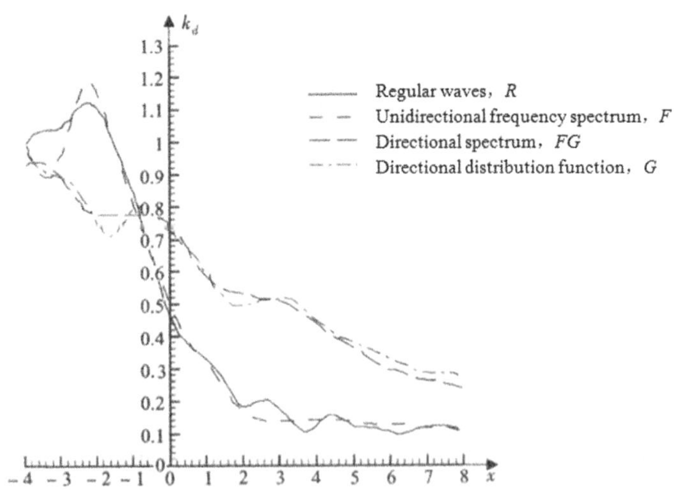

Fig. 4.14 Wave diffraction coefficients behind the single breakwater ($y/L = 6$)

directional spectrum (FG) are close. These indicate that the diffraction coefficients are not sensitive to the frequency, but the directional distribution function has a great effect on the diffraction coefficients.

4.4 Mathematical Models of Mild Slope Equations for Wave Refraction and Diffraction

When waves propagate from deep water to coastal regions near structures, the wave deformation cannot be regarded as a problem of pure refraction or diffraction. The combined effects of wave refraction and diffraction should be taken into account. Under assumptions of inviscid and irrotational small-amplitude water waves, and assuming that the bathymetry varies slowly, Berkhoff [1] initially proposed a mild slope equation describing combined wave refraction and diffraction in 1972. Subsequently, many scholars used different methods to derive the mild slope equation in different forms. The study contents have also been more enriched such as the effects of current and bottom friction and the parabolic approximation to the equation [14–17].

4.4.1 Elliptic Mild Slope Equation Model

1. Basic equations

Based on the linear small-amplitude wave theory, the potential function $\Phi(x, y, z, t)$ can be expressed as

$$\Phi(x, y, z, t) = Z(z)\phi(x, y)\mathrm{e}^{-i\omega t}$$

$$Z(z) = \frac{\cosh k(z+d)}{\cosh kd} \tag{4.4.1}$$

Under the condition of varying bathymetry, Φ satisfies

$$\nabla^2\Phi + \Phi_{zz} = 0, \quad -d \le z \le 0 \tag{4.4.2}$$

$$\Phi_z = -\frac{1}{g}\Phi_{tt}, \quad z = 0 \tag{4.4.3}$$

$$\frac{\partial\Phi}{\partial z} = \frac{\partial\Phi}{\partial x}\frac{\partial d}{\partial x} + \frac{\partial\Phi}{\partial y}\frac{\partial d}{\partial y}, \quad z = -d \tag{4.4.4}$$

The bottom boundary condition is identical to Eq. (2.1.25). Substituting Eq. (4.4.1) into Eqs. (4.4.2) and (4.4.4) and introducing the condition of bathymetry varying slowly, which means $\mathrm{O}([\nabla h]^2)$ and $\mathrm{O}(\nabla^2 h)$ can be ignored, yield

$$\nabla \cdot (cc_g \nabla \phi) + k^2 cc_g \phi = 0 \tag{4.4.5}$$

This is the basic form of the mild slope equation proposed by Berkhoff first [1] where c and c_g are the phase velocity and the group velocity, respectively.

$$c = \frac{\omega}{k} = \sqrt{\frac{g}{k}\tanh(kd)}$$

$$c_g = \frac{d\omega}{dk} = nc = \frac{1}{2}\left[1 + \frac{2kd}{\sinh(2kd)}\right] \cdot c$$

It is seen from Eq. (4.4.4) that c and c_g do not vary with spatial position when the water depth is constant. Then, Eq. (4.4.5) becomes similar to Eq. (4.3.10)

$$\nabla^2 \phi + k^2 \phi = 0$$

This is exactly the Helmholtz equation of the potential function Φ in the domain with constant water depth for a pure diffraction problem.

When using the mild slope equation in Eq. (4.4.5) to solve the wave refraction and diffraction, the hybrid element method is an effective method, which is normally to divide the water around the structure into the interior and the exterior parts by a circle. The mild slope equation can be solved by using a finite element method for the calculation in the interior domain. In the exterior domain where it is thought to have constant water depth, the wave diffraction problem is solved. The source intensity on the circumference and the solution in the interior domain can be obtained by means of the continuity conditions of pressure and the normal velocity on the circumference [8, 18].

Following Booij [19] added a dissipation term to the mild slope equation for consideration of the bottom friction effect. A concrete form of the friction factor had been given by Dalrymple [20]. The mild slope equation considering the friction resistance reads

$$\nabla \cdot (cc_g \nabla \phi) + k^2 cc_g \left(1 - \frac{iw}{kC_g}\right)\phi = 0 \tag{4.4.6}$$

where w is the total dissipation due to wave breaking and bottom friction. The calculation method on the dissipation term of wave breaking and bottom friction had been described in detail in the book by Dingemans [21].

The traditional mild slope equation was derived under an assumption of linear waves. Some researchers introduced nonlinearity into the mild slope equation in their studies, such as Karpman [22], Yue and Mei [23], and Dalrymple and Kirby [20]. The following gives a weakly nonlinear mild slope equation obtained by Liu and Tsay [17] using multiple perturbation expansion approach:

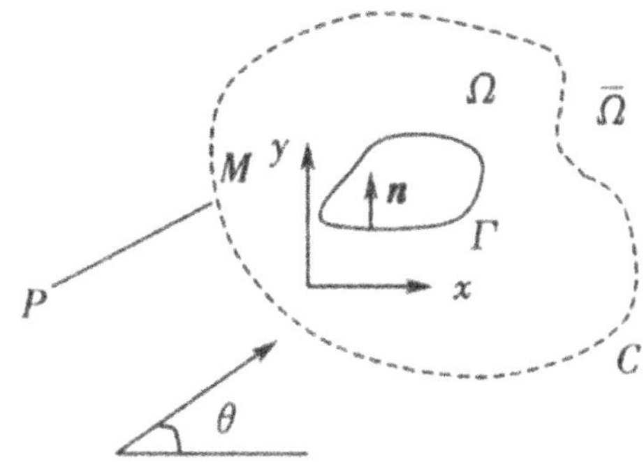

Fig. 4.15 Sketch of the interior and exterior domains

$$\nabla \cdot (cc_g \nabla \phi) + k^2 cc_g \phi - \chi \left(\frac{\omega}{g}\right)^2 |\phi|^2 \phi = 0 \tag{4.4.7}$$

where

$$\chi = \frac{k^4 c^2}{8 \sinh^4(kd)} \left[8 + \cosh(4kd) - 2 \tanh^2(kd)\right] \tag{4.4.8}$$

2. Hybrid element model for the elliptic mild slope equation

It is assumed that the complex bathymetry, large structures, varying coastline, and water depth can be confined within a closed curve C. The fluid region inside C is denoted by Ω and called the interior domain. The region outside C with an assumption of constant water depth is denoted by $\bar{\Omega}$ and called the exterior domain. See Fig. 4.15 for the sketch.

A finite element method is used to solve the elliptic mild slope equation in the interior domain, and a boundary element method is used to solve the pure diffraction problem in the exterior domain with constant water depth. Thus, it is called a hybrid element model.

Let

$$\phi = \phi_I + \phi_S \tag{4.4.9}$$

where ϕ_I is the incident wave potential function, and ϕ_S is the scattered wave potential function.

The angle between the incident wave direction and the x-axis is denoted by θ. Then, the incident wave potential function becomes

$$\phi_I = A_0 \mathrm{e}^{ik(x \cos\theta + y \sin\theta)} \tag{4.4.10}$$

where A_0 is a constant related to the incident wave height.

In the interior domain Ω,

$$\nabla \cdot (CC_g \nabla \phi) + \omega^2 \frac{C_g}{C} \phi = 0, \quad \text{in the } \Omega \text{ domain} \tag{4.4.11}$$

$$\frac{\partial \phi}{\partial n} + iRk\phi = 0, \quad \text{on the body surface} \tag{4.4.12}$$

where R is the reflection factor of the solid wall.

In the exterior domain $\bar{\Omega}$, the scattered wave potential function of the pure diffraction satisfies the Helmholtz equation and the Sommerfeld radiation condition,

$$\nabla^2 \bar{\phi}_S + k^2 \bar{\phi}_S = 0 \tag{4.4.13}$$

$$\lim_{r \to \infty} \sqrt{kr} \left(\frac{\partial}{\partial r} - ik \right) \phi_S = 0 \tag{4.4.14}$$

$\bar{\phi}_s$ can be expressed as

$$\bar{\phi}_S = \int_C \mu(s) \frac{1}{2i} H_0^{(2)}(kr) \mathrm{d}s \tag{4.4.15}$$

where $\mu(s)$ is the source intensity on the boundary C. $H_0^{(2)}(kr)$ is the zero-order Hankel function of the second kind.

$$\frac{\partial \bar{\phi}}{\partial n} = \frac{\partial \phi_I}{\partial n} - \mu(p) + \int_C \mu(s) \frac{\partial}{\partial n} \left[\frac{1}{2i} H_0^2(k_O r) \right] \mathrm{d}s \tag{4.4.16}$$

The total potential function in the exterior domain $\bar{\Omega}$ reads

$$\bar{\phi} = \bar{\phi}_I + \bar{\phi}_S$$

On the boundary C, the matching conditions are satisfied

$$\phi = \bar{\phi} \tag{4.4.17}$$

$$\frac{\partial \phi}{\partial n} = \frac{\partial \bar{\phi}}{\partial n} \tag{4.4.18}$$

The potential function in the interior domain Ω can be solved by using a finite element method through the matching conditions of the interior and exterior domains.

If a finite element method based on the variational principle is used for the solution in the interior domain, the function J can be expressed as

$$J = \frac{1}{2}\iint\limits_{\Omega}\left[cc_g(\nabla\phi)^2 - \frac{\omega^2 c_g}{c}\phi^2\right]\mathrm{d}x\mathrm{d}y$$
$$+\int\limits_{C} cc_g\left[\left(\frac{1}{2}\bar{\phi} - \phi\right)\frac{\partial\bar{\phi}}{\partial n} - \frac{1}{2}\bar{\varphi}\frac{\partial\bar{\phi}}{\partial n}\right]\mathrm{d}s + \frac{1}{2}\int\limits_{\Gamma} icc_g Rk\phi^2\mathrm{d}s \quad (4.4.19)$$

It can be proved that the extremum problem of the functional J is equivalent to the boundary problem of ϕ in the Ω domain and $\bar{\phi}_s$ in the $\bar{\Omega}$ domain [18], i.e., Equations (4.4.11) to (4.4.14), Eqs. (4.4.17), and (4.4.18).

The Ω domain is divided into finite elements, and the interpolation function is given at each element. Then, a set of linear algebraic equations expressed in terms of the potential functions at the element nodes can be obtained by solving the extremum of the functional J. In the numerical calculation, one-dimensional (1D) variable bandwidth storage technology for large coefficient matrix can be used to solve the complex potential function ϕ in the Ω domain. Finally, the refraction-diffraction coefficient can be obtained by

$$K_{\mathrm{rd}} = \frac{H_{\mathrm{rd}}}{H_I} = |\phi|$$

where H_{rd} is the wave height after the effect of refraction and diffraction, H_I is the incident wave height, and ϕ is determined by Eq. (4.4.9).

Figure 4.16 shows the calculated diffraction results of the breakwater gap by using the hybrid element model compared with the experimental results. In Fig. 4.16, (a) is for the regular waves, (b) for the unidirectional irregular waves, and (c) for the directional irregular waves [12].

4.4.2 Time-Dependent Mild Slope Equation

The following equation was given by Luke [2] when he used a variational principle to study potential waves:

$$\eta_t = -\nabla\cdot\left(\frac{cc_g}{g}\nabla\phi'\right) + \frac{\omega^2 - k^2 cc_g}{g}\phi' + \frac{\omega^2}{g}\left[R_1(\nabla d)^2 + R_2\nabla^2 d\right]\phi' \quad (4.4.20)$$

The potential function ϕ' is defined as the total potential function $\Phi(x, y, z, t) = Z(z)\phi'(x, y, t)$, where $Z(z)$ is defined by Eq. (4.4.1), and the other symbols in the equation are the same as above. R_1 and R_2 are the functions for consideration of the second-order derivatives of bathymetry.

Ignoring the second-order derivatives of bathymetry yields

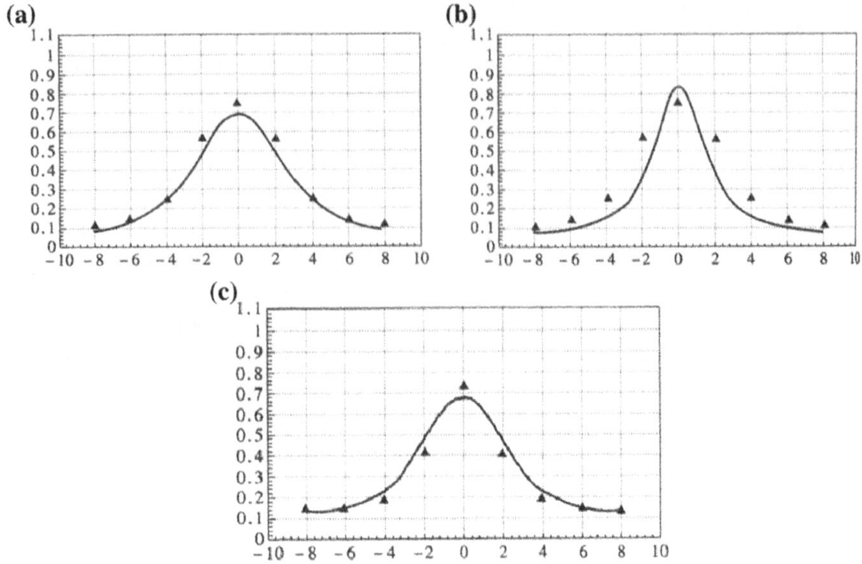

Fig. 4.16 Calculated diffraction coefficients of breakwater gap with 2 m gap width compared with the experimental results, **a** regular waves, at $y = 4$ m, **b** unidirectional irregular waves, at $y = 4$ m, **c** directional irregular waves, at $y = 2$ m. —— calculated results, ▲ experimental results

$$\eta_t = -\nabla \cdot \left(\frac{cc_g}{g} \nabla \phi' \right) + \frac{\omega^2 - k^2 cc_g}{g} \phi' \tag{4.4.21}$$

Considering the linear kinematic condition of free surface in Eq. (2.2.3),

$$\Phi_t = -g\eta$$

a time-dependent mild slope equation can be obtained by eliminating η or ϕ from Eqs. (4.4.20) and (4.4.21),

$$\phi'_{tt} = \nabla \cdot \left(cc_g \nabla \phi' \right) - \left(\omega^2 - k^2 cc_g \right) \phi' \tag{4.4.22}$$

or

$$\eta_{tt} = \nabla \cdot \left(cc_g \nabla \eta \right) - \left(\omega^2 - k^2 cc_g \right) \eta \tag{4.4.23}$$

Equation (4.4.22) or Eq. (4.4.23) is a time-dependent mild slope equation. For harmonic wave motion,

$$\phi'(x, y, t) = \phi(x, y) e^{i\omega t} \tag{4.4.24}$$

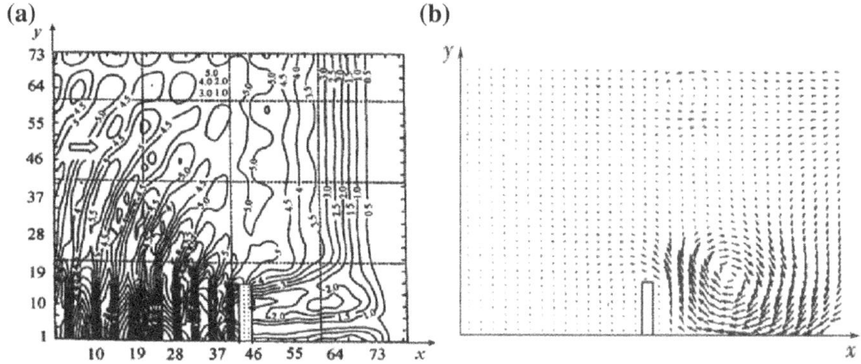

Fig. 4.17 Calculated wave height distribution (**a**) and velocity field (**b**) near an offshore breakwater

The elliptic mild slope Eq. (4.4.5) can be obtained by substituting Eq. (4.4.24) into Eq. (4.4.22). The time-dependent mild slope Eqs. (4.4.22) and (4.4.23) had also been derived by Dingemans [21] by using a variational principle based on the Hamiltonian method (the total energy method). The characteristics of the equation could be clearly understood through the derivation.

Figure 4.17 shows a diffracted wave height distribution and velocity field behind an offshore breakwater calculated by solving the time-dependent mild slope equation (only half domain is presented because of symmetry) [24].

4.4.3 Parabolic Mild Slope Equation Model

In the design of coastal engineering, it is usually needed to calculate the wave elements from deep sea to shallow water. The ray method for the wave refraction can be used if the bathymetry is gentle. However, the natural bathymetry is complex in some estuaries and coastal regions. The combined effects of refraction and diffraction should be taken into account simultaneously. Due to the large calculation area from more than a dozen kilometers to tens of kilometers, the computational workload is large with the elliptic mild slope equation or the time-dependent mild slope equation. Then, the parabolic equation can be used to simulate the wave field in a large area.

1. Governing equation

The basic assumptions of parabolic approximation to mild slope equation are that waves propagate in a principal direction and can be spitted into the transmitted waves propagating forward and the reflected waves propagating backward, and the reflected waves are negligible. The parabolic mild slope equation can be derived in many ways. Some key results given by Radder [3] are described here.

Based on the elliptic mild slope equation,

$$\nabla \cdot (cc_g \nabla \phi) + k^2 cc_g \phi = 0$$

introducing a transformation,

$$\xi = \sqrt{cc_g}\phi \tag{4.4.25}$$

then Eq. (4.4.5) is turned into the Helmholtz equation,

$$\nabla^2 \xi + k_c^2 \xi = 0 \tag{4.4.26}$$

$$k_c^2 = k^2 - \frac{\nabla^2(\sqrt{cc_g})}{\sqrt{cc_g}} \tag{4.4.27}$$

Substituting Eqs. (4.4.27) into (4.4.26) yields

$$\frac{\partial^2 \xi}{\partial x^2} + k_c^2 Q \xi = 0 \tag{4.4.28}$$

$$Q = \frac{1}{k_c^2}\frac{\partial^2}{\partial y^2} + 1 \tag{4.4.29}$$

The wave field is split into a transmitted wave field ξ^+ propagating forward and a reflected wave field ξ^- propagating backward,

$$\xi = \xi^+ + \xi^- \tag{4.4.30}$$

Introducing a small incident wave angle with an assumption of $\theta = \tan^{-1}(k_y/k) \ll 1$, neglecting the reflected wave field ξ^-, the equation for the transmitted wave field could be obtained through the derivation,

$$\frac{\partial \xi^+}{\partial x} = \left(ik - \frac{1}{2k}\frac{\partial k}{\partial x} + \frac{i}{2k}\frac{\partial^2}{\partial y^2}\right)\xi^+ \tag{4.4.31}$$

A parabolic mild slope equation which was proposed by Liu [25] by using the wave amplitude A as a variable and introducing weakly nonlinear item reads

$$2ikA_x + (cc_g A_y)y + \left[k(k - k_0) + \frac{1}{cc_g}(kcc_g)_x\right]A - \frac{D}{cc_g}|A^2|A = 0 \tag{4.4.32}$$

2. Higher-order approximate parabolic mild slope equation model

(1) Governing equation

Kirby [26] proposed a parabolic mild slope equation model with higher-order approximation by using Padé expansion and saddle point method for the improvements.

It allows waves propagating at certain angles to the principal direction and provides a good mathematical model for irregular wave propagation in a large area. A higher-order approximate parabolic mild slope equation reads

$$
\begin{aligned}
&c_g A_x + i \cdot (\bar{k} - \beta_0 k) c_g A + \frac{1}{2}(c_g)_x A + \frac{i}{\omega}\left(\beta_1 - \beta_2 \frac{\bar{k}}{k}\right)\left(c c_g A_y\right)_y \\
&- \frac{\beta_2}{\omega k}\left(c c_g A_y\right)_{yx} + \frac{\beta_2}{\omega}\left(\frac{k_x}{k^2} + \frac{(c_g)_x}{2kc_g}\right)\left(c c_g A_y\right)_y + \frac{i\omega k^2}{2} \cdot D \cdot |A|^2 A + \frac{W}{2} A = 0
\end{aligned}
\tag{4.4.33}
$$

where A is the complex wave amplitude function; i is the imaginary unit; k is the wave number; ω is the angular frequency; $\bar{k}$ is the characteristic wave number; c is the wave celerity; c_g is the group velocity;

$$c_g = nc$$

$$n = \frac{1}{2}\left(1 + \frac{2kh}{\sinh 2kh}\right)$$

where $h = d + \eta$ is the total water depth; β_0 , β_1 , β_2 in Eq. (4.4.33) are the coefficients for which different values correspond to different approximations to the equation; D is a factor for nonlinearity effect which can be calculated by

$$D = \frac{\cosh 4kh + 8 - 2\tanh^2 kh}{8\sinh^4 kh} \tag{4.4.34}$$

W is a factor for wave energy loss and generation which is expressed as

$$W = W_f + W_b + W_w \tag{4.4.35}$$

where W_f is the bottom friction factor, W_b is the wave breaking factor, and W_w is the wind effect factor.

W defined by the wave energy dissipation rate ε is

$$W = \frac{\varepsilon}{E} \tag{4.4.36}$$

where E is the wave energy, and ε is correspondingly expressed as

$$\varepsilon = \varepsilon_f + \varepsilon_b + \varepsilon_w \tag{4.4.37}$$

Calculation of the bottom friction factor W_f:

The bed shear stress can be approximately considered as

$$\tau_b = \frac{1}{2}\rho \cdot f_w U_b |U_b| \tag{4.4.38}$$

where $U_b = U_m \cos(\omega \cdot t)$ is the wave particle velocity above the bottom.

In the condition of turbulence, the time-averaged energy loss rate per unit seabed ε_f is

$$\varepsilon_f = \overline{\tau_b U_b} = \frac{1}{2T}\int_t^{t+T} \rho \cdot f_w U_b^2 |U_b| \mathrm{d}t = \frac{1}{2}\rho \cdot f_w U_m^3 \left(\frac{1}{T}\int_t^{t+T} \left|\cos^3(\omega \cdot t)\right| \mathrm{d}t\right) \tag{4.4.39}$$

where

$$U_m = \frac{\pi H}{T} \cdot \frac{1}{\sinh(kh)} \tag{4.4.40}$$

$$\frac{1}{T}\int_t^{t+T} \left|\cos^3 \omega t\right| \mathrm{d}t = \frac{4}{3\pi} \tag{4.4.41}$$

which gives

$$\varepsilon_f = \frac{16\pi^2}{3g} \frac{f_w \cdot H}{T^3 \sinh^3(kh)} E \tag{4.4.42}$$

In the parabolic mild slope equation,

$$W_f = \frac{16\pi^2}{3g} \frac{f_w H}{T^3 \sinh^3(kh)} \tag{4.4.43}$$

where f_w is the friction coefficient as a function of fluctuating Reynolds number and seabed roughness. Therefore, the friction coefficient varies with location. The roughness turbulence friction coefficient given by Jonsson [27] has been widely used but it is needed to use iteration for the solution. A convenient form of the friction coefficient was proposed by Swart [28],

$$f_w = \begin{cases} 0.00251 \exp[5.21(A/k_s)^{-0.19}] & A/k_s > 1.57 \\ 0.3 & A/k_s \leq 1.57 \end{cases} \tag{4.4.44}$$

In practical engineering, an empirical constant is commonly selected for f_w.

Calculation of the wind effect factor W_w:

The wind effect is taken into account by using an empirical expression (see Refs. [29, 28]),

$$W_w = -\frac{2c_g}{H_s}\frac{\mathrm{d}H_s}{\mathrm{d}x}\cos(\theta_w - \theta) \tag{4.4.45}$$

In order to calculate the derivative $\mathrm{d}H_s/\mathrm{d}x$, the wind wave growth curves given by Krylov or Wilson could be used as

$$\tilde{H}_s = \beta\left[1 - \frac{1}{\left(1+\alpha\sqrt{\tilde{x}_w}\right)^2}\right] \tag{4.4.46}$$

where $\tilde{H}_s = \frac{gH_s}{U_w^2}$, $\tilde{x}_w = \frac{gx_w}{U_w^2}$.
where H_s is the significant wave height, U_w is the wind speed, x_w is the wind fetch, θ_w is the wind direction angle, θ is the wave direction angle, and α and β are the coefficients.

(2) Boundary conditions

The incident boundary condition is given at $x = 0$. For normal incident waves,

$$A(0, y) = A_0 \tag{4.4.47}$$

where A_0 is the initial wave amplitude. For oblique incident waves,

$$A(0, y) = A_0 \cdot \mathrm{e}^{ily} \tag{4.4.48}$$

where $l = k\sin\theta$, θ is the incident wave angle defined by the angle between the wave propagation direction and the x-axis.

The lateral boundary conditions are the following (Let B be the width of the domain).

(1) Full reflection boundary condition:

$$\frac{\partial A}{\partial y} = 0, \quad \text{for } y = 0 \quad \text{or} \quad y = B$$

This boundary condition is applied for vertical walls of structures such as vertical wall breakwaters and sea walls, or used for large computational domain in which the boundary is far away from the concerned region.

(2) Transmission boundary condition:

$$\frac{\partial A}{\partial y} = ik\sin\theta \cdot A, \quad \text{for} \quad y = 0 \quad \text{or} \quad y = B$$

where k is the wave number. This boundary condition refers to that the waves can travel freely through the boundary.

3. Verification of the higher-order approximate parabolic mild slope equation model by laboratory data

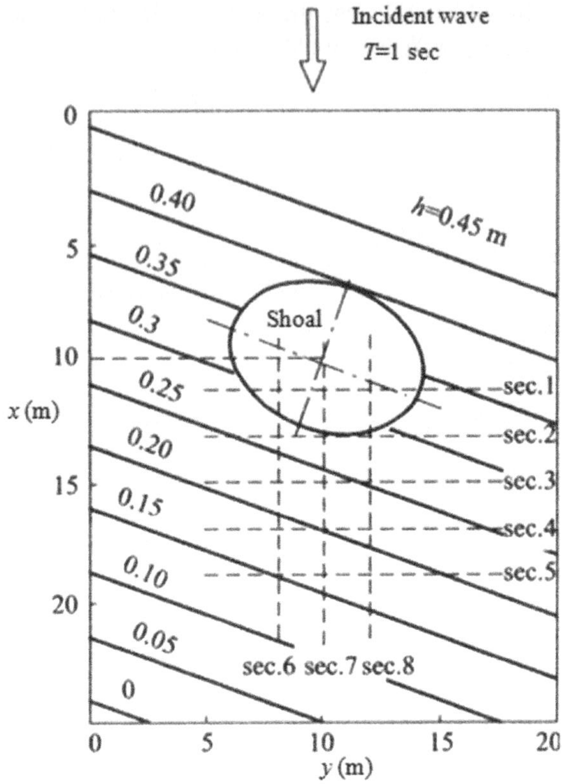

Fig. 4.18 Geometry of the submerged shoal experiment

In the laboratory experiments described by Berkhoff et al. [30], an elliptic shoal resting on a plane sloping bottom was arranged. The plane slope was turned at a certain angle to the incident wave. The geometry of the submerged shoal experiment is shown in Fig. 4.18.

The incident wave height was 0.046 m, and the wave period was 1 s. The deepest water depth was 0.45 m. Grid spacing was set to $\Delta x = \Delta y = 0.25$ m for the calculation. The refraction-diffraction calculations were based on three different models including the lowest-order approximate linear equation, the lowest-order approximate nonlinear equation, and the higher-order approximate nonlinear equation. Figure 4.19 shows comparisons of the numerical and experimental results at different sections. The nonlinear model results are in better agreement with the laboratory data than the linear model results. The higher-order model results are significantly better than the lower-order model results. The comparisons indicate that the nonlinear term plays a significant role in the calculation under the condition of complex bathymetry, and the results of higher-order approximation have more accuracy.

4. Calibration and verification of the higher-order approximate parabolic mild slope equation model by field measurements

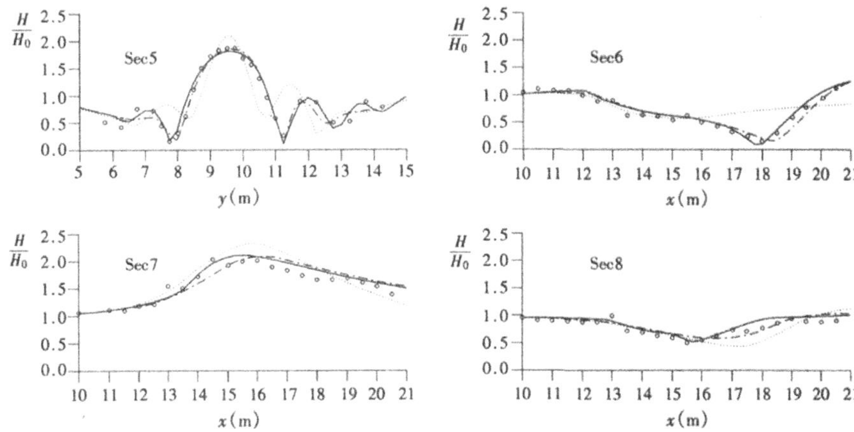

Fig. 4.19 Numerical and experimental relative amplitudes along transects for submerged shoal. ○: laboratory data, . . . the lowest-order linear approximation, - - - -: the lowest-order nonlinear approximation, ——: the higher-order nonlinear approximation

In the design of navigation channel regulation projects, such as the Yangtze River estuary deepwater channel regulation project and the Huanghua Port channel regulation project, it is necessary to calculate the wave field in existing bathymetry before the project for the determination of design wave parameters. Because of the large project area and complex bathymetry, the combined refraction-diffraction parabolic mild slope equation model can be used for simulations. However, the wave field after the project implementation is the basis of the structure design. Due to the existence of structures such as jetties and spur dikes, the wave reflection is not ignorable. The parabolic mild slope equation model is not valid anymore. Then the elliptic, time-dependant mild slope equation models or Boussineq model should be applied.

For the requirement of engineering design, field measurements of wave parameters at several stations and the analysis of historical data in the project area are carried out. The higher-order approximate parabolic mild slope equation model can be calibrated and verified by using the field measurements.

The bottom friction coefficient f_w is calibrated first. The wave measurements in the conditions of calm wind or small wind with different incident wave directions are selected for finding a reasonable f_w to make the simulated results be in good agreements with the measurements, which is called model calibration. Then, the model is verified by comparing the simulated results using the calibrated f_w with the measurements in the other wind conditions.

Table 4.1 presents an example of verified results at a measurement station. The calculated significant wave heights match the measurements in general. The calculation by mathematical model is deemed to be reasonable. The reliability of the model can be further proved by the verification with more measured data.

Table 4.1 Verification of wave height in different return periods at a measurement station

Incident wave direction	Return period (year)	$H_{1/10}$ (m)	H_s (m)	Calculated H_s (m)
NNE	50	4.40	3.46	3.99
	25	4.03	3.17	4.0
	5	3.21	2.52	2.81

4.5 Mathematical Model of Boussinesq Equations for Dispersive Shallow Water Waves

The numerical model based on Boussinesq equations is another type of wave model to simulate the wave refraction, diffraction, reflection, and shoaling due to complex bathymetry and the coastal engineering structures such as breakwaters and artificial islands in shallow water.

Boussinesq equations describe wave motion with dispersion and nonlinearity. The dispersion of Boussinesq equations is between the non-dispersive shallow water long waves and the Stokes dispersion waves (see Sect. 2.3.2).

4.5.1 *Governing Equations*

The basic Boussinesq equations are under the conditions of weak nonlinearity and weak dispersion, which means the relative wave height $\varepsilon = \frac{a}{d} \ll 1$, and the relative water depth $\mu = \frac{d}{L} \ll 1$. According to Eqs. (2.3.17) and (2.3.18), taking the surface elevation η and the depth-averaged velocity U as variables yields 1D Boussinesq equations for waves over flat bottom,

$$\frac{\partial \eta}{\partial t} + h\frac{\partial U}{\partial x} = 0 \tag{4.5.1}$$

$$\frac{\partial U}{\partial t} + U\frac{\partial U}{\partial x} + g\frac{\partial \eta}{\partial x} - \frac{h^2}{3}\frac{\partial^3 U}{\partial x^2 \partial t} = 0 \tag{4.5.2}$$

where η is the surface elevation above the still water level; $U = \frac{1}{h}\int_{-d}^{\eta} u dz$ is the depth-averaged velocity; $h = d + \eta$ is the total water depth; d is the still water depth.

The 2D conservative form of Boussinesq equations considers the effect of bottom friction for waves over flat bottom reads

$$\frac{\partial \eta}{\partial t} + \frac{\partial P}{\partial x} + \frac{\partial Q}{\partial y} = 0 \tag{4.5.3}$$

$$\frac{\partial P}{\partial t}+\frac{\partial}{\partial x}\left(\frac{P^2}{h}\right)+\frac{\partial}{\partial y}\left(\frac{PQ}{h}\right)+gh\frac{\partial \eta}{\partial x}+g\frac{\sqrt{P^2+Q^2}}{C^2h^2}P$$
$$=\frac{dh}{3}\left(\frac{\partial^3 P}{\partial x^2\partial t}+\frac{\partial^3 Q}{\partial x\partial y\partial t}\right) \tag{4.5.4}$$

$$\frac{\partial Q}{\partial t}+\frac{\partial}{\partial x}\left(\frac{PQ}{h}\right)+\frac{\partial}{\partial y}\left(\frac{Q^2}{h}\right)+gh\frac{\partial \eta}{\partial y}+g\frac{\sqrt{P^2+Q^2}}{C^2h^2}Q$$
$$=\frac{dh}{3}\left(\frac{\partial^3 P}{\partial x\partial y\partial t}+\frac{\partial^3 Q}{\partial y^2\partial t}\right) \tag{4.5.5}$$

where $p = Uh,\ Q = Vh$; U and V are the depth-averaged velocity components in x- and y-directions, respectively; C is the Chezy coefficient.

4.5.2 Boussinesq-Type Equations

The above basic Boussinesq Eqs. (4.5.1)–(4.5.5) are only for the description of waves with weak nonlinearity and weak dispersion. In recent years, many studies for improvements of Boussinesq equations on dispersion and nonlinearity have been undertaken. A series of higher-order Boussinesq equations and techniques to improve the dispersion have been proposed, which formed Boussinesq-type equations.

1. Improvements on linear dispersion

It has been proved practically that when waves propagate from shallow to deep water, the waves calculated by using the above basic Boussinesq equations are difficult to propagate. Why does this phenomenon occur? The phase velocities of the Boussinesq equation wave and the linear Stokes wave have been compared in Chap. 2. The phase velocity of the Boussinesq wave, which reflects wave motion characteristics, is a lower-order approximation to the phase velocity of the linear Stokes wave. If the approximation to the linear Stokes phase velocity becomes higher, the dispersion of the Boussinesq equations will be improved and the waves will be able to propagate from shallow to relatively deep water.

Among the improvements on dispersion, typical developments were made by Madsen et al. [31] and Nwogu [32]. In 1991, Madsen et al. proposed a new form of Boussinesq equations with improved linear dispersion by adding higher-order terms in the basic Boussinesq equations. This new form of Boussinesq equations being suitable for the sea waves over flat bottom includes an adjustable coefficient B and reads

$$\eta_t + P_x + Q_y = 0 \tag{4.5.6}$$

$$P_t + \left(\frac{P^2}{h}\right)_x + \left(\frac{PQ}{h}\right)_y + gd\eta_x - \left(B + \frac{1}{3}\right)d^2(P_{xxt} + Q_{xyt})$$
$$- Bgd^3(\eta_{xxx} + \eta_{xyy}) = 0 \tag{4.5.7}$$

$$Q_t + \left(\frac{PQ}{h}\right)_x + \left(\frac{Q^2}{h}\right)_y + gd\eta_y - \left(B + \frac{1}{3}\right)d^2(Q_{yyt} + P_{xyt})$$
$$- Bgd^3(\eta_{yyy} + \eta_{xxy}) = 0 \tag{4.5.8}$$

where $h = \eta + d$, $P = hU$, $Q = hV, U = \frac{1}{h}\int_{-d}^{\eta} u\mathrm{d}z$, $V = \frac{1}{h}\int_{-d}^{\eta} v\mathrm{d}z$, u and v are the velocity components in x- and y-directions, respectively; B is the correction coefficient.

The dispersion relation of the above equations reads

$$c^2 = \frac{\omega^2}{k^2} = gh\frac{1 + B(kh)^2}{1 + \left(B + \frac{1}{3}\right)(kh)^2} \tag{4.5.9}$$

When using $B = 1/15$ in Eqs. (4.5.7) and (4.5.8), the above dispersion relation including the term of $(kh)^4$ is equivalent to retaining three terms in the Padé expansion of the linear Stokes dispersion relation in Eq. (2.3.26). Figure 4.20 shows the ratio of phase velocity c/c^{Stokes}, where c is determined by Eq. (4.5.9) with different values of B, and c^{Stokes} is the linear Stokes (Airy waves) phase velocity. It reflects the dispersion characteristics of the new form of Boussinesq equations. It is seen from the figure that wave dispersion is improved significantly by using $B = 1/21$ which makes the phase velocity closer to c^{Stokes}, so that the waves could propagate to deeper water. In 1992, Madsen et al. further improved the Boussinesq equations which can be applicable to the wave propagation over a slowly varying bathymetry [33].

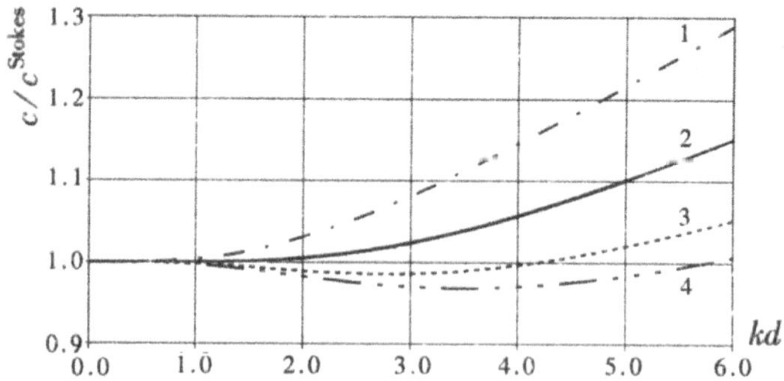

Fig. 4.20 Ratio of phase velocity c/c^{Stokes}. 1: $B = 1/10$; 2: $B = 1/15$; 3: $B = 1/21$; 3: $B = 1/25$

Nwogu [32] used the velocity $\bar{u}_\alpha$ at an arbitrary elevation z_a as the velocity variable instead of the commonly used depth-averaged velocity to obtain Boussinesq equations in the following form:

$$\eta_t + \nabla \cdot [(h + \varepsilon\eta)\bar{u}_\alpha] + \mu^2 \nabla \cdot \left\{ \left(\frac{z_\alpha^2}{2} - \frac{h^2}{6} \right) h \nabla (\nabla \cdot \vec{u}_\alpha) + \left(z_\alpha + \frac{h}{2} \right) h \nabla [\nabla \cdot (h\bar{u}_\alpha)] \right\} = 0 \quad (4.5.10)$$

$$\bar{u}_{\alpha t} + \nabla\eta + (\bar{u}_\alpha \cdot \nabla)\bar{u}_\alpha + \mu^2 \left\{ \frac{z_\alpha^2}{2} \nabla(\nabla \bar{u}_{\alpha t}) + z_\alpha \nabla[\nabla \cdot (h\bar{u}_{\alpha t})] \right\} = 0 \quad (4.5.11)$$

where z_a is a certain elevation. This Boussinesq model should be restricted to the range $0.0 < h/L_0 < 0.3$.

The embedded dispersion relation is

$$c^2 = \frac{\omega^2}{k^2} = gh \frac{1 - \left(\alpha + \frac{1}{3}\right)(kh)^2}{1 - \alpha(kh)^2} \quad (4.5.12)$$

When choosing $\alpha = -0.39$ corresponding to $z_a = 0.53h$, the dispersion relation is equivalent to retaining three terms in the Padé expansion of the linear Stokes dispersion relation in Eq. (2.3.26), which makes the waves propagate to relatively deep water.

2. Higher-order Boussinesq equations

In 1953, Serre [34] proposed a form of weakly dispersive, nonlinear Boussinesq equations with an accuracy of $O(\mu^2, \varepsilon\mu^2)$, which include the terms of $\varepsilon\mu^2$.

$$\eta_t + [(\eta + d)u]_x = 0 \quad (4.5.13)$$

$$u_t + g\eta_x + uu_x + g\eta_x + \frac{1}{h}\left[\frac{h^3}{3}GG + \frac{h^2}{2}M\right] = d_x\left[\left(\frac{h}{2}GG + M\right) - \frac{p_a}{p} + B_d^2[u_{xxt} + g\eta_{xxx}]\right] \quad (4.5.14)$$

$$GG = -u_{xt} - uu_{xx} + u_n^2, M = -d_x(u_t + uu_x) - u^2 d_{xx}$$

where d is the still water depth, and $p_a(x)$ is the surface pressure.

A comparison of the Serre equations and the basic Boussinesq Eqs. (2.3.17) and (2.3.18) had been made by Barthelemy et al. [35] who carried out a series of experiments on shoaling of solitary waves over the bottom with slopes varied within 1/76–1/60. Under the condition of small amplitude ($A/d = 0.08$–0.091), the results of different methods could hardly be distinguished. Under the condition of large amplitude with the slope of 1/60, the results by Serre equations were closer to the experimental results. With the increase of the amplitude near the breaking point, the differences between the results by the basic Boussinesq equations and the experimental results were greater. Therefore, the higher-order Boussinesq equations are better for simulating highly nonlinear waves.

Madsen and Schäffer [36] introduced a question of surface gravity waves with full nonlinearity and the parameter μ. By introducing an expansion of the velocity potential as a power series in the vertical coordinate, and together with the corresponding free surface and bottom boundary conditions, different higher-order Boussinesq-type equations in terms of different velocity variables were proposed. The equations in terms of the velocity at the still water level u_0, the depth-averaged velocity U, and depth-integrated flux P are accurate up to $O(\varepsilon^3\mu^2, \varepsilon^3\mu^4, \mu^6)$, $O(\varepsilon^3\mu^2, \varepsilon^2\mu^4, \mu^6)$, and $O(\mu^4)$, respectively.

Following the lower-order Boussinesq Eqs. (4.5.10) and (4.5.11) in terms of the velocity at an arbitrary z location derived by Nogu, Madsen and Schäffer [36] proposed a set of higher-order Boussinesq equations in terms of the velocity at an arbitrary z location with accuracy up to $O(\varepsilon^3\mu^2, \varepsilon^3\mu^4, \mu^6)$.

$$\begin{aligned}
&\eta_t + h\tilde{u}_x + \mu^2\left(\alpha + \frac{1}{3}\right)h^3\tilde{u}_{xxx} + \mu^4\frac{5}{6}\left(\alpha + \frac{2}{5}\right)^2 h^5\tilde{u}_{xxxxx} \\
&+ \varepsilon\frac{\partial}{\partial x}\left\{\eta\tilde{u} + \mu^2(\alpha h^2\eta\tilde{u}_{xx}) + \mu^4\left(\frac{5}{6}\alpha\left(\alpha + \frac{2}{5}\right)h^4\eta\tilde{u}_{xxxx}\right)\right\} \\
&- \varepsilon^2\frac{\partial}{\partial x}\left\{\frac{1}{2}\mu^2 h\eta^2\tilde{u}_{xx} + \frac{1}{2}\mu^4\left(\alpha + \frac{1}{3}\right)h^3\eta^2\tilde{u}_{xxxx}\right\} \\
&= O(\varepsilon^3\mu^2, \varepsilon^3\mu^4, \mu^6)
\end{aligned} \tag{4.5.15}$$

and

$$\begin{aligned}
&\tilde{u}_t + \eta_x + \mu^2(\alpha h^2\tilde{u}_{xxt}) + \mu^4\left(\frac{5}{6}\alpha\left(\alpha + \frac{2}{5}\right)h^4\tilde{u}_{xxxxt}\right) + \varepsilon\tilde{u}\tilde{u}_x + \varepsilon\frac{\partial}{\partial x} \\
&\left\{\mu^2\left(\alpha h^2\tilde{u}\tilde{u}_{xx} + \frac{1}{2}h^2(\tilde{u}_x) - h\eta\tilde{u}_{xt}\right) + \mu^4\left(\frac{5}{6}\alpha\left(\alpha + \frac{2}{5}\right)h^4\tilde{u}\tilde{u}_{xxxx}\right.\right. \\
&\left.\left.+\left(\alpha + \frac{1}{3}\right)h^4\tilde{u}_x\tilde{u}_{xxx} + \frac{1}{2}\alpha^2h^4(\tilde{u}_{xx})^2 - \left(\alpha + \frac{1}{3}\right)h^3\eta\tilde{u}_{xxxt}\right)\right\} \\
&+ \varepsilon^2\frac{\partial}{\partial x}\left\{\mu^2\left(-\frac{1}{2}\eta^2\tilde{u}_{xt} + h\eta(\tilde{u}_x^2 - \tilde{u}\tilde{u}_{xx})\right)\right. \\
&\left.+ \mu^4\left(\alpha h^3\eta(\tilde{u}_x\tilde{u}_{xxx} - \tilde{u}_{xx}^2) - \frac{1}{2}\alpha h^2\eta^2\tilde{u}_{xxxt} + \left(\alpha + \frac{1}{3}\right)h^3\eta(\tilde{u}_x\tilde{u}_{xxx} - \tilde{u}\tilde{u}_{xxxx})\right)\right\} \\
&= O(\varepsilon^3\mu^2, \varepsilon^3\mu^4, \mu^6)
\end{aligned} \tag{4.5.16}$$

where

$$\alpha = \frac{z}{h} + \frac{1}{2}\left(\frac{z}{h}\right)^2 \quad -\frac{1}{2} \le \alpha \le 0 \tag{4.5.17}$$

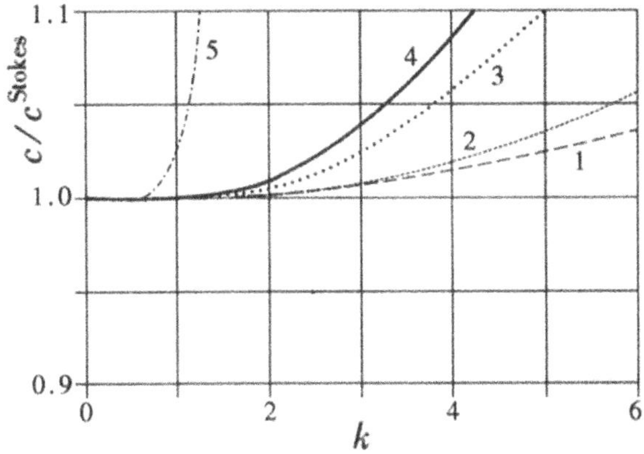

Fig. 4.21 Ratio of phase velocity c/c^{Stokes}. 1: $\alpha = -0.429648$; 2: $\alpha = -4/9$; 3: $\alpha = -2/5$; 4: $\alpha = -1/2$; 5: $\alpha = 0$

The dispersion relation is

$$\frac{\omega^2}{k^2 h} = \frac{1 - \left(\alpha + \frac{1}{3}\right)\kappa^2 + \frac{5}{6}\left(\alpha + \frac{2}{5}\right)^2 \kappa^4}{1 - \alpha\kappa^2 + \frac{5}{6}\alpha\left(\alpha + \frac{2}{5}\right)\kappa^4} \tag{4.5.18}$$

where $\kappa = \mu k h$.

Figure 4.21 shows the phase velocity ratio between Eq. (4.5.18) and Stokes's reference solution for different α. From the figure, it can be seen that the choice of $\alpha = -0.429648$ gives the most accurate linear dispersion characteristics.

In 2002, Kennedy et al. [37] derived a set of simplified higher-order Boussinesq equations with accuracy up to $O\left(\mu^4\right)$, in which the highest derivative in the terms is not greater than the third order. By choosing appropriate coefficients, the Boussinesq horizontal velocity profiles could match the linear stokes results.

There are many other forms of higher-order Boussinesq equations such as Eldeberty and Battjies [38, 39], Dingemans [40, 21], Gobbi and Kirby [41], Zou [42, 43], and some others.

3. Nonlinear characteristics

In the content of "Improvements on linear dispersion" in this section, the linear Stokes dispersion has always been compared to analyze the improvement of dispersion embedded in various Boussinesq equations. The nonlinearity characteristics of the Boussinesq-type equations can be understood by means of expanding the solutions of the equations using perturbation method and comparing them with the corresponding wave parameters and frequencies of the third-order Stokes waves. The weakly nonlinear solution of the Boussinesq equations in terms of surface elevation η and velocity U can be expressed by

$$\eta = a_1 \cos\theta + \varepsilon a_2 \cos 2\theta + \varepsilon^2 \cos 3\theta \tag{4.5.19a}$$

$$U = U_1 \cos\theta + \varepsilon U_2 \cos 2\theta + \varepsilon^2 U_3 \cos 3\theta \tag{4.5.19b}$$

$$\theta = \omega t - kx \tag{4.5.19c}$$

The surface elevation of the third Stokes waves reads

$$\eta = a_1 \cos\theta + \varepsilon a_2 \cos 2\theta + \varepsilon^2 \cos 3\theta \tag{4.5.20}$$

where a_1 is the first-order wave amplitude. The dispersion relation in the first-order approximation reads

$$\frac{\omega^2}{k^2 h} = \frac{\tanh\kappa}{\kappa}$$

where $\kappa = \mu k h$.

The third Stokes wave solution derived by Skjelbreia and Hendrickson [44] reads

$$a_2 = \frac{1}{4}\frac{a_1^2}{h}\kappa \coth\kappa\left(3\coth^2\kappa - 1\right) \tag{4.5.21a}$$

$$a_3 = \frac{3}{64}\frac{a_1^3}{h^2}\kappa^2\frac{1 + 8\cosh^6\kappa}{\sinh^6\kappa} \tag{4.5.21b}$$

$$\omega_{13} = \frac{1}{16}\frac{a_1^2}{h^2}\kappa^2\frac{9\tanh^4\kappa - 10\tanh^2\kappa + 9}{\tanh^4\kappa} \tag{4.5.21c}$$

where ω_{13} denotes a modification of the third-order amplitude on the first-order dispersion relation

$$\omega \equiv \omega\left(1 + \varepsilon^2\omega_{13}\right) \tag{4.5.22}$$

For the Boussinesq equations in terms of the velocity at an arbitrary z location, when choosing $\alpha = -0.429648$ in Eqs. (4.5.15) and (4.5.16), the expansions of a_2, a_3, and ω_{13} read

$$a_2 = \frac{3}{4}\frac{a_1^2}{h}\frac{1}{\kappa^2}\left(1 + \frac{2}{3}\kappa^2 + 0.169\kappa^4 + O\left(\kappa^6\right)\right) \tag{4.5.23a}$$

$$a_3 = \frac{27}{64}\frac{a_1^3}{h^2}\frac{1}{\kappa^4}\left(1 + \frac{5}{3}\kappa^2 + 1.496\kappa^4 + O\left(\kappa^6\right)\right) \tag{4.5.23b}$$

$$\omega_{13} = \frac{9}{16}\frac{a_1^2}{h^2}\frac{1}{\kappa^2}\left(1 + \frac{2}{9}\kappa^2 + 0.851\kappa^4 + O\left(\kappa^6\right)\right) \tag{4.5.23c}$$

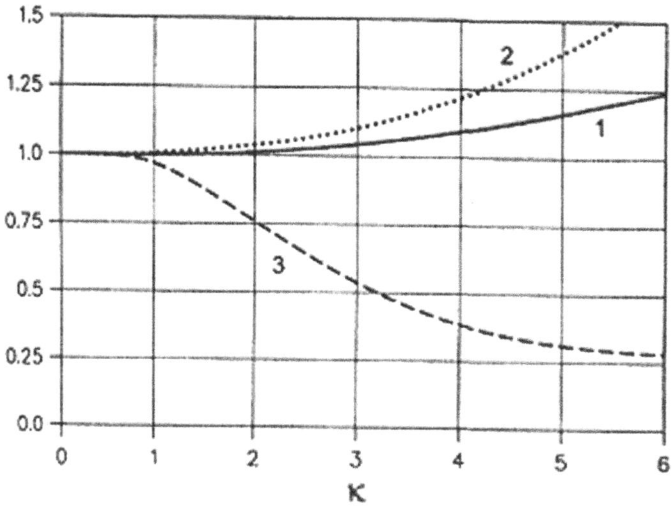

Fig. 4.22 Comparison of nonlinearity characteristics between the Boussinesq equations with $\alpha = -0.429648$ and the third-order Stokes nonlinear waves. 1: a_2/a_2^{Stokes}; 2: a_3/a_3^{Stokes}; 3: $\omega_{13}/\omega_{13}^{\text{Stokes}}$

Figure 4.22 shows a comparison of nonlinearity characteristics between the Boussinesq equations with $\alpha = -0.429648$ and the third-order Stokes nonlinear waves. It is seen from Figs. 4.21 and 4.22 that the Boussinesq equations with $\alpha = -0.429648$ have the optimal linear dispersion characteristics and also good nonlinearity characteristics.

4.5.3 Difference Scheme with Higher Accuracy for the Boussinesq Equations

A large amount of studies have been undertaken to improve the dispersion and nonlinearity of the Boussinesq equations as described above. These improvements to the equations are very important. When applying the equations, the selection of appropriate discretization method (such as difference scheme) will directly affect the accuracy of the final solution.

When the differential equations are solved by using a difference method, the accuracy of the difference equations must be at least up to the highest order of the derivatives in the differential equations. For instance, if the highest order of the derivatives in the Boussinesq equations is the third order, the accuracy of the difference for each item in the equations must be above the third order. The numerical dispersion occurs if some of the terms in the difference equations have only the second-order accuracy, which will "interfere" the physical dispersion of the original equations. Therefore, a difference scheme with higher-order accuracy for solving the

Fig. 4.23 Definition sketch

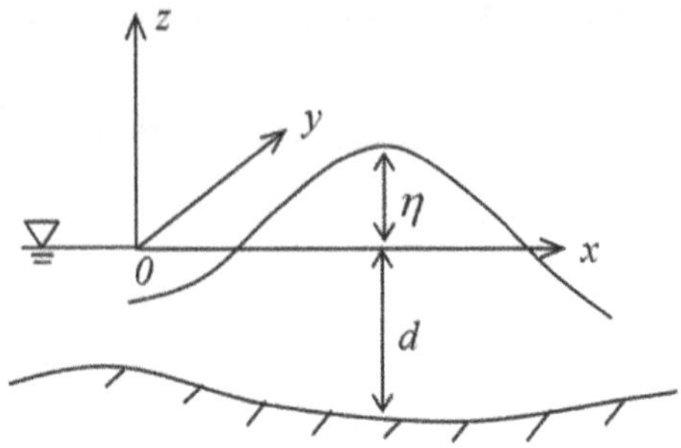

Boussinesq equations is very important, especially for the convection terms which affect the nonlinearity of the equations. The effects of the difference schemes for the Boussinesq equations on the dissipation, dispersion, nonlinearity, and shoaling by using the compact central difference scheme and the general central difference scheme are described below.

1. Governing equations

In the Cartesian coordinate system (x, y, z), the x- and y-axes are located on the still water surface, and the z-axis is perpendicular to the still water surface and positive in the upward direction. Figure 4.23 shows a sketch of the coordinate system.

The Boussniesq equations for waves over an uneven bottom derived by Schäffer and Madsen [45] read

$$\eta_t + P_x + Q_y = 0 \tag{4.5.24a}$$

$$\begin{aligned} &P_t + g(d+\eta)\eta_x + \left[\frac{P^2}{d+\eta}\right]_x + \left[\frac{PQ}{d+\eta}\right]_y + \left(\frac{1}{6} - B_2\right)d^3\left(\left[\frac{P}{d}\right]_{txx} + \left(\frac{Q}{d}\right)_{txy}\right) \\ &- \left(\frac{1}{2} + B_1\right)d^2\left(P_{txx} + Q_{txy}\right) - B_2 g d^3\left(\eta_{xxx} + \eta_{xyy}\right) \\ &- B_1 g d^2\left([d\eta_x]_{xx} + [d\eta_y]_{xy}\right) = 0 \end{aligned} \tag{4.5.24b}$$

$$\begin{aligned} &Q_t + g(d+\eta)\eta_y + \left[\frac{Q^2}{d+\eta}\right]_y + \left[\frac{PQ}{d+\eta}\right]_x + \left(\frac{1}{6} - B_2\right)d^3\left(\left[\frac{Q}{d}\right]_{tyy} + \left(\frac{P}{d}\right)_{txy}\right) \\ &- \left(\frac{1}{2} + B_1\right)d^2\left(Q_{tyy} + P_{txy}\right) - B_2 g d^3\left(\eta_{yyy} + \eta_{xxy}\right) \\ &- B_1 g d^2\left([d\eta_y]_{yy} + [d\eta_x]_{xy}\right) = 0 \end{aligned} \tag{4.5.24c}$$

where P and Q are the depth-integrated velocity components in x- and y-directions, respectively; η is the surface elevation; d is the still water depth; $h = d + \eta$ is the total water depth. By analyzing the dispersion and nonlinearity, the optimal coefficients $B_1 = 1/15$ and $B_2 \approx 0$ were found.

There are many methods to solve Eqs. (4.5.24a)–(4.5.24c) including explicit, implicit, or mixed alternating direction method, operator splitting method, finite volume method, and some others. The most widely and successfully used approaches are the alternating direction implicit (ADI) finite difference method and the finite volume method.

When using the ADI method to discrete the differential equations, one time step at any time Δt is decomposed into two steps. In the first half time step, the x-component momentum equation and the continuity equation are simultaneously solved for P and η by using implicit scheme to obtain P^{n+1} and $\eta^{n+1/2}$. In the second half time step, the y-component momentum equation and the continuity equation are simultaneously solved for Q and η by using implicit scheme to obtain $Q^{n+3/2}$ and η^{n+1}. As the equations are solved implicitly only in one direction in each half time step, the computation is very efficient and the stability of the scheme is also good.

The Boussniesq equations have terms of the third-order derivatives, which result in dispersion. In order to avoid the situation that the physical dispersion is masked by the non-physically numerical dispersion because of using a scheme with accuracy less than the third-order, a difference scheme with higher-order accuracy should be applied to discrete the Boussinesq equations. In the previous studies, the second-order central scheme was commonly used to discrete the temporal and spatial derivatives. After that, the Taylor expansion was applied for correction in order to obtain the difference equations with the third-order accuracy in both time and space. As the nonlinear terms in the governing equations are sensitive to the nonlinearity and dispersion, it is very important to increase the difference accuracy of the nonlinear terms. Normally, the improvement of the difference accuracy makes the number of grid points involved to increase, which increases the difficulty of solving the equations, especially for the treatment of boundary points. For this reason, a compact scheme had been introduced to discrete the nonlinear terms. By using this scheme, the difference accuracy on space can reach the fourth order but the number of the involved grid points is not increased.

1. Difference equations constructed with a compact scheme

A compact scheme is described briefly in the following.

Let $u(x)$ be a function. $u(x_i) = u_i$ is the function value at the grid point x_i. The cubic spline function $S_\Delta(u, x) = S_\Delta(x)$ satisfies the continuity of both the first-order and the second-order derivatives in the interval $[a, b]$. Meanwhile, it can be written as a cubic polynomial in each small interval $x_{i-1} \le x \le x_i$ and satisfies $S_\Delta(u_i, x_i) = u_i$. The function $u(x)$ is $C^4[a, b]$. It is assumed that $m_i = S'_\Delta(x_i)$ and $M_i = S''_\Delta(x_i)$. The two equations on the first- and second-order derivatives m and M can be derived, which are the two basic formulas for construction of the compact difference scheme (grid spacing is equidistant).

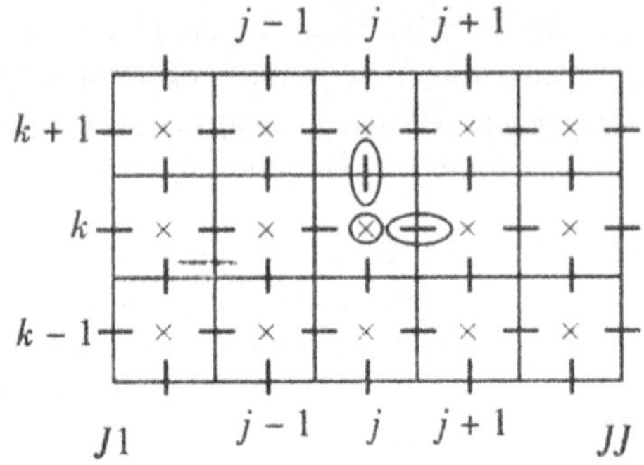

Fig. 4.24 Grid notation. —: P, |: Q, × : η and d

$$m_{i-1} + 4m_i + m_{i+1} = \frac{3}{\Delta x}(u_{i+1} - u_{i-1}) + O\big((\Delta x)^4\big) \tag{4.5.25a}$$

$$M_{i-1} + 4M_i + M_{i+1} = \frac{6}{(\Delta x)^2}(u_{i+1} - 2u_i + u_{i-1}) + O\big((\Delta x)^2\big) \tag{4.5.25b}$$

A staggered grid system is used as shown in Fig. 4.24 for the grid notation.

The continuity Eq. (4.5.24a) is discrete by using FTCS scheme and modified to be accurate up to the third order. The difference central point is located at (j, k), i.e., $\eta_{j,k}^{n+1/2}$. P and Q are located at $(j + 1/2, k)$ and $(j, k + 1/2)$, respectively, i.e., $P_{j+1/2,k}^{n+1/2}$ and $Q_{j,k+1/2}^{n+1/2}$ as shown in Fig. 4.24. The difference equation reads

$$A_{1j} P_{j,k}^{n+1} + B_{1j} \eta_{j,k}^{n+1/2} + C_{1j} P_{j+1,k}^{n+1} = D_{1j} \tag{4.5.26}$$

where A_{1j}, B_{1j}, C_{1j}, and D_{1j} are known coefficients.

The finite difference of the x-component momentum equation is taken as an example here. The temporal derivatives are discretized with the second-order central scheme and then modified to be accurate up to the third order. The central compact difference scheme is used for the spatial derivatives. The finite difference of Eq. (4.5.24b) is described in the following. The difference centers for the variables are $P_{j+1/2,k}^{n+1/2}$, $Q_{j,k+1/2}^{n+1/2}$, and $\eta_{j,k}^{n+1/2}$ as shown in Fig. 4.24.

In order to apply the compact difference scheme formulas Eqs. (4.5.25a) and (4.5.25b), performing the following combination of the momentum Eq. (4.5.24b),

$$(4.5.24\text{b})_{j+3/2,k} + 4 \times (4.5.24\text{b})_{j+1/2,k} + (4.5.24\text{b})_{j-1/2,k} = 0 \tag{4.5.27}$$

Substituting Eq. (4.5.24b) into Eq. (4.5.27) yields

$$
\begin{aligned}
&\left(\frac{P^{n+1}-P^{n}}{\Delta t}\right)_{j+3/2,k}+4\left(\frac{P^{n+1}-P^{n}}{\Delta t}\right)_{j+1/2,k}+\left(\frac{P^{n+1}-P^{n}}{\Delta t}\right)_{j-1/2,k}\\
&+(F_x)_{j+3/2,k}+4(F_x)_{j+1/2,k}+(F_x)_{j-1/2,k}+(G_y)_{j+3/2,k}\\
&+4(G_y)_{j+1/2,k}+(G_y)_{j-1/2,k}+H_1+\left(\frac{1}{6}-B_2\right)[H_2-H_3+H_4-H_5]\\
&+\left(\frac{1}{2}-B_1\right)[H_6-H_7+H_8-H_9]-B_2g[H_{10}+H_{11}]\\
&-B_1g[H_{12}+H_{13}]=0
\end{aligned}
\tag{4.5.28}
$$

where $F = P^2/(d+\eta)$; $G = PQ/(d+\eta)$; H_1 to H_{13} denote the finite differences of the second-order and third-order terms.

Based on Eq. (4.5.25a), the three terms related to F in Eq. (4.5.28) can be rewritten as

$$
\begin{aligned}
(F_x)_{j+3/2,k}+4(F_x)_{j+1/2,k}+(F_x)_{j-1/2,k}=\frac{3}{\Delta x}\Bigg(&\frac{P^{n}_{j+3/2,k}\cdot P^{n+1}_{j+3/2,k}}{(d+\eta)_{j+3/2,k}}\\
&-\frac{P^{n}_{j-1/2,k}\cdot P^{n+1}_{j-1/2,k}}{(d+\eta)_{j-1/2,k}}\Bigg)
\end{aligned}
\tag{4.5.29}
$$

and by substituting Eq. (4.5.25a) or Eq. (4.5.25b) into the other remaining terms in Eq. (4.5.28), the following difference equations can be obtained.

For the interior points, the central compact difference scheme is used for H_1, and the difference equation reads

$$
\begin{aligned}
&S_{\text{moj}}\eta^{n+1/2}_{j-1,k}+P_{\text{moj}}P^{n+1}_{j-1/2,k}+A_{\text{moj}}\eta^{n+1/2}_{j,k}+B_{\text{moj}}P^{n+1}_{j+1/2,k}+C_{\text{moj}}\eta^{n+1/2}_{j+1,k}\\
&+R_{\text{moj}}P^{n+1}_{j+3/2,k}+T_{\text{moj}}\eta^{n+1/2}_{j+2,k}=D_{\text{moj}}
\end{aligned}
\tag{4.5.30}
$$

where S_{moj}, P_{moj}, A_{moj}, B_{moj}, C_{moj}, R_{moj}, T_{moj}, and D_{moj} are known coefficients.

On the left boundary $j = J1$, the difference center is located at $J1 + 1/2$. The eccentric compact difference scheme is used for H_1, and the difference equation reads

$$
P_{\text{moj}}P^{n+1}_{j-1/2,k}+A_{\text{moj}}\eta^{n+1/2}_{j,k}+B_{\text{moj}}P^{n+1}_{j+1/2,k}+C_{\text{moj}}\eta^{n+1/2}_{j+1,k}+R_{\text{moj}}P^{n+1}_{j+3/2,k}+T_{\text{moj}}\eta^{n+1/2}_{j+2,k}=D_{\text{moj}}
\tag{4.5.31}
$$

where P_{moj}, A_{moj}, B_{moj}, C_{moj}, R_{moj}, T_{moj}, and D_{moj} are known coefficients.

Similar to the left boundary, on the right boundary $j = JJ$, $JJ - 3/2$ is the difference center. The difference equation reads

$$
\begin{aligned}
&S_{\text{moj}}\eta^{n+1/2}_{j-1,k}+P_{\text{moj}}P^{n+1}_{j-1/2,k}+A_{\text{moj}}\eta^{n+1/2}_{j,k}+B_{\text{moj}}P^{n+1}_{j+1/2,k}+C_{\text{moj}}\eta^{n+1/2}_{j+1,k}\\
&+R_{\text{moj}}P^{n+1}_{j+3/2,k}=D_{\text{moj}}
\end{aligned}
\tag{4.5.32}
$$

where S_{moj}, P_{moj}, A_{moj}, B_{moj}, C_{moj}, R_{moj}, and D_{moj} are known coefficients.

P and η are the variables of the Eqs. (4.5.26), (4.5.30), (4.5.31), and (4.5.32). By eliminating η through the continuity equation, the five-diagonal equations on P can be obtained so that P at each point is solved. Then, η at each point is obtained by substituting P into the continuity equation.

Next, the variables and parameters in the Eqs. (4.5.26), (4.5.30), (4.5.31), and (4.5.32) are altered in the following ways:

$P \to Q, \Delta x \to \Delta y, \Delta y \to \Delta x, n \to n+1/2, n+1/2 \to n+3/2, n-1/2 \to n, j \to k, k \to j, x \to y$. Then, the difference equations for y-direction can be obtained. According to the above difference equations, the convective terms can be accurate up to the fourth order. With correction on the equations by means of the Taylor expansion, each term of the equations has accuracy up to the third- or fourth-order in time and space.

3. Numerical results of the Boussinesq equations in compact difference scheme and analysis

Numerical calculations of the Boussinesq equations in compact difference scheme had been carried out. The improvements on dissipation, dispersion, and nonlinearity had been analyzed and compared with the results using the general central difference scheme [10].

(1) Dissipation

The calculation case is wave propagation in an area with constant water depth. The still water depth is $d = 10$ m, the incident wave height is $H = 1$ m, and the wave period is $T = 10$ s. The calculation domain is 25 km × 10 km. Figure 4.25 shows the calculated surface elevation compared with the analytical solution. In the 25 km long calculation domain, there is no dissipation in the wave propagation in general.

(2) Dispersion

Figure 4.26 shows the ratios of the phase velocity to the Airy wave phase velocity by using different methods, including the Airy wave theory, the dispersion relation

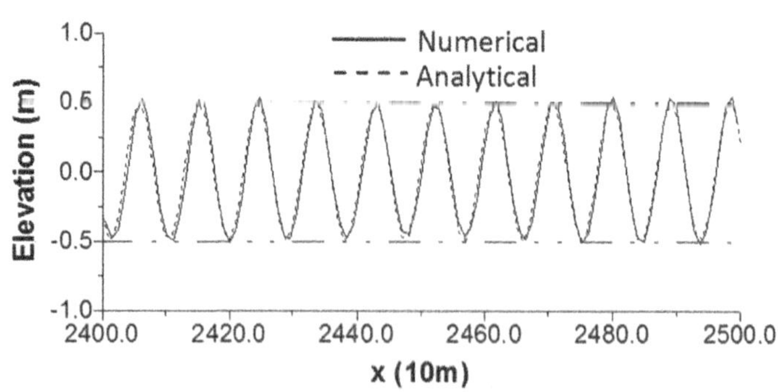

Fig. 4.25 Numerical surface elevation compared with the analytical solution in the local area of 24–25 km

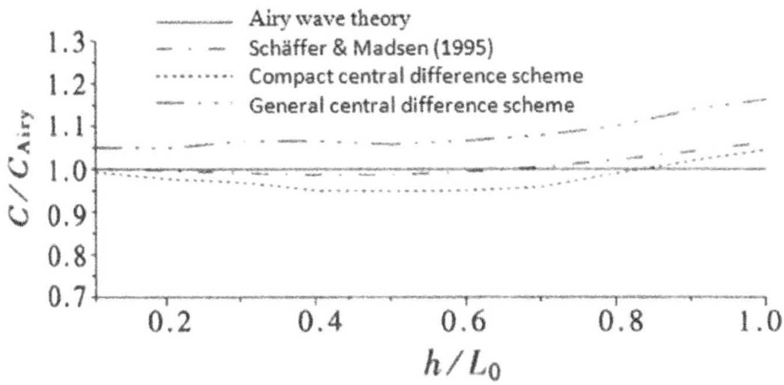

Fig. 4.26 Comparison of C/C_{Airy}

embeds in the Boussinesq equations derived by Schäffer and Madsen [45], and the two numerical results by solving the Boussinesq equations with the compact central difference scheme and the general central difference scheme. The numerical results in the figure are obtained by using numerical models to simulate a sinusoidal wave propagating in a long distance with constant water depth, counting the wavelength and then solving the wave celerity. It is seen from the figure that there is a significant difference between the two curves of the numerical models using the compact central difference scheme and the general central difference scheme. By comparing these two results with the solution of the Airy wave theory, it is clear that the dispersion of the numerical model using the compact central difference scheme is better.

(3) Nonlinearity

The calculation case is wave propagation over a submerged breakwater as shown in Fig. 4.27a. The incident waves propagate from the left to the right. The wave height is 2 cm, and the wave period is 2 s. Seven sections have been picked before and after the breakwater along the incident wave direction. Figure 4.27b–h show comparisons of the time series of surface elevation at each section by different methods including the experimental results given by Beji and Battjes [46], the two numerical results by solving the Boussinesq equations using the compact central difference scheme and the general central difference scheme, and a linear wave model. Due to the bathymetry change caused by the submerged breakwater, the nonlinearity and the wave shape change significantly as the wave propagates. As shown in Fig. 4.27b for the initial section of the breakwater, the numerical results of the surface elevation simulated by the linear and the nonlinear models are almost the same. As shown in Fig. 4.27c–h, there are significant differences between the linear model and the nonlinear models when the bathymetry changes. The time series of surface elevation simulated by the nonlinear models matches with the experimental results. By further comparison of the nonlinear models using the two different schemes, it is found that the model results using the compact central scheme are closer to the experimental results than using the general central scheme.

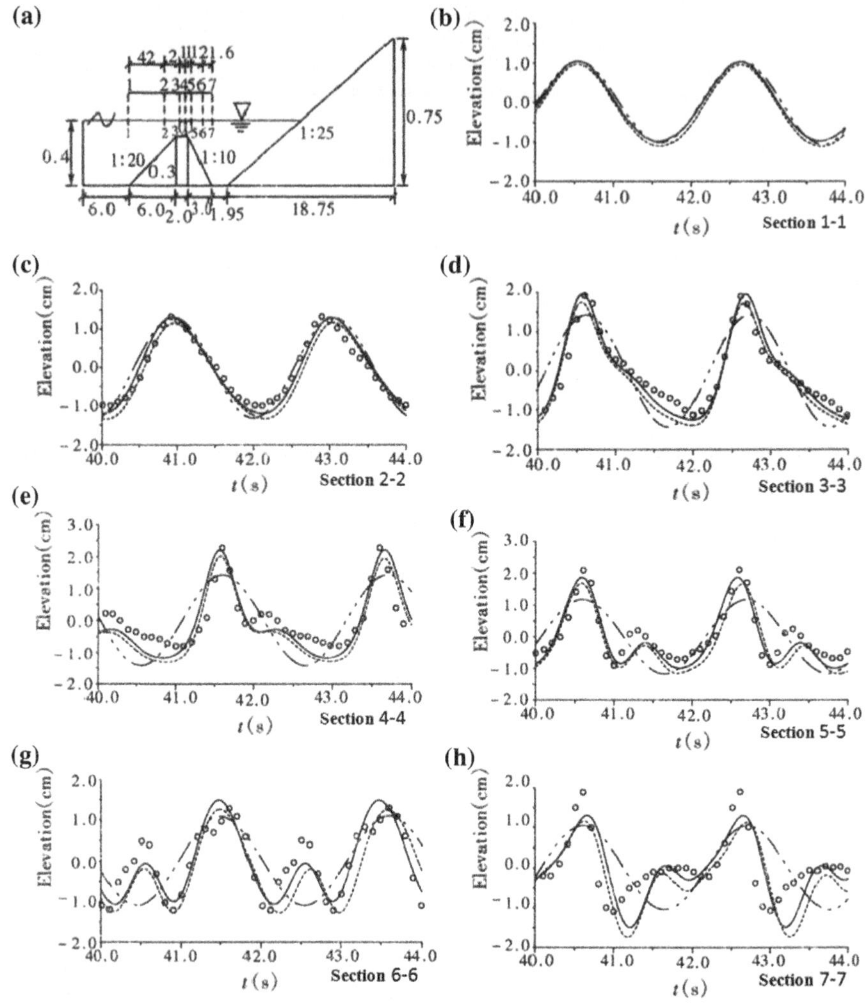

Fig. 4.27 Time series of surface elevation at different locations before and after the submerged breakwater. ○: Experimental results by Beji and Battjes [46]; ——: Numerical results in the compact central difference scheme;- - - -: Numerical results in the general central difference scheme; _._.Linear wave model results

References

1. Berkhoff JCW. Computation of combined refraction diffraction. Proceedings of the 13th international coastal engineering conference. New York: ASCE; 1972. p. 471–90.
2. Luke JC. A variational principle for a fluid with a free surface. J. Fluid Mech. 1967;27:393–7.
3. Radder AC. On the parabolic equation method for water-wave propagation. J. Fluid Mech. 1979;95(1):159–76.
4. Boussinesq MJ. Théorie des ondes et des ramous qui se propagent le long d'un canal rectangulaire horizontal, en communiquant au liquide contenu dans ce canal des vitesses sensiblement

pareilles de la surface au found. Journal de Mathématique Pures et Appliquées, deuxième sérrie. 1872;17:55–108.
5. Peregrine DH. Long waves on a beach. J. Fluid Mech. 1967;27:815–27.
6. Stoker JJ. Water waves. New York: InterScience Publishers Inc.; 1957.
7. Numerical wave modelling study on outer channel regulation project of Huanghua Port for Shenhua Group. Technical report in Department of Mechanics. Tianjin: Tianjin University; 2004.
8. Tao J, Long W. The critical angle of wave refraction and multiple refraction by harbor approach channel and basin. In: 29th IAHR Conference, Sept 2001. Beijing. p. 472–478.
9. Sommerfeld A. Partial differential equations in physics. New York: Academic Press; 1949.
10. Qin W. Improvement of mathematical model of the Boussinesq equations and the engineering applications. Dissertation for Master degree, Tianjin: Tianjin University; 2003.
11. Goda Y. Random seas and design of maritime structures. Tokyo: University of Tokyo Press; 1985.
12. Xia Z. Numerical simulation and experimental study on the combined refraction-diffraction of irregular waves. Dissertation for Master degree, Tianjin: Tianjin University; 1994.
13. Tao J, Zhao J. Mathematical model and graphical display of wave diffraction. Technical report in Department of Mechanics. Tianjin: Tianjin University; 1988.
14. Copeland GJM. A practical alternative to the mild-slope wave equation. Coast. Eng. 1985;9:125–49.
15. Kirby JT, Dalrymple RA. A parabolic equation for the combined refraction-diffraction of Stokes waves by mildly varying topography. J. Fluid Mech. 1983;136:453–66.
16. Kirby JT, Dalrymple RA. Verification of a parabolic equation for propagation of weakly-nonlinear waves. Coast. Eng. 1984;8:219–32.
17. Liu PL-F, Tsay TK. Refraction-diffraction model for weakly nonlinear water wave. J. Fluid Mech. 1984;141:265–74.
18. Jianhua Tao, Jincheng Zhao. Calculation of the wave diffraction in the harbor by using boundary element method. Port Eng. Technol. 1987;1987(4):17–22.
19. Booij N. Gravity waves on water with non-uniform depth and current. Report 81-1, Department of Civil Engineering, Delft University of Technology; 1981.
20. Dalrymple RA, Kirby JT, Paul AH. Wave diffraction due to areas of energy dissipation. J. Waterway Port Coast. Ocean Eng. 1984;110(1):67–78.
21. Dingemans MW. Water wave propagation over uneven bottoms. World Scientific Published; 1997.
22. Karpman VI. Non-lincar waves in dispersive media. Oxford: Pergamonn Press; 1975. p. 183.
23. Yue DK, Mei CC. Forward diffraction of Stokes' waves by a thin wedge. J. Fluid Mech. 1980;99:33–55.
24. Zhang H. Mathematical study on the medium-term coastal evolution of sand beach. Dissertation for Master degree. Tianjin: Tianjin University; 1997.
25. Liu PL-F. Wave transformation, in the sea. In: Le Me Haute B, Hanes DM, editors, vol. 9. Wiley Interscience; 1990. p. 27–63.
26. Kirby JT. Higher-order approximations in the parabolic equation method for water waves. J. Geophys. Res. 1986;91(C1):933–52.
27. Jonsson IG, Carlsen NA. Experimental and theoretical investigations in an oscillatory rough turbulent boundary layer. J. Hydraul. Res. 1976;14:45.
28. Swart DH. Offshore sediment transport and equilibrium beach profiles. Publication No. 131, Delft Hydraulics Laboratory, Delft, The Netherlands; 1974.
29. Gao Q, Radder AC. A refraction-diffraction model for irregular waves. In: Proceedings of 26th coastal engineering conference ASCE. Copenhagen; 1998. p. 367–379.
30. Berkhoff JCW, Booij N, Radder AC. Verification of numerical wave propagation models for simple harmonic water waves. Coast. Eng. 1982;6:255–79.
31. Madsen PA, Murray R, Sørensen OR. A new form of the Boussinesq equations with improved linear dispersion characteristics. Coast. Eng. 1991;1991(15):371–88.

32. Nwogu OG. Alternative form of Boussinesq equations for nearshore wave propagation. J. Waterway, Port, Coast. Ocean Eng. 1993;119(6):618–38.
33. Madsen PA, Sorensen OR. A new form of the Boussinesq equations with improved linear dispersion characteristics. Part 2: a slowly-varying bathymetry. Coast. Eng. 1992;18:183–204.
34. Serre F. Contribution à L'étude des écoulements permanents et variables dans les canaux. La Houille Blanche. 1953;8:374–88.
35. Bathelemy E, Guibourg S. A theoretical and experimental comparison of the Boussinesq and Serre equations. MAST G8M overall workshop, Grenoble; 1993.
36. Madsen PA, Schäffer HA. Higher-order Boussinesq-type equations for surface gravity waves: derivation and analysis. Phil. Trans. R. Soc. Lond. A. Printed Great Britain. 1998;356:3123–84.
37. Kennedy AB, Kirby JT, Gobbi MF. Simplified higher-order Boussinesq equations I: linear simplifications. Coast. Eng. 2002;44:205–29.
38. Eldeberky Y, Battjes JA. Phase lock in waves passing over a bar. In: Issacson M, Quick M, editors. Proceedings of international symposium: waves—physical and numerical modelling, vol. II. Vancouver; 1994. p. 1086–95.
39. Eldeberky Y, Battjies JA. Spectral modeling of wave breaking: application to Boussinesq equations. J Geophys Res. 1996;102:1253–64.
40. Dingmans MW. Comparison of computations with Boussinesq-like models and laboratory measurements. MAST-G8M note, H1684, Delft Hydraulics; 1994. 32 p.
41. Gobbi MF, Kirby JT. A forth order Boussinesq-type model. In: Proceedings of 25 international conference on coastal engineering. Orlando. p. 1116–1129.
42. Zhili Zou. Higher-order Boussinesq equations with strong currents. Acta Oceanol Sin. 2000;22(4):41–50.
43. Zou Z. Higher-order Boussinesq equations for rapidly varying topography. Acta Oceanologica Sinica. 2001;23(1):109–19.
44. Skjelbreia L, Henderson JA. Fifth order gravity wave theory. In: Proceedings of 7th conference on coastal engineering. ASCE; 1961. p. 184–196.
45. Schäffer HA, Madsen PA. Further enhancement of Boussinesq-type equations. Coast. Eng. 1995;26:1–15.
46. Beji S, Battjies JA. Experimental investigations of wave propagateion over a bar. Coast. Eng. 1993;19:151–62.

Chapter 5
Numerical Simulation of Wave Run-up and Breaking on Beach

When waves propagate from deep sea to coastal regions, wave shoaling occurs and eventually waves will break into breaking waves. The area between the breaker line and the limit of wave uprush is called the surf zone, where the current and wave motions are complex. Wave setup, wave run-up, and nearshore currents (including longshore current and rip currents) are the dominant factors affecting sediment and pollutant transports there. Hence, it is very important to reproduce wave and current fields and determine the location of the breaker line in numerical models. However, because of the complexity of wave breaking process, mechanisms of wave breaking are not fully understood. Wave breaking is accompanied by a large amount of wave energy dissipation, and various assumptions are normally applied for the models to simulate wave breaking at present.

Wave run-up and wave breaking on the beach are critical issues in the coastal engineering design. Two important matters have to be resolved in the numerical simulation of the two physical processes. One is the treatment of moving boundary on the beach, and the other is the simulation of energy dissipation due to the wave breaking.

The wave run-up on the beach can be classified into two cases: non-breaking and breaking. When waves propagate over a steep slope, they do not break but reflect. If the bottom friction is ignored, the wave run-up in front of the vertical wall is equal to the incident wave height, and the total wave height becomes twice of the incident wave height. If the beach slope is gentle, the waves may break. In this case, it is necessary to deal with the moving boundary and also to simulate wave breaking. Therefore, normally it is needed to simulate the wave breaking and wave run-up on the beach simultaneously.

J. Tao, *Numerical Simulation of Water Waves*, Springer Tracts in Civil Engineering, https://doi.org/10.1007/978-981-15-2841-5_5

5.1 A Treatment of Moving Boundary on Beach—The Slot Method

The slot method was initially proposed by Tao Jianhua (the author of this book) when she worked at Danish Hydraulic Institute (DHI) as a visiting scholar in 1982. It was for the treatment of moving boundary in the wave model packages S21 and MK8 (Boussinesq equation-based models). This method allows the two numerical models to simulate wave propagation and breaking in the near shore [1, 2].

The slot method was applied to simulate the 2D flow with transient boundary by He and Wang [3] in China. Wang et al. [4] used this method to treat the moving boundary and moving sources in the numerical simulations of tidal current and pollutant distribution in Bohai Bay. Madsen and Sorensen [5, 6, 7] applied the slot method to the treatment of moving boundary for the advanced Boussinesq equations. Kennedy et al. [8] improved the slot expression for the treatment of moving boundary for the Boussinesq equations. As an approach to dealing with the moving boundary of wave motion, the slot method has been widely used.

5.1.1 *Basic Idea of the Slot Method*

The basic idea of the slot method is assuming that a very narrow "slot" existing in unit width of the beach has the water depth consistent with the coastal water depth or below the lowest water level, which makes the water in the slots connect with the coastal water. Therefore, the boundary of the computational domain can be located in the narrow slots inside the beach, so as to make the moving boundary be a fixed boundary with a certain water depth. With the existence of narrow slots, the water has been extended into the slots. Then it is not needed to judge and change the boundary location on the beach. Meanwhile, because of the very narrow width of the slots inside the beach, it has been proved that the existence of slots has little impact on the wave activities on the beach [1].

Figure 5.1 shows a sketch of the narrow slot. The origin of the z-coordinate is located on the still water level. An exponential function is used to define the slot width in order to ensure that the cross section is continuously smooth.

$$f(\eta) = \begin{cases} \varepsilon + (1-\varepsilon)\mathrm{e}^{\alpha(\eta - z^*)}, & \eta \le z^* \\ 1, & \eta \ge z^* \end{cases} \tag{5.1.1}$$

where $f(\eta)$ is the width coefficient (the slot width in unit width of beach); η is the surface elevation; z^* is the elevation of the seabed, and the slot width is 1 for the level above z^*; ε is the slot width parameter (the slot width for the infinitely deep); α is the shape smoothing parameter; d is the still water depth.

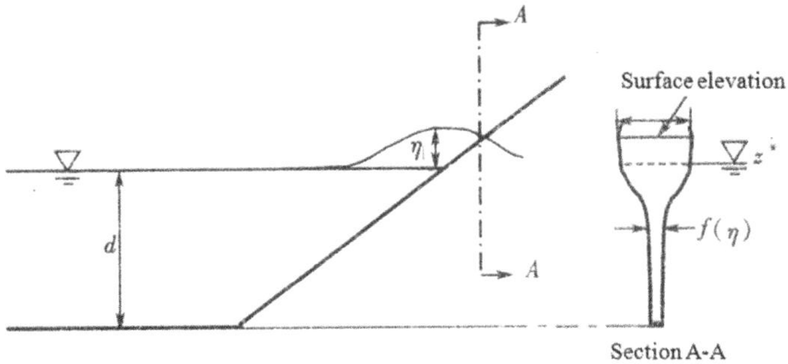

Fig. 5.1 Sketch of a narrow slot

According to this width definition, the cross-sectional area of the slot is obtained as

$$A(\eta) = \int_{-d}^{\eta} f(\eta)\mathrm{d}\eta = \begin{cases} \varepsilon(\eta + d) + \dfrac{1-\varepsilon}{\alpha}\mathrm{e}^{-\alpha(d+z^*)}\left[\mathrm{e}^{\alpha(\eta+d)} - 1\right] & \eta \le z^* \\ \eta - z^* + \varepsilon\left(d + z^*\right) + \dfrac{1-\varepsilon}{\alpha}\left[1 - \mathrm{e}^{-\alpha(z^*+d)}\right] & \eta \ge z^* \end{cases} \tag{5.1.2}$$

The continuity and momentum equations need to be changed to comply with the narrow slots. Here, the 1D Boussinesq equations with narrow slots are presented.

The continuity equation reads

$$\frac{\partial A}{\partial t} + \frac{\partial P}{\partial x} = 0 \tag{5.1.3}$$

where P is the discharge per unit width; A is the wetted area.

$$\frac{\partial A}{\partial t} = \frac{\partial}{\partial t}\int_{-d}^{\eta} f(\eta)\mathrm{d}\eta = F\frac{\partial h}{\partial t} \tag{5.1.4}$$

$$F = f(\eta) \tag{5.1.5}$$

Thus, the continuity equation with narrow slots reads

$$F\frac{\partial h}{\partial t} + \frac{\partial P}{\partial x} = 0 \tag{5.1.6}$$

The momentum equation:

As the shape of water in the original unit width near the slope is no longer regular with the existence of narrow slots, the water depth h is replaced by the slot cross-sectional area A in the Boussinesq Eq. (4.5.4) to make it turn into

$$\frac{\partial P}{\partial t} + \frac{\partial}{\partial x}\left(\frac{P^2}{A}\right) + gA\frac{\partial \eta}{\partial x} + g\frac{|P|P}{AC^2R} = E\frac{\partial^2 P}{\partial x^2} + \frac{1}{3}A^2\frac{\partial^3 P}{\partial x^2 \partial t} \tag{5.1.7}$$

where C is the Chezy number, R is the hydraulic radius, E is the comprehensive viscosity coefficient.

5.1.2 Selection of the Slot Parameters

1. The slot width parameter ε

To demonstrate the applicability of this approach, the effects of the slots on the wave activities need to be investigated.

Figure 5.2 shows a process of solitary wave reflection by a vertical wall with the existence of narrow slots calculated by a Boussinesq equation model with narrow slots. It is found that the surface elevation at the vertical wall has the maximum value (around twice of the incident wave height) at the calculation time step $N = 69$, meanwhile the wave propagates into the slots. It is clear that the solitary wave is reflected by the vertical wall at $N = 90$, and at the same time, the wave propagates forward in the slots. These indicate that the characteristics of the water wave reflection in front of the vertical wall remain unchanged with the narrow slots.

Figure 5.3 shows a comparison of the reflected wave surface elections without and with narrow slots (with different ε) in the vertical wall. As ε decreases, the reflected wave gradually approaches to the case without slot. When $\varepsilon = 0.02$, the reflected wave surface election is very close to the result without slot. It clearly demonstrates that the existence of narrow slots does not affect the wave activities near the beach (and on the beach) when the slot width parameter ε is small enough.

2. The slot length

The slot length (the grid number in the slot for the calculation) is also an important parameter. The long slot length means a large grid number, which needs increasing

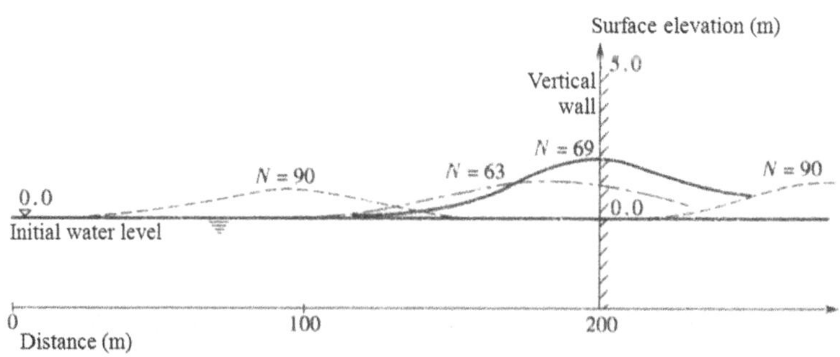

Fig. 5.2 Solitary wave reflection by a vertical wall and wave propagation in the narrow slots

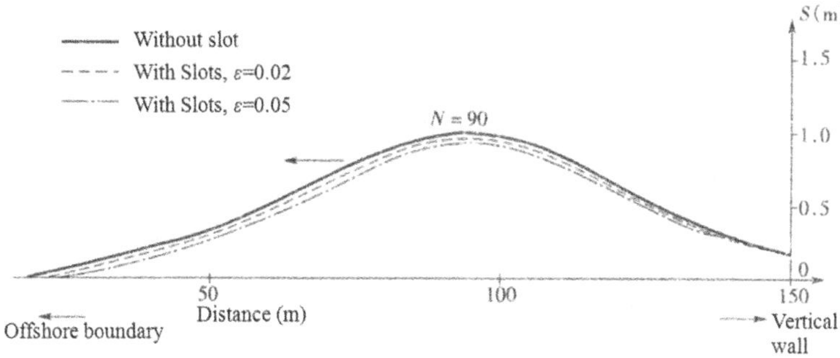

Fig. 5.3 Comparison of the reflected wave surface elections for different ε

computation time. While the grid number is too small, the wave in the slots might be reflected by the boundary in the slots, which affects the wave motion significantly. Therefore, the slot length should be properly chosen. Normally, sponge layers can be applied at the slot boundary as a numerical wave absorber to reduce the wave reflection in the slots.

3. Shape smoothing parameter

The determination of the shape smoothing parameter α is to ensure a smooth transition of the cross-sectional shape from a unit width to a narrow slot, and it should make the cross-sectional area below the beach elevation reduced quickly. From the results of numerous numerical experiments, it is found that $\alpha = 2.3$ is fairly appropriate.

The slot width was been modified by Kennedy et al. [8]; the new expression is

$$f(\eta) = \begin{cases} 1 & \eta \geq z^* \\ \varepsilon + (1-\varepsilon)\exp[\beta_{\mathrm{r}}(\eta - z^*)/d] & \eta \leq z^* \end{cases} \tag{5.1.8}$$

where β_{r} is equivalent to α in Eq. 5.1.1.

The difference between Eqs. 5.1.8 and 5.1.1 is that the exponential term divided by the still water depth d makes it turn to a dimensionless term. The expression of the corresponding cross-sectional area of the slot reads

$$A(\eta) = \begin{cases} (\eta - z^*) + \varepsilon(z^* + d) + \frac{(1-\varepsilon)d}{\beta_r}\left[1 - \mathrm{e}^{-\beta_r(1+z^*/d)}\right] & \eta \geq z^* \\ \varepsilon_\gamma(\eta + d) + \frac{(1-\varepsilon)d}{\beta_r}\mathrm{e}^{\beta_r(\eta - z^*)/d}\left[1 - \mathrm{e}^{-\beta_r(1+\eta/d)}\right] & \eta \leq z^* \end{cases} \tag{5.1.9}$$

The slot parameters in the above expressions had been studied by Qin [9], who presented a comparison of the slot cross-sectional profiles between different ε_r and β_r for $d = -10.0$ m and $z^* = -4.0$ m as shown in Fig. 5.4. The transition of the cross-sectional shape is smooth for $\beta_r = 20$, which corresponds to the case of $\alpha = 2.0$ in Eq. 5.1.1, and is closed to $\alpha = 2.3$ proposed by Tao (the writer of this book) [1]. If α is too large, the rapid change of the cross section from a unit width to a narrow slot may cause instability.

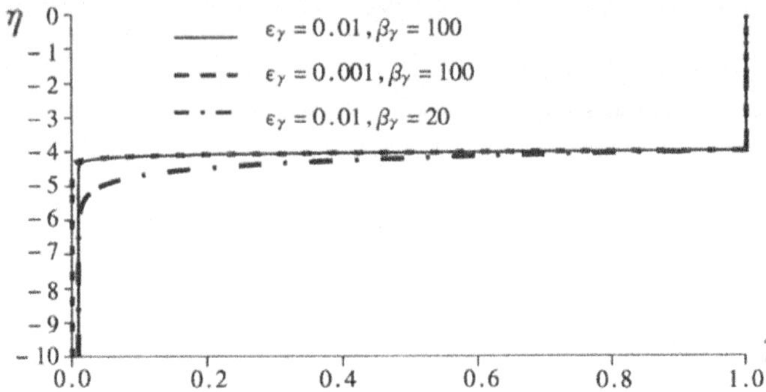

Fig. 5.4 Slot cross-sectional profiles for different ε_r and β_r

5.1.3 Numerical Simulation of Wave Run-up on a Beach

Figure 5.5 shows an example of wave run-up on a beach using the slot method. The incident wave is a solitary wave, and the beach slope is 1/10. The process of solitary wave run-up and run-down has been clearly shown by the numerical results (the wave motion in the slots is not shown here).

For the 2D wave problems, it can be assumed that the slots in the x- and y-directions are independent of each other, and the wave motion in the slots is one dimensional. Similar to Eq. (5.1.7), the 2D Eqs. (4.5.24a)–(4.5.24c) should be modified correspondingly for the existence of narrow slots [1].

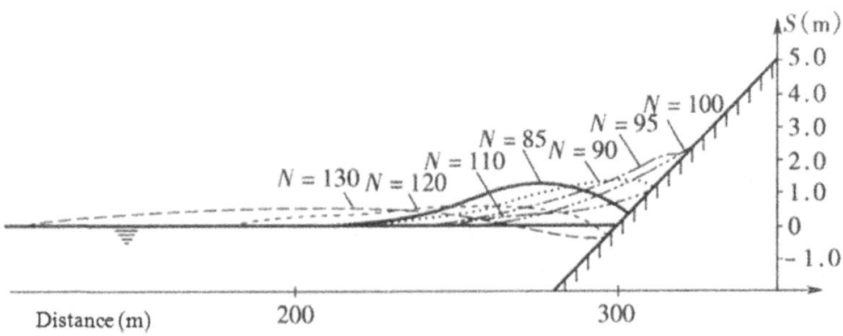

Fig. 5.5 The process of solitary wave run-up on the beach

5.2 Wave Breaking Criteria

5.2.1 Classical Wave Breaking Criterion

Stokes [10] proposed that, when the horizontal particle velocity on the wave crest is equal to the phase velocity of wave propagation, the waveform cannot be stable and the wave breaks. Accordingly, the wave breaks when the crest angle reaches 120°, which is the so-called classical wave breaking criterion. Based on this criterion, Michell [11] proposed that, when the deepwater wave steepness *H/L* reaches 0.142, wave breaking occurs. Thereafter, Miche [12] suggested a wave breaking criterion on the wave steepness for shallow water waves,

$$(H/L)_b = 0.142 \tanh kd \tag{5.2.1}$$

Li and Dong [13] made a further study on the wave breaking criteria and proposed that the constant of 0.142 in Eq. (5.2.1) should be 0.128 for irregular waves when the seabed slope is less than 1/50.

5.2.2 Criterion of Ultimate Wave Height

In shallow water areas, the relative wave height is commonly used as a wave breaking index, and it is called the criterion of ultimate wave height. McGowan [14] derived a wave breaking index on a beach using the solitary wave theory,

$$\gamma_b = \frac{H_b}{d_b} = 0.78 \tag{5.2.2}$$

where the subscript "*b*" denotes the breaking point. According to a study by Goda [15], the ultimate wave height can be expressed as

$$\gamma_b = \frac{H_b}{d_b} = B\left\{1 - \exp\left[-1.5\frac{\pi d_b}{L_0}\left(1 + 15(\tan\theta)^{4/3}\right)\right]\right\} \Big/ \left(\frac{d_b}{L_0}\right) \tag{5.2.3}$$

where θ is the beach slope angle; L_0 is the wavelength in deep water. For regular waves, Goda [15] obtained $B = 0.17$. For irregular waves, Li Yucheng et al. proposed $B = 0.15$ for spilling breakers and $B = 0.12$ for significant waves.

5.3 Turbulence Model for Breaking Waves

A viscosity term is introduced into the Boussinesq Eq. (5.1.7) to simulate the energy loss due to wave breaking. Taking the 1D problem as an example, the Boussinesq equations with the existence of narrow slots read

$$F\frac{\partial h}{\partial t} + \frac{\partial p}{\partial x} = 0 \tag{5.3.1}$$

$$\frac{\partial p}{\partial t} + \frac{\partial}{\partial x}(p^2/A) + gA\frac{\partial \eta}{\partial x} + g\frac{|p|p}{A^2C^2R} = E\frac{\partial^2 p}{\partial x^2} + \frac{1}{3}A^2\frac{\partial^3 p}{\partial x^2 \partial t} \tag{5.3.2}$$

where $E\frac{\partial^2 p}{\partial x^2}$ is the viscosity term; E is the comprehensive viscosity coefficient (viscosity coefficient for short). In the surf zone, this term includes not only the turbulent energy loss but also the contribution from the non-uniform flow velocity distribution over the depth (see Chap. 3). The role of the viscosity term in the simulation of energy dissipation will be illustrated by some numerical results in the following.

5.3.1 *Constant Viscosity Coefficient*

Various constant viscosity coefficients are applied first to illustrate the effect of the viscosity coefficient on the wave run-up and wave breaking. Figure 5.6 shows a comparison of calculated surface elevation for a solitary wave propagating on the beach with slope of 1/50 by using three different viscosity coefficients including $E = 0$, $E = 2$, and $E = 10$. When $E = 0$, i.e., the viscosity term is eliminated, the wave front becomes steep and the wave breaks at the time step $N = 150$; as a result, the calculation breaks down. When $E = 2$, the maximum wave height is reduced by half, but the wave front is still steep and the wave breaks at $N = 150$. When $E = 10$, the wave height is further reduced, the wave front is no longer steep, and the non-breaking wave propagates forward. These results signify the importance of viscosity

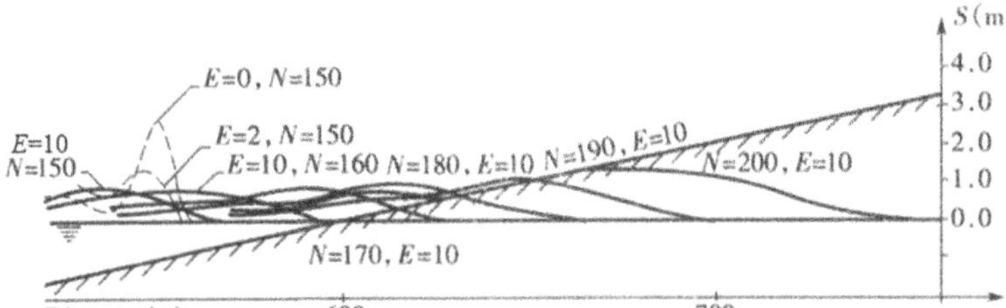

Fig. 5.6 Comparison of surface elevation for a solitary wave propagating on the beach with slope of 1/50 by using $E = 0$, $E = 2$, and $E = 10$

coefficient in the calculation of wave breaking. But the constant viscosity coefficient in the whole wave field is not physically realistic. The viscosity coefficient should have a more realistic representation, e.g., from a model.

5.3.2 k-Equation Model

With the assumption that the main energy loss in the surf zone is caused by the turbulence, the viscosity term can be determined by a turbulence model. Let ν_t denote the eddy viscosity coefficient, then the viscosity coefficient $E \approx \nu_t$.

The kinetic energy per unit mass in the turbulence reads

$$\bar{k} = \frac{1}{2}\Sigma\bar{u}_i'^2 \tag{5.3.3}$$

$\bar{k}$ reflects the turbulence intensity at a certain point. The relation between $\bar{k}$ and the eddy viscosity coefficient can be obtained with dimensional analysis. From Boussinesq assumption

$$\nu_t = C'_\mu \sqrt{\bar{k}}L \tag{5.3.4}$$

where C'_μ is an empirical coefficient; L is the length scale. Equation (5.3.4) is the Kolmogorov–Prandtl formula.

In the k-equation model, $\bar{k}$ is determined by the following equation,

$$\frac{\partial \bar{k}}{\partial t} + \bar{u}_i\frac{\partial \bar{k}}{\partial x_i} = \frac{\partial}{\partial x_j}\left(\frac{\nu_t}{\sigma_\kappa}\frac{\partial \bar{k}}{\partial x_i}\right) + \nu_i\left(\frac{\partial \bar{u}_i}{\partial x_j} + \frac{\partial \bar{u}_j}{\partial x_i}\right)\frac{\partial \bar{u}_i}{\partial x_j} - C_D\frac{\bar{k}^{3/2}}{L} \tag{5.3.5}$$

where C_D and σ_κ are empirical coefficients.

The 1D depth-integrated k-equation reads

$$\frac{\partial \bar{k}}{\partial t} + U\frac{\partial \bar{k}}{\partial x} = \frac{\nu_t}{\sigma_\kappa}\frac{\partial^2 \bar{k}}{\partial x^2} + \frac{1}{\sigma_\kappa}\left(\frac{\partial \nu_i}{\partial x}\frac{\bar{k}}{x}\right) + 2\nu_t\left(\frac{U}{x}\right)^2 - C_D\frac{\bar{k}^{3/2}}{L} + P_{kU} \tag{5.3.6}$$

where $P_{kU} = \frac{n^2 g U^2}{h}$, U is the depth-averaged velocity, P_{kU} is an additional term arising from the non-uniformity of the flow velocity distribution over the depth, n is the Manning number.

$\bar{k}$ can be obtained from the k-equation with an explicit difference scheme. ν_t is obtained from Eq. (5.3.4). After that, $E = \nu_t$ is substituted into Eq. (5.3.2).

The k-equation model has been demonstrated to be appropriate in the calculation of wave run-up and wave breaking. The empirical coefficients are obtained by trial calculations as

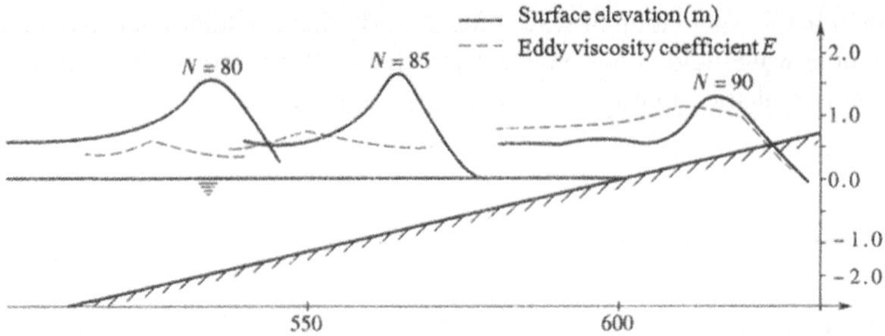

Fig. 5.7 Solitary wave run-up on the beach and the distribution of eddy viscosity coefficient (k-equation model, slope of 1/50, $\varepsilon = 0.05$)

$$\sigma_\kappa = 1.0,\ L = 4\,\text{m},\ C_D = 0.08,\ \sigma'_\mu = 1.0.$$

Figure 5.7 shows the calculated solitary wave results using the k-equation model. The eddy viscosity coefficient increases correspondingly with the wave run-up and wave breaking.

The k-ε model was used in the above case, but the results were not satisfactory. This may be due to the significant contribution to the viscosity term by the non-uniformity of the flow velocity distribution over the depth.

5.4 Surface Roller Model for Breaking Waves

The basic concept of the surface roller model is to consider the velocity distribution and the energy dissipation in the isolated surface roller after wave breaking and in the lower main wave area separately.

5.4.1 Roller Model for Breaking Waves in the Boussinesq Equations

The energy dissipation in the surface roller after wave breaking is considered by an additional convective momentum term in the Boussinesq equations. Here, a 1D problem is taken as an example to demonstrate the model.

1. Roller model for breaking waves by Schäffer and Madsen [16]

   Basic equations:

$$\eta_t + p_x = 0 \tag{5.4.1}$$

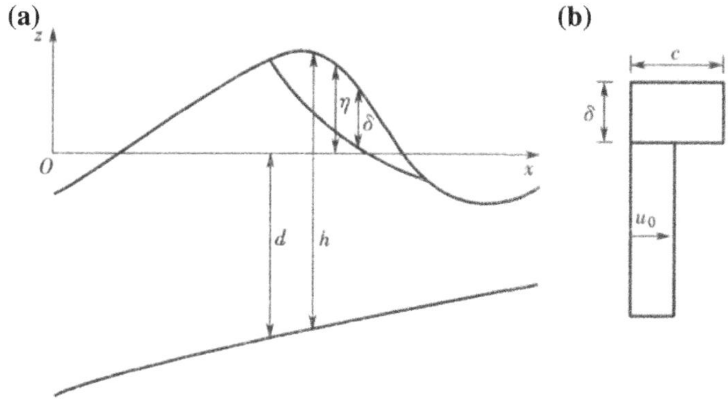

Fig. 5.8 Definition sketch: surface roller of a breaking wave (**a**) and vertical profile of the horizontal particle velocity (**b**)

$$p_t + \left(\frac{p^2}{h}\right)_x + gh\eta_x + \frac{d^2}{6}\left(\frac{p}{h}\right)_{xxt} - \frac{d^2}{2}p_{xxt} + M_{0,b} = 0 \tag{5.4.2}$$

where η is the surface elevation above the still water level; $p = Uh$, U is the depth-averaged velocity, $h = \eta + d$, d is the still water depth; $M_{0,b} = R_x$ is the excess momentum term derived by Schäffer.

As shown in Fig. 5.8 for the definition sketch, the assumed vertical distribution of the horizontal particle velocity under the surface roller reads

$$u(\eta) = \begin{cases} c & \eta - \delta \le z \le \eta \\ u_0 & -h \le z \le \eta - \delta \end{cases} \tag{5.4.3}$$

where δ is the thickness of the surface roller.

$$p = \int_{-d}^{\eta} u\mathrm{d}z \tag{5.4.4}$$

Let $\frac{p^2}{h} = \int_{-d}^{\eta} u^2\mathrm{d}z = M$.

By introducing the above velocity distribution under the surface roller, p and M read

$$p = u_0 h + (c - u_0)\delta \tag{5.4.5}$$

$$M = \frac{p^2}{h} = u_0^2 h + \left(c^2 - u_0^2\right)\delta \tag{5.4.6}$$

The excess term R due to the non-uniform velocity distribution reads

$$R = M - \frac{p^2}{h} \tag{5.4.7}$$

Finally, it yields

$$R = \delta\left(c - \frac{p}{h}\right)^2\left(1 - \frac{\delta}{h}\right)^{-1} \tag{5.4.8}$$

$M_{0,b}$ in the momentum Eq. (5.4.2) is obtained by

$$M_{0,b} = \frac{\partial R}{\partial x} \tag{5.4.9}$$

The key to calculate the excess momentum term $M_{0,b}$ is the determination of the thickness of the surface roller. The method presented in Ref. [16] is complicated. Based on the experimental study by Stive [17], Karambas [18] proposed that

$$\delta = \frac{U - u_0}{u_\eta - u_0} \cdot \frac{h}{0.45} \tag{5.4.10}$$

where $U = \frac{1}{h}\int u\mathrm{d}z$, u_η is the velocity at the surface, u_0 is the velocity below the surface roller. The results by Karambas are in agreement with the experimental data.

2. Roller model for breaking waves by Skotner and Apelt [19, 20]

This model is based on the Boussinesq-type equations with improved dispersion characteristics derived by Nwogu [21]. The relative wave height $\delta' = H_0/d$ is used as a nonlinear parameter in the model. The speed of surface roller is $c(x)\delta'$, in which $c(x)$ is the dimensionless wave speed.

The modified Boussinesq equations read

$$\eta_t + (u_\alpha h)_x + M_{bc} = 0 \tag{5.4.11}$$

$$u_{\alpha t} + u_\alpha u_{\alpha x} + g\eta_x + \frac{1}{2}z^2 d u_{\alpha xxt} + z_a (d u_{\alpha t})_{xx} + M_{bm} = 0 \tag{5.4.12}$$

where u_α is the velocity at the specific depth z_α. The other symbols are identical to the ones in Eqs. (4.5.10) and (4.5.11).

The excess term in the continuity equation reads

$$M_{bc} = [d(c - u_\alpha)]_x + \left[\left(\frac{1}{2}z_\alpha^2 - \frac{1}{6}d^2\right)du_{xxx} + \left(z_\alpha + \frac{1}{2}d\right)d(du_\alpha)_{xx}\right]_x \tag{5.4.13}$$

The excess term in the momentum equation reads

$$M_{bm} = \frac{[d(c - u_\alpha)]_t}{h} + \frac{dcc_x}{h} + \frac{(dc)_x(c - u_0)}{h} \tag{5.4.14}$$

5.4.2 Roller Model for Breaking Waves in the Parabolic Mild Slope Equation

Tao and Han [22] applied the roller model in the higher-order approximate parabolic mild slope Eq. (4.4.33) to determine the wave breaking factor W_b.

Figure 5.9 shows a sketch of a surface roller. A is the area of the surface roller. The surface roller travels with the wave speed c. The energy dissipation occurs at the shear layer between the surface roller and the wave below. With the surface roller, the energy dissipation includes two parts contributed by the wave part energy E_w and the surface roller energy E_r.

$$E = E_w + E_r \tag{5.4.15}$$

The wave energy equation can be written as

$$\frac{\partial(E_w c_g)}{\partial x} + \frac{\partial(E_r c_g)}{\partial x} = -\varepsilon \tag{5.4.16}$$

where ε is the wave energy dissipation rate.

The linear wave theory gives

$$E_w = \frac{1}{8}\rho_w g H^2 \tag{5.4.17}$$

where ρ_w is the density of seawater.

$$E_r = \frac{1}{2}\rho_r c^2 \frac{A}{L} \tag{5.4.18}$$

where A is the area of the surface roller; L is the wavelength; ρ_r is the density of the surface roller and taken as $\rho_r \approx \rho_w = \rho$. With an assumption of fully developed surface roller (the roller formed on the whole wave front), an approximate expression

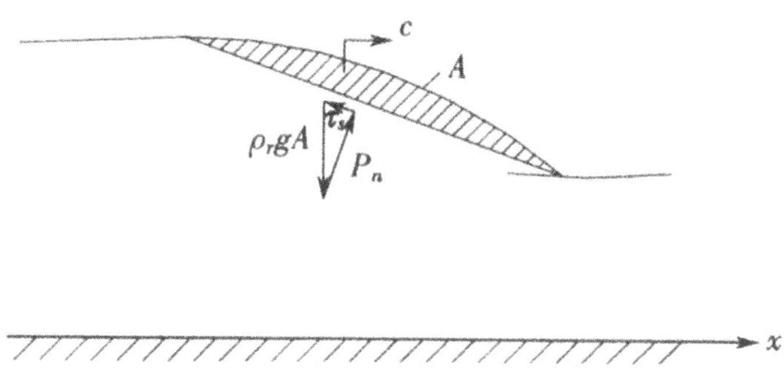

Fig. 5.9 Sketch of a surface roller

of A was derived by Engelund [23],

$$A = \frac{H_b^3}{4h \tan \sigma} \tag{5.4.19}$$

where σ is the angle between the wave and the stress vector of the roller τ_s; H_b is the wave height at breaking. The wave speed is expressed by

$$c = f_m \cdot L \tag{5.4.20}$$

where $f_m = \frac{1}{T}$ is the frequency.

Substituting Eqs. (5.4.19) and (5.4.20) into Eq. (5.4.18) yields

$$E_r = \frac{1}{8} \rho \, c \, f_m \frac{H_b^3}{h \tan \sigma} \tag{5.4.21}$$

It is assumed that the dissipation caused by the bottom friction is negligible compared with the dissipation due to the surface roller, and the maximum energy loss caused by the wave and the surface roller occurs at the interface between the wave and the roller. Then $\varepsilon_b = \varepsilon_r$ can be obtained from the work done by the shear between the roller bottom boundary and the wave surface boundary,

$$\varepsilon_r = \frac{\tau_s c}{L} \tag{5.4.22}$$

It is assumed that the interface between the wave and the roller is linear as shown in Fig. 5.9, then

$$\tau_s = \rho \, g \, A \sin \sigma \tag{5.4.23}$$

Substituting Eqs. (5.4.19) and (5.4.23) into Eq. (5.4.22) yields

$$\varepsilon_r = \frac{1}{4} \rho g \, f_m \frac{H_b^3}{h} \cos \sigma \tag{5.4.24}$$

For irregular waves, according to the studies by Thornton and Guza [24] and Whitford [25], using the Rayleigh distribution to describe the wave heights both inside and outside the surf zone, and adding an energy dissipation term into the wave evolution equation yield

$$\begin{aligned} W_b = {} & \frac{3\sqrt{\pi}}{2} f_m M \cdot \frac{H_{rms}^3}{h} \cos \sigma \left[1 - \frac{1}{(1+M)^{\frac{5}{2}}} \right] \\ & + \frac{3\sqrt{\pi}}{4} \cdot \frac{f_m}{\tan \sigma} \cdot \frac{1}{H_{rms}^2} \frac{\partial}{\partial x} \left\{ M H_{rms}^3 \left(1 - \frac{1}{(1+M)^{\frac{5}{2}}} \right) \right\} \end{aligned} \tag{5.4.25}$$

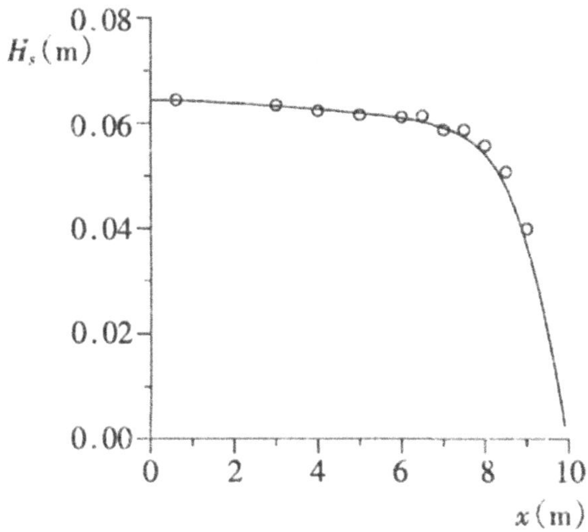

Fig. 5.10 Calculated significant wave height of irregular waves on a plane beach (curve) compared with the experimental data (scattered points)

where $M = \left(\frac{H_{rms}}{\gamma h}\right)^2$, γ is an adjustable parameter taken as $\gamma = 0.3$–0.6, $\sigma = 5°$–$30°$; H_{rms} is the root-mean-square wave height.

It can be seen from Eq. (5.4.25) that the energy dissipation term tends to be zero when the water depth goes to infinity. The energy dissipation is small in deep water. When the water becomes shallow, the effect of this term is significant. It decreases gradually again with the water depth increase. Therefore, the bathymetry change can be adapted in this model, and it is not necessary to judge when the wave breaks and recovers. Moreover, this model can simulate wave breaking more than once. The feasibility of this model will be shown in the following cases.

Figure 5.10 shows a comparison of significant wave height of irregular waves propagating over a beach with constant slope between the numerical results and the experimental data by Cox et al. [26]. The greatest water depth is 0.47 m, and the beach slope is 1:20. The input waves are generated on the basis of a target spectrum of Pierson–Moskowitz type with a peak frequency of 1.0 Hz and a significant wave height of 6.45 cm.

5.4.3 Calculated Irregular Breaking Waves on a Circular Shoal Compared with Experimental Data

This test case is based on the experiment of irregular breaking waves by Chawla et al. [27]. The experiment was carried out in the directional wave basin at the University of Delaware. The wave basin is approximately 18 m long and 18.2 m wide. The

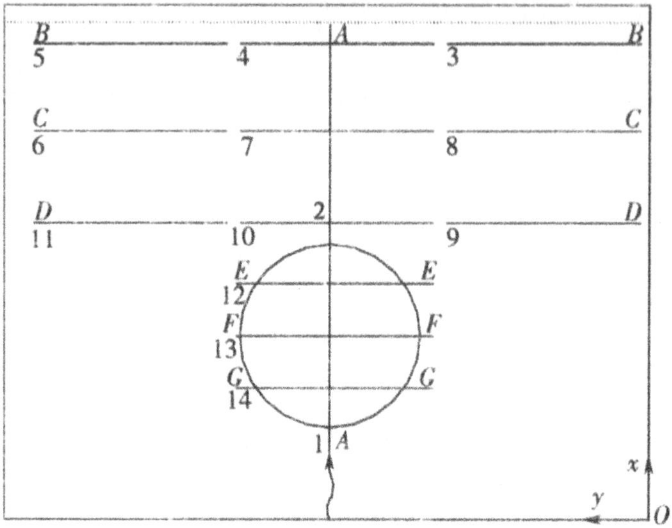

Fig. 5.11 Sketch of the circular shoal experiment setup

bottom is flat except for a circular shoal in the center and a stone beach at the far end to minimize reflections as shown in Fig. 5.11. There are seven transects selected for the wave height measurements, including one longitudinal transect (A-A) going over the shoal and six transversal transects (B-B, C-C, D-D, E-E, F-F, and G-G) behind and on top of the shoal. The input irregular sea state is described by a TMA spectrum. Figure 5.12 shows comparisons of the normalized significant wave height along the transects between the numerical results and the experimental data. The parameters $\gamma = 0.6$ and $\sigma = 10°$ are used in the numerical calculation. The numerical results match the experimental data well. It could be seen from the calculations that the value of σ within the range has no significant impact on the numerical results which mainly depend on the value of γ.

5.5 Energy Dissipation Model for Breaking Waves

5.5.1 Multiple-Breaking Model for Regular Waves

The breaking wave model developed by Dally et al. [28] is applied in the parabolic mild slope equation. It is assumed that the energy dissipation rate ε_b is proportional to the difference between the local energy flux E and the steady energy flux E_s,

$$\varepsilon_b \propto \frac{[Ec_g - E_s c_g]}{h} \tag{5.5.1}$$

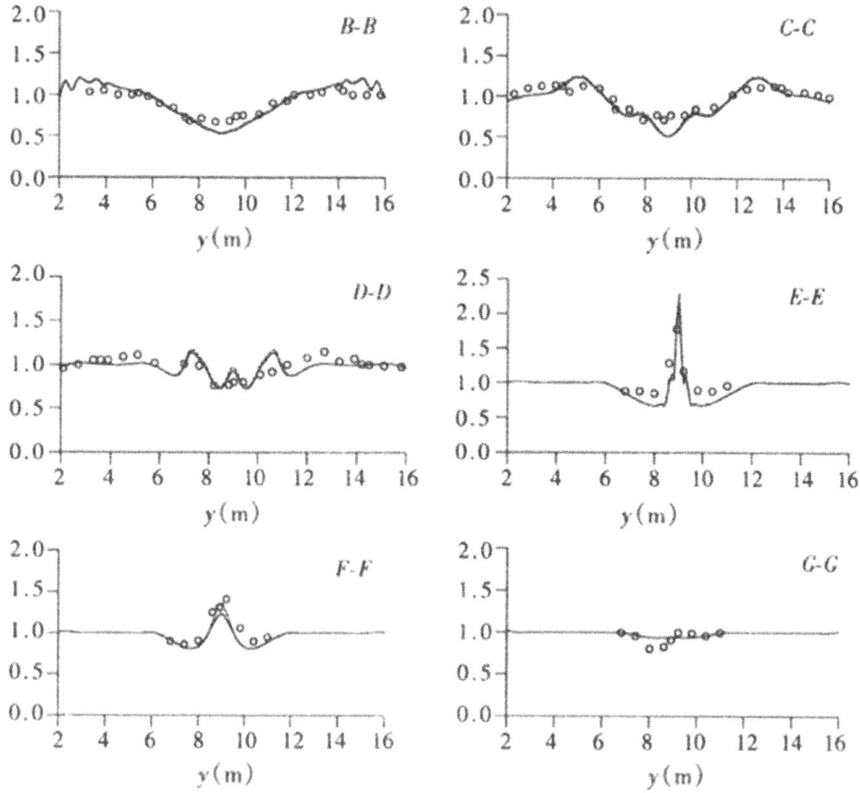

Fig. 5.12 Comparisons of the significant wave height normalized with the input significant wave height along the different transects between the numerical results (curve) and the experimental data (scattered points). The input significant wave height is 1.39 m, and the peak period is 0.73 s

Let K_d denote the proportional coefficient (decaying coefficient), which gives

$$\varepsilon_b = \frac{K_d c_g}{h}[E - E_s] \tag{5.5.2}$$

Let H_s be the stable wave height and Γ be the stability factor, the linear wave theory gives

$$E_s = \frac{1}{8}\rho g H_s^2 = \frac{1}{8}\rho g(\Gamma h)^2 \tag{5.5.3}$$

Generally, K_d is taken as 0.15 and Γ is taken as 0.4. Then W_b in the parabolic mild slope equation reads

$$W_b = \frac{\varepsilon_b}{E} = \frac{K_d c_g}{h}\left[1 - \frac{E_s}{E}\right] \tag{5.5.4a}$$

Substituting Eq. (5.5.3) into Eq. (5.5.4b) yields

$$W_b = \frac{K_d c_g}{h}\left[1 - \left(\frac{\Gamma h}{H}\right)^2\right] \tag{5.5.4b}$$

The advantage of this model is that it is able to stop the calculation of breaking dissipation when the breaking waves reach the recovery area. In the calculation, wave breaking is judged first according to the wave breaking index, and then the breaking wave energy is calculated. If the calculated energy flux is less than the steady energy flux, K_d is taken as 0. Multiple wave breaking can be simulated by this model. When the wave height breaking index $H_b = \gamma_b d_b$ is used in the model, $\gamma_b = 0.78$ can be taken, and accordingly $K_d = 0.15$. The numerical results were compared with the experimental data for three different bathymetries by Sun et al. [29], and the numerical results and the experimental data are in good agreement.

5.5.2 Energy Dissipation Model of Irregular Breaking Waves

For regular waves, the breaker line is clear, and the wave energy loss starts to occur from the breaker line. But it is different for irregular waves. On each transect parallel to the coastline, the different wave heights are distributed with a certain probability, which corresponds to different breaking points. In offshore areas, only a small part of waves break. As the waves approach the coast and the water depth decreases gradually, more and more individual wave components start to break until almost all the individual wave components break inside the surf zone. Therefore, there is no obvious breaker line for irregular waves.

Based on the regular wave breaking theory, it is assumed that the mean energy dissipation rate of the irregular breaking waves is proportional to the difference between the local mean energy density and the steady-state energy density. By introducing a wave breaking rate factor Q_b, the mean energy dissipation rate of the irregular breaking waves is expressed as

$$\overline{D_B} = \frac{K_1 Q_b C_p}{h}[E_m - E_s] \tag{5.5.5}$$

where C_p is the wave speed corresponding to the peak period T_p of the irregular waves, h is the water depth, K_1 is a constant which can be taken as 0.1. E_m is the local mean energy density and E_s is the steady-state energy density. They can be obtained by

$$E_m = \frac{1}{8}\rho g H_{rms}^2 \tag{5.5.6}$$

$$E_s = \frac{1}{8}\rho g H_s^2 = \frac{1}{8}\rho g(\Gamma_i h)^2 \tag{5.5.7}$$

where H_{rms} is the root-mean-square wave height, H_s is the steady-state wave height, Γ_i is the steady-state wave factor of irregular waves. Equation (5.5.5) can be rewritten as

$$\overline{D}_B = \frac{K_1 Q_b c_p \rho g}{8h}\left[H_{rms}^2 - (\Gamma_i h)^2\right] \tag{5.5.8}$$

Thus, the energy dissipation factor for irregular wave breaking is written as

$$\bar{W} = \frac{\bar{D}_B}{E_m} = \frac{K_1 Q_b C_p}{h}\left[1 - \left(\frac{\Gamma_i h}{H_{rms}}\right)^2\right] \tag{5.5.9}$$

The steady-state wave factor Γ_i is expressed by the following empirical formula,

$$\Gamma_i = \exp\left[K_2\left(-0.36 - 1.25\frac{h}{\sqrt{L_p H_{rms}}}\right)\right] \tag{5.5.10}$$

where L_p is the wavelength corresponding to the peak period; K_2 is a coefficient which can be taken as 0.1.

Compared with the regular waves, the irregular breaking waves have no specific breaker line. Therefore, a parameter Q_b denoting the probability of wave breaking is introduced in the calculation of energy loss. The irregular wave breaking is also judged by this parameter. It is determined by the assumed wave height probability density function. The expression of Q_b proposed by Battjes and Janssen [30] reads

$$\frac{1 - Q_b}{-\ln Q_b} = \left(\frac{H_{rms}}{H_b}\right)^2 \tag{5.5.11}$$

where H_b is the maximum possible wave height which could be determined by the regular wave breaking criterion. The breaking point of the wave component with the maximum possible wave height is located farthest from the coastline. The energy dissipation region induced by the irregular wave breaking is much larger than that caused by the regular wave breaking. The above Eq. (5.5.11) is an implicit expression, not easy to solve. An explicit expression for Q_b proposed by Rattanapitikon and Shibayama [31] through multiple regression analysis reads

$$Q_b = \begin{cases} 0, & \frac{H_{rms}}{H_b} \le 0.43 \\ -0.738\left(\frac{H_{rms}}{H_b}\right) - 0.28\left(\frac{H_{rms}}{H_b}\right)^2 + 1.785\left(\frac{H_{rms}}{H_b}\right)^3 + 0.235, & \frac{H_{rms}}{H_b} > 0.43 \end{cases} \tag{5.5.12}$$

The wave height at the breaking H_b can still be determined by the wave breaking criterion Eq. (5.2.1) for irregular waves.

5.6 Wave-Induced Radiation Stress

5.6.1 Concept of Radiation Stress

The fluid motion in the surf zone is very complex. The energy loss due to wave breaking can be calculated by turbulence models and energy dissipation models described in the above section.

The contribution of waves, a kind of surface perturbation, to the nearshore currents in the surf zone can be calculated by the radiation stress. Waves influence the bottom friction of the seabed, which has been described earlier. Here, only the concept and the calculation method on the radiation stress are presented.

In the 1960s, Bowen [32] and Longuet-Higgins and Stewart [33] proposed the concept of radiation stress at almost the same time. They believed that the radiation stress should be added to the momentum equations to calculate the fluid motion in the surf zone.

When the x- and y-coordinates are parallel and perpendicular to the incident wave direction, respectively, and the z-coordinate is upward, the horizontal momentum fluxes induced by the waves read

$$\left.\begin{aligned} M_{xx} &= \int_{-h}^{\eta} (P + \rho u^2)\mathrm{d}z - \int_{-h}^{0} P_0 \mathrm{d}z \\ M_{xy} &= M_{yx} = \int_{-h}^{\eta} \rho uv \mathrm{d}z \\ M_{yy} &= \int_{-h}^{\eta} (P + \rho v^2)\mathrm{d}z - \int_{-h}^{0} P_0 \mathrm{d}z \end{aligned}\right\} \tag{5.6.1}$$

where P is the pressure in the wave field; P_0 is the atmospheric pressure.

Dynamically, the horizontal momentum fluxes are equivalent to the stresses. The horizontal momentum fluxes from waves can be treated as certain horizontal stresses called radiation stresses. Furthermore, since they are functions of time t, they are averaged over time to obtain the time-averaged $\bar{M}_{xx}$, $\bar{M}_{xy}$, $\bar{M}_{yx}$ and $\bar{M}_{yy}$ denoted as S_{11}, S_{12}, S_{21} and S_{22}, respectively. Here, "—" denotes a time average within a wave period, i.e., $\frac{1}{T}\int_0^T (\ )\mathrm{d}t$. Thus, we have

$$\left.\begin{aligned} S_{11} &= \overline{\int_{-h}^{\eta} (P + \rho u^2)\mathrm{d}z - \int_{-h}^{0} P_0 \mathrm{d}z} \\ S_{12} = S_{21} &= \overline{\int_{-h}^{\eta} \rho u v \mathrm{d}z} \\ S_{22} &= \overline{\int_{-h}^{\eta} (P + \rho v^2)\mathrm{d}z - \int_{-h}^{0} P_0 \mathrm{d}z} \end{aligned}\right\} \tag{5.6.2}$$

For an individual wave propagating over constant water depth, let

$$S_{11} = S_{11}^{(1)} + S_{11}^{(2)} + S_{11}^{(3)}$$

where

$$S_{11}^{(1)} = \overline{\int_{-h}^{\eta} \rho u^2 \mathrm{d}z}, \ S_{11}^{(2)} = \overline{\int_{-h}^{0} (P - P_0)\mathrm{d}z}, \ S_{11}^{(3)} = \overline{\int_{0}^{\eta} P \mathrm{d}z} \tag{5.6.3}$$

In the expression of $S_{11}^{(1)}$, the integral upper limit $z = \eta$ is approximated by $z = 0$, which yields

$$S_{11}^{(1)} = \overline{\int_{-h}^{0} \rho u^2 \mathrm{d}z} = \int_{-h}^{0} \rho \overline{u^2} \mathrm{d}z, \ S_{11}^{(2)} = \int_{-h}^{0} (\bar{P} - P_0)\mathrm{d}z$$

At $z = 0$, the vertical momentum flux $\overline{P + \rho w^2} = -\rho g z = P_0$($w$ is the vertical velocity component) makes $\bar{P} - P_0 = -\rho \overline{w^2}$, which leads to

$$S_{11}^{(1)} + S_{11}^{(2)} = \int_{-h}^{0} \rho (\overline{u^2 - \omega^2})\mathrm{d}z \tag{5.6.4}$$

Based on the small-amplitude wave theory,

$$\overline{u^2 - w^2} = \frac{1}{T} \int_{0}^{T} (u^2 - w^2)\mathrm{d}t = \frac{a^2 \omega^2}{2 \sinh^2 (kh)} \tag{5.6.5}$$

Thus,

$$S_{11}^{(1)} + S_{11}^{(2)} = \frac{1}{2}\frac{\rho a^2\omega^2 h}{\sinh^2(kh)} \tag{5.6.6}$$

From the dispersion relation $\omega^2 = gk\tanh(kh)$, we have

$$S_{11}^{(1)} + S_{11}^{(2)} = \frac{\rho g a^2 kh}{\sinh(2kh)}$$

Because of $P = \rho g(\eta - z)$,

$$S_{11}^{(3)} = \frac{1}{2}\rho g\overline{\eta^2} = \frac{E}{2}$$

where E is the wave energy per unit area, $E = \frac{\rho g a^2}{2}$. Thus, we have

$$S_{11} = E\left(\frac{2kh}{sh(2kh)} + \frac{1}{2}\right) \tag{5.6.7}$$

Similarly, it can be derived that $S_{12} = S_{21} = 0$, and

$$S_{22} = E\frac{kh}{\sinh(2kh)} \tag{5.6.8}$$

5.6.2 Radiation Stress Tensor

When waves propagate in the x-direction, S_{11}, S_{12}, S_{21}, and S_{22} are the radiation stress components of the radiation stress in tensor form S,

$$S = E\cdot\begin{bmatrix} \frac{2kh}{\sinh(2kh)} + \frac{1}{2} & 0 \\ 0 & \frac{kh}{\sinh(2kh)} \end{bmatrix} \tag{5.6.9}$$

Under the condition of uneven water depth, the wave refraction and diffraction result in the wave direction change. Let α be the angle between the wave direction and the x-coordinate, which yields

$$\begin{aligned} S_{xx} &= S_{11}\cos^2\alpha + S_{22}\sin^2\alpha \\ &= \frac{1}{2}E[2n(1+\cos^2\alpha) - 1] \end{aligned} \tag{5.6.10}$$

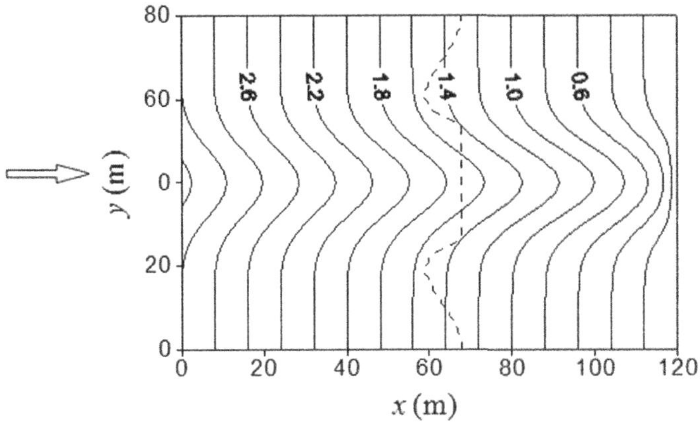

Fig. 5.13 Bathymetry (dash line shows the breaker line)

$$S_{xy} = S_{11} \sin\alpha\cos\alpha - S_{22}\sin\alpha\cos\alpha = \frac{E}{2} n \sin 2\alpha \tag{5.6.11}$$

$$S_{yy} = S_{11}\sin^2\alpha + S_{22}\cos^2\alpha = \frac{1}{2}E[2n(1+\sin^2\alpha) - 1] \tag{5.6.12}$$

$$S = \begin{bmatrix} S_{xx} & S_{xy} \\ S_{yx} & S_{yy} \end{bmatrix} = E\begin{bmatrix} n(1+\cos^2\alpha) - \frac{1}{2} & \frac{n}{2}\sin 2\alpha \\ \frac{n}{2}\sin 2\alpha & n(1+\sin^2\alpha) - \frac{1}{2} \end{bmatrix} \tag{5.6.13}$$

5.6.3 Calculation of Radiation Stress—An Example

The bathymetry for the calculation is shown in Fig. 5.13. The water depth in the figure is expressed by

$$h\,(x, y) = Jx\left\{1 + A_0 \exp\left[-3(x/20)^{\frac{1}{3}}\right]\sin^{10}\left[\left(\frac{\pi}{\lambda}\right)(y - x\tan\varepsilon)\right]\right\} \tag{5.6.14}$$

where J is the mean beach slope taken as 0.025; λ is the length of the coastline taken as 80 m; A_0 is the maximum amplitude of the bathymetry variation taken as 20 m. ε is the inclination of the bathymetry variation taken as 0. The wave propagates from the left with the incident wave height of 1 m and wave period of 3.5 s. The wave field is calculated by a higher-order approximate parabolic mild slope equation with grid spacing of $\Delta x = \Delta y = 1$ m. Figure 5.14 shows the calculated radiation stress components. It is found that the bathymetry has a great influence on the distribution of radiation stress.

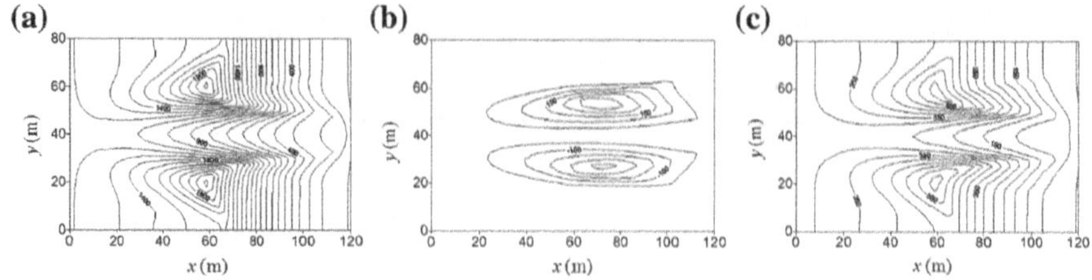

Fig. 5.14 Calculated radiation stress components. **a** S_{xx}; **b** S_{xy}; **c** S_{yy}

References

1. Tao JH. Computation of wave rup-up and wave breaking. In: Internal report. Danish Hydraulic Institute; 1983. 40 pp.
2. Jianhua Tao. Numerical simulation of wave run-up and wave breaking on the beach. Acta Ocenologica Sinica (Chinese version). 1984;6(5):692–700.
3. Shaoling He, Liangxiang Wang. Application of the slot method in the computation of two dimensional flow with transient boundary. J Hydraul Eng. 1986;12:11–9.
4. Zeliang Wang, Rixin Wang, Jianhua Tao. Numerical simulation of tidal current and pollutant distribution of Bohai Bay. Oceanologia Et Limnologia Sinica. 1999;30(2):224–30.
5. Madsen PA, Sorensen OR. Bound waves and triad interaction in shallow water. Ocean Eng. 1993;20:359–88.
6. Madsen PA, Sorensen OR, Schaffer HA. Surf zone dynamics simulated by a Boussinesq type model. Part I: model description and cross-shore motion of irregular waves. Coast Eng. 1997;32:255–87.
7. Madsen PA, Sorensen OR, Schaffer HA. Surf zone dynamics simulated by a Boussinesq type model. Part II: surf beat and swash oscillations for wave groups and irregular waves. Coast Eng. 1997;32:289–319.
8. Kennedy AB, Chen Q, Kirby JT, et al. Boussinesq modeling of wave transformation breaking, and runup. I: 1D. J Waterw Port Coast Ocean Eng. 2000;126(1):39–47.
9. Qin W. Improvement of Boussinesq equation mathematical model and its engineering application. In: Dissertation for master degree. Tianjin: Tianjin University; 2003.
10. Stokes GG. Considerations relative to the greatest height of oscillatory irrotational waves which can be propagated without change of form. Math Phys Pap. 1880;I:225–8.
11. Michell JH. On the highest waves in water. Philos Mag. 1893;36:430–7 5th series.
12. Miche R. Mouvements ondulatoires de la mer en profundeur constante on decroissante. Ann. des Ponts et Claussees; 1944. 114 pp.
13. Li Y, Dong G. Breaker indices of irregular waves on gentle beach. J Hydrodyn Ser A. 1993;8(1):21 7.
14. McGowan J. On the highest wave of permanent type. Lond Edinburgh Dublin Phil Mag J Sci. 1894;38:351.
15. Goda Y. Random seas and design of maritime structures. Tokyo: University of Tokyo Press; 1985.
16. Schaffer HA, Madsen PA, Deigaard R. A Boussinesq model for waves breaking in shallow water. Coast Eng. 1993;20:185–202.
17. Stive MJF. Energy dissipation in waves breaking on gentle slopes. Coast Eng. 1980;8:99–127.
18. Karambas TV. A numerical method for integral properties of periodic non-linear breaking and no-breaking waves. J Hydrosci Hydraul Eng. 1999;17(1):171–85.
19. Skotner C, Apelt CJ. Application of a Boussinesq model for the computation of breaking waves. Part 1: development and verification. Ocean Eng. 1999;26:927–47.

20. Skotner C, Apelt CJ. Application of a Boussinesq model for the computation of breaking waves. Part 2: wave-induced setdown and setup on a submerged coral reef. Ocean Eng. 1999;26:927–47.
21. Nwogu OG. Alternative form of Boussinesq equations for nearshore wave propagation. J Waterw Port Coast Ocean Eng. 1993;119(6):618–38.
22. Tao JH, Han G. Numerical simulation of breaking wave based on higher-order mild slope equation. China Ocean Eng. 2001;15(2):269–80.
23. Engelund F. A simple theory of weak hydraulic jumps. In: Progress report no. 54. Institute of Hydrodynamics and Hydaulic Engineering, ISVA, Technical University of Denmark; 1981. p. 29–32.
24. Thornton EB, Guza RT. Transformations of wave height distribution. J Goephys Res. 1983;88:5925–38.
25. Whitford DJ, Thornton EB. Bed shear stresses for longshore currents over a barred beach during superduck. Coast Eng. 1988;27:243–62.
26. Cox DT, Mase H, Sakai T. An experiment on the effect of fluid acceleration on seabed stability. In: Report no. 91-HY-01. Japan: Kyoto University; 1991.
27. Chawla A, Ozkan-Haller HT, Kirby JT. Spectral model for wave transformation and breaking over irregular bathymetry. J Waterw Port Coast Ocean Eng. 1998;124(4):189–98.
28. Dally WR, Dean RG, Dalrymple RA. Wave height variation across beaches of arbitrary profile. J Geophys Res. 1985;90(C6):11917–27.
29. Sun T, Han G, Tao J. Numerical simulation of wave-induced long-shore currents and experimental verification. J Hydraul Eng. 2002;33(11):1–7.
30. Battjes JA, Janssen JPFM. Energy loss and set-up due to breaking of random waves. In: Proceedings of 16th international conference coastal engineering ASCE; 1978. p. 569–587.
31. Rattanapitikon K, Shibayama T. Energy dissipation model for regular and irregular breaking waves. Coast Eng. 1998;40(4):327–46.
32. Bowen AJ. The generation of longshore currents on a plane beach. J Mar Res. 1969;27:206–15.
33. Longuet-Higgins MS, Stewart RW. Radiation stresses in water waves: a physical discussion, with applications. Deep-Sea Res. 1964;11:529–62.

Chapter 6
Numerical Simulation of Wave Forces on Structures

6.1 Introduction

Wave forces are the main external loads on marine engineering structures. The wave force calculation is one of the key issues in the design of coastal and marine engineering.

The fluid–structure interaction is complex. It depends on the Reynolds number of the flow, the K_C number, and the scale, shape and the relative roughness of the structure. The early determination of wave forces on marine structures mainly relied on semiempirical formulas and physical model tests.

Since the 1970s, with rapid developments of computer and computational methods, great progress has been made in numerical calculation of wave forces. At present, the calculation of wave forces on various marine structures using the linear wave theory is mature. However, the calculation using a nonlinear wave theory is still imperfect due to issues associated with the free surface boundary and the scattered wave condition at infinity. There have not been many studies on the numerical simulation of wave forces on marine structures using the N-S equations or the Reynolds equations [1]; hence, this approach is less commonly applied. The content in this aspect will be described in Chap. 9.

The calculation of wave forces on marine structures can be summarized into two types; one for small-scale structures and the other for large-scale structures depending on the ratio of the characteristic scale of the structure D to the wavelength L (D/L). According to the main aspects of wave action on structures, different algorithms are applied in the wave force calculations.

When the ratio of the characteristic scale of the structure to the wavelength is relatively small, $D/L \leq 0.15$, the structure is classified as small-scale structure. It is assumed that the existence of a small-scale structure does not affect the incident wave field, and the dominant forces acting on the structure in the waves are the viscous force and the inertia force. When the characteristic scale of the structure is

J. Tao, *Numerical Simulation of Water Waves*, Springer Tracts in Civil Engineering, https://doi.org/10.1007/978-981-15-2841-5_6

large with $D/L > 0.15$, the structure is then classified as large-scale structure and the existence of the structure affects the incident wave field. The structure is treated as a disturbance source in the wave field, and it can be calculated based on the diffraction theory without consideration of the viscosity effect. If the scale of structure is just in between small- and large-scale structures, all the viscosity effect, the inertia effect, and the disturbance of the structure on the wave field need to be taken into account. In this case, it is more appropriate to use a model based on the basic equations of fluid dynamics, such as the N-S equations or the Reynolds equations.

6.2 Wave Forces on a Small-Scale Structure

Although it can be assumed that a small-scale structure does not affect the wave field, when the fluid bypasses the structure the flow pattern is very complex due to the separation of the boundary layer, the development of the vortices in the wake, the formation of the vortex street and wake oscillation. Numerical models can be used to simulate the wave action on small-scale structures and obtain the wave forces. However, it is difficult to use numerical models in engineering design because of the great workload and the demand of high numerical accuracy. At present, the most widely used approach is the semiempirical method for determining the force coefficients by means of theoretical analyses combined with experiments. In order to better understand the semiempirical formulas, the wave forces on a small-scale structure are described starting with the forces on a structure in steady current in the following.

6.2.1 Flow Around a Cylinder in Steady Current and the Forces

When steady current flows around a cylinder, the flow regime depends on the ratio of the viscosity force to the inertia force acting on the cylinder, so that it is closely related to the Reynolds number (Re) which is defined as

$$\mathrm{Re} = \frac{UD}{\nu}$$

in which U is the inflow velocity, D is the diameter of the cylinder, ν is the kinematic viscosity of the fluid. Figure 6.1 presents the various regimes of flow around a smooth circular cylinder in steady current [2].

Figure 6.1 shows that, when $\mathrm{Re} < 5$, the flow is symmetric laminar flow without boundary layer separation. With the increase of the Reynolds number, the boundary layer starts to separate and turns from the laminar boundary layer to the turbulent

(a) 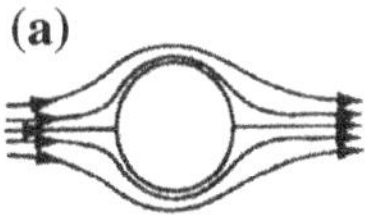	No separation. Laminar flow.	$Re<5$
(b) 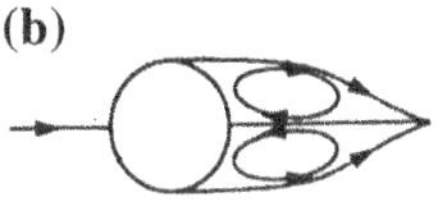	A fixed pair of symmetric vortices	$5<Re<40$
(c) 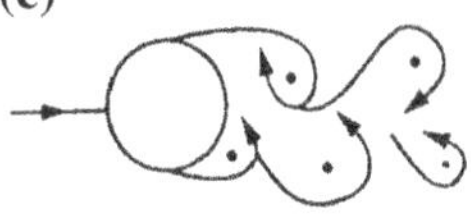	Laminar vortex street	$40 < Re < 200$
(d) 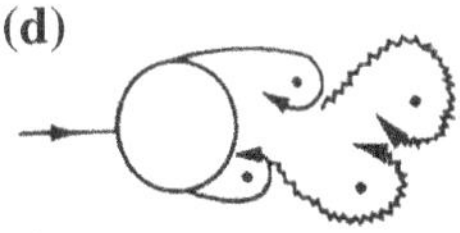	Transition to turbulence in the wake	$200 < Re < 300$
(e) 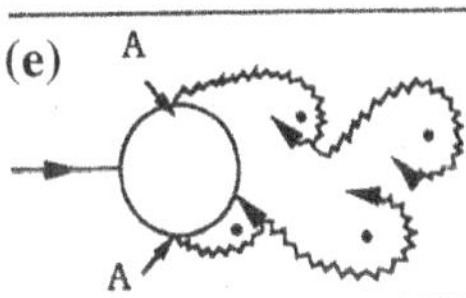	Complete turbulence A: Laminar boundary layer separation	$200 < Re < 300$
(f) 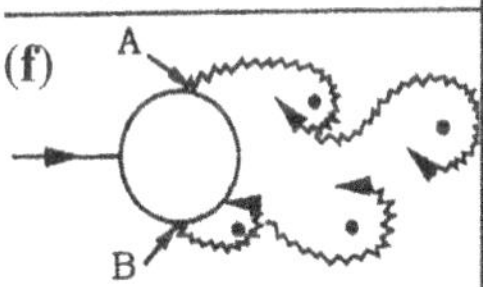	A: Laminar boundary layer separation B: Turbulent boundary layer separation	$300 < Re < 3.5 \times 10^5$
(g) 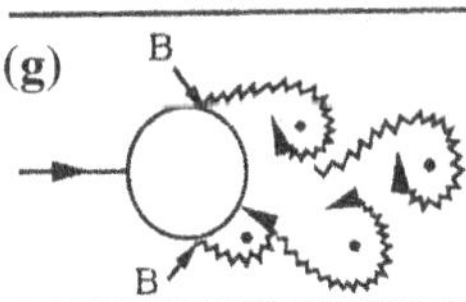	B: Turbulent boundary layer separation	$3 \times 10^5 < Re < 3.5 \times 10^5$
(h) 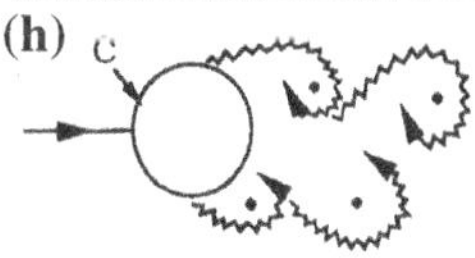	C: Boundary layer completely turbulent at one side	$1.5 \times 10^6 < Re < 4 \times 10^6$
(i) 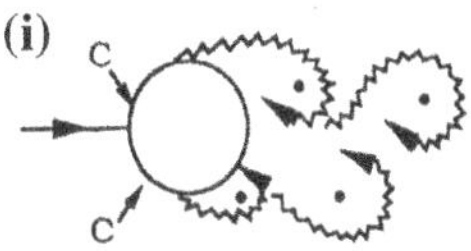	C: Boundary layer completely turbulent at two sides	$Re > 4 \times 10^6$

Fig. 6.1 Regimes of flow around a smooth circular cylinder in steady current

boundary layer. Then the vortex street is formed behind the cylinder and the flow becomes asymmetric.

In steady current, the in-line force parallel to the flow direction on unit length of the cylinder F_D is called the drag force, which consists of the in-line pressure force on the cylinder and the in-line friction force,

$$F_D = F_p + F_f \tag{6.2.1}$$

$$F_p = \int_0^{2\pi} p \cos \beta r_0 \mathrm{d}\phi \tag{6.2.2}$$

$$F_f = \int_0^{2\pi} \tau_0 \sin \beta r_0 \mathrm{d}\phi \tag{6.2.3}$$

in which F_p and F_f are the in-line pressure force and in-line friction force on unit length of the cylinder, respectively; p is the pressure at a point on the cylinder; β is the angle between the normal direction at this point and the flow direction; r_0 is the radius of the cylinder; τ_0 is the friction stress at this point.

p and τ_0 can be obtained by experiments or numerical calculation. In general, the drag force on a structure in steady current can be calculated by the following semiempirical formula

$$F_D = C_D \frac{1}{2} \rho D U^2 \tag{6.2.4}$$

in which C_D is the drag coefficient; ρ is the fluid density; D is the projected area per unit length of the cylinder perpendicular to the flow direction, which is equal to the diameter of the cylinder; U is the current speed.

When the Reynolds number is large, the pressure distribution on the cylinder becomes asymmetric, there exists a nonzero component on the cylinder in the direction perpendicular to the flow direction, which is called the lift force and denoted by F_L,

$$F_L = C_L \frac{1}{2} \rho D U^2 \tag{6.2.5}$$

in which C_L is the lift coefficient.

Figure 6.2 shows time series of the drag coefficient C_D and the lift coefficient C_L from the measurements by Orescher in 1956 [3]. It is found that C_D changes periodically in time oscillating a little around the mean value $\overline{C}_D$, and C_L changes significantly and periodically in time. The oscillating period of the lift force is basically in agreement with the vortex-shedding period.

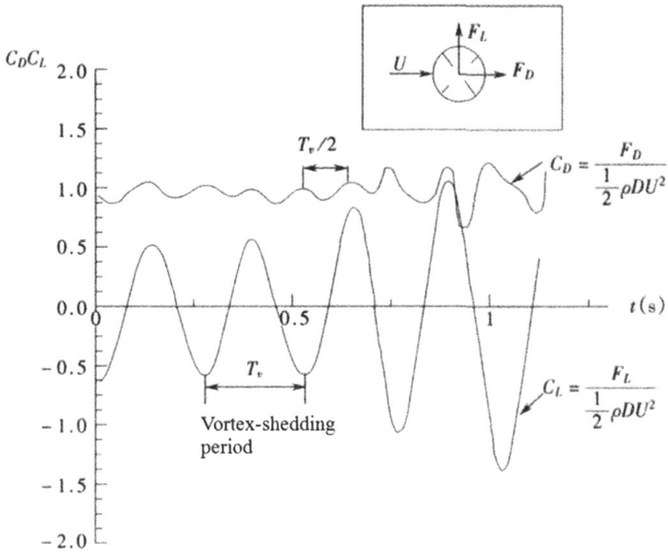

Fig. 6.2 Time series of the drag coefficient C_D and the lift coefficient C_L in the case of flow around a cylinder in steady current

6.2.2 *Flow Around a Cylinder in Oscillatory Flow and the* K_C *Number*

As the particle orbits under waves are circles or ellipses, the wave motion can be regarded as a kind of oscillatory flow. Besides the Reynolds number Re, another important dimensionless parameter to describe the action of the oscillatory flow on a cylinder is the Keulegan–Carpenter number, the K_C number for short, which is defined as

$$K_C = \frac{U_m T_w}{D} \tag{6.2.6}$$

in which U_m is the maximum velocity of the oscillatory flow; T_w is the period of the oscillatory flow. The flow velocity can be given by

$$U = U_m \sin(\omega t) \tag{6.2.7}$$

in which the maximum velocity is $U_m = a\omega = \frac{2\pi a}{T_w}$ by the linear wave theory, ω is the angular frequency of the oscillatory flow, and a is the amplitude of the oscillatory flow.

According to the linear wave theory, the K_C number can be written as

$$K_C = \frac{2\pi a}{D} \tag{6.2.8}$$

The physical meaning of the K_C number can be interpreted as the ratio of the scale of the particle orbit to the diameter of the cylinder. A small K_C number means that the range of the water particle motion is smaller than the diameter of the cylinder. The wave height $H = 2a$ and the maximum velocity $U_m = \frac{\pi H}{T_w}$ lead to

$$K_C = \frac{\pi H}{D}. \tag{6.2.9}$$

6.2.3 Forces on a Small-Scale Cylinder in Regular Waves—The Morison Equation

1. Horizontal forces on a cylinder

Morison et al. [4] proposed that, the horizontal wave force f_H acting on a cylinder at any position (z level above the seabed) consists of two components: one is the force (the horizontal drag force f_D) caused by the horizontal particle velocity u of the wave motion and the other is the force (the horizontal inertia force f_I) caused by the horizontal particle acceleration $\frac{du}{dt}$ of wave motion. The horizontal wave force on unit length of a vertical cylinder at any position z reads

$$\begin{aligned} f_H = f_D + f_I &= \frac{1}{2}\rho\, C_D D|u|u + \rho\, \overline{V}_0 \frac{du}{dt} + \rho\, C_m \overline{V}_0 \frac{du}{dt} \\ &= \frac{1}{2}\rho\, C_D D|u|u + \rho\, C_M \overline{V}_0 \frac{du}{dt} \end{aligned} \tag{6.2.10}$$

This equation is known as the Morison equation, in which C_D is the coefficient of the drag force perpendicular to the axis of the cylinder, also called the drag coefficient; $C_M = C_m + 1$, where C_M is the inertia coefficient, C_m is the added mass coefficient; $\overline{V}_0 = \frac{\pi D^2}{4}$ is the volume per unit length of the cylinder, where D is the diameter of the cylinder.

Similar to the drag force on a cylinder in steady current, the first term on the right side of Eq. (6.2.10) is the resistance on the cylinder which is also called the drag force. Since the flow direction may change, U^2 in steady current is replaced by $|u|u$ in order to maintain the positive and negative properties of the drag force. The second term on the right side of Eq. (6.2.10) is the inertia force including the hydrodynamic mass force and the Froude–Krylov force which are related to the acceleration. The hydrodynamic mass force is the inertial force from the acceleration of the fluid in the immediate surroundings of the fixed body, which is equivalent to the force by

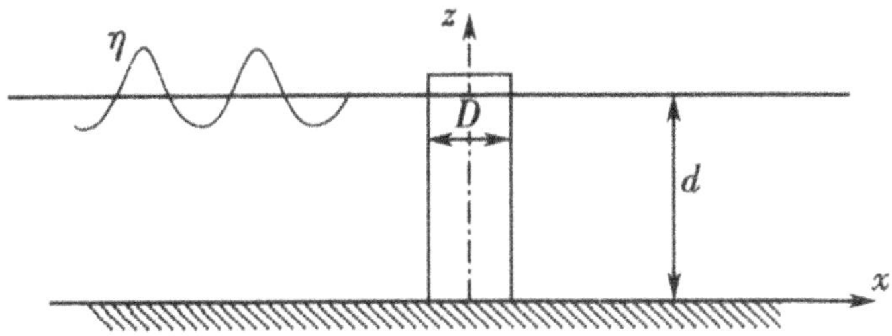

Fig. 6.3 Definition sketch for a small-scale cylinder

the cylinder moving with the acceleration in the opposite direction in still water. The Froude–Krylov force is caused by the pressure gradient generated by the accelerated motion of the fluid in the outer-flow region.

Since the small-scale cylinder has no significant influence on the wave motion, u and $\frac{du}{dt}$ in Eq. (6.2.10) can be approximated by the corresponding horizontal velocity u and the acceleration $\frac{\partial u}{\partial t}$ at the location of the cylinder center in the water without cylinder, respectively.

The horizontal wave force f_H on unit length of a vertical cylinder can be rewritten as

$$f_H = \frac{1}{2}\rho\, C_D D|u|u + \rho\, C_M \frac{\pi D^2}{4}\frac{\partial u}{\partial t} \tag{6.2.11}$$

In the coordinate system shown in Fig. 6.3, the horizontal wave force on unit length dz of the cylinder reads

$$\mathrm{d}f_H = \frac{1}{2}\rho\, C_D D|u|u\mathrm{d}z + \rho\, C_M \frac{\pi D^2}{4}\frac{\partial u}{\partial t}\mathrm{d}z \tag{6.2.12}$$

Integrating Eq. (6.2.12) from $z_1 = 0$ to $z_2 = d + \eta$ yields the total horizontal wave force on the whole cylinder

$$F_H = \int_0^{d+\eta}\left(\frac{1}{2}\rho\, C_D D|u|u\right)\mathrm{d}z + \int_0^{d+\eta}\left(\rho\, C_M \frac{\pi D^2}{4}\frac{\partial u}{\partial t}\right)\mathrm{d}z \tag{6.2.13}$$

Similarly, the moment of the total horizontal wave force to the seabed can be given as

$$M_H = \int_0^{d+\eta} z f_H \mathrm{d}z = \int_0^{d+\eta}\left(\frac{1}{2}\rho\, C_D D|u|u\right)z\mathrm{d}z + \int_0^{d+\eta}\left(\rho\, C_M \frac{\pi D^2}{4}\frac{\partial u}{\partial t}\right)z\mathrm{d}z \tag{6.2.14}$$

The distance from the action point of the total horizontal wave force to the seabed is

$$e = \frac{M_H}{F_H} \tag{6.2.15}$$

u and $\frac{\partial u}{\partial x}$ in the Morison equation vary with different wave theories. Thus, the key issue to correctly calculate the horizontal wave force F_H on a vertical cylinder, and the corresponding moment of force M_H is to select a suitable wave theory to calculate u and $\frac{\partial u}{\partial x}$ based on the conditions of bathymetry, the design wave height and wave period in the area where the cylinder is located. Meanwhile, the drag coefficient C_D and the inertia coefficient C_M should be selected properly.

In the coordinate system shown in Fig. 6.3, the linear wave theory gives

$$u = \frac{H\omega}{2}\frac{\cosh kz}{\sinh kd}\cos\theta \tag{6.2.16}$$

$$\frac{\partial u}{\partial t} = \frac{H}{2}\omega^2\frac{\cosh kz}{\sinh kd}\sin\theta \tag{6.2.17}$$

where $\theta = kx - \omega t$. In deep water, we have $\omega^2 = gk$. Then the horizontal force on unit length of the vertical cylinder reads

$$f_H = \frac{1}{2}\rho\, C_D D\left(\frac{\pi H}{T}\frac{\cosh kz}{\sinh kd}\right)^2 |\cos(\theta)|\cos(\theta) + \rho\, C_M\frac{\pi D^2}{4}\frac{2\pi^2 H}{T^2}\frac{\cosh kz}{\sinh kd}\sin(\theta) \tag{6.2.18}$$

The total horizontal force in the direction of wave propagation reads

$$F_H = \int_0^{d+\eta} f_H \mathrm{d}z = C_D\frac{\gamma DH^2}{2}K_1\cos(\theta)|\cos(\theta)| + C_M\frac{\gamma\pi D^2 H}{8}K_2\sin(\theta) \tag{6.2.19}$$

in which

$$\gamma = \rho g$$

$$K_1 = \frac{2k(d+\eta) + \sinh 2k(d+\eta)}{8\sinh 2kd} \tag{6.2.20}$$

$$K_2 = \frac{\sinh k(d+\eta)}{\cosh kd} \tag{6.2.21}$$

The total moment of the wave force on the cylinder reads

$$M_H = \int_0^{d+\eta} z f_H \mathrm{d}z \tag{6.2.22}$$

The linear wave theory gives

$$M_H = C_D \frac{\gamma}{2\pi} \frac{DH^2L}{} K_3 \cos(\theta)|\cos(\theta)| + C_M \frac{\gamma}{16} \frac{D^2HL}{} K_4 \sin(\theta) \tag{6.2.23}$$

where L is the wavelength of the small-amplitude wave, and

$$K_3 = \frac{1}{32 \sinh 2kd}\left[2k^2(d+\eta)^2 + 2k(d+\eta)\sinh 2k(d+\eta) - \cosh 2k(d+\eta) + 1\right] \tag{6.2.24}$$

$$K_4 = \frac{1}{\cosh kd}[k(d+\eta)\sinh k(d+\eta) - \cosh k(d+\eta) + 1] \tag{6.2.25}$$

At present, in engineering design the wave forces on small- and medium-scale cylinders are mainly calculated from the Morison equation. The inertia coefficient C_M and the drag coefficient C_D are determined by experiments. There are abundant experimental data on mono-, multi-, and oblique cylinders for references [5].

In recent years, numerical methods are also used to solve the wave forces on cylinders. Mostly the N-S equations or Reynolds equations are solved using the finite element method or the finite difference method, which can simulate the vortex street behind the cylinder properly. However, a higher-order computing scheme is required and the computing workload is heavy when the Reynolds number is high.

2. Lift force on a cylinder

When a small-scale cylinder is in an oscillatory flow or waves, it may undergo a lift force. If the flow around the cylinder is unseparated (very small K_C numbers), the lift force is zero. As the K_C number increases, the lift force fluctuates. The maximum lift force can be expressed as

$$F_{L\max} = \frac{1}{2}\rho C_{L\max} D U_m^2 \tag{6.2.26}$$

in which $C_{L\max}$ is the coefficient of the maximum lift force.

3. Determination of the inertia coefficient C_M, the drag coefficient C_D, and the lift coefficient C_L.

In the Morison equation, the horizontal force on a small-scale cylinder consists of two parts—the drag force and the inertia force. The key to correctly calculate the drag force and the inertia force is the determination of the drag coefficient C_D and the inertia coefficient C_M.

Based on dimensional analysis, the wave force on a cylinder can be expressed as

$$\frac{2F}{\rho LDu_{\max}^2} = f\left(\frac{u_{\max}T}{D}, \frac{u_{\max}D}{\nu}, \frac{\Delta}{D}, \frac{t}{T}\right) \tag{6.2.27}$$

in which D is the characteristic length of the cylinder; Δ is the relative roughness of the cylinder. The inertia coefficient C_M, the drag coefficient C_D, and the lift coefficient C_L can be expressed as

$$C_M = f_2\left(\mathrm{K_C}, \mathrm{Re}, \frac{\Delta}{D}, \frac{t}{T}\right) \tag{6.2.28}$$

$$C_D = f_1\left(\mathrm{K_C}, \mathrm{Re}, \frac{\Delta}{D}, \frac{t}{T}\right) \tag{6.2.29}$$

$$C_L = f_3\left(\mathrm{K_C}, \mathrm{Re}, \frac{\Delta}{D}, \frac{t}{T}\right) \tag{6.2.30}$$

C_M, C_D, and C_L are considered to be independent of time in engineering applications, which yields

$$C_M = f_2\left(\mathrm{K_C}, \mathrm{Re}, \frac{\Delta}{D}\right) \tag{6.2.31}$$

$$C_D = f_1\left(\mathrm{K_C}, \mathrm{Re}, \frac{\Delta}{D}\right) \tag{6.2.32}$$

$$C_L = f_3\left(\mathrm{K_C}, \mathrm{Re}, \frac{\Delta}{D}\right) \tag{6.2.33}$$

In wave motion, the Reynolds number Re can be replaced by the frequency number β which is defined as

$$\beta = \frac{D^2}{\nu T} = \frac{\mathrm{Re}}{\mathrm{K_C}} \tag{6.2.34}$$

It gives $\mathrm{Re} = \beta\,\mathrm{K_C}$, i.e., β is the ratio of Re to $\mathrm{K_C}$. Therefore, the inertia coefficient C_M, the drag coefficient C_D and the lift coefficient C_L can be expressed as

$$C_M = f_2\left(\mathrm{K_C}, \beta, \frac{\Delta}{D}\right) \tag{6.2.35}$$

$$C_D = f_1\left(\mathrm{K_C}, \beta, \frac{\Delta}{D}\right) \tag{6.2.36}$$

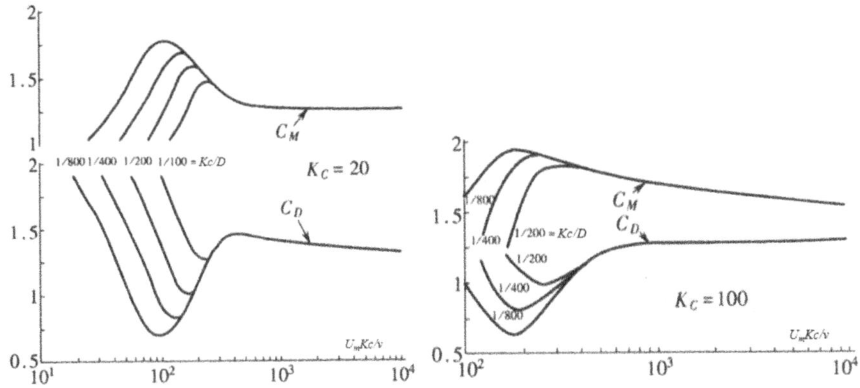

Fig. 6.4 C_D and C_M versus Kc and Re [5]

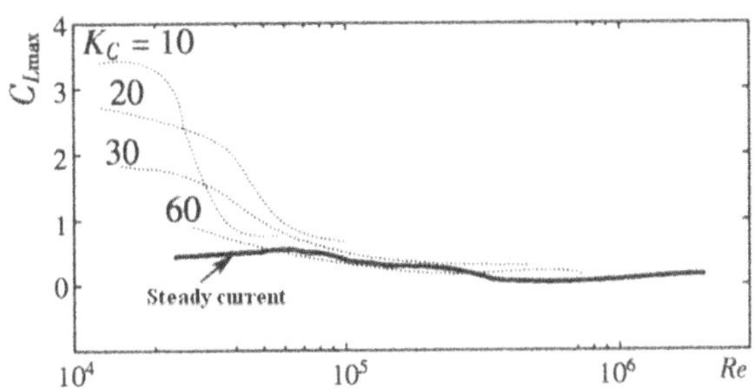

Fig. 6.5 $C_{L\max}$ versus $\mathrm{K_C}$ and Re [5]

$$C_L = f_3\left(\mathrm{K_C}, \beta, \frac{\Delta}{D}\right) \tag{6.2.37}$$

For cylindrical structures, the inertia coefficient C_M and the drag coefficient C_D can be obtained from the diagram according to the $\mathrm{K_C}$ number, β (or Re) and the surface roughness Δ. However, for structures with special section shape, they should be determined by experiments. Figure 6.4 shows the inertia coefficient C_M and the drag coefficient C_D versus $\mathrm{K_C}$ and Re when the Reynolds number Re is large. Figure 6.5 shows the maximum lift coefficient $C_{L\max}$ versus $\mathrm{K_C}$ and Re.

6.3 Wave Forces on a Large-Scale Circular Cylinder

In coastal engineering, when the ratio of geometric dimension of the structure or pile D to the wavelength L, $D/L > 0.15$, the structure affects the wave field. When the length of the water particle trajectory in one wave period is smaller than the scale of the structure, the separation of flow on the structure surface and the fluid viscosity can be neglected. In this case, the wave diffraction theory can be applied.

6.3.1 Problem of Linear Wave Diffraction

It is assumed that the fluid is inviscid, irrotational, and potential in the coordinate system shown in Fig. 6.6. The velocity potential is denoted by $\Phi(x, y, z, t)$. Then the potential function of linear water wave diffraction $\Phi(x, y, z, t)$ satisfies the following governing equation and boundary conditions,

$$\nabla^2\Phi = 0, \qquad \text{in the domain} \tag{6.3.1}$$

$$\frac{\partial^2\Phi}{\partial t^2} + g\frac{\partial\Phi}{\partial z} = 0, \quad z = 0 \tag{6.3.2}$$

$$\eta = -\frac{1}{g}\left(\frac{\partial\Phi}{\partial t}\right), \quad z = 0 \tag{6.3.3}$$

$$\frac{\partial\Phi}{\partial z} = 0, \quad z = -d \tag{6.3.4}$$

$$\frac{\partial\Phi}{\partial n} = 0, \quad \text{on the body surface} \tag{6.3.5}$$

Radiation condition, at infinity

in which z is the vertical coordinate based on the reference of still water level; η is the free surface elevation above the still water level; d is the still water depth; a is the radius of the circular cylinder.

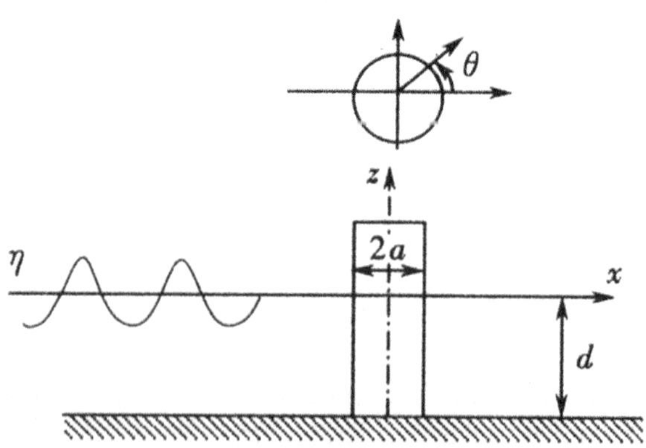

Fig. 6.6 Definition sketch for a large-scale cylinder

By the linear wave theory, the potential function Φ can be decomposed into the incident wave potential function Φ_I and the scattered wave potential function Φ_S,

$$\Phi = \Phi_I + \Phi_s \tag{6.3.6}$$

The incident wave potential function Φ_I refers to the potential function of the small-amplitude wave without obstacle in the fluid. The scattered wave potential function Φ_S is the potential function caused by the existence of the obstacle, which is scattered outward and should satisfy the radiation condition at infinity ($r \to \infty$).

Assuming that,

$$\Phi(x, y, z, t) = \phi(x, y, z)\mathrm{e}^{-i\omega t} \tag{6.3.7}$$

$$\phi = \phi_I + \phi_s \tag{6.3.8}$$

Sommerfeld [6] proposed a radiation condition at infinity for the existent and unique solution of the scattered wave potential function Φ_S in the linear diffraction problem. It reads

$$\lim_{r\to\infty} r^{1/2}\left(\frac{\partial \phi_s}{\partial r} - ik\phi_s\right) = 0 \tag{6.3.9}$$

in which r is the radial distance from the point in the domain to the obstacle. Thus, the governing equation and the conditions of fixed solution for the linear scattered wave potential read,

$$\nabla^2\phi_s = 0 \quad \text{in the domain} \tag{6.3.10}$$

$$\frac{\partial \phi_s}{\partial z} - \frac{\omega^2}{g}\phi_s = 0 \quad z = 0 \tag{6.3.11}$$

$$\frac{\partial \phi_s}{\partial z} = 0 \quad z = -d \tag{6.3.12}$$

$$\frac{\partial \phi_s}{\partial r} = -\frac{\partial \phi_I}{\partial r} \quad \text{on the body surface } r = a \tag{6.3.13}$$

$$\lim_{r\to\infty} r^{1/2}\left(\frac{\partial \phi_s}{\partial r} - ik\phi_s\right) = 0 \quad \text{at infinity} \tag{6.3.14}$$

where ϕ_I is the incident wave potential function, and it satisfies Eqs. (6.3.1)–(6.3.5) and can be expressed as

$$\phi_I = -i\frac{gH}{2\omega}\frac{\cosh[k(z+d)]}{\cosh(kd)}\mathrm{e}^{ikx}. \tag{6.3.15}$$

6.3.2 Wave Forces on a Large-Scale Cylinder

It is assumed that $\phi = \phi(r, \theta, z)$ in a cylindrical coordinate system with three coordinates of r, θ, and z. Then the Laplace equation in the domain reads

$$\frac{\partial^2\phi}{\partial r^2} + \frac{1}{r}\frac{\partial\phi}{\partial\theta} + \frac{1}{r^2}\frac{\partial^2\phi}{\partial\theta^2} + \frac{\partial^2\phi}{\partial z^2} = 0 \tag{6.3.16}$$

By means of $e^{ikx} = e^{ikr\cos\theta} = \cos(kr\cos\theta) + i\sin(kr\cos\theta)$, without obstacle in the fluid, the series expansion of the incident wave potential function ϕ_I of the small-amplitude wave in the cylindrical coordinate system can be derived as

$$\phi_I = -i\frac{gH}{2\omega}\frac{\cosh[k(z+d)]}{\cosh(kd)}\left[\sum_{m=0}^{\infty}\beta_m J_m(kr)\cos(m\theta)\right] \tag{6.3.17}$$

$$\beta_m = \begin{cases} 1 & m = 0 \\ 2i^m & m \geq 1 \end{cases}$$

where $J_m(kr)$ is the Bessel function of the first kind, k is the wave number which satisfies the dispersion relation,

$$\omega^2 = gk \cdot \tanh(kh) \tag{6.3.18}$$

Next, the expression of the scattered potential function ϕ_s is sought, and it should satisfy the Laplace equation, the free surface and bottom boundary conditions, and the Sommerfeld radiation condition. It is assumed that the scattered potential function ϕ_s has a form similar to Eq. (6.3.17) for the incident wave potential function ϕ_I. An unknown coefficient B_m is introduced, and the Bessel function in Eq. (6.3.17) is replaced by the Hankel function of the first kind, then the scattered wave potential function ϕ_s is expressed as

$$\phi_s = -i\frac{gH}{2\omega}\frac{\cosh[k(z+d)]}{\cosh(kd)}\left[\sum_{m=0}^{\infty}\beta_m B_m H_m^{(1)}(kr)\cos(m\theta)\right] \tag{6.3.19}$$

It can be proved that Eq. (6.3.19) satisfies the Laplace equation, the free surface and bottom boundary conditions, and the Sommerfeld radiation condition at infinity. Substituting Eqs. (6.3.17) and (6.3.19) into Eq. (6.3.13) yields

$$B_m = -\frac{J'_m(ka)}{H_m^{(1)'}(ka)}$$

Thus, the total potential function of the linear wave around a large-scale circular cylinder is

$$\Phi = -i\frac{gH}{2\omega}\frac{\cosh[k(z+d)]}{\cosh(kd)}\left[\sum_{m=0}^{\infty}\beta_m\left(J_m(kr) - \frac{J_m'(ka)}{H_m^{(1)'}(ka)}H_m^{(1)}(kr)\right)\cos(m\theta)\right]e^{-i\omega t} \tag{6.3.20}$$

Once the potential function is determined, all other physical quantities can be obtained. The following identity should be noted in solution finding,

$$J_m(ka) - \frac{J_m'(ka)}{H_m^{(1)'}(ka)}H_m^{(1)}(ka) = \frac{2i}{\pi ka H_m^{(1)'}(ka)}$$

According to the Bernoulli equation, the pressure on the cylinder surface is given as

$$\begin{aligned} p|_{r=a} &= -\rho gz - \rho \left.\frac{\partial\Phi}{\partial t}\right|_{r=a} \\ &= -\rho gz + \rho gH\frac{\cosh(k(z+d))}{\cosh(kd)}\sum_{m=0}^{\infty}\frac{i\beta_m\cos(m\theta)}{\pi ka H_m^{(1)}(ka)}e^{-i\omega t} \end{aligned} \tag{6.3.21}$$

The horizontal resulting force on the cylinder reads

$$\begin{aligned} F_x &= -\int_{-d}^{0}\mathrm{d}z\int_{0}^{2\pi} a\cdot p\cos\theta\,\mathrm{d}\theta \\ &= 2\rho gHad\frac{A(ka)}{ka}\frac{\tanh(kd)}{kd}\cos(\omega t - \delta) \end{aligned} \tag{6.3.22}$$

where

$$A(ka) = \left[J_1'^2(ka) + Y_1'^2(ka)\right]^{-\frac{1}{2}}$$

$$\delta = -\tan^{-1}\left[Y_1'(ka)/J_1'(ka)\right]$$

It is found from Eqs. (6.3.20) and (6.3.21) that, both the diffraction potential function and the wave pressure for the linear wave around a large-scale cylinder can be expressed in a form of analytic solution. Similarly, other physical quantities such as the moment of the force on the large-scale cylinder can also be obtained. See the reference [5] for the detail.

6.4 Wave Forces on Large-Scale Circular Multiple Cylinders

6.4.1 Coordinate System and Basic Equations

When cylinders have circular sections, a polar coordinate system is used with the plane of (r, α) coinciding with the still water plane, and the z-coordinate being vertical upward. Equations (6.3.1)–(6.3.6) are the governing equations and the boundary conditions describing the wave diffraction of multi-cylinders.

It is assumed that there are N_c vertical circular cylinders appearing out of the water in a wave field. The locations of any two cylinders and their interaction can be described by the following quantities. See Fig. 6.7 for a sketch of two cylinders.

a_i the radius of the ith circular cylinder;
R_i the distance between the center of the ith circular cylinder and the pole;
ψ_i the angle between the polar axis and the Oi line;
R_{ij} the distance between the centers of the ith and jth circular cylinders;
ψ_{ij} the angle between the polar axis and the ij line.

The potential function Φ for the wave diffraction of a single cylinder is expressed in Eq. (6.3.20), and it has analytic solution. The potential function for the wave diffraction of multi-cylinders can be obtained by a superposition of the potential functions for the individual cylinders. However, the transformation relation between the coordinate systems should be considered.

The incident wave potential function is

$$\Phi_I^o = -f(z, t)\exp[ikr\cos(\theta - \alpha)] \tag{6.4.1}$$

where θ is the angle between the incident wave direction and the x-axis (the polar axis).

$$f(z, t) = \frac{gH}{2\omega}\frac{\cosh[k(z + d)]}{\cosh(kd)}e^{-i\omega t} \tag{6.4.2}$$

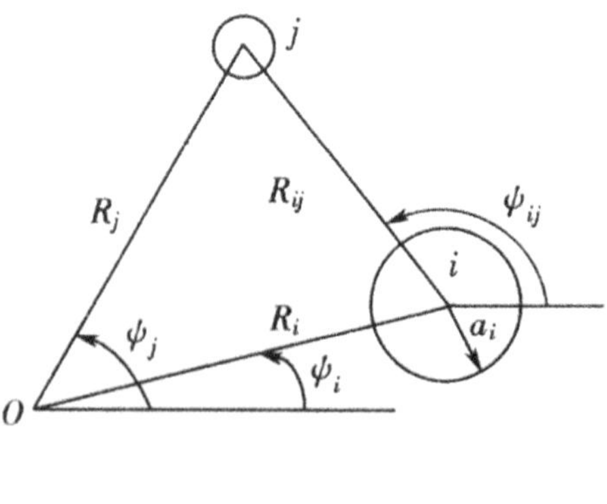

Fig. 6.7 Sketch of two cylinders

The superscript "O" of Φ_I^o denotes that the values of the independent variable in space are in the (r, α) coordinates with the pole of "O". ω and k are the angular frequency and the wave number, respectively, which satisfy the dispersion relation,

$$\omega^2 = gk\tanh(kd)$$

Separating the independent variables in Eq. (6.4.1) yields

$$\Phi_I^o = -f(z,t)\sum_{n=-\infty}^{+\infty} J_n(kr)\exp\left[in\left(\theta - \alpha + \frac{\pi}{2}\right)\right] \quad (6.4.3)$$

Firstly, the incident wave potential function in the jth coordinates (r_j, α_j) is given by

$$\Phi_I^j = -f(z,t)\sum_{n=-\infty}^{\infty} J_n(kr_j)\exp\left[ikR_j\cos(\Psi_j - \alpha) + in\left(\theta - \alpha_j + \frac{\pi}{2}\right)\right] \quad (6.4.4)$$

For linear problems, the total potential function can be expressed by a superposition of the components,

$$\Phi = \Phi_I + \sum_{j=1}^{N_C} \Phi_{Sj} \quad (6.4.5)$$

where Φ_{Sj} is the scattered wave potential under the condition of the only one jth cylinder existing in the wave field. By using the variable separation method, the general form for Φ_{Sj} (in the jth coordinates) can be obtained as

$$\Phi_{Sj} = f(z,t)\sum_{m=-\infty}^{+\infty} A_{mj}H_m(kr_j)\exp(im\alpha_j) \quad (6.4.6)$$

where A_{mj} is unknown complex coefficient determined by the boundary condition on the cylinder surface; H_m is the mth-order Hankel function of the first kind.

The total potential function can be obtained by a superposition of the scattered wave potential components. It should be noted that various coordinate systems have been used in solution finding. In order to implement the superposition easily, a common coordinate system is used for the expressions first. Here the addition theorem for Bessel functions is utilized.

By means of the transformation through the addition theorem for Bessel functions, the scattered wave potential of the ith cylinder in the jth coordinate system can be obtained as

$$\Phi_{si}(r_j, \alpha_j) = f(z,t) \sum_{n=-\infty}^{+\infty} J_{-n}(kr_j) \exp(in\alpha_j)$$
$$- \sum_{n=-\infty}^{+\infty} A_{mj}(kR_{ji}) \exp\left[i(m\psi_{ij} - n\psi_{ji})\right] \quad (6.4.7)$$

Eventually, with diffraction of the Nc cylinders, the total potential function of the ith cylinder in the polar coordinate system having the center of the jth cylinder as the pole can be expressed by

$$\Phi_j(r_j, \alpha_j) = f(z,t) \sum_{n=-\infty}^{+\infty} \phi_{nj}(kr_j) \exp\left[in\alpha_j\right] \quad (6.4.8)$$

in which

$$\Phi_{nj}(kr_j) = -J_n(kr_j) \exp\left[in\left(\frac{\pi}{2} - \theta\right) + ikR_j \cos(\psi_j - \theta)\right] + A_{nj} H_n(kr_j)$$
$$+ \sum_{\substack{i=1 \\ i \neq j}}^{N_C} \sum_{m=-\infty}^{+\infty} A_{mj} J_{-n}(kr_j) H_{m-n}(kR_{ji}) \exp\left[i(m\psi_{ij} - n\psi_{ji})\right] \quad (6.4.9)$$

where the unknown complex coefficients A_{mj} are determined by the boundary condition on the cylinder surface; H_m is the mth-order Hankel function of the first kind.

Using the condition of the zero normal velocity on the cylinder surface required by the boundary,

$$\Phi_{nj}(kr_j) = 0, n = 0, \pm 1, \pm 2, \ldots, j = 1, 2, \ldots, N_C \quad (6.4.10)$$

The unknown complex coefficients A_{mj} can be obtained.

6.4.2 Wave Forces on Multiple Circular Cylinders

After the potential function, Φ is determined, and the Bernoulli equation gives the pressure

$$p = -\rho \frac{\partial \Phi}{\partial t} \quad (6.4.11)$$

By integrating Eq. (6.4.11) over the cylinder surface, the total wave force on the cylinder can be calculated

$$F_{xj} = -\int_{-d}^{0} \mathrm{d}z \int_{0}^{2\pi} \rho \frac{\partial \Phi}{\partial t} \cos(\theta - \alpha_j) a_j \mathrm{d}\alpha_j$$
$$= i\pi\rho g a_j \frac{H}{2k} \tanh(kd)[\phi_{-1j}(ka_j)\mathrm{e}^{-i\theta} + \phi_{1j}(ka_j)\mathrm{e}^{i\theta}]\mathrm{e}^{-i\omega t} \quad (6.4.12)$$

$$F_{yj} = -\int_{-d}^{0} \mathrm{d}z \int_{0}^{2\pi} \rho \frac{\partial \Phi}{\partial t} \sin(\theta - \alpha_j) a_j \mathrm{d}\alpha_j$$
$$= \pi\rho g a_j \frac{H}{2k} \tanh(kd)[\phi_{-1j}\mathrm{e}^{-i\theta} + \phi_{1j}\mathrm{e}^{i\theta}]\mathrm{e}^{-i\omega t} \quad (6.4.13)$$

$$F_{xj} = \frac{\pi\rho g H D_j^2}{8} \frac{\tanh(kd)}{ka_j} \overline{C}_{xj} \mathrm{e}^{-i\omega t} \quad (6.4.14)$$

$$F_{yj} = \frac{\pi\rho g H D_j^2}{8} \frac{\tanh(kd)}{ka_j} \overline{C}_{yj} \mathrm{e}^{-i\omega t} \quad (6.4.15)$$

where

$$\overline{C}_{xj} = i(\phi_{-1j}\mathrm{e}^{-i\theta} + \phi_{1j}\mathrm{e}^{i\theta})$$

$$\overline{C}_{yj} = i(\phi_{-1j}\mathrm{e}^{-i\theta} - \phi_{1j}\mathrm{e}^{i\theta})$$

The position of the action point of the wave force is given by

$$z_f/d = \frac{kd\sinh(kd) - \cosh(kd) + 1}{kd\cosh(kd)}. \quad (6.4.16)$$

6.4.3 Random Wave Forces on Large-Scale Multiple Cylinders

The calculation of random wave forces involves three parts: (1) determination of the incident wave spectrum and calculation of the corresponding characteristic quantities; (2) determination of the transfer function; (3) calculations of the out force spectrum and the corresponding characteristic quantities.

1. Input wave spectrum

There are two kinds of input wave spectrum: frequency spectrum and directional spectrum. Frequency spectrum is denoted by $S(\omega)$. Taking JONSWAP spectrum as an example

$$S(\omega) = \frac{\alpha g^2}{\omega^5} \exp\left[-1.25\left(\frac{\omega}{\omega_0}\right)^{-4}\right]\gamma^a \tag{6.4.17}$$

in which

$$\alpha = \exp\left[-\frac{(\omega - \omega_0)^2}{2\sigma^2\omega_0^2}\right] \tag{6.4.18}$$

where $S(\omega)$ is the spectral density function; α is a parameter related to the wind speed and fetch; T is the wave period; σ is a constant; ω_0 is the peak angular frequency; ω is the angular frequency.

$$\omega = \frac{2\pi}{T} \tag{6.4.19}$$

$$\gamma = 3.3.$$

When $\omega = \omega_0$, the peak spectrum is

$$S_p = \frac{\alpha g^2}{\omega_0^5} e^{-1.25}\gamma \tag{6.4.20}$$

In the following, the superscript "′" is added to denote the dimensional quantities. By using the peak spectrum S_P and the corresponding peak frequency ω_0 as the parameters for non-dimensionalization, the non-dimensional results can be obtained

$$S(\omega) = \frac{S'(\omega)}{S_p} = \omega^{-5} \exp\left[-\frac{5}{4}(\omega^{-4} - 1)\right]\gamma^a \tag{6.4.21}$$

$$a = \exp\left[-\frac{(\omega - 1)^2}{2\sigma^2}\right] - 1 \tag{6.4.22}$$

where $\omega = \omega'/\omega_0$

$$\upsilon = \begin{cases} 0.07, & \text{for } \omega \leq 1 \\ 0.09, & \text{for } \omega > 1 \end{cases}$$

The ith-order moment of the spectrum is defined by

$$m_i = \int_0^\infty \omega^i S(\omega) \mathrm{d}\omega \tag{6.4.23}$$

The relations between the spectral moments and the statistical characteristic quantities of waves, such as the significant wave height H_s and the mean zero-crossing frequency ω_c, read,

$$\omega_c = \omega_0 \sqrt{\frac{m_2}{m_0}} \tag{6.4.24}$$

$$H_s = 4(S_p \omega_0 m_0)^{1/2} \tag{6.4.25}$$

The wave directional spectrum of waves is defined by the product of the frequency spectrum and the normalized directional distribution function

$$S(\omega, \theta) = S(\omega) G(\theta) \tag{6.4.26}$$

Taking the square cosine type of the directional distribution function as an example

$$G(\theta) = G_0 \cos^2(\theta - \theta_0), \quad \theta - \theta_0 \in \left[-\frac{\pi}{2}, \frac{\pi}{2}\right] \tag{6.4.27}$$

where θ_0 is the main wave direction, G_0 is obtained by the normalization relation as $G_0 = \frac{2}{\pi}$.

2. Transfer function

It is assumed that the input wave frequency spectrum and the output wave force spectrum are $S_I(\omega)$ and $S_o(\omega)$, respectively. For linear systems, frequencies remain unchanged in the transfer process. If the input is a harmonic excitation signal with unit amplitude and a frequency of ω, the output of the system is a response with the same frequency and an amplitude of $H(\omega)$. If a random signal of frequency spectrum $S_I(\omega)$ is added to the input, the output force spectrum becomes

$$S_o(\omega) = |H(\omega)|^2 S_I(\omega) \tag{6.4.28}$$

This equation can be regarded as the relation between the input and output spectra, also the definition of the transfer function. $|H(\omega)|^2$ is defined as the transfer function of the system

$$|H(\omega)|^2 = \frac{S_o(\omega)}{S_I(\omega)} \tag{6.4.29}$$

$H(\omega)$ can be obtained from the wave forces of the regular wave with a single frequency acting on a single or multiple cylinders.

The transfer function of the directional spectrum is defined as $H(\omega, \theta)$, which satisfies

$$|H(\omega)|^2 = \int_{-\frac{\pi}{2}}^{\frac{\pi}{2}} H(\omega,\theta)^2 G(\theta) \mathrm{d}\theta \tag{6.4.30}$$

Once the transfer function is solved, the output force spectrum denoted by S^o_{aj} can be calculated

$$S^o_{\alpha j}(\omega,\theta) = \left|H_{\alpha j}(\omega,\theta)\right|^2 S_i(\omega,\theta)$$

where $\alpha = (x, y)$, $j = 1, 2, \ldots, N_c$, N_c is the number of the cylinders.

By means of Eqs. (6.4.14) and (6.4.15), the wave forces of the regular wave with a single frequency acting on multiple cylinders can be obtained by

$$\left.\begin{aligned} P_{xj} &= \frac{H}{2} H_j C_{xj} \\ P_{yj} &= \frac{H}{2} H_j C_{yj} \end{aligned}\right\}, \tag{6.4.31}$$

in which H is the wave height, and

$$\left.\begin{aligned} H_j &= \frac{\pi}{4}\rho g D_j^2 \\ C_{xj} &= \overline{C}_{xj}\frac{\tanh(kd)}{ka_j} \\ C_{yj} &= \overline{C}_{yj}\frac{\tanh(kd)}{ka_j} \end{aligned}\right\}, \tag{6.4.32}$$

The wave force on a single cylinder has only one component, parallel to the main wave direction,

$$P_j^0 = \frac{H}{2} H_j \frac{4\tanh(kd)}{\pi\left(ka_j\right)^2}\left[J_1'^2(ka_j) + Y_1'^2(ka_j)\right]^{-\frac{1}{2}}$$

According to the definition, the transfer functions reads

$$\left.\begin{aligned} \left|H_{xj}(\omega,\theta)\right|^2 &= H_j^2 C_{xj}^2 \\ \left|H_{yj}(\omega,\theta)\right|^2 &= H_j^2 C_{yj}^2 \end{aligned}\right\} \tag{6.4.33}$$

The transfer functions for a single cylinder read

$$\left|H_{xj}^0(\omega,\theta)\right|^2 = H_j^2 C_j^{02} \cos^2(\theta - \theta_0) \tag{6.4.34}$$

$$\left|H_{yj}^{0}(\omega,\theta)\right|^{2}=H_{j}^{2}C_{j}^{02}\sin^{2}(\theta-\theta_{0}) \tag{6.4.35}$$

For a circular cylinder, an analytic solution of C_j^0 exists

$$C_{j}^{0}=\frac{4}{\pi}\frac{\tanh(kd)}{ka_{j}}[J_{1}^{\prime 2}(ka_{j})+Y_{1}^{\prime 2}(ka_{j})] \tag{6.4.36}$$

In the case that only frequency spectrum is considered, the definition of the transfer function in Eq. (6.4.3) gives

$$\left|H_{xj}(\omega)\right|^{2}=\int_{-\frac{\pi}{2}}^{\frac{\pi}{2}}H_{xj}(\omega,\theta)^{2}G(\theta)\mathrm{d}\theta \tag{6.4.37}$$

Several other transfer functions also have similar relations, especially the transfer functions for single cylinder, which have analytic results

$$\left.\begin{aligned}\left|H_{xj}^{0}(\omega)\right|^{2}&=\frac{3}{4}H_{j}^{2}C_{j}^{02}\\ \left|H_{yj}^{0}(\omega)\right|^{2}&=\frac{1}{4}H_{j}^{2}C_{j}^{02}\end{aligned}\right\}, \tag{6.4.38}$$

3. Output force spectrum

The dimensional output force spectrum reads

$$S'_{\alpha j}(\omega,\theta)=\left|H_{\alpha j}(\omega,\theta)\right|^{2}S'(\omega,\theta) \tag{6.4.39}$$

Extracting the non-dimensional part yields

$$S'_{\alpha j}(\omega,\theta)=S_{pj}S_{aj}(\omega,\theta) \tag{6.4.40}$$

in which

$$S_{pj}=G_{0}S_{p}H_{j}^{2} \tag{6.4.41}$$

$$S_{aj}(\omega,\theta)=C_{\alpha j}^{2}S(\omega,\theta)\cos^{2}(\theta-\theta_{0}) \tag{6.4.42}$$

where $\alpha=(x,y)$ and $j=1,2,\ldots,N_{c}$.

The output effective force is

$$P_{aj}=2H_{j}\sqrt{G_{0}S_{p}\omega_{0}}\sqrt{(m_{0})_{aj}} \tag{6.4.43}$$

The mean zero-crossing frequency is

$$\omega_{aj}^{0} = \omega_0 \sqrt{(m_2)_{aj}/(m_0)_{aj}} \tag{6.4.44}$$

where

$$(m_k)_{aj} = \int_0^{\infty} \int_{-\frac{\pi}{2}}^{\frac{\pi}{2}} \omega^k S_{aj}(\omega, \theta) \mathrm{d}\theta \mathrm{d}\omega \tag{6.4.45}$$

The cylinder group effect coefficient K_{aj} is defined by the ratio of the wave force on multiple cylinders to the wave force on a single cylinder,

$$K_{aj} = \frac{P_{aj}}{P_{xj}^{0}} \tag{6.4.46}$$

where P_{xj}^{0} is the effective force in the main wave direction under the same wave condition when the only jth cylinder exists,

$$P_{xj}^{0} = 2H_j \sqrt{G_0 S_p \omega_0} \sqrt{(m_0)_j} \tag{6.4.47}$$

$$(m_0)_j = \frac{3}{4} \int_0^{\infty} C_j^{02}(\omega) S(\omega) \mathrm{d}\omega \tag{6.4.48}$$

It can be seen from the above that, the key to solving the random wave force problem is to calculate the transfer function $|H^2|$ and the numerical integral Eqs. (6.4.45), (6.4.48), and (6.4.23).

6.4.4 Diagrams of Cylinder Group Effect Coefficients for Irregular Wave Forces on Multiple Cylinders

Xinyu et al. [7] proposed the diagrams for cylinder group effect coefficients using the above calculation method for the irregular wave forces on multiple cylinders.

For large-scale multiple cylinders, when the distances between the centers of the cylinders are not more than $4D$ ($l \leq 4D$), the total horizontal wave force is the product of the wave force on the individual cylinder and the cylinder group effect coefficient K_{aj}.

Figures 6.8, 6.9, 6.10, 6.11, 6.12, 6.13, 6.14, 6.15, 6.16 and 6.17 show the cylinder group effect coefficients for the irregular wave forces on two cylinders, three cylinders, and four cylinders. In the diagrams, the parameter for different curves is kD, in which $k = 2\pi/L_p$, L_p is the wavelength calculated by $T_p = 1.2\,\overline{T}$, $\overline{T}$ is the mean period.

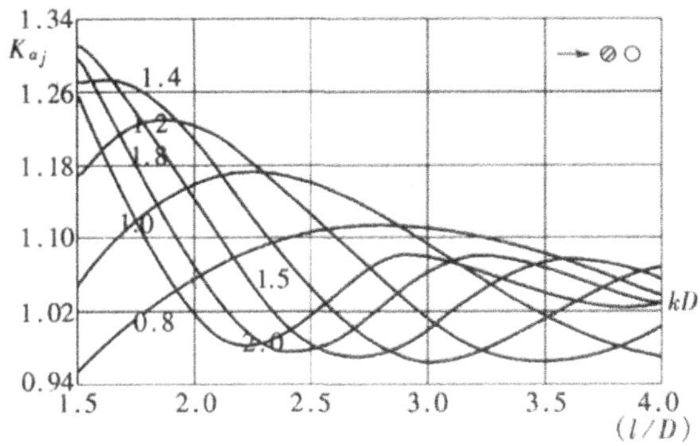

Fig. 6.8 Diagram of the effect coefficients for the first of the two tandem cylinders

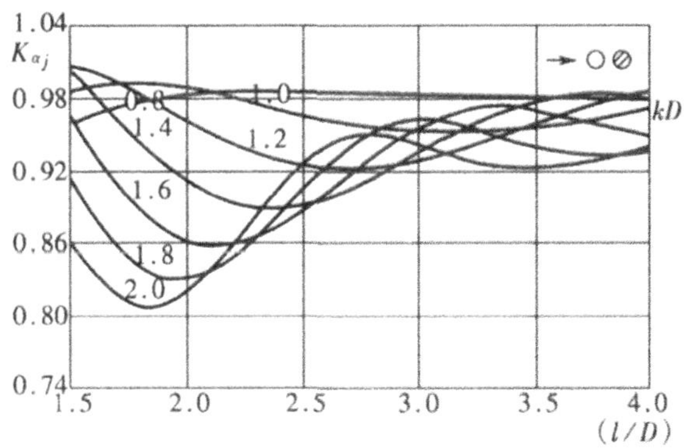

Fig. 6.9 Diagram of the effect coefficients for the second of the two tandem cylinders

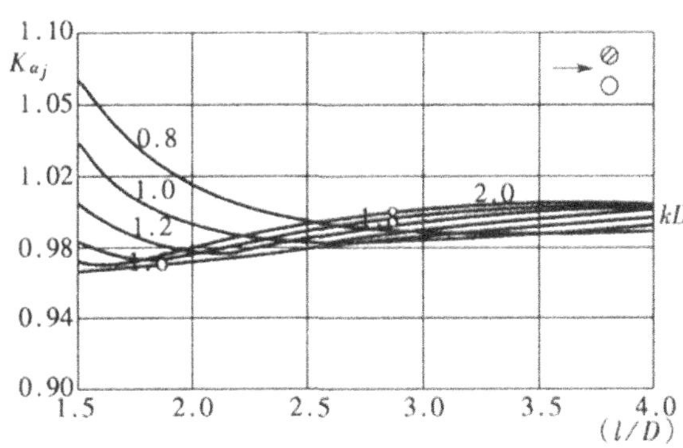

Fig. 6.10 Diagram of the effect coefficients for the first of the two parallel cylinders

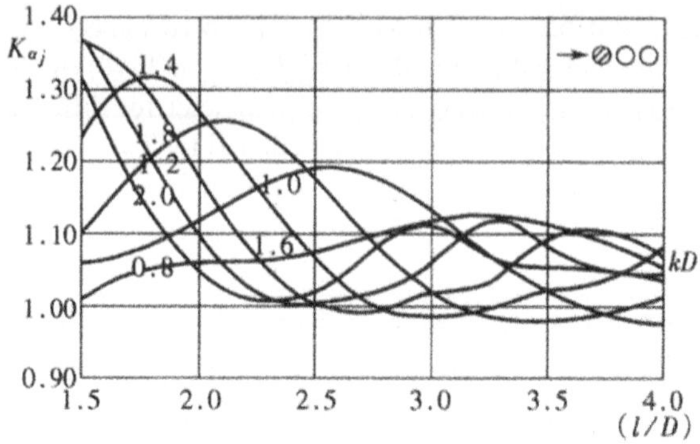

Fig. 6.11 Diagram of the effect coefficients for the first of the three tandem cylinders

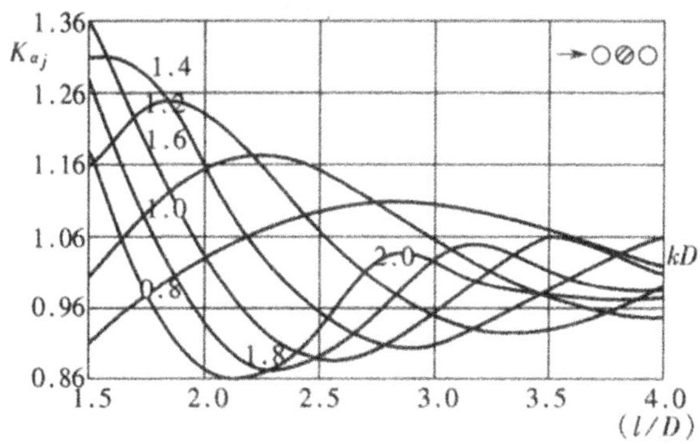

Fig. 6.12 Diagram of the effect coefficients for the second of the three tandem cylinders

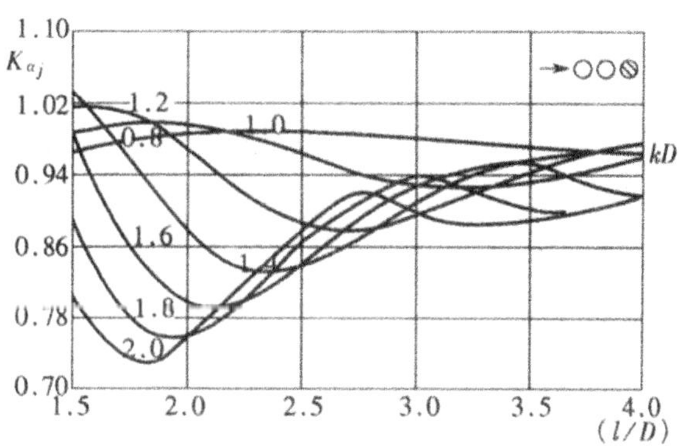

Fig. 6.13 Diagram of the effect coefficients for the third of the three tandem cylinders

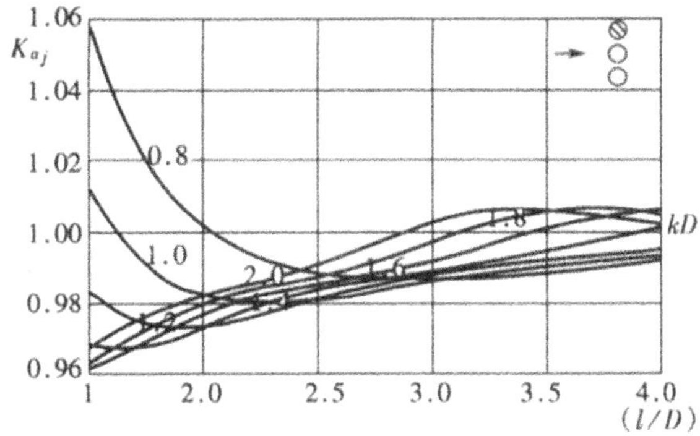

Fig. 6.14 Diagram of the effect coefficients for the first of the three parallel cylinders

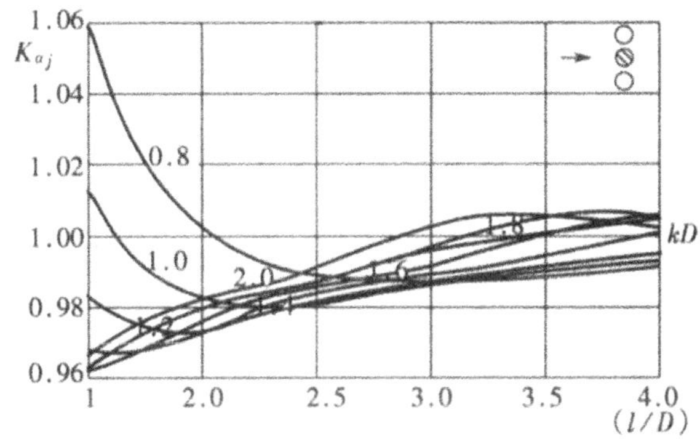

Fig. 6.15 Diagram of the effect coefficients for the second of the three parallel cylinders

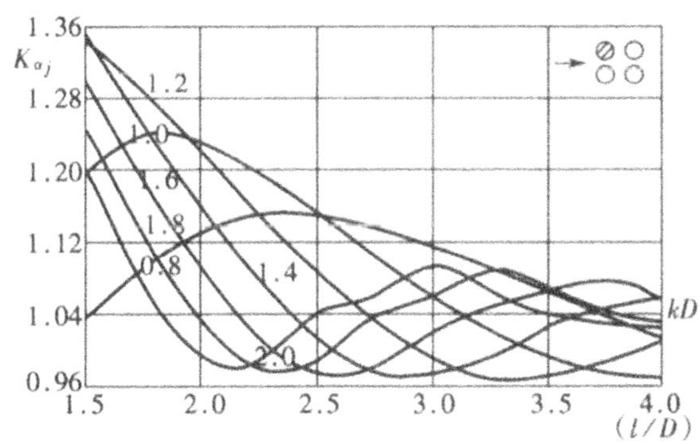

Fig. 6.16 Diagram of the effect coefficients for the front one of the four cylinders

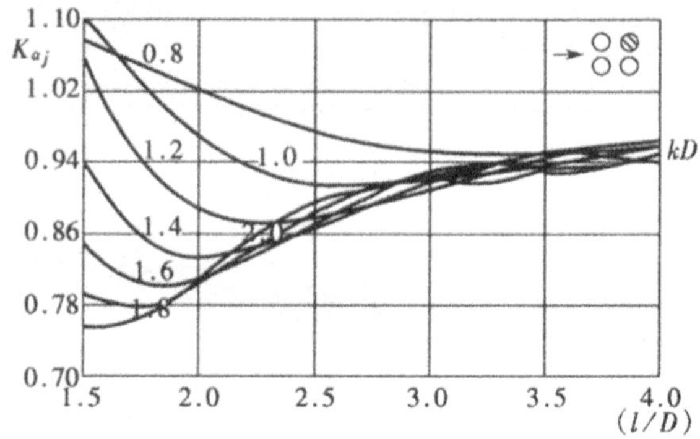

Fig. 6.17 Diagram of the effect coefficients for the rear one of the four cylinders

6.5 Wave Forces on 2D Large-Scale Structures

6.5.1 Calculation of Wave Forces on a 2D Large-Scale Structure Using Boundary Element Method

The wave forces on a large circular cylinder have been described in Sect. 6.6.3, and an analytic solution of wave forces is given by Eq. (6.3.22). When the cross section of the structure is not circular, the analytic solution does not exist anymore, and numerical approaches are needed. A numerical method for calculating the wave forces on a pillar with constant non-circular cross section over the water depth is described in the following. The diffraction of the pillar with constant cross section over the water depth can be calculated as a 2D horizontal diffraction problem, as shown in Fig. 6.18 for a sketch.

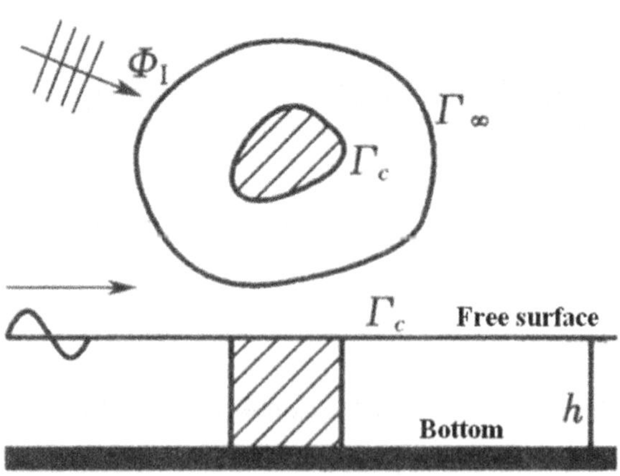

Fig. 6.18 Sketch of 2D diffraction of a pillar with constant cross section

1. Basic equations

Here the velocity potential for the linear wave diffraction still satisfies the equations and the boundary conditions from Eqs. (6.3.1)–(6.3.5). For a 2D problem, the potential function $\Phi(x, y, z, t)$ can be expressed by separation of the variables

$$\Phi(x, y, z, t) = Z(z)\phi(x, y)e^{-i\omega t} \tag{6.5.1}$$

$$Z(z) = \frac{-gH}{2\omega}\frac{\cosh[k(z+d)]}{\cosh kd} \tag{6.5.2}$$

$$\phi(x, y) = \phi_I + \phi_s \tag{6.5.3}$$

where H is the incident wave height; ω is the incident wave angular frequency; k is the wave number; d is the still water depth.

The dispersion relation for the linear small-amplitude waves reads

$$\omega^2 = gk\tanh(kd) \tag{6.5.4}$$

When there is no structure, the known incident wave potential function reads

$$\phi_I(x, y) = e^{ik(x\cos\theta + y\sin\theta)} \tag{6.5.5}$$

where θ is the angle between the incident wave direction and the x-axis.

By substituting Φ in Eqs. (6.5.3) into (6.3.1) and separating the incident wave from the scattered wave, the scattered wave potential ϕ_S satisfies the Helmholtz equation, the problem becomes a horizontal 2D problem on ϕ_S.

$$\frac{\partial^2\phi_S}{\partial x^2} + \frac{\partial^2\phi_S}{\partial y^2} + k^2\phi_S = 0 \quad \text{in the domain} \tag{6.5.6}$$

$$\frac{\partial\phi_S}{\partial n} = -\frac{\partial\phi_I}{\partial n} \quad \text{on the solid wall } \Gamma_c \tag{6.5.7}$$

$$\frac{\partial\phi_S}{\partial r} - ik\phi_s = 0 \quad \text{at infinity } \Gamma_\infty \tag{6.5.8}$$

2. Solving the scattered wave potential ϕ_S and wave forces

Here, the subscript "S" is omitted in the scattered wave potential ϕ_s which can be solved with the boundary element method.

For a boundary value problem of the Helmholtz equation, the direct boundary integral equation of the boundary element method reads

$$c_i\phi_i + \int_\Gamma \left(\frac{\partial\phi^*}{\partial n}\phi + \phi^*\frac{\partial\phi}{\partial n}\right)d\Gamma = 0 \tag{6.5.9}$$

where Γ is the sum of the boundary of the structure cross section Γ_c and the boundary at infinity Γ_∞, $\Gamma = \Gamma_c + \Gamma_\infty$.

The fundamental solution of the Helmholtz equation reads

$$\phi^* = \frac{i}{4} H_0^1(kr) \tag{6.5.10}$$

where $r = [(x-\xi)^2 + (y-\eta)^2]^{1/2}$, H_0^1 is the zero-order Hankel function of the first kind.

The boundary integral Eq. (6.5.9) can be further written as

$$c_i\phi_i + \int_{\Gamma_c} \left(\frac{\partial \phi^*}{\partial n}\phi + \phi^* \frac{\partial \phi}{\partial n} \right) \mathrm{d}\Gamma + \int_{\Gamma_\infty} \left(\frac{\partial \phi^*}{\partial n} - ik\phi^* \right) \phi \mathrm{d}\Gamma = 0 \tag{6.5.11}$$

The asymptotic expression of the Hankel function at infinity is

$$\phi^* \approx \frac{i}{4}\sqrt{\frac{2}{\pi k r}} \exp\left[i\left(kr - \frac{4}{\pi} \right) \right] \quad r \to \infty \tag{6.5.12}$$

Substituting Eq. (6.5.12) into Eq. (6.5.11) yields that the last term of the integral over Γ_∞ in Eq. (6.5.11) is zero. The boundary integral in Eq. (6.5.11) is thus only implemented over the boundary Γ_c,

$$c_i\phi_i + \int\limits_{\Gamma_c} \left(\frac{\partial \phi^*}{\partial n}\phi + \phi^* \frac{\partial \phi}{\partial n} \right) d\Gamma = 0 \tag{6.5.13}$$

In the case that there is an obtuse angle α on the boundary as shown in Fig. 6.19, we have $c_i = 1 - \alpha/2\pi$.

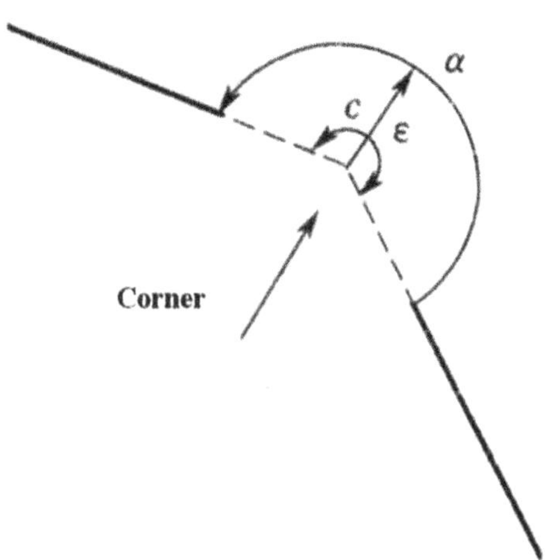

Fig. 6.19 Sketch of an obtuse angle on the boundary

In numerically solving Eq. (6.5.13), the pillar section is divided into M segments. The discrete form of the equation reads

$$c_i\phi_i + \sum_{j=1}^{M} \int_{\Gamma_j} \frac{\partial \phi^*}{\partial n}[N]\mathrm{d}\Gamma\{\phi_j^n\} = -\sum_{j=1}^{M} \int_{\Gamma_j} \phi^*[N]\mathrm{d}\Gamma \left\{\frac{\partial \phi}{\partial n}\right\}_j^n \tag{6.5.14}$$

where N is an interpolation function.

Using linear element for each segment, j yields

$$\phi_j = [N]\{\phi_i^n\} = N_1\phi_1^n + N_2\phi_2^n \tag{6.5.15}$$

$$\left(\frac{\partial \phi}{\partial n}\right)_j = [N]\left\{\frac{\partial \phi}{\partial n}\right\}_j^n = N_1\left(\frac{\partial \phi}{\partial n}\right)_1^n + N_2\left(\frac{\partial \phi}{\partial n}\right)_2^n \tag{6.5.16}$$

where N is a linear interpolation function.

After obtaining M linear algebraic equations by Eq. (6.5.14), ϕ_s at each node can be solved. Then ϕ is obtained by Eq. (6.5.3).

From the linear Bernoulli equation, the pressure on the structure is

$$p = -\rho g z - \rho\frac{\partial \Phi}{\partial t} \tag{6.5.17}$$

The total force acting on the pillar can be obtained by the integral over the surface of the pillar.

6.5.2 *A Case Study of Numerical Solution of Wave Forces on Multiple Cylinders by Using Boundary Element Method [8]*

A quay steering platform at a harbor is supported by four circular cylinders each with a diameter of 18 m. Wave loads should be provided for the design of the cylinders. Figure 6.20 shows elevation and plan views of the cylinders. The seabed is flat in this case study. The 2D boundary element method is used to calculate the wave forces acting on the circular cylinders.

By dividing the boundary of the cross section of each circular cylinder into several elements (line segments), the wave forces on each cylinder and the total wave force on multiple cylinders can be obtained. The influences of the incident wave angle and the distance between the circular cylinders on wave loads are investigated.

Table 6.1 presents comparisons of the numerical and experimental wave force results. Figure 6.21 shows the calculated waves forces on the cylinders. The maximum

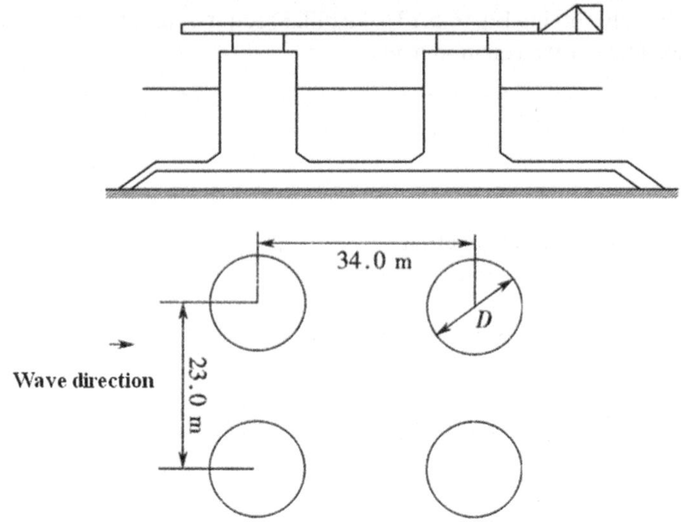

Fig. 6.20 Sketch of the quay steering platform supported by four cylinders

Table 6.1 Comparisons of the numerical and experimental wave force results

Period (s)	Maximum wave forces on the front cylinder (*t*)		Maximum wave forces on the rear cylinder (*t*)		Maximum wave forces on multiple cylinders (*t*)	
	Numerical results	Experimental results	Numerical results	Experimental results	Numerical results	Experimental results
8.8	2033.529	2496.3	1767.678	1407.5	3514.463	3825.53
10.0	1764.736	2240.0	1522.826	1810.6	3776.091	3465.53
11.0	1579.922	1807.9	1363.913	1641.3	3874.544	4860.76
13.0	1325.114	1401.5	1178.658	1416.5	3939.965	3699.89
14.2	1216.936	1195.8	1100.080	1258.1	3838.300	3646.0

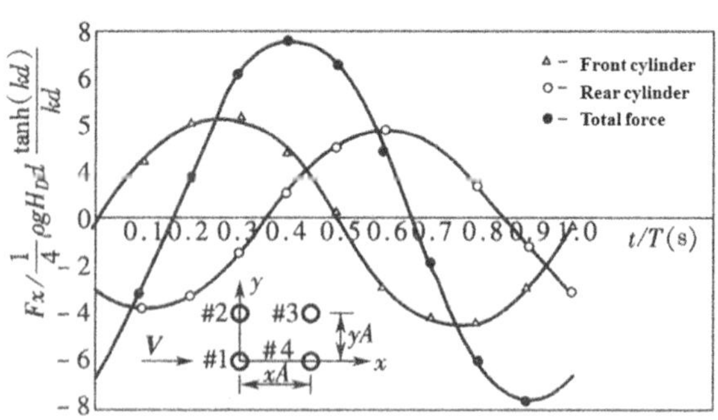

Fig. 6.21 Calculated waves forces on the cylinders in one period

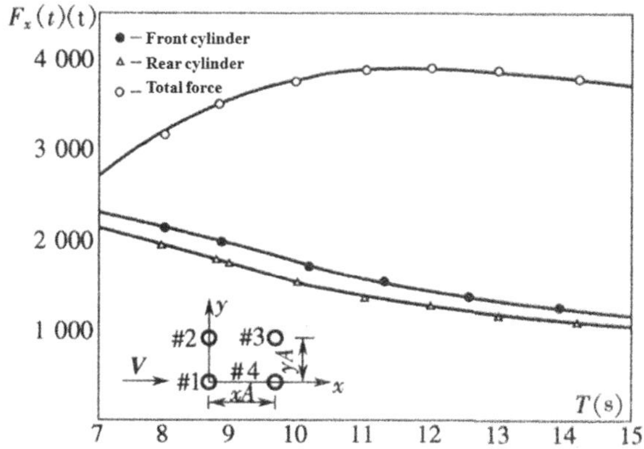

Fig. 6.22 Total wave force on the four cylinders varying with the period

value of the total wave force on multiple cylinders does not occur at the phase with the maximum wave force on a single cylinder, but occurs at a certain phase when the wave forces on both the front and the rear cylinders are positive values.

The total wave force on the four cylinders varying with the period is shown in Fig. 6.22. The maximum total wave force occurs at the wave period of 12 s.

6.6 Nonlinear Wave Forces on a Semicircular Breakwater

Semicircular breakwater is a type of elongated structure with semicircular cross section, and it becomes a 2D structure in a vertical plane when taking unit length for study. In engineering design, mostly, the crest elevation of the semicircular breakwater is close to the design water level. The crest is thus commonly in the state of being submerged and emerged alternately. When waves pass over the crest of the breakwater, free surface vortex and wave breaking may occur sometimes, and the state of the free surface is complex. To simulate the wave forces on such structures, some particular numerical techniques are required in addition to an appropriate mathematical model.

6.6.1 Physical Phenomena of the Interaction of Nonlinear Waves and a Semicircular Breakwater

Experiments and observations demonstrate that when waves pass over a semicircular breakwater, there are three typical hydrodynamic statuses as shown in Fig. 6.23. (1): The water level is relatively high, and the breakwater is submerged. In some extreme wave cases, the waves break near the crest of the breakwater, surface roller

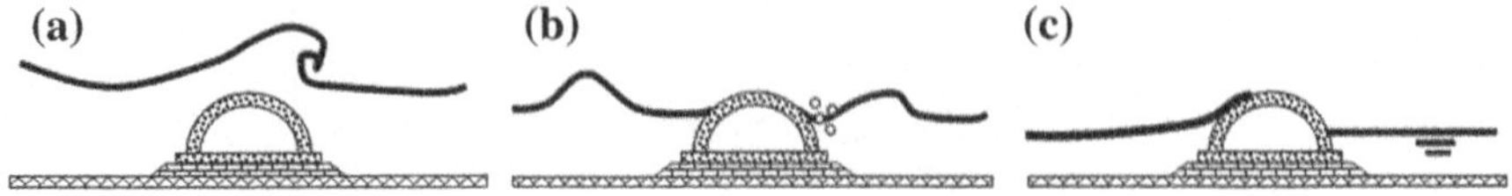

Fig. 6.23 Three typical hydrodynamic statuses for semicircular breakwater

and vortex may occur (in Fig. 6.23a). (2): The water level is close to the crest of the breakwater (in Fig. 6.23b), the waves break and vortex occur obviously near the crest of the breakwater. (3): The crest of the breakwater is emerged above the water surface (in Fig. 6.23c); the waves run up and down along the surface of the breakwater. Sometimes wave overtopping may occur and waves break near the breakwater. Wave breaking results in a huge amount of energy dissipation, and it also leads to discontinuity of the free surface. Furthermore, when waves approach a semicircular breakwater, the wave deformation and reflection are significant in all cases. Wave reflection is more significant in the cases of low water level and the water level close to the crest of the structure.

The energy losses caused by surface roller and wave breaking need to be taken into account in the mathematical model. Numerical techniques to capture the free surface and to deal with the non-reflected open boundary for nonlinear waves are required.

6.6.2 *Mathematical Model*

For the local hydrodynamic phenomena of the interaction of waves with structures, the traditional method is that the viscosity is ignored, and the potential flow theory is then applied. However, for the interaction of nonlinear waves with semicircular breakwaters, experimental data together with a large amount of numerical tests already showed that wave breaking may still occur near the breakwater even in the case of the submerged breakwater and cause severe vortex motion and energy dissipation. Therefore, the energy dissipation of vortex cannot be neglected. Yuan et al. [9] modified the traditional method of the potential flow theory by introducing a dissipative term on the energy dissipation caused by wave breaking and established a numerical model to simulate the interaction of nonlinear waves with a semicircular breakwater. The numerical results agreed well with the experimental data.

1. Governing equation and boundary conditions

The definition sketch for the mathematical model for solving a vertical 2D wave problem is shown in Fig. 6.24. The horizontal and vertical coordinates are denoted by x and z, respectively. η is the free surface elevation above the still water level. The free surface is denoted by S_f. S_B is the solid boundary (including the surface of the breakwater and the seabed). S_L and S_R are the left and right vertical boundaries.

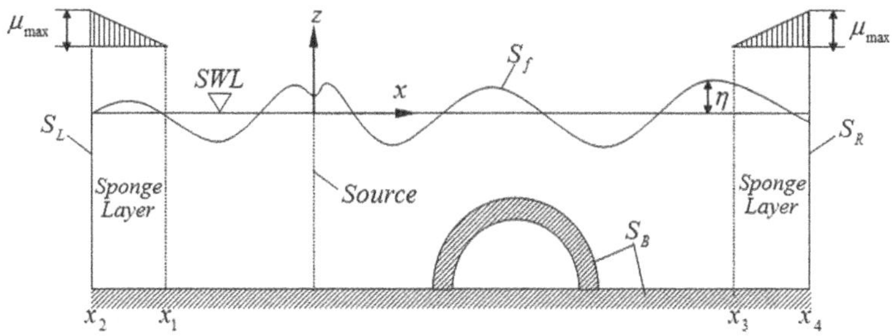

Fig. 6.24 Definition sketch for the mathematical model

In the potential flow theory, the velocity potential in the domain satisfies the Laplace equation

$$\nabla^2 \phi = 0 \tag{6.6.1}$$

The kinematic boundary condition at the free surface reads

$$\frac{\partial \eta}{\partial t} + u\frac{\partial \eta}{\partial x} - w = 0, \quad \text{for} \quad z = \eta \tag{6.6.2}$$

where t is time; u and w are velocity components in x- and z-directions, respectively.

The dynamic boundary condition at the free surface reads

$$\frac{\partial \phi}{\partial t} + \frac{1}{2}|\nabla \phi|^2 + gz + \frac{P_a}{\rho} = C(t), \quad \text{for} \quad z = \eta \tag{6.6.3}$$

where P_a is the atmospheric pressure, ρ is the density of water, $C(t)$ is a function with time, and it is normally set to 0.

The solid boundary condition reads

$$\frac{\partial \phi}{\partial n} = 0 \quad \text{on } S_B \tag{6.6.4}$$

where n is the exterior normal direction of the solid boundary.

The scale of ocean is very large. However, the model domain is limited, so that it involves artificially delimited calculation boundaries, i.e., open boundaries, in which the waves should propagate freely without reflection. In order to make waves pass through the open boundaries without reflection, a lot of studies have been carried out on the open boundary issue. For small-amplitude linear waves, the Sommerfeld radiation condition [10] can be used as the open boundary condition. For nonlinear waves, since there is no theoretically perfect radiation condition available, numerical techniques are employed to absorb the outgoing waves. The numerical techniques on wave absorption will be described in detail later.

2. Wave breaking model

When highly nonlinear waves propagate to the vicinity of a structure, the waves are deformed severely and even break due to the wave reflection and shoaling. Under the different wave level conditions as shown in Fig. 6.23, when the incident wave height is relatively high, wave breaking near the breakwater and the presence of violent vortex and energy dissipation were observed in the physical experiments. When the waves break near the structure, the energy dissipation by the wave breaking affects the wave motion significantly. Hence, the potential flow theory cannot describe the interaction of the waves with the structure correctly. Furthermore, the free surface is no longer continuous, which poses a challenge to solving the governing equation in differential form.

To simulate the energy dissipation by the wave breaking, a dissipative term is introduced into the dynamic free surface boundary condition in Eq. (6.6.3)

$$\frac{\partial \phi}{\partial t} + \frac{1}{2}|\nabla \phi|^2 + gz + \frac{P_a}{\rho} + \varepsilon \frac{\partial^2 \phi}{\partial x^2} = C(t) \tag{6.6.5}$$

where ε is the dissipative coefficient expressed by

$$\varepsilon = C_e \frac{H}{C_z} \sqrt{g\left(u_s^2 + w_s^2\right)} \tag{6.6.6}$$

To keep the model stable, the dissipative coefficient ε near the breaking point is smoothed by

$$\varepsilon = \begin{cases} \frac{L - D_{\text{ist}}}{L} \varepsilon_{\text{BRK}} & , D_{\text{ist}} < L \\ 0 & , D_{\text{ist}} \geq L \end{cases}$$

where ε_{BRK} is the dissipative coefficient ε at the breaking point; D_{ist} represents the horizontal distance between the breaking point and the other point on the free surface; L is the wavelength; C_e= 0.1~10; C_z is the Chezy coefficient; u_s and w_s are the velocity components in the x- and z-directions on the free surface; H is the wave height. In this approach, wave breaking is treated as energy dissipation, which avoids the direct simulation on wave breaking.

6.6.3 Numerical Nonlinear Wave Generation and Boundary Treatments

1. Numerical nonlinear wave generation method

Numerical wave generation method is a key technique in numerical modeling of water waves. It should generate nonlinear waves with expected wave elements at specified locations in the model domain. The two common wave generation methods are (1) giving certain water level or velocity at the incident wave boundary to generate waves and (2) adding a source term in the governing equation to generate waves. The former is easy to implement, but it is difficult to deal with the wave reflection at the incident wave boundary. The method of generating waves with a source term has advantage in eliminating waves at the boundaries and can effectively avoid the interference of the wave reflection from the boundaries in the wave field. This method is widely used in the mathematical model based on the potential wave theory.

In order to eliminate the influence of the wave reflection from the boundaries on the calculation, the method of adding a source term to the governing equation is adopted in the mathematical model. The location of the wavemaker can thus be set in the middle of the computational domain, and the left and right boundaries can be set to the open boundaries with wave absorber (as in Fig. 6.24). Therefore, the governing Eq. (6.6.2) in the computational domain should be changed to

$$\nabla^2\phi = S_{rc} \tag{6.6.7}$$

where $S_{rc} = S_{rc}(x, z)$ is the source term. According to the approach given by Bronson and Larsen [11], it reads

$$S_{rc} = 2w\delta(x - x_s) \tag{6.6.8}$$

in which x_s is the horizontal coordinate at the wavemaker; $w = w(x_s, z, t)$ is the vertical velocity component at the wavemaker; δ is the Dirac delta function.

2. Open boundary treatment

As mentioned above, for linear small-amplitude waves, the Sommerfeld radiation condition can be used to approximate the open boundary condition. For nonlinear waves, there is no theoretically perfect radiation condition at present. Larsen and Dancy [12] proposed an approach of absorbing boundary conditions by sponge layers to attenuate the waves in the sponge layers so as to absorb the reflected waves.

Although the open boundary treatment by sponge layers does not have theoretical explanations so far, good results have been obtained in practical applications. At present, it has become a common approach to deal with the open boundary conditions for nonlinear waves.

In order to effectively eliminate the wave reflection from the left and right boundaries and avoid the influence of the reflected waves on the wave field in the computational domain, the sponge layers and the Sommerfeld radiation boundary condition can be combined at the left and right boundaries to absorb the outgoing waves (see Fig. 6.24).

The combined conditions for wave absorbing read

in sponge layers (left side):

$$\frac{\partial\phi}{\partial t} + \frac{1}{2}|\nabla\phi|^2 + gz + \frac{P_a}{\rho} + \varepsilon\frac{\partial^2\phi}{\partial x^2} - \mu\phi - \int_{x_1}^{x} \mu_x\phi \mathrm{d}x = C(t) \tag{6.6.9a}$$

At the left boundary:

$$\frac{\partial \phi}{\partial x} = \frac{1}{c}\left(\frac{\partial \phi}{\partial t} + \mu\phi + \int_{x_1}^{x_2} \mu_x \phi \mathrm{d}x\right) \tag{6.6.9b}$$

At the right boundary:

$$\frac{\partial \phi}{\partial x} = \frac{1}{c}\left(\frac{\partial \phi}{\partial t} + \mu\phi - \int_{x_3}^{x_4} \mu_x \phi \mathrm{d}x\right) \tag{6.6.9c}$$

where μ is the damping coefficient, which is assumed to be a linear function of the x-coordinate in the sponge layers (see Fig. 6.2); μ_x is the derivative of μ to x; c is the phase velocity of waves at the left and right boundaries.

3. Moving boundary treatment

In the two statuses shown in Fig. 6.23b, c, the waves run up and down along the surface of the breakwater. The intersection of the free surface and the breakwater surface varies with time, presenting a complex moving boundary.

When the water level is relatively low (Fig. 6.23c), the waves cannot pass over the structure to reach the leeside of the structure, but only run up and down along the waveward surface of the breakwater. In this case, Dekui [13] used the method of "tracing point of intersection" to treat the moving boundary, which means the boundary of the computational domain and the elements vary with the intersection of the structure surface and the free surface.

When the water level is close to the crest of the breakwater (Fig. 6.23b), the crest of the breakwater is alternately submerged and emerged. The waves run up and down along the surface of the breakwater. Also, the phenomenon of water body splitting and merging on the left and right sides of the structure takes place. For this case, a combination of "tracing point of intersection" and "thin layer of water" can be applied to treat the moving boundary of the alternately submerged and emerged semicircular breakwater.

6.6.4 *Establishment, Discritization, and Numerical Solution of the Boundary Integral Equation*

Considering the complexity of the structure boundary, the boundary element method with a good adaptability to the boundary is used to discretize the governing equation in the domain, and the free surface boundary is solved by the finite difference method. An iteration method is applied to ensure the time marching of the governing equation in the domain being at the same rate as the boundary conditions.

1. Mathematical model in boundary integral form

Implementing the appropriate treatments mentioned above (adding the source term of wave generation, the sponge absorbing boundary, and the energy dissipation term), the basic equations become

$$\nabla^2\phi = S_{rc} \quad \text{in the domain} \tag{6.6.10}$$

$$\frac{\partial\phi}{\partial t} + \frac{1}{2}|\nabla\phi|^2 + gz + \frac{P_a}{\rho} + \varepsilon\frac{\partial^2\phi}{\partial x^2} - \mu\phi - \int_{x_1}^{x}\mu_x\phi\mathrm{d}x = C(t) \quad , z = \eta \tag{6.6.11}$$

$$\frac{\partial\eta}{\partial t} + u\frac{\partial\eta}{\partial x} - w = 0 \quad , z = \eta \tag{6.6.12}$$

$$\frac{\partial\phi}{\partial n} = 0 \quad \text{on the solid boundary} \tag{6.6.13}$$

$$\frac{\partial\phi}{\partial x} = \frac{1}{c}\left(\frac{\partial\phi}{\partial t} + \mu\phi + \int_{x_1}^{x_2}\mu_x\phi\mathrm{d}x\right) \quad \text{at the left boundary} \tag{6.6.14a}$$

$$\frac{\partial\phi}{\partial x} = \frac{1}{c}\left(\frac{\partial\phi}{\partial t} + \mu\phi - \int_{x_3}^{x_4}\mu_x\phi\mathrm{d}x\right) \quad \text{at the right boundary} \tag{6.6.14b}$$

The boundary value problem described by Eqs. (6.6.10)–(6.6.14) can be solved by the finite element method or the boundary element method. For the problem of a semicircular breakwater, there is a difficulty in mesh generation and reconstruction for the complex computational domain using the finite element method. This can be avoided by using the boundary element method. The boundary element theory gives that the integral equation for the boundary value problem described by the above governing equation and boundary conditions can be written as

$$\gamma\phi(P_P) = \int_S\left[\phi(P_Q)\frac{\partial G_r}{\partial n} - G_r\frac{\partial\phi(P_Q)}{\partial n}\right]\mathrm{d}S + \int_\Omega S_{rc}\mathrm{d}\Omega \tag{6.6.15}$$

where $S = S_B + S_f + S_L + S_R$ is the whole boundary of the computational domain; Ω is the computational domain; $G_r(P_P, P_Q)$ is the fundamental solution of the Laplace equation; $P_Q = (x_Q, z_Q)$ is the "source" point on the boundary; $P_P = (x_P, z_P)$ is the "observation" point. In case of P_P being an interior point in the domain, γ is equal to 2π. In case of P_P being a boundary point, γ is equal to the interior angle (radian) at P_P on the boundary. The fundamental solution or Green's function G_r reads

$$G_r(P_P, P_Q) = \ln \frac{1}{r} \tag{6.6.16}$$

where $r = \sqrt{(x_P - x_Q)^2 + (z_P - z_Q)^2}$ is the distance between P_P and P_Q.

2. Discretization of the equation

In order to treat the nonlinear terms in the governing equation, the equation is discretized and solved with a time-marching technique. At each time step, the governing equation is solved in the computational domain by using the boundary element method, and the time-dependent terms are solved by the finite difference method. In the following description, the superscripts n, $n + 1$, etc., denote the number of time step in the solving process. For instance, ϕ^{n+1} denotes the potential function at time $t = (n + 1)\Delta t$, in which Δt is the time step. The boundary of the whole domain is discretized into N line elements. In this model, the boundary integral equation is solved by using linear elements, which means ϕ and $\partial\phi/\partial n$ vary linearly at each element. The detailed discretization formula can be found in Reference [13].

6.6.5 *Procedure of Solving the Equations and the Pressure Formula*

Obviously, except for the solid boundary condition in Eq. (6.6.14), the other four boundary conditions and the continuity equation in the domain are coupled, so that they need to be solved simultaneously. In this model, the five equations are discretized and solved separately at the same time step. The synchronous solutions can be achieved by iterating at this time step. The main procedure includes the following steps

(1) Determining the source term S_{rc} in Eq. (6.6.10), and solving the boundary integral Eq. (6.6.15).
(2) Solving Eq. (6.6.11) to obtain the potential function at the new time step.
(3) Updating the boundary conditions with the new potential function obtained from Step (2).
(4) Judging whether a given iteration accuracy is reached. If so, marching a time step forward and going back to Step (1). Otherwise, going back to Step (2) for the next iteration until the given accuracy is reached.

In Step (2), the boundary conditions at the first iteration of each time step can be estimated by using the values at the last time step.

At each time step, after solving the potential function in the domain and the free surface boundary condition, the pressure on the structure surface can be obtained by the Bernoulli equation

$$\frac{\partial \phi}{\partial t} + \frac{1}{2}|\nabla \phi|^2 + gz + \frac{p}{\rho} = C(t) \tag{6.6.17}$$

where p is the pressure at any point in the computational domain.

This equation can be discretized in finite difference scheme as

$$\frac{p_i^{n+1}}{\rho} = \frac{\phi_s^{n+1} - \phi_i^{n+1} - \phi_s^n + \phi_i^n}{\Delta t} + \frac{1}{2}(|\nabla \phi_s|^2 - |\nabla \phi_i|^2)^{n+1/2} + g(\eta_i^{n+1/2} - z) \tag{6.6.18}$$

where the subscript "s" denotes a point on the free surface. The pressure on the free surface can be taken as 0 for convenience.

6.6.6 Verification of Numerical Model

A large amount of numerical calculations were carried out by Yuan [9] in order to verify the reliability of the above model. Figure 6.25 shows the locations of the measurement points on the surface of a semicircular breakwater. Figure 6.26 shows the comparison between the numerical and experimental pressure results. The numerical results of the maximum total horizontal wave force compared with the experimental results are shown in Fig. 6.27. Figure 6.28 shows the comparison of the numerical and experimental maximum total vertical wave forces.

The comparisons of the numerical and experimental results suggest that the mathematical model, the numerical method, and techniques used can be applied to calculate wave forces.

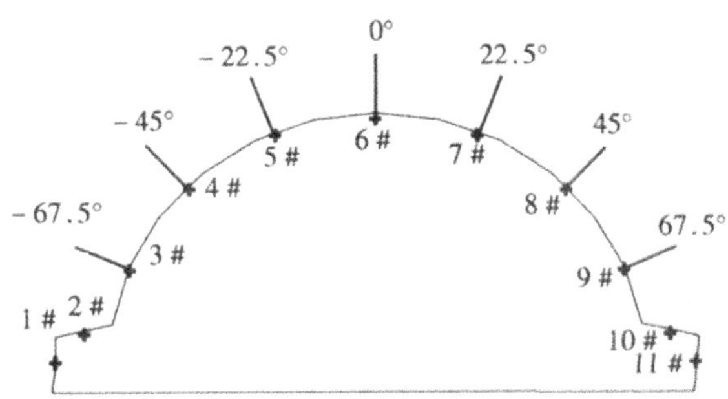

Fig. 6.25 Distribution of the measurement points (radius = 4.0 m)

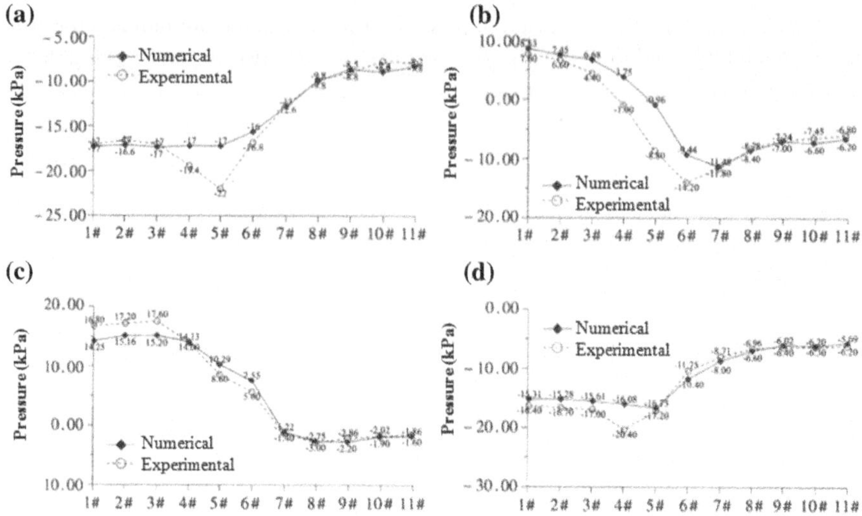

Fig. 6.26 Comparison of the numerical results and the experimental data of the pressure along the surface of the breakwater. **a** At the moment with the minimum total horizontal wave force at extreme high water level; **b** At the moment with the maximum total vertical wave force at extreme high water level; **c** At the moment with the maximum total horizontal wave force at design high water level; **d** The minimum pressure at each point, at design high water level

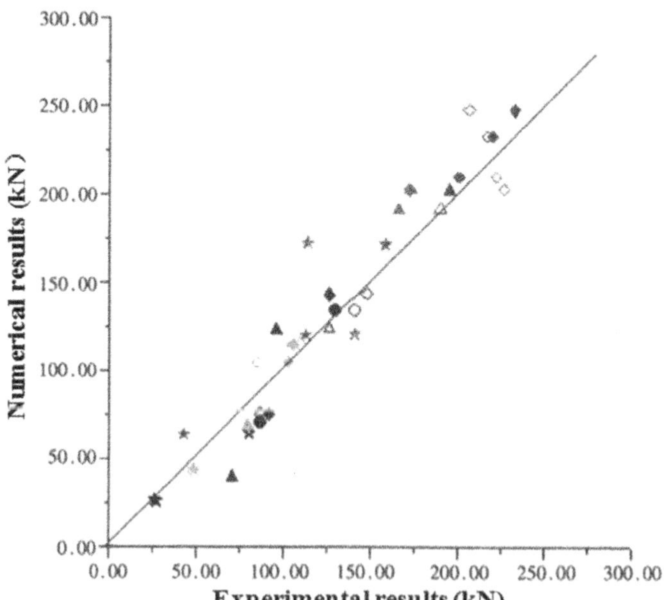

Fig. 6.27 Comparison of the numerical results and the experimental data of the maximum total horizontal wave forces for different radiuses of the semicircular breakwater at different water levels. Solid symbols: regular waves; Hollow symbols: irregular waves

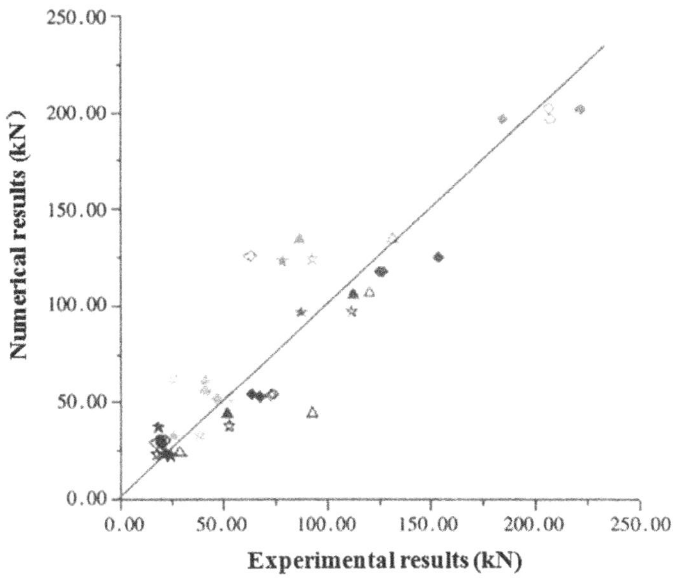

Fig. 6.28 Comparison of the numerical results and the experimental data of the maximum total vertical wave forces for different radiuses of the semicircular breakwater at different water levels. Solid symbols: regular waves; Hollow symbols: irregular waves

6.7 Wave Forces on 3D Large-Scale Structures

6.7.1 Solutions of Wave Forces by Linear Diffraction Theory

The wave forces acting on arbitrarily shaped 3D large-scale structures can be obtained by solving the potential function of a 3D diffraction problem. The indirect boundary element method [14], namely complex Green's function, for solving the potential function, is described in the following.

1. Potential function

It is assumed that the potential function is $\Phi(x, y, z, t) = \phi(x, y, z)\mathrm{e}^{-i\omega t}$, where $\phi(x, y, z) = \phi_I + \phi_S$, ϕ_I is the incident wave potential function without structure in the water, and ϕ_S is the scattered wave potential function due to the presence of the structure.

In Green's function method, the scattered wave potential function at the point (x, y, z) in the computational domain can be expressed as

$$\phi_S(x, y, z) = \frac{1}{4\pi} \iint f(\xi, \eta, \zeta) G(x, y, z, \xi, \eta, \zeta) \mathrm{d}S \tag{6.7.1}$$

where $f(\xi, \eta, \zeta)$ is the source intensity at the point (ξ, η, ζ) on the body surface; $G(x, y, z, \xi, \eta, \zeta)$ is Green's function, which is not a fundamental solution of the Laplace equation (the fundamental solution can be called the simple Green's function), but a function satisfying the linear free surface condition, the bottom boundary condition, the radiation condition, and satisfying the Laplace equation except for the point (ξ, η, ζ) and is thus called complex Green's function; S is the whole surface of the structure. If Green's function satisfying the above conditions is found, the boundary integral will only be done on the structure surface to save workload. However, Green's function is commonly very complex.

The source intensity f can be obtained by solving the following equation

$$-f(x, y, z) + \frac{1}{2\pi}\iint_S f(\xi, \eta, \zeta)\frac{\partial}{\partial n}G(x, y, z, \xi, \eta, \zeta)\mathrm{d}S = 2\frac{\partial \phi_I}{\partial n} \tag{6.7.2}$$

where $\frac{\partial G}{\partial n}$ is the directional derivative of Green's function in the exterior normal direction of the structure surface S; $\frac{\partial \phi_I}{\partial n}$ is the directional derivative of the incident wave potential function ϕ_I in the exterior normal direction of the structure surface S, which is a known value.

$$\frac{\partial G}{\partial n} = \nabla G \cdot \boldsymbol{n} = \frac{\partial G}{\partial x}n_x + \frac{\partial G}{\partial y}n_y + \frac{\partial G}{\partial z}n_z \tag{6.7.3}$$

where $\boldsymbol{n} = \boldsymbol{i}n_x + \boldsymbol{j}n_y + \boldsymbol{k}n_z$ is the exterior normal vector. Equation (6.7.1) and Eq. (6.7.2) are the two basic equations for solving the potential function ϕ_S. The source intensity f is obtained by solving Eq. (6.7.2) first, and then ϕ_S is solved by Eq. (6.7.1).

2. Green's function

(1) Green's function in integral form

It was proposed by John [15],

$$G = \frac{1}{R} + \frac{1}{R'} + 2\text{P.V.}\int_0^\infty \frac{(\mu + \nu)\mathrm{e}^{-\mu d}\cosh[\mu(\zeta + d)]\cosh[\mu(z + d)]}{\mu\sinh(\mu d) - \nu\cosh(\mu d)}J_0(\mu r)\mathrm{d}\mu$$
$$+ i\frac{2\pi(k^2 - \nu^2)\cosh[k(\zeta + d)\cosh[k(z + d)]}{k^2 d - \nu^2 d + \nu}J_0(kr) \tag{6.7.4}$$

where

$$R = [(x - \xi)^2 + (y - \eta)^2 + (z - \zeta)^2]^{1/2}$$
$$R' = [(x - \xi)^2 + (y - \eta)^2 + (z + 2d + \zeta)^2]^{1/2}$$
$$r = [(x - \xi)^2 + (y - \eta)^2]^{1/2}$$
$$\nu = k\tanh(kd) = \omega^2/g$$

"P.V." denotes the principal value integral.

(2) Green's function in series form

Since Green's function in integral form is complex, it can be calculated in series form

$$G = \frac{1}{2}\frac{\nu^2 - k^2}{k^2 d - \nu^2 d + \nu}\cosh[k(z+d)\cosh[k(\zeta+d)][Y_0(kr) - iJ_0(kr)]$$
$$+ \frac{1}{\pi}\sum_{m=1}^{\infty}\frac{(\mu_m^2 + \nu^2)}{\mu_m^2 d + \nu^2 d - \nu}\cos[\mu_m(z+d)]\cos[\mu_m(\zeta+d)]K_0(\mu_m r) \quad (6.7.5)$$

where J_0 and Y_0 are the zero-order Bessel functions of the first and second kinds; K_0 is the zero-order modified Bessel function of the second kind; μ_m is the real positive solution to the following equation

$$\mu_m \tan(\mu_m d) + \nu = 0 \quad (6.7.6)$$

The numerical calculations indicated that Green's function in series form has better convergence, and it is simple and efficient. However when $r \to 0$, $K_0(\mu_k r) \to \infty$ has singularities, and Green's function in series cannot be applied directly. It is usually needed to use Green's function locally in integral form instead of Green's function in series form. In addition to calculating the principal value integral, using Green's function in integral form also needs to calculate the integral of $1/R$ over the panel surface. When the shape of the structure is complex and there are lots of types of element division, the calculation is very complicated. Resorting to the treatment of singularity for Green's function in series form, the local Green's function in integral form can be avoided and Green's function in series form is adopted only, so that the calculation is simplified.

(3) Green's function in asymptotic form [16]

$$G = \frac{1}{R} + \frac{1}{R_1} + \sum_{n=1}^{\infty}\left[\frac{1}{R_{2n}} + \frac{1}{R_{3n}} + \frac{1}{R_{4n}} + \frac{1}{R_{5n}}\right], \omega \to 0$$

where

$$R = [r^2 + (z-\zeta)^2]^{1/2}$$
$$R_1 = [r^2 + (z+\zeta)^2]^{1/2}$$
$$R_{2n} = [r^2 + (z - 2nd - \zeta)^2]^{1/2}$$
$$R_{3n} = [r^2 + (z + 2nd + \zeta)^2]^{1/2}$$
$$R_{4n} = [r^2 + (z + 2nd - \zeta)^2]^{1/2}$$
$$R_{5n} = [r^2 + (z - 2nd + \zeta)^2]^{1/2}$$

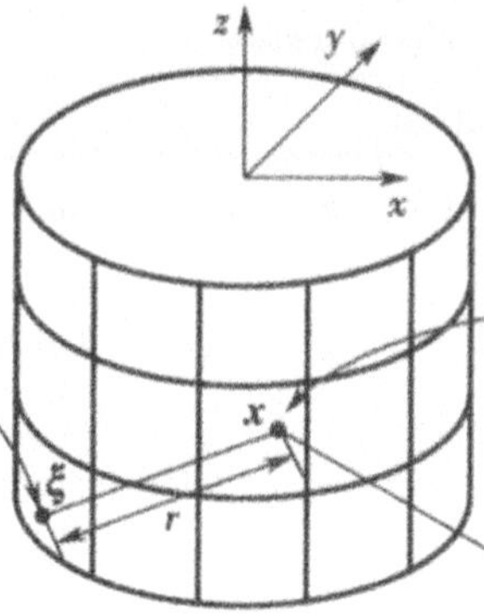

Fig. 6.29 Definition sketch on the structure surface for a 3D diffraction problem

$$r = [(x-\xi)^2 + (y-\eta)^2]^{1/2}$$

Green's function in asymptotic form only performs real operations with simple calculation but less accuracy.

2. Numerical calculation method

For numerical calculation, the surface of the structure is divided into N elements, and each element has area of ΔS_j ($j = 1,2, \ldots, N$). The centroid point of each element is used as the control point with the coordinates of (x_i, y_i, z_i). Figure 6.29 shows a definition sketch on the structure surface. Then, Eqs. (6.7.1) and (6.7.2) can be written as

$$\phi_s(x_i, y_i, z_i) = \frac{1}{4\pi}\sum_{j=1}^{N}\iint_{\Delta S_j} f_j(\xi,\eta,\zeta)\cdot G(x_i, y_i, z_i, \xi, \eta, \zeta)\mathrm{d}S \tag{6.7.7}$$

$$-f_i + \frac{1}{2\pi}\sum_{j=1}^{N}\iint_{\Delta S_j} f_j(\xi,\eta,\zeta)\frac{\partial}{\partial n}G(x_i, y_i, z_i, \xi, \eta, \zeta)\mathrm{d}S = 2\frac{\partial \phi_I}{\partial n} \tag{6.7.8}$$

If constant elements are used, i.e., assuming that the source intensity on the surface of each element f_i is constant and represented by the value at the centroid point (x_i, y_i, z_i), it can be put outside the integral sign in Eqs. (6.7.1) and (6.7.2). Then the basic equations are discretized as

$$\phi_s = \boldsymbol{\beta}_{ij} f_j \tag{6.7.9}$$

$$-f_i + \alpha_{ij} f_j = 2\frac{\partial \phi_I}{\partial n} \quad i = 1, 2, \ldots, N; \quad j = 1, 2, \ldots, N \tag{6.7.10}$$

where

$$\boldsymbol{\beta}_{ij} = \frac{1}{4\pi} \iint_{\Delta S_j} G(x_i, y_i, z_i, \xi, \eta, \zeta)\mathrm{d}S \tag{6.7.11}$$

$$\boldsymbol{\alpha}_{ij} = \frac{1}{2\pi} \iint_{\Delta S_j} \frac{\partial}{\partial n} G(x_i, y_i, z_i, \xi, \eta, \zeta)\mathrm{d}S \tag{6.7.12}$$

Since Green's function $G(x, y, z, \xi, \eta, \zeta)$ is known, when the generation of ΔS_j on the structure surface is completed, the normal direction $\boldsymbol{n}$ of each element is also known. Thus it is not difficult to find out the coefficient matrices $\boldsymbol{\beta}_{ij}$ and $\boldsymbol{\alpha}_{ij}$. After getting $\boldsymbol{\alpha}_{ij}$, the source intensity f_i can be solved by the linear algebraic Eq. (6.7.10). Then the scattered wave potential ϕ_S at each element on the structure surface is obtained by Eq. (6.7.9).

3. Wave forces

After getting the scattered wave potential, the potential function can be obtained by

$$\phi = \phi_I + \phi_S \tag{6.7.13}$$

The Bernoulli equation gives the pressure on the structure,

$$p = -\rho g z + i\omega\rho\phi \mathrm{e}^{-i\omega t} \tag{6.7.14}$$

The total wave force is obtained by the integral over the structure surface

$$F = -\int_S \boldsymbol{p} \cdot \boldsymbol{n}\mathrm{d}S \tag{6.7.15}$$

The total wave force moment reads

$$M = \int_S \boldsymbol{p} \cdot (\boldsymbol{r} \times \boldsymbol{n})\mathrm{d}S \tag{6.7.16}$$

The procedure of using linear diffraction theory to calculate the wave forces on a 3D large structure includes that, generating the structure surface elements ΔS_j; getting the coordinates of the centroid point and the exterior normal direction for each element; solving the coefficient matrices $\boldsymbol{\beta}_{ij}$ and $\boldsymbol{\alpha}_{ij}$ by Eqs. (6.7.11) and (6.7.12) through the integrals of Green's function and its derivative; calculating the potential function $\phi = \phi_I + \phi_S$ at the centroid point of each element by Eq. (6.7.9); then obtaining the wave force and moment by Eqs. (6.7.15) and (6.7.16).

4. Treatment of singularity

The key to getting the linear wave forces on a 3D structure is to solve the two linear algebraic equations (6.7.9) and (6.7.10) for the scattered wave potential ϕ_S. To solve these two equations, the coefficient matrices $\boldsymbol{\beta}_{ij}$ and $\boldsymbol{\alpha}_{ij}$ need to be obtained first. The calculations of the two matrices $\boldsymbol{\beta}_{ij}$ and $\boldsymbol{\alpha}_{ij}$ are put down to the integrations of the Neumann function Y_0 and the Bessel function K_0, i.e., $\int_0^a Y_0(Br)\mathrm{d}r$ and $\int_0^a K_0(Br)\mathrm{d}r$.

where

$$B = \begin{cases} k, & \text{for non infinite series terms} \\ \mu_m, & \text{for infinite series terms} \end{cases}$$

However, when $r \to 0$, the two functions Y_0 and K_0 tend to infinity, so that singularities appear. Yan et al. [17] treated the integral singularity by using the following method.

When $r \to 0$, the asymptotic property of $Y_0(kr)$ and $K_0(kr)$ gives

$$Y_0(r) \to -\frac{2}{\pi}\ln\frac{2}{r}$$

$$K_0(r) \to \ln\frac{2}{r}$$

The integral $\int_0^a K_0(r)\mathrm{d}r$ is taken as an example. The asymptotic expression is deducted from the integral function first, so that the singularity can be eliminated. After that the integral of the asymptotic expression is re-added.

$$\int_0^a K_0(r)\mathrm{d}r = \int_0^a \left[K_0(r) - \ln\frac{2}{r}\right]\mathrm{d}r + \int_0^a \ln\frac{2}{r}\mathrm{d}r$$

In the above equation, the first term on the right side can be integrated by a numerical method. The integral of the asymptotic expression can be calculated by analytical integral, which gives

$$\int_0^a \ln\frac{2}{r}\mathrm{d}r = r\ln\frac{2}{r}\Big|_0^a + \int_0^a \mathrm{d}r = a\left(\ln\frac{2}{a} + 1\right)$$

In order to verify the feasibility of this method, calculations are carried out within the possible practical range of the integral bound. Table 6.2 [18] presents the calculated results with the treatment of singularity compared with the accurate values and the results obtained by using the four-point Gaussian integral directly without the treatment of singularity. The comparisons indicate that this treatment is effective.

Table 6.2 Comparison of the calculated results with the treatment of integral singularity

a	Accurate values	Without the treatment of singularity		With the treatment of singularity	
		Values	Errors (%)	Values	Errors (%)
0.5	0.927103	0.912444	1.6	0.926873	0.025
1.0	1.242510	1.211272	2.5	1.241079	0.12
1.5	1.394582	1.347641	3.4	1.392010	0.18
5.0	1.567387	1.409871	10.05	1.552013	0.98
10.0	1.570779	1.241437	20.1	1.556119	0.93

6.7.2 *Numerical Calculation of Wave Forces on an Offshore Gravity Platform [19]*

Here we use a case that an offshore gravity platform consists of a caisson, four support cylinders and an upper platform, as shown in Fig. 6.30. Wave loads mainly act on the cylinders and the caisson.

The hydrodynamic properties of the gravity platform allow that the wave forces on the gravity platform can be separated into two parts. The upper cylinders are

Fig. 6.30 Sketch of the offshore gravity platform

Fig. 6.31 Definition sketch. a Elevation view; b Plan view

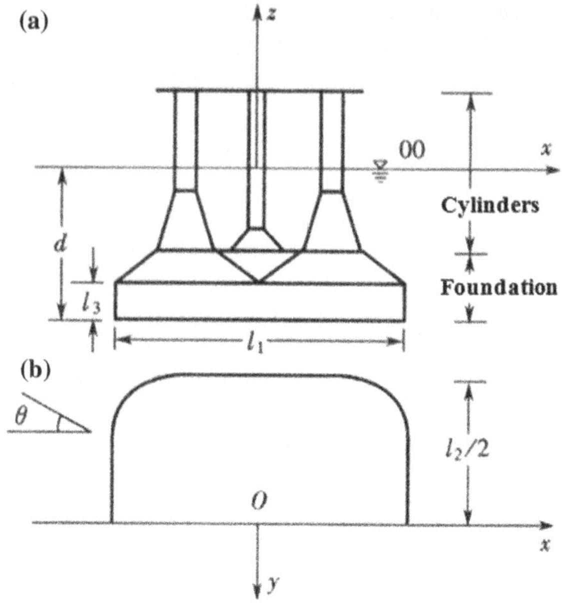

treated as small-scale structures in which the wave forces can be calculated by the Morison equation using the fifth-order Stokes wave theory. On the bottom caisson as a large-scale structure, the wave forces can be numerically calculated by the 3D linear diffraction theory, which has been verified by the physical model in a wave flume.

1. Calculation of wave forces on the support cylinders

The Cartesian coordinate system is used in the calculations, as shown in Fig. 6.31 for the definition sketch. As the distances between the upper cylinders are large, the mutual influence is not taken into account. When $t = 0$, the wave crest just passes the central position of the caisson. For a small-scale structure, only the action of incident waves is taken into account. Then the wave force on the structure reads

$$f = \frac{1}{2}C_D\rho Du|u| + C_M\rho\frac{\pi D^2}{4}\frac{\mathrm{d}u}{\mathrm{d}t} \tag{6.7.17}$$

where C_D and C_m are the drag coefficient and the inertia coefficient, respectively; ρ is the seawater density; D is the diameter of the circular cylinder; u and $\frac{du}{dt}$ are the velocity and the acceleration, respectively.

When the incident waves are the fifth-order Stokes waves, the horizontal wave force acting on unit height of each cylinder is the sum of the forces by the waves of all the orders

$$f = \sum_{m=1}^{5} \left[\frac{1}{2} C_D \rho D u_m |u_m| + C_m \rho \frac{\pi D^2}{4} \left(\frac{\partial u}{\partial t} \right)_m \right] \tag{6.7.18}$$

in which m is the order of the Stokes waves and let $\left(\frac{du}{dt}\right)_m = \left(\frac{\partial u}{\partial t}\right)_m$;

$$u_m = k\lambda_m m \cosh[mk(z+d)] \cos[m(kx\cos\theta + ky\sin\theta - \omega t)] \tag{6.7.19}$$

$$\left(\frac{\partial u}{\partial t} \right)_m = k\omega m^2 \lambda_m \cosh[mk(z+d)] \sin[m(kx\cos\theta + ky\sin\theta - \omega t)] \tag{6.7.20}$$

where k is the wave number; ω is the angular frequency; θ is the incident wave angle.

$$c^2 = \frac{g\tanh(kd)}{k}[1 + \lambda^2 C_1 + \lambda^4 C_2]$$

$$\lambda_1 = \frac{c}{k}(\lambda A_{11} + \lambda^3 A_{13} + \lambda^5 A_{15});\ \lambda_2 = \frac{c}{k}(\lambda^2 A_{22} + \lambda^4 A_{24});$$

$$\lambda_3 = \frac{c}{k}(\lambda^3 A_{33} + \lambda^5 A_{35});\ \lambda_4 = \frac{c}{k}(\lambda^4 A_{44});\ \lambda_5 = \frac{c}{k}(\lambda^5 A_{55});$$

in which the parameters λ, A_{11}, A_{13}, A_{15}, A_{22}, A_{24}, A_{33}, A_{35}, A_{44}, A_{55}, C_1, and C_2 are only related to kd. See Ref. [10] for the detail.

As each cylinder has varying diameter, the cylinder is divided into n segments. The horizontal wave force acting on the whole cylinder with varying diameter is the sum of the forces on all the segments

$$F = \sum_{i=1}^{n} \sum_{m=1}^{5} \left[\frac{1}{2} C_D \rho D_i u_{mi} |u_{mi}| l_i + C_m \rho \frac{\pi D_i^2}{4} \left(\frac{\partial u}{\partial t} \right)_{mi} l_i \right] \tag{6.7.21}$$

where n is the number of the cylinder segments; l_i is the height of the ith cylinder segment; D_i is the diameter of the ith cylinder segment. Using the coordinate system shown in Fig. 6.31, the wave force on each cylinder can be calculated by Eq. (6.7.18). Then the total wave force on the whole cylinders can be obtained according to the coordinates of the cylinders and their phase differences.

2. Calculation of wave forces on the bottom caisson

Green's function in series form expressed by Eq. (6.7.5) is used in the calculation. The treatment of singularity described above is applied. The coefficient matrices $\boldsymbol{\beta}_{ij}$ and $\boldsymbol{\alpha}_{ij}$ are solved first by Eqs. (6.7.11) and (6.7.12). Then the source intensity f_i and the scattered wave potential ϕ_{Si} can be obtained by Eqs. (6.7.9) and (6.7.10).

The total potential function:

$$\Phi = (\phi_I + \phi_S)e^{-i\omega t} \tag{6.7.22}$$

The pressure on the caisson:

$$p = -\rho g z + \frac{1}{\rho}\frac{\partial \Phi}{\partial t} \tag{6.7.23}$$

The total wave force:

$$F = \int_S \boldsymbol{p} \cdot \boldsymbol{n} \mathrm{d}S \tag{6.7.24}$$

where $\boldsymbol{n} = (\boldsymbol{n}_x, \boldsymbol{n}_y)$ is the unit normal vector at a point, $\boldsymbol{n}_x$ and $\boldsymbol{n}_y$ are the unit vector components in the x- and y-directions.

The total horizontal force:

$$F_H = \int_S \boldsymbol{p} \cdot \boldsymbol{n}_x \mathrm{d}x \tag{6.7.25}$$

The total vertical force:

$$F_V = \int_S \boldsymbol{p} \cdot \boldsymbol{n}_y \mathrm{d}y \tag{6.7.26}$$

The total horizontal force coefficient:

$$f_H = \frac{F_H}{\frac{1}{2}\rho g H l^2} \tag{6.7.27}$$

The total vertical force coefficient:

$$f_V = \frac{F_V}{\frac{1}{2}\rho g H l^2} \tag{6.7.28}$$

where H is the wave height, l is the characteristic length of the caisson.

3. Comparison of the calculated and experimental results

In order to verify the validity of the above numerical method, overall physical model tests on the gravity platform had been carried out in a wave flume at the hydraulic laboratory in Tianjin University [19]. The wave flume was 60 m long, 3 m wide, and 2 m high.

Figure 6.32 shows time series of the calculated and experimental results of the relative surface elevation η/H, the total horizontal force coefficient f_H, and the total vertical force coefficient f_V caused by the normal incident wave on both the cylinders and the caisson. The calculated relative surface elevation was obtained by the fifth-order Stokes wave theory. Figure 6.33 shows time series of the calculated horizontal

Fig. 6.32 Time series of η/H, total f_H and total f_V by normal incident wave

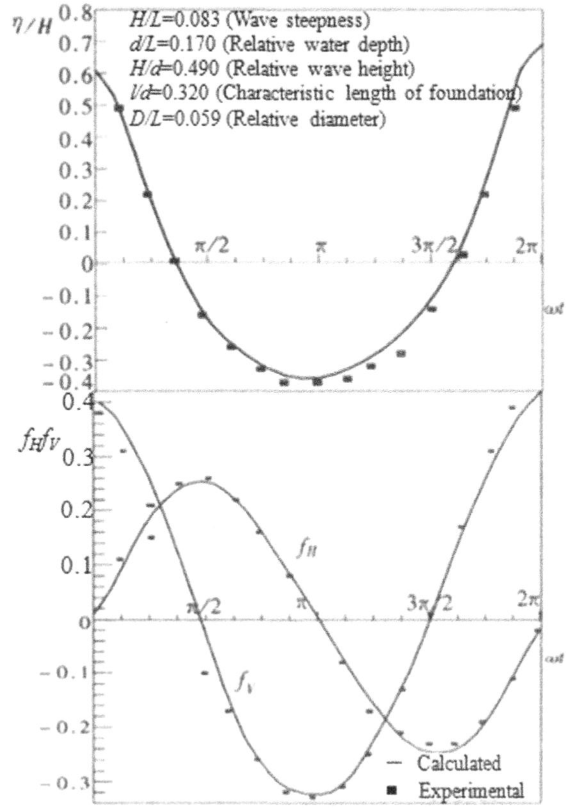

and vertical force coefficients f_H and f_V of the caisson caused by the normal incident wave by using Green's functions in series form and in asymptotic form respectively. Green's functions in different forms result in significant difference in the vertical force coefficient. The maximum deviation is about 14%. As Green's function in asymptotic form only performs real operations, the computing time using this form is only 1/3 of the time using series form. Thus, Green's function in asymptotic form can be used for the preliminary estimation of wave forces. Figure 6.34 shows the maximum total horizontal wave force coefficient f_{Hmax} and the maximum total vertical wave force coefficient f_{Vmax} versus the relative water depth d/L. The calculated and experimental results matched well.

In this case, the wave forces on the caisson and the cylinders are calculated separately. For the cylinders treated as small-scale structures, the Morison equation and the fifth-order wave theory are used. For the caisson, the 3D source distribution method is applied. Compared with the experimental results, the deviation of the calculated total horizontal force is only about 1%. The computing time by taking the caisson and cylinders separately is only 1/4 of the time by taking the caisson and cylinders as an integrate structure. In this case, the wave force on the cylinders is

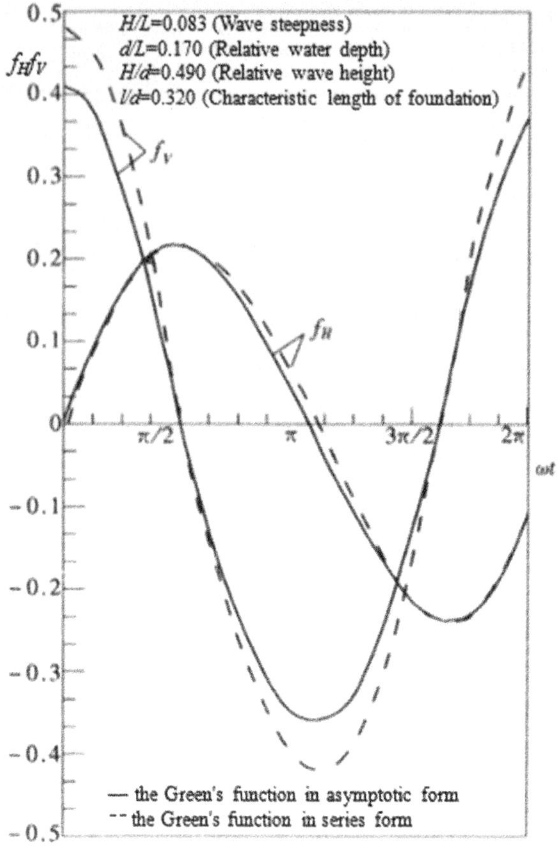

Fig. 6.33 Time series of f_H and f_V of the caisson by normal incident wave

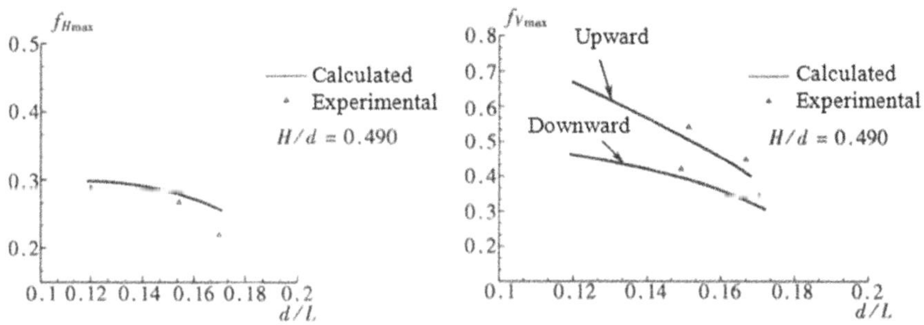

Fig. 6.34 $f_{H\max}$ and $f_{V\max}$ versus d/L

about 14% of the total wave force. Therefore, it is reasonable to calculate the caisson and cylinders separately for simplicity.

6.8 Second-Order Wave Forces on a Large-Scale Cylinder Using Green's Function Method

The calculation of wave forces on 3D structures using complex Green's function method of linear diffraction theory has been described above. When *H/d* is large, the linear theory is not valid, and a nonlinear diffraction theory and corresponding calculation method need be sought. Here a method for calculating the wave forces on a cylinder in second-order nonlinear Stokes waves is described [20].

6.8.1 *Basic Equations*

The coordinate system is established as shown in Fig. 6.35. Here, H is the wave height; d is the still water depth; η is the surface elevation; a is the radius of the circular cylinder. The potential function is denoted by $\Phi(x, y, z, t)$. $r = (x^2 + y^2)^{1/2}$.

The perturbation expansions of the potential function Φ and the surface elevation η read,

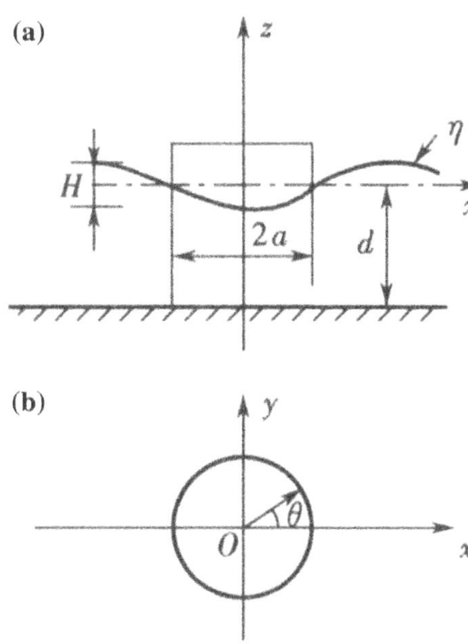

Fig. 6.35 Definition sketch. **a** Elevation view; **b** Plan view

$$\Phi = \varepsilon\Phi_l + \varepsilon^2\Phi_q + \cdots \tag{6.8.1}$$

$$\eta = \varepsilon\eta_l + \varepsilon^2\eta_q + \cdots \tag{6.8.2}$$

where ε is a perturbation parameter, $\varepsilon = \frac{1}{2}kH$, in which k is the incident wave number; Φ_l and η_l are the first-order potential function and surface elevation, respectively; Φ_q and η_q are the second-order potential function and surface elevation. By means of perturbation expansion, the nonlinear problem is decomposed into two fixed solution problems. One of them is a first-order fixed solution problem, i.e., a linear wave problem, for which the method described above can be applied. The second-order equation and boundary conditions are given in the following.

$$\nabla^2\Phi_q = 0 \quad r > a \tag{6.8.3}$$

$$\frac{\partial^2\Phi_q}{\partial t^2} + g\frac{\partial\Phi_q}{\partial z} = -\frac{\partial}{\partial t}|\nabla\Phi_l|^2 + \frac{1}{g}\frac{\partial\Phi_l}{\partial t}\frac{\partial}{\partial z}\left[\frac{\partial^2\Phi_l}{\partial t^2} + g\frac{\partial\Phi_l}{\partial z}\right], r \geq a \text{ and on } z = 0 \tag{6.8.4}$$

$$\frac{\partial\Phi_q}{\partial z} = 0, z = -d \tag{6.8.5}$$

$$\frac{\partial\Phi_q}{\partial r} = 0, r = a \text{ and } -d \leq z \leq \eta \tag{6.8.6}$$

Radiation condition, at infinity

where $\Phi_l = \Phi_l^I + \Phi_l^S$, $\Phi_q = \Phi_q^I + \Phi_q^S$.

Φ_l^I and Φ_l^S are the first-order incident and scattered wave potentials, respectively. Φ_q^I and Φ_q^S are the second-order incident and scattered wave potentials. The Stokes wave theory gives

$$\Phi_l^I = \text{Re}(\phi_l^I e^{-i\omega t}) \quad \Phi_l^S = \text{Re}(\phi_l^S e^{-i\omega t}) \tag{6.8.7}$$

$$\Phi_q^I = \text{Re}(\phi_q^I e^{-2i\omega t}) \quad \Phi_q^S = \text{Re}(\phi_q^S e^{-2i\omega t}) \tag{6.8.8}$$

where "Re" denotes the real part of the complex.

On the basis of obtaining the first-order scattered wave potential, the second-order fixed solution problem becomes a problem of solving ϕ_q^S.

When solving the first-order scattered wave potential Φ_l^S, the well-known Sommerfeld condition (the radiation condition at infinity) makes the linear solution unique. However, the radiation condition at infinity for the general nonlinear scattered wave potential is not perfect yet. As a condition at infinity, the radiation condition has a physical meaning of waves propagating only outward without reflection. The calculation of the second-order wave forces is described in the following.

6.8.2 The Fixed Solution Problem of Second-Order Scattered Waves

Omitting the subscript "q" for simplicity, then the fixed solution problem of the second-order scattered wave potential reads

$$\nabla^2\Phi^S = 0, \quad \text{in the domain } r > a \tag{6.8.9}$$

$$\frac{\partial\Phi^S}{\partial z} - \frac{4\omega^2}{g}\Phi^S = f, \quad z = 0 \tag{6.8.10}$$

$$\frac{\partial\Phi^S}{\partial z} = 0, \quad z = -d \tag{6.8.11}$$

$$\frac{\partial\Phi^S}{\partial r} = -\frac{\partial\Phi^I}{\partial r}, \quad r = a \tag{6.8.12}$$

$$\text{Radiation condition,} \quad r \to \infty \tag{6.8.13}$$

in which,

$$\begin{aligned} f = &-\frac{\partial\Phi^I}{\partial z} + \frac{4\omega^2}{g}\Phi^I \\ &- \frac{i\omega}{zg}(\Phi_l^I + \Phi_l^S)\frac{\partial}{\partial z}\left[\frac{\partial}{\partial z}(\Phi_l^I + \Phi_l^S) - \frac{\omega^2}{g}(\Phi_l^I + \Phi_l^S) + \frac{i\omega}{g}(\nabla\Phi_l^I + \nabla\Phi_l^S)^2\right] \end{aligned} \tag{6.8.14}$$

The second-order incident wave potential Φ^I is

$$\varepsilon^2\Phi^I = \frac{3}{8}\left(\frac{H}{2}\right)^2\omega\frac{\cosh[2k_2(z+d)]}{\sinh^4(k_2d)}\sin[2(k_2x\cos\theta + k_zy\sin\theta) - \omega t] \tag{6.8.15}$$

where k_2 satisfies,

$$4\omega^2 = gk_2\tanh(k_2d) \tag{6.8.16}$$

It is assumed that the second-order scattered wave potential is composed of a particular solution φ and a general solution ϕ_2,

$$\Phi^S = \varphi + \phi_2 \tag{6.8.17}$$

The particular solution ϕ satisfies the following equation and boundary conditions,

$$\nabla^2\varphi = 0, \quad \text{in the domain} \tag{6.8.18}$$

$$\frac{\partial \varphi}{\partial z} - \frac{4\omega^2}{g}\varphi = f, \quad z = 0 \tag{6.8.19}$$

$$\frac{\partial \varphi}{\partial z} = 0, \quad z = -d \tag{6.8.20}$$

$$\lim_{r\to\infty} \sqrt{r}\left(\frac{\partial \varphi}{\partial r} - ik_2\varphi\right) = 0, \quad \text{at infinity} \tag{6.8.21}$$

The radiation condition for the first-order scattered wave potential is applied here. The numerical results indicate that the nonlinear effect is negligible in the region with the distance further than 2–3 times of the wavelength from the structure.

The general solution ϕ_2 satisfies the following equation and boundary conditions

$$\nabla^2\phi_2 = 0, \quad \text{in the domain} \tag{6.8.22}$$

$$\frac{\partial \phi_2}{\partial z} - \frac{4\omega^2}{g}\phi_2 = 0, \quad z = 0 \tag{6.8.23}$$

$$\frac{\partial \phi_2}{\partial r} = -\frac{\partial \phi^I}{\partial r} - \frac{\partial \phi}{\partial r}, \quad r = a \tag{6.8.24}$$

$$\frac{\partial \phi_2}{\partial z} = 0, \quad z = -d \tag{6.8.25}$$

$$\lim_{r\to\infty} \sqrt{r}\left(\frac{\partial \phi_2}{\partial r} - ik_2\phi_2\right) = 0, \quad \text{at infinity.} \tag{6.8.26}$$

6.8.3 Solving the Particular Solution φ and the General Solution Φ_2 by Using Green's Function Method

A method similar to solving the linear scattered wave potential can be applied to solve the general solution ϕ_2. Green's function utilized is similar to Green's function for the linear scattered wave potential, in which k is replaced by k_2.

The method for solving the particular solution φ is the focus here. Omitting the terms of higher than the second order in the particular solution φ, then φ can be written as

$$\varphi = \varphi_2 \exp(-2i\omega t)$$

Let,

$$\varphi_2 = \frac{1}{2\pi} \iint\limits_{r \geq a} G(x, y, z, \xi, \eta, 0) f(\xi, \eta, 0) \mathrm{d}\xi \, \mathrm{d}\eta \tag{6.8.27}$$

where f is the source intensity on the surface, which is determined by Eq. (6.8.14). The integral is performed on the free surface except for the occupied part of the structure.

The problem now is to find Green's function G so that φ can be the solution to the fixed solution problem. Because the free surface boundary condition is non-homogeneous, the form of Green's function for the first-order solution cannot be used. Based on Green's function for the first-order solution, Green's function for the second-order solution is presented. It reads in integral form

$$\begin{aligned} G(x, y, z, \xi, \eta, 0) = {} & \frac{1}{R} + \frac{1}{R'} + 2\text{P.V.} \int\limits_0^\infty \frac{\nu e^{-\mu d} \cosh(\mu d) \cosh[\mu(z+d)]}{\mu \sinh(\mu d) - \nu \cosh(\mu d)} J_0(\mu r) \mathrm{d}\mu \\ & + \text{P.V.} \int\limits_0^\infty \frac{\mu e^{-2\mu d} \cosh[\mu(z+d)]}{\mu \sinh(\mu d) - \nu \cosh(\mu d)} J_0(\mu r) \mathrm{d}\mu \\ & - i \frac{\pi (k_2^2 - \nu^2) \cosh[k_2(z+d)] \cosh k_2 d}{k_2^2 d - \nu^2 d + \nu} J_0(k_2 r) \end{aligned} \tag{6.8.28}$$

in which,

$$R = [(x-\xi)^2 + (y-\eta)^2 + z^2]^{1/2}; \quad R' = [(x-\xi)^2 + (y-\eta)^2 + (z+2d)^2]^{1/2};$$
$$r = [(x-\xi)^2 + (y-\eta)^2]^{1/2}; \quad \nu = \frac{4\omega^2}{g}.$$

Green's function in series form reads

$$\begin{aligned} G(x, y, z, \xi, \eta, 0) = {} & \pi \frac{(k_2^2 - \nu^2) \cosh[k_2(z+d)] \cdot \cosh(k_2 d)}{k_2^2 d - \nu^2 d + \nu} [Y_0(k_2 r) - i J_0(k_2 r)] \\ & + 2 \sum_{k=1}^{\infty} \frac{\mu_k^2 + \nu^2}{\mu_k^2 d + \nu^2 d - \nu} \cos[\mu_k(z+d)] \cos(\mu_k d) K_0(\mu_k r) \end{aligned} \tag{6.8.29}$$

where μ_k is the kth positive solution to the equation $\mu_k \tan(\mu_k d) + \nu = 0$. In Eq. (6.8.29), J_0 and Y_0 are the zero-order Bessel functions of the first and second kinds, respectively. K_0 is the zero-order modified Bessel function of the second kind.

As above Green's function G satisfies the Laplace equation, the bottom boundary condition, and the radiation condition, and φ_2 satisfies them as well. φ_2 will be proved here to satisfy the non-homogeneous free surface boundary condition in Eq. (6.8.19). On the free surface, we have

$$\frac{\partial \varphi_2}{\partial z} - \nu\varphi_2|_{z=0} = \frac{1}{2\pi}\iint_{r\geq a}\left(\frac{\partial G}{\partial z} - \nu G\right)|_{z=0} f(\xi, \eta, 0)d\xi\, \mathrm{d}\eta \tag{6.8.30}$$

and

$$\begin{aligned}\frac{\partial G}{\partial z} - \nu G|_{z=0} = {} & \frac{\partial}{\partial z}\left(\frac{1}{R} + \frac{1}{R'}\right) - \nu\left(\frac{1}{R} + \frac{1}{R'}\right) \\ & + 2\mathrm{P.V.}\int_0^\infty \nu \mathrm{e}^{-\mu d}\cosh(\mu\, d)J_0(\mu\, r)\mathrm{d}\mu \\ & + \mathrm{P.V.}\int_0^\infty \mu \mathrm{e}^{-2\mu d} J_0(\mu\, r)\mathrm{d}\mu \\ & + i\frac{\pi(\nu^2 - k_2^2)\cosh(k_2 d)}{k_2^2 d - \nu^2 d + \nu} J_0(k_2 r)[k_2\sinh(k_2 d) - \nu\cosh(k_2 d)]\end{aligned} \tag{6.8.31}$$

By means of

$$\left.\begin{aligned}\frac{1}{R} &= \mathrm{P.V.}\int_0^\infty \mathrm{e}^{\mu z} J_0(\mu\, r)\mathrm{d}\mu \\ \frac{1}{R'} &= \mathrm{P.V.}\int_0^\infty \mathrm{e}^{-\mu(z+2d)} J_0(\mu\, r)\mathrm{d}\mu\end{aligned}\right\}, \tag{6.8.32}$$

which gives

$$\frac{\partial G}{\partial Z} - \nu G|_{z\to 0} = \frac{\partial}{\partial z}\left(\frac{1}{R}\right) \tag{6.8.33}$$

Thus, we have

$$\frac{\partial \varphi_2}{\partial z} - \nu\varphi_2|_{z\to 0} = \lim_{z\to 0}\frac{1}{2\pi}\iint_{r\geq a}\left(\frac{\partial G}{\partial z} - \nu G\right) f(\xi, \eta, 0)\mathrm{d}\xi\, \mathrm{d}\eta \tag{6.8.34}$$

Taking an infinitesimal surface s which contains the point $(x, y, 0)$, and dividing the above integral into two parts

$$\frac{\partial \varphi_2}{\partial z} - \nu \varphi_2|_{z\to 0} = \lim_{\substack{z\to 0\\ s\to 0}} \frac{1}{2\pi} \iint\limits_{r\geq a-s} \frac{\partial}{\partial z}\left(\frac{1}{R}\right) f(\xi,\eta,0)\mathrm{d}\xi\,\mathrm{d}\eta$$

$$+ \lim_{\substack{z\to 0\\ s\to 0}} \frac{1}{2\pi} \iint\limits_{s} \frac{\partial}{\partial z}\left(\frac{1}{R}\right) f(\xi,\eta,0)\mathrm{d}\xi\,\mathrm{d}\eta$$

$$= \lim_{\substack{z\to 0\\ s\to 0}} \frac{1}{2\pi} \iint\limits_{s} \frac{\partial}{\partial z}\left(\frac{1}{R}\right) f(\xi,\eta,0)\mathrm{d}\xi\,\mathrm{d}\eta \tag{6.8.35}$$

By the potential theory, this integral value is $f(x, y, 0)$, thus φ_2 satisfies the non-homogeneous free surface boundary condition.

6.8.4 Second-Order Wave Forces and Moment on a Large-Scale Cylinder

With the particular solution φ and the general solution ϕ_2, the second-order scattered wave potential ϕ_q^S can be solved by Eq. (6.8.17). Then the total potential function Φ is obtained by Eq. (6.8.1). The pressure on the structure is determined by

$$p = -\rho\frac{\partial \Phi}{\partial t} - \rho\frac{|\nabla\Phi|^2}{2} - \rho\, gz \tag{6.8.36}$$

The expression of the pressure with accuracy up to the second-order is

$$\begin{aligned} p &= \rho\,\omega\,\varepsilon\,\mathrm{Re}\{i\phi_I \mathrm{e}^{-i\omega t}\} + 2\rho\,\omega\,\varepsilon^2 \mathrm{Re}\{i\phi_q \mathrm{e}^{-i2\omega t}\} \\ &\quad - \frac{1}{4}\rho\,\varepsilon^2\,\mathrm{Re}\left\{\left[\left(\frac{\partial\phi_I}{\partial x}\right)^2 + \left(\frac{\partial\phi_I}{\partial y}\right)^2 + \left(\frac{\partial\phi_I}{\partial z}\right)^2\right]\mathrm{e}^{-i2\omega t}\right\} \\ &\quad - \frac{1}{4}\rho\,\varepsilon^2\left[\left(\frac{\partial\phi_I}{\partial x}\right)\left(\frac{\partial\bar{\phi}_I}{\partial x}\right) + \left(\frac{\partial\phi_I}{\partial y}\right)\left(\frac{\partial\bar{\phi}_I}{\partial y}\right) + \left(\frac{\partial\phi_I}{\partial z}\right)\left(\frac{\partial\bar{\phi}_I}{\partial z}\right)\right] - \rho\, gz \end{aligned} \tag{6.8.37}$$

where "—" denotes the complex conjugate. The total force on the structure is

$$F = -\iint\limits_{S} \boldsymbol{p}\cdot\boldsymbol{n}\mathrm{d}S \tag{6.8.38}$$

The total moment is

$$M = -\iint\limits_{s} p\boldsymbol{r}' \times \boldsymbol{n}\mathrm{d}S \tag{6.8.39}$$

where $\boldsymbol{n}$ is the normal vector on the surface, $\boldsymbol{r}'$ is the position vector of the force acting position to the certain point; S is the wetted surface area of the structure.

When the wave forces on the structure are calculated using above second-order Green's function method, the integral radius in the numerical calculation, such as for Eq. (6.8.27), can be chosen as 2–3 times of the wave length. For the integral of second-order Green's function, the singularity problem can be solved with a method similar to the treatment on the singularity in the linear diffraction problem [17].

6.8.5 Calculation Cases

The second-order wave forces on a fixed vertical circular cylinder are calculated to verify the numerical method of the second-order wave forces described above [20].

Figures 6.36 and 6.37 show the calculated wave forces compared with the results of the other methods for the two cases of $h/a = 1.16$, $H/2a = 0.19$, and $h/a = 1.16$, $H/2a = 0.131$. The calculated results using the method described above matched the experimental data. Compared with the nonlinear results in Reference [21], the results using the method described above are closer to the experimental data in the low frequency, and the curve obtained using this method had better agreement with the curve from experimental data. Obviously, the second-order results using the method described above were closer to the experimental data than the results in Ref. [21].

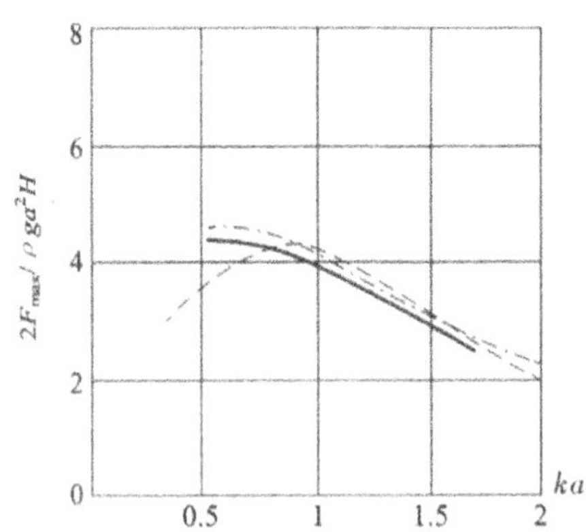

—— Second-order nonlinear results[16];
-·-·- Experimental curve; ---- Nonlinear results[7].

Fig. 6.36 Comparison of the calculated and experimental results of wave forces on a large-scale circular cylinder. $h/a = 1.16$, $H/2a = 0.19$

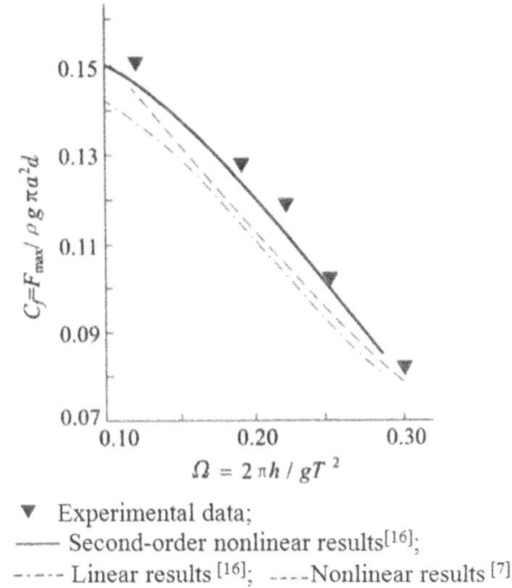

Fig. 6.37 Comparison of the calculated and experimental results of wave forces on a large-scale circular cylinder. $h/a = 1.16, H/2a = 0.131$

6.9 Mathematical Model of Wave–Current Forces on a Submerged Structure Near the Free Surface

The calculation methods of pure wave forces on marine structures have been described in the above sections in this chapter. However, waves and currents coexist in many coastal, marine engineering and marine environments. It is very important to study the wave–current interaction on marine structures. Since the regimes of the interaction among waves, currents, and structures are very complex, it is a challenge to find a commonly effective calculation method. At present, wave forces and current forces are mostly calculated separately and then add them together.

Under the condition of weak currents, the wave–current interaction on a structure can be regarded as a diffraction problem. The potential flow theory can be used to study the current effect on the diffraction around the structure and the wave–current forces on the structure.

For the problem of wave diffraction with presence of current, the current effects on the free surface and radiation conditions should be taken into account. Even under the condition of linear waves, it is much more complex than the pure wave action.

In this section, the effect of steady current on the wave diffraction of a submerged body near the water surface will be described. Here the submerged body is a 2D circular cylinder near the free surface. In order to avoid introducing the complicated free surface condition and radiation condition at the same time, the method of the internal and external fields matching is used to study the effect of steady current on the wave diffraction of a submerged body near the water surface [22].

6.9.1 Wave-Making Problem in Steady Current

When a steady current rounds a submerged body near the water surface, the free surface may fluctuate near the body. Figure 6.38 shows a definition sketch for this type of wave-making problem. The steady current in the direction of the negative x axis flows around a 2D body submerged below the free water surface, which generates wave motion behind the body. The x-axis coincides with the still water surface, and the z-axis is vertical upward.

Let the velocity potential of steady current around the body be

$$\Phi_0(x, z) = -Ux + \varphi_v(x, z) \tag{6.9.1}$$

where $\varphi_v(x, z)$ is the potential function of the wave generated by the steady current around a submerged body.

1. Boundary value problem of the linear wave potential generated by steady current

Based on the potential flow theory and the linearized free surface condition, the equation and the boundary conditions satisfied by the wave potential generated by the steady current φ_v read,

$$\frac{\partial^2 \varphi_v}{\partial x^2} + \frac{\partial^2 \varphi_v}{\partial z^2} = 0, \quad \text{in the domain} \tag{6.9.2}$$

$$\frac{\partial^2 \varphi_v}{\partial x^2} + K_0 \frac{\partial \varphi_v}{\partial z} = 0, \quad z = 0 \tag{6.9.3}$$

$$\frac{\partial \varphi_v}{\partial n} = U n_x, \quad \text{on the body surface} \tag{6.9.4}$$

$$\frac{\partial \varphi_v}{\partial z} = 0, \quad z \to -\infty \tag{6.9.5}$$

Radiation condition:

$$\frac{\partial \varphi_v}{\partial x} = \begin{cases} 0 & x \to +\infty \\ w(x, z) & x \to -\infty \end{cases} \tag{6.9.6}$$

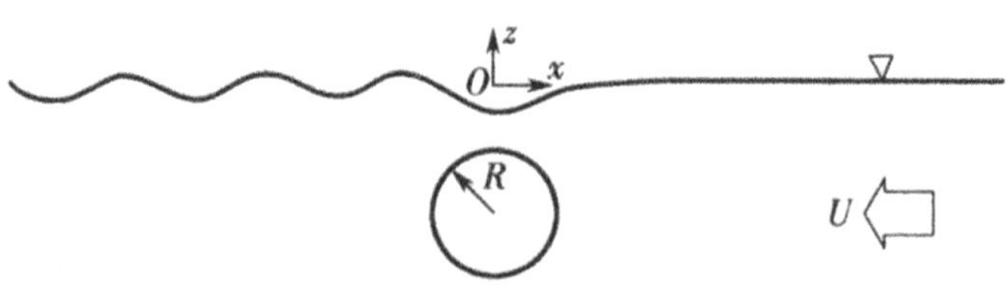

Fig. 6.38 Definition sketch for the wave-making problem of steady current

where

$$K_0 = \frac{g}{U^2}, \tag{6.9.7}$$

and $w(x, z)$ denotes a standing wave.

2. Solution of the wave potential generated by the steady current φ_v using Green's function method

Green's function (Kelvin sources) of the above boundary value problem reads [14],

$$\begin{aligned} G(x, z, \xi, \zeta) &= \ln r + \ln r_1 + 2\mathrm{Re}[\mathrm{e}^{\tau_0} E_1(\tau_0)] \\ &+ \begin{cases} 0, & x - \xi > 0 \\ 4\pi \mathrm{e}^{K0(z+\zeta)} \cdot \sin K_0(x - \xi), & x - \xi < 0 \end{cases} \end{aligned} \tag{6.9.8}$$

where $P(x, z)$ is the calculation point, $Q(\xi, \eta)$ is the source point.

$$\begin{aligned} r &= \sqrt{(x-\xi)^2 + (z-\zeta)^2}, r_1 = \sqrt{(x-\xi)^2 + (z+\zeta)^2} \\ \tau_0 &= K_0[z + \zeta + i(x - \xi)], \end{aligned}$$

$E_1(z)$ is the complex exponential integral defined as

$$E_1(z) = \int\limits_z^\infty \frac{\mathrm{e}^{-t}}{t} \mathrm{d}t, \quad |\arg z| < \pi,$$

"Re" denotes the real part of the complex.

When Green's function is known, the wave potential φ_v can be obtained by

$$\varphi_v(x, z) = \int\limits_{S_0} \sigma(\xi, \zeta) G(x, z, \xi, \zeta) \mathrm{d}S \tag{6.9.9}$$

where $\sigma(\xi, \zeta)$ is the source intensity, it satisfies,

$$-\pi\sigma(x, z) + \int\limits_{S_0} \sigma(\xi, \zeta) \frac{\partial G(x, z, \xi, \zeta)}{\partial n} \mathrm{d}S = U n_x \tag{6.9.10}$$

in which, S_0 is the body surface, and the point $P(x, z)$ is on the body surface S_0.

As to the procedure for the numerical solution of $\varphi_v(x, z)$, the method in Sect. 6.7in this chapter can be taken as a reference. After the steady generated wave potential is obtained, the velocity potential of the steady current around the body is given by

$$\Phi_0(x, z) = -Ux + \varphi_v(x, z) \tag{6.9.11}$$

Then the pressure or the other interesting physical quantities can be obtained by the Bernoulli equation.

6.9.2 Mathematical Model of Diffraction by a Submerged Body Near Water Surface Under Wave–Current Interaction

Figure 6.39 shows a definition sketch in the plane of (x, z). The origin is located on the still water level; the horizontal x-axis is perpendicular to the cylinder axis, and the z-axis is vertical upward; d is the submerged depth. Under the assumption of small-amplitude waves, when the steady current flows opposite to the x-axis direction with a speed of U, the velocity potential $\Phi(x, z, t)$ can be separated into

$$\Phi(x, z, t) = -Ux + \phi_v(x, z) + \mathrm{Re}\{\varphi(x, z)\mathrm{e}^{-i\omega t}\} \tag{6.9.12}$$

where φ_v is the steady perturbation potential; ϕ is the unsteady part of the spatial velocity potential.

In the case of low current velocity, the effect of the steady perturbation on the free surface is only significant in the vicinity of the body. In the area far away from the body, the velocity of the steady perturbation is much smaller than the incoming current velocity, so that it is feasible to take the effect of the steady current only into account for the free surface condition in that area. Under this assumption, the whole flow field is divided by a control surface S_J into two parts: an internal filed Ω which contains the body, and an external field $\overline{\Omega}$ which is far away from the body (as shown in Fig. 6.39).

1. Solution in the external field

The solution in the external field $\phi_2(x, z)$ can be decomposed into two parts, i.e., the incident wave potential ϕ_I and the scattered wave potential ϕ_S,

$$\phi_2 = \phi_I + \phi_S \tag{6.9.13}$$

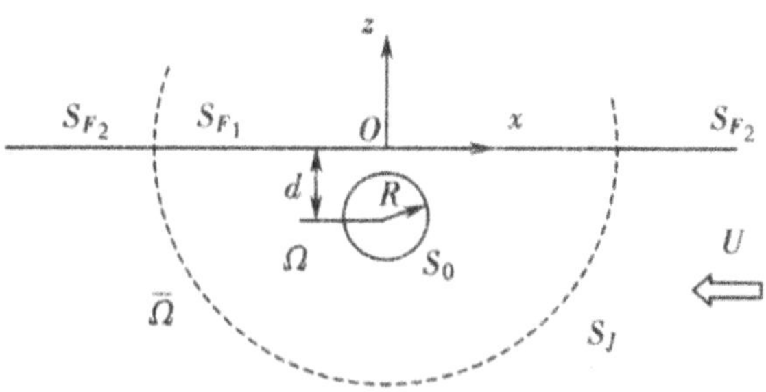

Fig. 6.39 Definition sketch

Considering that the velocity of the steady perturbation is much smaller than the incoming current velocity in the external field, only the effect of steady current is taken into account on the free surface. With accuracy of $O(U)$, the control equation and conditions satisfied by the scattered wave potential ϕ_S read,

$$\frac{\partial^2\phi_S}{\partial x^2} + \frac{\partial^2\phi_S}{\partial z^2} = 0, \quad \text{in } \overline{\Omega} \tag{6.9.14a}$$

$$-\omega^2\phi_S + i2U\omega\frac{\partial\phi_S}{\partial x} + g\frac{\partial\phi_S}{\partial z} = 0, \quad z = 0 \tag{6.9.14b}$$

$$\nabla\phi_S = 0, \quad z \to -\infty \tag{6.9.14c}$$

$$\text{Matching condition,} \quad \text{on } S_J \tag{6.9.14d}$$

$$\text{Radiation condition,} \quad \text{at infinity} \tag{6.9.14e}$$

Green's function for the above problem $G(x, z, \xi, \zeta)$ reads [23]

$$\begin{aligned} G(x, z, \xi, \zeta) = {} & \ln r - \ln r_1 - \frac{1}{1-2\tau}\mathrm{e}^{t_2}E_1(t_2) - \frac{1}{1+2\tau}\mathrm{e}^{t_4}E_1(t_4) \\ & - \frac{2\pi i}{1-2\tau}H(x-\xi)\mathrm{e}^{t_2} - \frac{2\pi i}{1+2\tau}H(\xi - x)\mathrm{e}^{t_4} \end{aligned} \tag{6.9.15}$$

in which,

$$\begin{aligned} & r = \sqrt{(x-\xi)^2 + (z-\zeta)^2}, \quad r_1 = \sqrt{(x-\xi)^2 + (z+\zeta)^2} \\ & t_2 = k_2[z + \zeta + i(x-\xi)], \quad t_4 = k_4[z + \zeta - i(x-\xi)] \\ & k_2 = \frac{\omega^2}{g}\frac{1}{1-2\tau}, \quad k_4 = \frac{\omega^2}{g}\frac{1}{1+2\tau}, \quad \tau = \frac{U\omega}{g} \\ & E_1(z) = \int_z^\infty \frac{\mathrm{e}^{-t}}{t}dt, \quad |\arg z| < \pi \\ & H(x) = \begin{cases} 1, & x > 0 \\ 0, & x < 0 \end{cases} \end{aligned}$$

Using Green's theorem,

$$\int_{\overline{\Omega}} (\varphi_1\nabla^2\varphi_2 - \varphi_2\nabla^2\varphi_1)dV = \int_S \left(\varphi_1\frac{\partial\varphi_2}{\partial n} - \varphi_2\frac{\partial\varphi_1}{\partial n}\right)dS \tag{6.9.16}$$

where $S = S_J + S_{F_2} + S_b + S_{+\infty} + S_{-\infty}$. Let $\varphi_1 = \phi_S$ and $\varphi_2 = G(x, z, \xi, \zeta)$. If the point $P(x, z)$ is in the field of $\overline{\Omega}$, Eq. (6.9.16) gives

$$\phi_S(x,z) = \int\limits_S \left(\phi_S \frac{\partial G}{\partial n} - G \frac{\partial \phi_S}{\partial n} \right) dS \tag{6.9.17}$$

This integral is based on the source point $Q(\xi, \zeta)$ being taken as the integral variable.

It is seen from Green's function expression in Eq. (6.9.15) that, comparing $G|_{\zeta=0}$, $-\frac{\partial G}{\partial \xi}|_{\zeta=0}$ and $\frac{\partial G}{\partial \zeta}|_{\zeta=0}$ with $G|_{z=0}$, $\frac{\partial G}{\partial x}|_{z=0}$ and $\frac{\partial G}{\partial z}|_{z=0}$, respectively, they have the same form except that ζ is replaced by z. Therefore, Green's function $G(x, z; \xi, \zeta)$ satisfies the following boundary condition on the free surface $\zeta = 0$,

$$-\omega^2 G - i2U\omega \frac{\partial G}{\partial \xi} + g \frac{\partial G}{\partial \zeta} = 0, \quad \zeta = 0 \tag{6.9.18}$$

Using the above boundary condition and the following boundary condition satisfied by the velocity potential $\phi_S(\xi, \zeta)$,

$$-\omega^2 \phi_S + i2U\omega \frac{\partial \phi_S}{\partial \xi} + g \frac{\partial \phi_S}{\partial \zeta} = 0, \quad \zeta = 0 \tag{6.9.19}$$

and letting

$$\left.\begin{aligned}
J_{S_{F_2}} &= \int\limits_{S_{F_2}} \left(\phi_S \frac{\partial G}{\partial \zeta} - G \frac{\partial \phi_S}{\partial \zeta} \right) \mathrm{d}S = \frac{i2U\omega}{g} [\phi_S G]_{(-\infty,0)}^{(+\infty_c,0)} - \frac{i2U\omega}{G} [\phi_S G]_{(-R_c,0)}^{(+R_{cc},0)} \\
J_{S_{-\infty}} &= \frac{i2U\omega}{g} \phi_S G|_{(-\infty,0)} \\
J_{S_{+\infty}} &= -\frac{i2U\omega}{g} \phi_S G|_{(+\infty_c,0)}
\end{aligned}\right\} \tag{6.9.20}$$

it can be proved that [17]

$$J_{S_{F_2}} + J_{S_{-\infty}} + J_{S_{+\infty}} = \frac{i2U\omega}{g} [\phi_S G]_{(-R_c,0)}^{(R_c,0)} \tag{6.9.21}$$

where R_c is the radius of the control surface S_J. According to the bottom condition, the integral over the bottom surface S_b is zero, i.e.,

$$J_{S_b} = 0 \tag{6.9.22}$$

Substituting the boundary integral values into Eq. (6.9.17) yields the scattered wave potential in the external field by the integral over the interface S_J between the internal and the external fields

$$\phi_S(x,z) = \int\limits_{S_J} \left(\phi_S \frac{\partial G}{\partial n} - G \frac{\partial \phi_S}{\partial n} \right) \mathrm{d}S - \frac{i2U\omega}{g} [\phi_S G]_{(-R_c,0)}^{(R_c,0)} \tag{6.9.23}$$

It is assumed that ϕ_{in} is an imaginary solution in the internal field, which satisfies

$$\frac{\partial^2 \phi_{\text{in}}}{\partial x^2} + \frac{\partial^2 \phi_{\text{in}}}{\partial z^2} = 0, \quad \text{in } \Omega$$

$$-\omega^2 \phi_{\text{in}} + i2U\omega \frac{\partial \phi_{\text{in}}}{\partial x} + g \frac{\partial \phi_{\text{in}}}{\partial z} = 0, \quad z = 0$$

$$\nabla \phi_{\text{in}} = 0, \quad z \to -\infty$$

Matching condition $\phi_{\text{in}} = \phi_S$, on S_J.

For a point $p(x, z)$ in the external field $\overline{\Omega}$, using the same approach as above yields

$$0 = \int\limits_{S_J} \left(\phi_{\text{in}} \frac{\partial G}{\partial n} - G \frac{\partial \phi_{\text{in}}}{\partial n} \right) \mathrm{d}S - \frac{i2U\omega}{g} [\phi_{\text{in}} G]_{(-R_c,0)}^{(R_c,0)} \tag{6.9.24}$$

Subtracting Eq. (6.9.24) from Eq. (6.9.23) gives

$$\begin{aligned} \phi_S(x,z) = & \int\limits_{S_J} \left[(\phi_S - \phi_{\text{in}}) \frac{\partial G}{\partial n} + \left(\frac{\partial \phi_{\text{in}}}{\partial n} - \frac{\partial \phi_S}{\partial n} \right) G \right] \mathrm{d}S \\ & - \frac{i2U\omega}{g} [(\phi_S - \phi_{\text{in}}) G]_{(-R_c,0)}^{(R_c,0)} \end{aligned} \tag{6.9.25}$$

By means of the matching condition on S_J, Eq. (6.9.25) turns into

$$\phi_s(x,z) = \int\limits_{S_J} \left(\frac{\partial \phi_{\text{in}}}{\partial n} - \frac{\partial \phi_S}{\partial n} \right) G \mathrm{d}S \tag{6.9.26}$$

On the control surface S_J, let

$$\sigma(\xi, \zeta) = \frac{\partial \phi_{\text{in}}}{\partial n} - \frac{\partial \phi_S}{\partial n} \tag{6.9.27}$$

Expressed by indirect boundary element, Eq. (6.9.23) becomes

$$\phi_S(x,z) = \int\limits_{S_J} \sigma(\xi, \zeta) G \mathrm{d}S \tag{6.9.28}$$

Substituting Eq. (6.9.28) back into Eq. (6.9.13) yields the solution in the external field expressed in integral form

$$\phi_2(x, z) = \varphi_I + \int_{S_J} \sigma(\xi, \zeta) G \mathrm{d}S \tag{6.9.29}$$

2. Solution in the internal field

For a body near the water surface, the wave generated by the steady current is not negligible in the near field. Let the solution in the internal field be

$$\phi_1(x, z, t) = \phi_1 \mathrm{e}^{-i\omega t} \tag{6.9.30}$$

The boundary value problem satisfied by ϕ_1 is described as

$$\frac{\partial^2 \phi_1}{\partial x^2} + \frac{\partial^2 \phi_1}{\partial z^2} = 0, \quad \text{in } \Omega \tag{6.9.31a}$$

$$-\omega^2 \phi_1 - i2\omega \frac{\partial \varphi_0}{\partial x} \frac{\partial \phi_1}{\partial x} - i\omega \frac{\partial^2 \varphi_0}{\partial x^2} \phi_1 + g \frac{\partial \phi_1}{\partial z} = 0, \quad z = 0 \tag{6.9.31b}$$

$$\frac{\partial \phi_1}{\partial n} = 0, \quad \text{on } S_0 \tag{6.9.31c}$$

$$\nabla \phi_1 = 0, \quad z \to -\infty \tag{6.9.31d}$$

where $\varphi_0 = -Ux + \varphi_v(x, z)$ is obtained by Eq. (6.9.1).

The above fixed solution problem can be solved by matching its solution with the solution in the field expressed in Eq. (6.9.29). The matching conditions read

$$\phi_1 = \phi_2, \quad \text{on } S_J \tag{6.9.32}$$

$$\frac{\partial \phi_1}{\partial n} = \frac{\partial \phi_2}{\partial n}, \quad \text{on } S_J \tag{6.9.33}$$

3. Matching method for the solution in the internal field

Appling Green's theorem to the internal field yields

$$-2\pi \phi_1(x, z) = \int_{S_0 + S_{F_1} + S_J} \left(\phi_1 \frac{\partial}{\partial n} \ln r - \frac{\partial \phi_1}{\partial n} \ln r \right) \mathrm{d}S, \quad P \text{ point in the field} \tag{6.9.34}$$

$$-\pi\phi_1(x,z) = \int\limits_{s_0+s_{F_1}+s_J} \left(\phi_1 \frac{\partial}{\partial n}\ln r - \frac{\partial \phi_1}{\partial n}\ln r\right)\mathrm{d}S, \quad P \text{ point on the boundary} \tag{6.9.35}$$

where r is the distance between the points $P(x, z)$ and $Q(\xi, \zeta)$.

As long as one of $\partial\phi_1/\partial n$ and ϕ_1 is known or the relation between $\partial\phi_1/\partial n$ and ϕ_1 is known on all the boundaries, $\partial\phi_1/\partial n$ and ϕ_1 on the boundaries can be solved by the boundary integral Eq. (6.9.35). Then ϕ_1 at the any point in the field can be obtained using the boundary integral Eq. (6.9.34). In fact, on the boundary S_0, $\partial\phi_1/\partial n$ has been given by the condition on the body surface in Eq. (6.9.31c). Also on the boundary S_{F_1}, the relation between $\partial\phi_1/\partial n$ and ϕ_1 is given by the free surface condition in Eq. (6.9.31b). On the control surface S_J, the relation between $\partial\phi_1/\partial n$ and ϕ_1 can be gotten by the matching conditions Eqs. (6.9.32) and (6.9.33) and the solution in the external field expressed in Eq. (6.9.29).

Substituting the solution in the external field Eq. (6.9.29) into the matching conditions Eqs. (6.9.32) and (6.9.33) yields

$$\phi_1 = \phi_I + \int\limits_{S_J} \sigma G \mathrm{d}S, \quad \text{on } S_J \tag{6.9.36}$$

$$\frac{\partial \phi_1}{\partial n} = \frac{\partial \phi_I}{\partial n} - \pi\sigma + \int\limits_{S_J} \sigma \frac{\partial G}{\partial n}\mathrm{d}S, \quad \text{on } S_J \tag{6.9.37}$$

6.9.3 Numerical Solution

1. Discretization of the equations

The control surface S_J is discretized into N elements. Then the above matching conditions can be discretized as

$$(\phi_1)_i = (\phi_I)_i + \boldsymbol{\beta}_{ij}\sigma_j, \quad i, j = 1, 2, \ldots, N, \tag{6.9.38}$$

$$\left(\frac{\partial \phi_1}{\partial n}\right)_i = \left(\frac{\partial \phi_I}{\partial n}\right)_i - \pi\sigma_i + \boldsymbol{\alpha}_{ij}\sigma_j, \quad i, j = 1, 2, \ldots, N, \tag{6.9.39}$$

where

$$\boldsymbol{\alpha}_{ij} = \int\limits_{\Delta S_j} \frac{\partial G}{\partial n}\mathrm{d}S, \quad i, j = 1, 2, \ldots, N; \quad \boldsymbol{\beta}_{ij} = \int\limits_{\Delta S_j} G\mathrm{d}S, \quad i, j = 1, 2, \ldots, N$$

As Green's function G is known, the coefficient matrixes $\boldsymbol{\alpha}_{ij}$ and $\boldsymbol{\beta}_{ij}$ can be obtained. Eliminating the source intensity σ_j from Eqs. (6.9.38) and (6.9.39) yields the relation between $\partial\phi_1/\partial n$ and ϕ_1 on the control surface S_J. Then $\partial\phi_1/\partial n$ and ϕ_1 on the control surface S_J can be solved by Eq. (6.9.35).

2. wave–current forces

The forces acting on the body can be obtained by integrating the pressure on the body surface

$$F_j = -\int_{S_0} p n_j \mathrm{d}S, \quad j = 1, 2 \tag{6.9.40}$$

Under the assumption of small-amplitude waves, substituting Eq. (6.9.12) into the Bernoulli equation and omitting the second-order small quantities yield the wave pressure expressed as

$$p(x, z) = -\rho(-i\omega + \boldsymbol{V} \cdot \nabla)\phi \tag{6.9.41}$$

where

$$\boldsymbol{V} = \nabla(-Ux + \varphi_v)$$

Substituting Eq. (6.9.40) into Eq. (6.9.41) and using the theory proposed by Tuck [14] can yield the forces acting on the body

$$F_j = -\rho\int_{S_0} (i\omega n_j + m_j)\phi \mathrm{d}S, \quad j = 1, 2, \tag{6.9.42}$$

where,

$$\boldsymbol{m} = -\boldsymbol{n} \cdot \nabla(\nabla\varphi_0)$$

3. Comparisons of numerical results and the experimental data [22]

Figure 6.40 shows comparisons of numerical results and the experimental data for the diffraction effect caused by a steady current rounding a circular cylinder near the water surface. The experiments were carried out in a wave flume at the hydraulic laboratory in Tianjin University. In Fig. 6.40, the numerical wave force results are compared with the experimental data for the cases of the relative submerged depth *d*/*R* being 1.5 and 1.25 in the condition of the incident wave direction same as the current direction. Meanwhile, the numerical results of the cases with and without considering the local steady perturbation are both shown in the figure. Without considering the local steady perturbation, the calculation was based on the diffraction of a submerged object in deep water, i.e., the steady current was uniform at the free surface. In

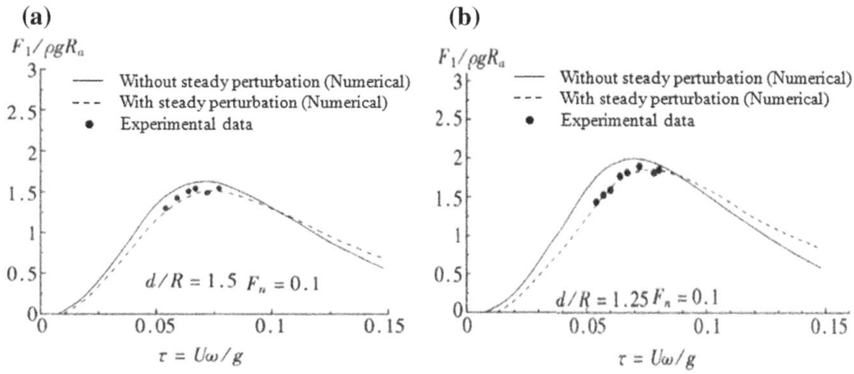

Fig. 6.40 Comparisons of numerical results and the experimental data ($F_n = U/\sqrt{gR} = 0.1$)

Fig. 6.40, the solid line and the dash line represent the numerical results without and with considering the local steady perturbation, respectively. The numerical results with the local perturbation effect of the steady current using the method of matching external and internal fields agreed with the experimental results better.

6.10 Mathematical Model of Wave–Current Forces on a Large-Scale Cylinder

6.10.1 Diffraction of a Large-Scale Vertical Cylinder Under the Wave–Current Interaction

The mathematical model for the diffraction of a submerged body near the water surface under the wave–current interaction has been described in the above section. That is a 2D problem in a vertical plane. However, the diffraction of a large-scale cylinder under the wave–current interaction is a 3D problem. As the interaction among the waves, current and the structure is very complex, especially in the vicinity of the cylinder, physical model or approximate calculation methods are commonly used in engineering at present. An approximate calculation method is to separate the wave forces and current forces; calculate wave forces using linear diffraction theory with the wave elements affected by current; calculate current forces according to the drag force of current around a cylinder; and then add them together to get the total forces under the wave–current interaction. This method does not consider the wave–current interaction and creates some differences with the experimental data. In ship engineering, the problem of ships moving at a uniform velocity in waves has been deeply studied. This problem is equivalent to the diffraction problem in wave–current field. Under the condition of linear theory, considering the effect of current on the free surface condition, and introducing the thin-ship assumption, then the problem can be solved by means of 3D moving oscillating source distribution on the body surface. This approach is feasible, but the calculation is heavy. Especially it is more difficult in finite water depth. In this section, a method for calculating wave-current forces under the wave–current interaction is described. In this method, it is assumed that the wave–current interaction on the cylinder is a potential flow problem. The velocity potential is divided into the steady potential and the unsteady potential. The steady potential consists of the double-model potential which is the potential by current around a cylinder without free surface, and the potential of the wave generated by the steady current around the cylinder. The generated wave potential is neglected because it is much smaller than the others. Considering the effect of the double-model potential on the local non-uniformity, i.e., the effect on the free surface boundary condition of the unsteady wave potential, the free surface condition along the streamline and the radiation condition for the wave–current field are introduced. The wave–current forces on a large-scale vertical circular cylinder are obtained using the direct boundary element method (simple Green's function method). The wave–current force curves presented in this section can be used directly for engineering practice [24, 18].

1. Basic equations

In order to obtain the wave–current field under the wave–current interaction, this case is assumed to be a problem of potential flow with inviscid, irrotational, and incompressible fluid.

Figure 6.41 shows a definition sketch for a large-scale cylinder in wave–current field. Under the wave–current interaction, the fixed solution problem described by the potential function $\Phi(x, y, z, t)$ reads

$$\nabla^2\Phi = 0, \quad -d < z < \eta \tag{6.10.1}$$

$$\frac{\partial\Phi}{\partial t} + g\eta + \frac{1}{2}|\nabla\Phi|^2 = d(t), \quad z = \eta \tag{6.10.2}$$

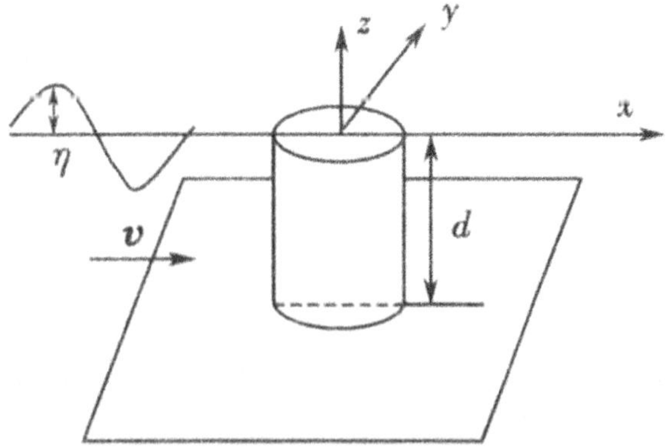

Fig. 6.41 Definition sketch for a large-scale cylinder in wave–current field

$$\frac{\partial\eta}{\partial t}+\frac{\partial\eta}{\partial x}\frac{\partial\Phi}{\partial x}+\frac{\partial\eta}{\partial y}\frac{\partial\Phi}{\partial y}=\frac{\partial\Phi}{\partial z},\quad z=\eta \tag{6.10.3}$$

$$\frac{\partial\Phi}{\partial n}=0\quad r=a \tag{6.10.4}$$

$$\frac{\partial\Phi}{\partial z}=0\quad z=-d \tag{6.10.5}$$

Radiation condition, On the open boundary S_c (6.10.6)

where $d(t)=\frac{1}{2}U^2$, U is the velocity of the uniform flow, and d is the still water depth; η is the free surface elevation; a is the radius of the circular cylinder.

By combining the above kinematic and dynamic free surface conditions and eliminating η, the unified free surface condition is given as

$$\frac{\partial^2\Phi}{\partial t^2}+g\frac{\partial\Phi}{\partial z}+\frac{1}{2}\nabla\Phi\cdot\nabla(\nabla\Phi\cdot\nabla\Phi)+2\nabla\Phi\cdot\nabla\frac{\partial\Phi}{\partial t}=0 \tag{6.10.7}$$

This free surface condition is nonlinear. Although the Laplace equation is linear, due to the nonlinearity of the free surface condition, the fluid dynamics problem with the existence of free surface is nonlinear essentially. Furthermore, since the calculation must be carried out in a limited domain, the problem of open boundary conditions must be dealt with.

2. Free surface condition

With the existence of a body, the potential function of the uniform flow in the vicinity of the body cannot be expressed by Ux.The free surface condition without existence of a body is thus not valid anymore. Now it is assumed that the total potential function is composed of three parts

$$\Phi(x,y,z,t)=\hat{\Phi}(x,y,z,t)+\phi(x,y,z)+\phi_v(x,y,z) \tag{6.10.8}$$

where $\hat{\Phi}$ is the unsteady potential; ϕ is the double-model potential, i.e., the potential by current around a cylinder without free surface; ϕ_v is the potential of the wave generated by the steady current around the cylinder.

Here the focus is on the weak current case, the generated wave potential is much smaller than the unsteady potential. It was proved by Lianwu [24, 18] that, when $\frac{U}{c}<0.2$, the generated wave potential is just about 1% of the unsteady potential. Then, the total potential becomes

$$\Phi(x,y,z,t)=\hat{\Phi}(x,y,z,t)+\phi(x,y,z) \tag{6.10.9}$$

Although the double-model potential ϕ does not contribute to the loads on the body, its effect on the local nonlinearity can be taken into account, so that the free surface boundary condition of the unsteady potential is affected.

By substituting Eq. (6.10.9) into the free surface boundary condition Eq. (6.10.7) and omitting the nonlinear terms of $\hat{\Phi}$, the free surface condition can be rewritten as

$$\frac{\partial^2 \Phi}{\partial t^2} + 2\Phi_x \hat{\Phi}_{xt} + 2\Phi_y \hat{\Phi}_{yt} + \Phi_z \hat{\Phi}_{zt} + \frac{1}{2}\Phi_x(\Phi_x^2 + \Phi_y^2 + \Phi_z^2)_x$$
$$+ \frac{1}{2}\Phi_y(\Phi_x^2 + \Phi_y^2 + \Phi_z^2)_y + g\Phi_z = 0 \qquad (6.10.10)$$

The non-uniformity of flow in the vicinity of the cylinder is significant. Thus, the free surface condition can be expressed as a differential form along the streamlines.

By the properties of the streamlines, any function B satisfies

$$\phi_x B_x + \phi_y B_y = \phi_L B_L \qquad (6.10.11)$$

where the subscript "L" denotes the differentiation along the streamlines. Therefore, the free surface condition in the differential form along the streamlines becomes

$$\frac{\partial^2 \hat{\Phi}}{\partial t^2} + 2\phi_L \hat{\Phi}_{Lt} + 2\phi_L \phi_{LL} \hat{\Phi}_L + \phi_L^2 \hat{\Phi}_{LL} + \phi_L^2 \phi_{LL} + g\hat{\Phi}_z = 0 \qquad (6.10.12)$$

Let $\hat{\Phi} = \mathrm{Re}[\hat{\Phi} e^{-i\omega t}]$,

$$\omega = \omega_0 + kU \qquad (6.10.13)$$

where ω_0 is the natural frequency without current, ω is the encountering frequency with current, k is the wave number. Then the free surface condition with the existence of the cylinder is

$$-\omega^2 \hat{\Phi} - 2i\omega\phi_L \hat{\Phi}_L + 2\phi_L \phi_{LL} \hat{\Phi}_L + \phi_L^2 \hat{\Phi}_{LL} + \phi_L^2 \phi_{LL} + g\hat{\Phi}_z = 0 \qquad (6.10.14)$$

where "L" denotes the differentiation along the streamlines as mentioned above.

3. Radiation condition under the wave–current interaction

The wave structure by the wave–current interaction on a cylinder is much more complicated than that without current. The oscillation of the diffraction potential in the field is not caused by the body oscillation, but generated by the oscillation of the incident wave velocity potential instead. However the result is equivalent to that by the body oscillation. Therefore, it can be represented by using the method of moving oscillating source.

It is assumed that the velocity potential of the moving oscillating source at finite water depth is $\Phi = \mathrm{Re}\{Ge^{-i\omega t}\}$. Then G satisfies the following conditions

$$\nabla^2 G(P, Q) = \delta(P - Q), \quad \text{in the domain} \tag{6.10.15}$$

$$-\omega^2 G - 2i\omega U \frac{\partial G}{\partial x} + U^2 \frac{\partial^2 G}{\partial x^2} + g\frac{\partial G}{\partial z} = 0, \quad z = 0 \tag{6.10.16}$$

$$\frac{\partial G}{\partial z} = 0, \quad z = -d \tag{6.10.17}$$

Radiation condition at infinity

Equation (6.10.16) is the linear free surface condition under the wave–current interaction.

In order to satisfy the bottom boundary condition, let

$$G = \frac{1}{r} + \frac{1}{r_d} + G^*(P, Q) \tag{6.10.18}$$

in which

$$r = [(x-\xi)^2 + (y-\eta)^2 + (z-\zeta)^2]^{\frac{1}{2}} \tag{6.10.19}$$

$$r_d = [(x-\xi)^2 + (y+2d+\eta)^2 + (z-\zeta)^2]^{\frac{1}{2}} \tag{6.10.20}$$

By means of Fourier transform method, the solution satisfying the free surface condition is obtained

$$G^* = \frac{2}{\pi}\int_0^\infty dk \int_{-\pi}^{\pi} d\theta \frac{\mathrm{e}^{-kd}[(\omega + Uk\cos\theta)^2 + gk]\cosh[k(\eta+d)]}{-(\omega+Uk\cos\theta)^2\cosh kd + gk\sinh kd} \cdot \cosh[k(y+d)]\cos[k(z-\zeta)\sin\theta]\cdot \mathrm{e}^{ik(x-\xi)\cos\theta} \tag{6.10.21}$$

There is a singular point at the denominator $gk\sinh kd - (\omega + Uk\cos\theta)^2\cosh kd$ in the integral kernel. It is assumed that k_1 and k_2 are the two roots of θ within $\left[0, \frac{\pi}{2}\right]$,$k_3$ and k_4 are the two roots of θ within $\left[\frac{\pi}{2}, \pi\right]$, $k_1 < k_2$ and $k_3 < k_4$. The properties of the roots are shown in Fig. 6.42.

By integrating G^*, G is obtained

$$G = \frac{1}{r} + \frac{1}{r_d} + \frac{2}{\pi}\int_0^\pi d\theta \cdot \text{P.V.}\int_0^\infty dk F(\theta, k) + \frac{2}{\pi}\int_0^{\frac{\pi}{2}} (\pi\, i\text{Res}_1 - \pi\, i\text{Res}_2)d\theta + \frac{2}{\pi}\int_{\frac{\pi}{2}}^{\pi} (\pi\, i\text{Re}\, s_3 + \pi\, i\text{Re}\, s_4)\mathrm{d}\theta \tag{6.10.22}$$

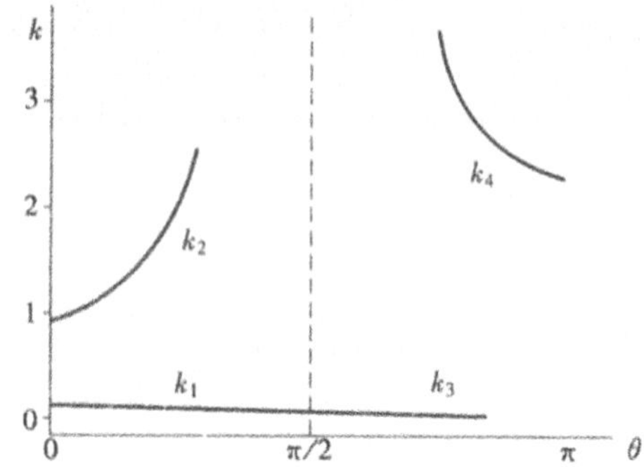

Fig. 6.42 Variation of k_1, k_2, k_3 and k_4 with θ

where

$$F(\theta, k) = \frac{e^{-kd}[(\omega + Uk\cos\theta)^2 + gk]\cosh[k(\eta + d)]}{gk\sinh(kd) - (\omega + Uk\cos\theta)^2\cosh kd} \cdot e^{ik(x-\xi)\cos\theta} \cdot \cosh[k(y + d)]\cos[k(z - \zeta)\sin\theta] \tag{6.10.23}$$

$$\mathrm{Res}_i = \frac{gk_i\cosh[k_i(\eta + d)] \cdot \cosh[k(y + d)]\cos[k_i(z - \zeta)\sin\theta]e^{ik_i(x-\zeta)\cos\theta}}{\cosh^2(k_i d)[g\tanh(k_i d) + gk_i d(1 - \tanh^2 k_i d) - 2(\omega + Uk_i\cos\theta)U\cos\theta]}$$
$$i = 1, 2, 3, 4 \tag{6.10.24}$$

By means of the analyses on the integration path and results, it is known that, k_1 and k_3 denote the waves with a radiation property, k_2 and k_4 denote the waves with certain moving property. In the calculation, U/c is relatively small. k_1 and k_3 are much smaller than k_2 and k_4. Therefore, the waves of k_2 and k_4 can be omitted, which creates an error less than 3%.

Then, the approximate radiation condition can be obtained

$$\frac{\partial\Phi_s}{\partial r} = ik_1(\theta)\Phi_s \quad \theta \le \frac{\pi}{2} \tag{6.10.25}$$

$$\frac{\partial\Phi_d}{\partial r} = ik_3(\theta)\Phi_d \quad \theta > \frac{\pi}{2} \tag{6.10.26}$$

k_1, k_2, k_3, and k_4 are determined by

$$gk\tanh kd = (\omega + Uk\cos\theta)^2 \tag{6.10.27}$$

Let,

$$\Phi = \Phi_I + \Phi_S \tag{6.10.28}$$

where Φ_I is the incident wave potential and Φ_S is the scattered wave potential. The radiation condition expressed by the velocity potential Φ reads

$$\frac{\partial(\Phi - \Phi_I)}{\partial r} = ik(\theta)(\Phi - \Phi_I) \tag{6.10.29}$$

$$\frac{\partial \Phi}{\partial r} = ik(\theta) \cdot (\Phi - \Phi_I) + \frac{\partial \Phi_I}{\partial r}. \tag{6.10.30}$$

6.10.2 Numerical Solution of Wave–Current Forces on a Large-Scale Circular Cylinder

1. Governing equation and boundary conditions

It is assumed that the unsteady potential is $\hat{\Phi}_S = \mathrm{Re}\{\varphi\, \mathrm{e}^{-i\omega t}\}$. Then the boundary value problem of φ is described as

$$\nabla^2 \varphi = 0 \tag{6.10.31}$$

$$-\omega^2\varphi - 2i\omega\phi_L\varphi_L + 2\phi_L\phi_{LL}\varphi_L + \phi_L^2\phi_{LL} + g\varphi_z = 0, \quad z = 0 \tag{6.10.32}$$

$$\frac{\partial \varphi}{\partial z} = 0 \quad z = -d \tag{6.10.33}$$

$$\frac{\partial \varphi}{\partial r} = 0 \quad r = a \tag{6.10.34}$$

$$\frac{\partial \varphi}{\partial r} = \begin{cases} ik_1(\theta)(\varphi - \varphi_I) + \frac{\partial \varphi_I}{\partial r} & \theta \le \frac{\pi}{2} \\ ik_3(\theta)(\varphi - \varphi_I) + \frac{\partial \varphi_I}{\partial r} & \theta > \frac{\pi}{2} \end{cases} \tag{6.10.35}$$

The problem of the unsteady scattered potential is turned into a boundary value problem in spatial domain, which can be solved directly by the boundary integral equation method.

The boundary element formula gives

$$\varphi(\boldsymbol{x}) = \frac{1}{\alpha(\boldsymbol{x})} \int\!\!\int_S \left[G(\boldsymbol{x}, \boldsymbol{\xi}) \frac{\partial \varphi(\boldsymbol{\xi})}{\partial n} - \varphi(\boldsymbol{\xi}) \frac{\partial G(\boldsymbol{x}, \boldsymbol{\xi})}{\partial n} \right] \mathrm{d}S \tag{6.10.36}$$

$$S = S_b + S_f + S_c \tag{6.10.37}$$

where S_b is the body surface, S_f is the free surface, S_c is the open boundary.

When the curved surface is smooth at the $\boldsymbol{x}$ point, $\alpha = 2\pi$. Here, in order to satisfy the symmetric conditions of the computational domain and at the bottom, Green's function is selected as

$$G = \frac{1}{r_1} + \frac{1}{r_2} + \frac{1}{r_3} + \frac{1}{r_4} \tag{6.10.38}$$

in which

$$r_1 = [(x-\xi)^2 + (y-\eta)^2 + (z-\zeta)^2]^{\frac{1}{2}} \tag{6.10.39}$$

$$r_2 = [(x-\xi)^2 + (y-\eta)^2 + (z+\zeta+2d)^2]^{\frac{1}{2}} \tag{6.10.40}$$

$$r_3 = [(x-\xi)^2 + (y+\eta)^2 + (z+\zeta+2d)^2]^{\frac{1}{2}} \tag{6.10.41}$$

$$r_4 = [(x-\xi)^2 + (y+\eta)^2 + (z-\zeta)^2]^{\frac{1}{2}} \tag{6.10.42}$$

2. Solution of wave forces using boundary element method

The cylinder surface S_b, the free surface S_f, and the open boundary S_c are divided into N_b, N_f, and N_c elements, respectively. Discretizing Eq. (6.10.36) yields

$$\varphi(\boldsymbol{x}_i) = \frac{1}{2\pi} \sum_{j=1}^{N_b+N_c+N_f} \left\{ \int_{\Delta S_j} G(\boldsymbol{x}, \boldsymbol{\xi}) \frac{\partial \varphi(\boldsymbol{\xi})}{\partial n} dS - \int_{\Delta S_j} \varphi(\boldsymbol{\xi}) \frac{\partial G(\boldsymbol{x}, \boldsymbol{\xi})}{\partial n} \mathrm{d}S \right\} \tag{6.10.43}$$

where N_b, N_c, and N_f are the element number on S_b , S_f and S_c, respectively.

By using constant elements in calculation, the linear algebraic equations are obtained

$$\sum_{j=1}^{N} \left[\boldsymbol{a}_{ij}\varphi_j + \boldsymbol{b}_{ij} \left(\frac{\partial \varphi}{\partial n} \right)_j \right] + \varphi(\boldsymbol{x}_i) = 0 \tag{6.10.44}$$

in which

$$N = N_b + N_c + N_f \tag{6.10.45}$$

$$\boldsymbol{a}_{ij} = \frac{1}{2\pi} \int_{\Delta S_j} \left(\frac{\partial G}{\partial n} \right)_{ij} \mathrm{d}S; \tag{6.10.46a}$$

$$\boldsymbol{b}_{ij} = -\frac{1}{2\pi} \int_{\Delta S_j} G_{ij} \mathrm{d}S \tag{6.10.46b}$$

The free surface condition gives

$$\frac{\partial \varphi}{\partial z} = \frac{\omega^2}{g}\varphi + C \tag{6.10.47}$$

in which,

$$C = \frac{2\phi_L i\omega}{g}\varphi_L - \frac{2}{g}\phi_L\varphi_{LL} - \frac{1}{g}\phi_L\varphi_{LL} - \frac{1}{g}\phi_L^2\varphi_{LL} \tag{6.10.48}$$

Since the relation between $\frac{\partial \varphi}{\partial z}$ and φ cannot be derived directly, the iterative method is used in the calculation. The initial value of φ is given first, and then C can be regarded as a known quantity. After that, we have,

$$\begin{aligned}&\sum_{j=1}^{N_b} \boldsymbol{a}_{ij}\varphi_j + \sum_{j=N_b+1}^{N_b+N_f}\left(\boldsymbol{a}_{ij} + \boldsymbol{b}_{ij}\tfrac{\omega^2}{g}\right)\varphi_j + \sum_{j=N_b+N_f+1}^{N}(\boldsymbol{a}_{ij} + ik(\theta)\boldsymbol{b}_{ij})\varphi_j + \varphi(\boldsymbol{x}_i)\\ &= -\sum_{j=N_b+1}^{N_b+N_f} \boldsymbol{b}_{ij}C_{ij} - \sum_{j=N_b+N_f+1}^{N} \boldsymbol{b}_{ij}\left[-ik(\theta)\varphi_I + \tfrac{\partial \varphi_I}{\partial n}\right]\end{aligned} \tag{6.10.49}$$

where

$$\varphi_I = \frac{4g}{i\omega}\cdot\frac{\cosh[k(z+d)]}{\cosh kd}\mathrm{e}^{ikx}.$$

To solve the above equations: find φ; obtain the value of C; and repeat the above procedure until the result converges.

The formula for the wave–current forces is derived from the Bernoulli equation

$$p = -\rho\left[\frac{\partial \Phi}{\partial t} + gz + \frac{1}{2}|\nabla\Phi|^2 - d(t)\right] \tag{6.10.50}$$

It is linearized as

$$p = -\rho[-i\omega\,\Phi + \phi_L\frac{\partial \Phi}{\partial L}]\mathrm{e}^{-i\omega t} \tag{6.10.51}$$

$$F = \iint_{S_b} \boldsymbol{p}\cdot\boldsymbol{n}_x \mathrm{d}S \tag{6.10.52}$$

$$M = \iint_{S_b} (\boldsymbol{p}\cdot\boldsymbol{n})(z+d)\mathrm{d}S \tag{6.10.53}$$

The dimensionless force coefficient is

$$C_F = \frac{F}{\rho\, g H\, a\, d \tanh(kd)/2kd} \tag{6.10.54}$$

The moment coefficient is

$$C_M = \frac{M}{\rho g\, H\, a\, d^2 \left[\frac{kd \sinh(kd)+1-\cosh(kd)}{2(kd)^2 \cosh(kd)}\right]} \tag{6.10.55}$$

where F is wave–current force, a is the radius of the circular cylinder, d is the water depth, H is the wave height, ρ is the fluid density, k is the wave number, and g is the gravitational acceleration.

3. Wave–current force on a large-scale circular cylinder

Figure 6.43 shows the variation of the force coefficient C_F with ka for a large-scale circular cylinder ($D/L > 0.2$) calculated by the mathematical model and the numerical method described above. The parameter is the ratio of the current velocity U to the wave celerity c, $c = \sqrt{gh}$. In the figure, the positive U/c represents that the wave direction and current direction are same, and the negative value represents that the wave and current directions are opposite. Due to the assumption of weak current and linearization in the model establishment, the above method and model results are applicable under the conditions of $D/L > 0.2$, $d/L > 0.16$ and $U/c < 0.2$. After finding C_F from Fig. 6.43, the wave–current force and moment can be calculated by

$$F = \frac{C_F \rho g \frac{H}{2} a \tanh(kd)}{k} \tag{6.10.56}$$

$$M = C_F \rho g \frac{H}{2} a \left[\frac{kd \sin(kd) + 1 - \cosh(kd)}{k^2 \cosh(kd)}\right] \tag{6.10.57}$$

It is seen from Fig. 6.43 that, when $U/c = 0$, i.e., for the case without existence of current, the result is identical to the analytical solution of linear diffraction theory

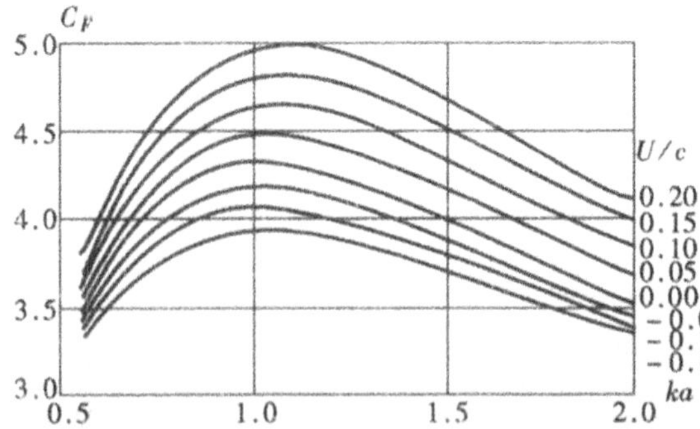

Fig. 6.43 Variation of C_F with ka

in Sect. 6.6.3. This verifies the validity of the above mathematical model and the calculation method.

It is also seen from Fig. 6.43 that, as the positive ratio U/c increases (the opposite ratio decrease), the effects of the current on the wave–current force increase. When $U/c = 0.1$, the wave–current force is larger than the wave force without existence of current by 10%.

Due to the lack of experimental data, only limited amount of experimental data were compared with the numerical results. The experimental and numerical results have a good agreement [24].

Presently, in the engineering design, the wave–current forces acting on the structure are mostly calculated by the method of linear diffraction theory, similar to the methods described above in Sects. 6.6.3, 6.6.5, and 6.6.6, but using the wave elements (such as wave height H and wave celerity c) affected by the wave–current interaction. Yanbao et al. [25] carried out an experimental study of wave–current forces on a large-scale circular cylinder under the wave–current interaction in the laboratory and calculated the wave forces on the cylinder using the linear diffraction theory with the wave elements affected by the wave–current interaction.

Figure 6.44 shows the ratio of the experimental wave force F_p to the calculated wave force F_L by the linear diffraction theory, F_p/F_L, versus the relative water depth d/L for different U/c. The wave-current forces calculated by the linear diffraction theory are significantly different from the experimental results, and the difference increases with the increase of U/c. With the decrease of d/L, the effects of nonlinearity in shallow water could not be neglected anymore. There is a big error in the calculation of wave forces even in the absence of current by means of the linear diffraction theory.

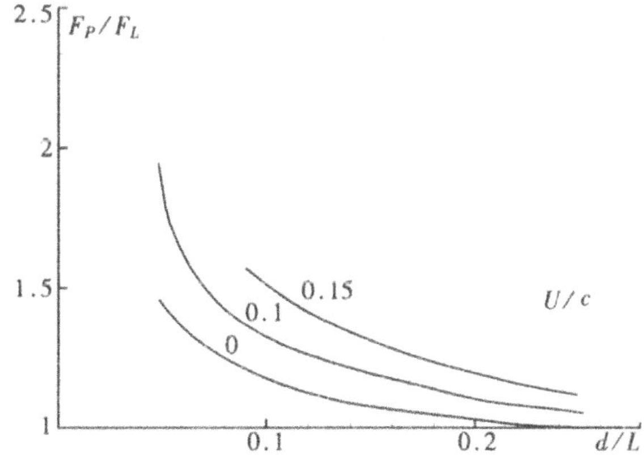

Fig. 6.44 Variation of F_p/F_L with d/L

References

1. Changgen L. Using unsteady Reynolds Navier-Stokes equations to simulate the water wave interaction with coastal structures. Dissertation for PhD degree, Tianjin University, Tianjin; 2003.
2. Mutlu SB, Fredsoe, J. Hydrodynamics Around Cylindrical Structures. World Scientific; 1999.
3. Drescher H. Messung der auf querangeströmte zylinder ausgeübten zeithlich veränderten Drücke. Z f Fluguiss. 1956;4(112):17–21.
4. Morison JR, Johnson JW, Schaaf SA. The force exerted by surface wave on piles. Petrol Trans AIME. 1950;189:149–54.
5. Sarpkaya T, Isaacson M. Mechanical of wave forces on offshore structures; 1981.
6. Sommerfeld A. Partial Differential Equation in Physics. New York: Academic Press; 1949.
7. Jianhua T, Xinyu D. Group effect coefficients of circular cylinders with large diameters. Serial documents of National Standard Technical Committee of water transport engineering, 1989;013: 52–66.
8. Jianhua T, Xinyu D. Numerical calculation of wave forces on multiple cylinders of a dock steering platform. Technical report. Department of Mechanics, Tianjin University, Tianjin; 1988.
9. Yuan DK, Tao JH. Wave forces on submerged, alternately submerged and emerged semi-circular breakwaters. Coastal Eng. 2003;48:75–93.
10. Skjelbreia L, Hendrickson JA, Fifth Order Gravity Wave Theory. In: Proceedings of 7th Coastal Engineering Conference, The Hague, 1960. p. 184–96.
11. Brorsen M, Larsen J. Source generation of nonlinear gravity waves with the boundary integral equation method. Coast Eng. 1987;11:93–113.
12. Larsen J, Dancy H. Open boundaries in short wave simulations—a new approach. Coastal Eng. 1983;7:285–97.
13. Dekui Y. The theory, method and application of numerical simulation on hydrodynamic characteristics of semicircular breakwaters. Dissertation for PhD degree, Tianjin University, Tianjin; 2004.
14. Ogilvie TF, Tuck, EO. A rational strip theory for ship motions, part 1. Rept. No. 013, Dname, University of Michigan; 1969.
15. John F. On the motion of the floating bodies, II. Commun Pure Appl Math. 1950;3: 45–101.
16. Garrison CJ. Wave Loading on North Sea Gravity Platforms A comparison of Theory and Experiment Proceedings of 9th OTC, 1977;(1):2794.
17. Jianhua T, Yan W. The treatment of singular integrals in the calculation of wave forces acting on a large scale structure with 3D source distribution method. Advances in Hydrodynamics, 1987;2(4):16–22.
18. Jianhua T, Lianwu L. Forces on a large circular cylinder by waves and current. Serial documents of Chinese Engineering Construction Standardization Association Water Transport Engineering Committee 1992;025: 61–72.
19. Jianhua T, Yanrong Z, Minghua Y. The numerical computation and experimental verification of wave forces on a gravity platform in the sea. J Tianjin Univ. 1987;4:85–94.
20. Jianhua T, Xu D. Second order wave loads on large structures. Acta Mech Sin. 1988;20(5):385–92.
21. Molin B. Second order diffraction loads upon three-dimensional bodies. App Ocean. 1979;1:197–202.
22. Yan W, Jianhua T. The effects of steady current on the diffraction for a body near water surface. Acta Oceanol Sin. 1997;19(2):133–40.
23. Yan W. The effects of steady current on the radiation and diffraction problems. Dissertation for PhD degree, Tianjin University, Tianjin, 1991.

24. Jianhua T, Lianwu L. Wave-current forces on a large vertical cylinder. J Hydrodyn. 1993;8(3):265–72.
25. YanBo L, Qi N. Forces on a Large Circular Cylinder by the Interaction of Nonlinear Waves and Current. Department of water resources and harbor engineering, Tainjin University.

Chapter 7
Numerical Simulation of Pollutant Transport Under Waves and Tidal Currents in Coastal Regions

7.1 Introduction

The pollutant transport and dispersion in seawaters is closely linked with ocean dynamics. In coastal regions, the flows under the combined action of waves and tidal currents are complex, especially in nearshore shallow waters with mild seabed slope. The nearshore currents induced by wave deformation and breaking make the pollutant transport and dispersion completely different from that in the coastal waters with steep slope.

The wave impacts on the pollutant transport and dispersion had not been paid enough attention for a long time. Recent studies have indicated that waves play an important role in the pollutant transport and diffusion in nearshore shallow waters with mild slope [8, 9].

In this chapter, a mathematical model used to study the pollutant transport and dispersion in the presence of waves in coastal regions with mild seabed slope will be introduced. The model was verified by a large amount of experimental data in wave basins. The mathematical model includes three sub-models: a wave model, a flow model regarding waves and tidal currents, and a model for the pollutant transport and dispersion. They are described separately in the following.

7.2 Wave Model

As the pollutant field in coastal water with mild seabed slope is relatively large, the wave model should be able to simulate the wave propagation, deformation, and breaking in a large coastal area. Here, the higher-order approximate parabolic mild slope equation model described in Chap. 4 is used. If the model domain is not very large, the wave field can be calculated using the elliptic mild slope equation, the time-dependent mild slope equation, or the Boussinesq equations.

J. Tao, *Numerical Simulation of Water Waves*, Springer Tracts in Civil Engineering, https://doi.org/10.1007/978-981-15-2841-5_7

7.2.1 Regular Waves

Based on the higher-order approximate parabolic mild slope Eq. (4.4.33), a wave model is established. Let the *x*-axis be the wave propagation direction. The surface elevation is expressed as,

$$\eta = \mathrm{Re}\left\{A(x, y) \cdot \mathrm{e}^{\mathrm{i}(\bar{k}x - \omega t)}\right\} \tag{7.2.1}$$

where Re{} denotes the real part of a complex; $A(x, y)$ is the complex wave amplitude function; "*i*" is the imaginary unit; k is the wave number; ω is the angular frequency; $\bar{k}$ is the characteristic wave number. The governing equation with the complex wave amplitude $A(x, y)$ reads,

$$\begin{aligned} & c_g A_x + \mathrm{i} \cdot (\bar{k} - \beta_0 k) c_g A + \frac{1}{2}(c_g)_x A + \frac{\mathrm{i}}{\omega}(\beta_1 - \beta_2 \frac{\bar{k}}{k})(c c_g A_y)_y \\ & - \frac{\beta_2}{\omega k}(c c_g A_y)_{yx} + \frac{\beta_2}{\omega}\left(\frac{k_x}{k^2} + \frac{(c_g)_x}{2k c_g}\right)(c c_g A_y)_y + \frac{\mathrm{i}\omega k^2}{2} \cdot D \cdot |A|^2 A + \frac{W}{2} A = 0 \end{aligned} \tag{4.4.33}$$

where c is the wave celerity; c_g is the wave group velocity; $\beta_0, \beta_1, \beta_2$ are the coefficients of wave approximation; D is a factor for the nonlinear effect as expressed in Eq. (4.4.34); W is a factor for wave energy loss and generation expressed as,

$$W = W_f + W_b + W_w \tag{7.2.2}$$

where W_f is the bottom friction factor, W_b is the wave breaking factor, and W_w is the wind effect factor. The relation between W and the wave energy dissipation rate ε has been given by Eq. (4.4.36),

$$W = \frac{\varepsilon}{E} \tag{4.4.36}$$

where $E = \frac{1}{8}\rho H^2$.

The model considers the effects of nonlinear terms, bottom friction, wave breaking, and wind action. The bottom friction factor W_f can be calculated by Eq. (4.4.43),

$$W_f = \frac{16\pi^2}{3g} \frac{f_w H}{T^3 \sinh^3(kh)}$$

where f_w is the friction coefficient, and it can be calculated by Eq. (4.4.44). The multiple-breaking model described in Chap. 5 can be used. The wave breaking factor is expressed by Eq. (5.5.4b),

$$W_b = \frac{K_d cn}{h}\left[1 - \left(\frac{\Gamma h}{H}\right)^2\right]$$

where $n = c_g/c$ is the ratio of the wave group velocity to the wave celerity; h is the water depth; H is the wave height. In general, K_d is taken as 0.15, and Γ is taken as 0.4. The wind effect factor W_w can be calculated by Eq. (4.4.45) in Chap. 4.

In the calculation, once the wave height, the wave period, and the wave incident angle at the incident wave boundary are given, the wave field can be obtained by a marching method.

7.2.2 Irregular Waves

By the irregular wave theory, the irregular waves can be regarded as a superposition of an infinite number of cosine waves with different amplitudes, frequencies, initial phases, and different propagation directions in the plane of (x, y). Based on Eq. (4.4.33) for an individual regular wave, the irregular wave model can be established by the linear superposition principle.

It is assumed that the frequency spectrum and the directional distribution function of irregular incident waves can be divided into some intervals. There is a representative frequency ω_i and a representative directional function θ_j in each interval. The wave with a certain frequency and a certain direction function is regarded as a regular wave component, which can be described by the parabolic mild slope equation. The propagations of these individual wave components are calculated on the grid nodes simultaneously. The statistical characteristics of the irregular waves can be obtained by the linear superposition of the individual wave components on each grid node. In the calculation of irregular waves, the wave surface elevation is expressed as,

$$\eta = \sum_i \sum_j \eta_{ij} = \mathrm{Re}\left\{\sum_{i=1}^{N_\omega} \sum_{j=1}^{N_\theta} A_{ij}(x, y) \cdot \mathrm{e}^{\mathrm{i}(\bar{k}_i x - \omega_i t)}\right\} \tag{7.2.3}$$

where the subscripts "i" and "j" represent for frequency and direction respectively; $A_{ij}(x, y)$ denotes the complex wave amplitude of the individual wave component at the ith frequency in the jth direction; N_ω and N_θ are the partition numbers of frequency and direction, respectively; ω_i is the angular frequency of the ith wave component.

$$\omega_i = gk_i \tanh k_i d$$

The characteristic wave number at the ith frequency is defined as,

$$\overline{k_i}(x) = \frac{1}{B}\int_0^B k_i(x, y)dy \tag{7.2.4}$$

where B is the bandwidth.

It is assumed that the refraction and diffraction of each wave component can be described by the parabolic mild slope Eq. (4.4.33). The complex wave amplitude of each individual wave component on each row of the wave propagation direction can be calculated simultaneously. The statistical quantities describing the random waves, such as the significant wave height and the root-mean-square wave height, can be obtained by a superposition of wave components on each grid node. The significant wave height denoted by $H_s(x, y)$ is given by,

$$H_s(x, y) = \left(8\sum_{i=1}^{N_\omega}\sum_{j=1}^{N_\theta}\left|A_{ij}(x, y)\right|^2\right)^{\frac{1}{2}} \tag{7.2.5}$$

The root mean square wave height denoted by $H_{\text{rms}}(x, y)$ is obtained as,

$$H_{\text{rms}}(x, y) = \left(4\sum_{i=1}^{N_\omega}\sum_{j=1}^{N_\theta}\left|A_{ij}(x, y)\right|^2\right)^{\frac{1}{2}} \tag{7.2.6}$$

The other statistical quantities such as the frequency spectrum $S(\omega)$, the directional spectrum $S(\omega, \theta)$, and the mean direction angle $\bar{\theta}$ can also be obtained. The frequency spectrum is calculated by,

$$S(\omega_i) = \frac{\sum_{j=1}^{N_\theta}\left|A_{ij}(x, y)\right|^2}{2(\Delta\omega_i)} \tag{7.2.7}$$

where $i = 1, \ldots, N_\omega$; $\Delta\omega_i$ is the bandwidth of each frequency interval.

The directional spectrum is,

$$S(\omega_i, \theta_b) = \frac{\sum_{j=\psi 1}^{\psi 2}\left|A_{ij}(x, y)\right|^2}{2\Delta\omega_i\Delta\theta} \tag{7.2.8}$$

The wave direction angle of each individual wave component on each grid node can be obtained by the complex wave amplitude,

$$\theta_{ij} = \arctan\left[\frac{\frac{\partial \mathrm{Im}(\ln A_{ij})}{\partial y}}{\frac{\partial \mathrm{Im}(\ln A_{ij})}{\partial x} + \bar{k}_i}\right] \quad (7.2.9)$$

where Im() denotes the imaginary part of a complex.

7.3 Model of Nearshore Currents Under Waves

When waves propagate toward the coast, a complex current system called nearshore current system is formed under the effect of bathymetry and accompanied with the transmission of mass, momentum, and energy. Nearshore currents are directly related to the action of waves and mainly occur in the surf zone between the breaker-line and the shoreline and the adjacent waters. The nearshore current system mainly consists of a longshore current (a current parallel to the shoreline), shoreward currents, and seaward currents.

Under the condition of shallow water with gentle seabed slope, the nearshore current induced by wave deformation and wave breaking is strong. Also, due to the existence of shoreward and seaward currents, the effect of the nearshore current is not only restricted to the surf zone. Knowledge's on the generation and distribution of the nearshore current can help understand the velocity distribution, further the pollutant transport and dispersion in this type of coastal waters.

7.3.1 *Mathematical Model of Nearshore Currents*

The mathematical model of nearshore currents under action of waves is established on the basis of the method proposed by Longuet-Higgins [5] and Thornton (1970).

It is assumed that the coastline is straight. The coordinate origin is set at the still water level. The x- and y-axes are parallel and perpendicular to the coastline, respectively. Figure 7.1 shows a definition sketch. η is the surface elevation, and d is the still water depth.

The horizontal two-dimensional (2D) hydrodynamic equations read,

Fig. 7.1 Definition sketch

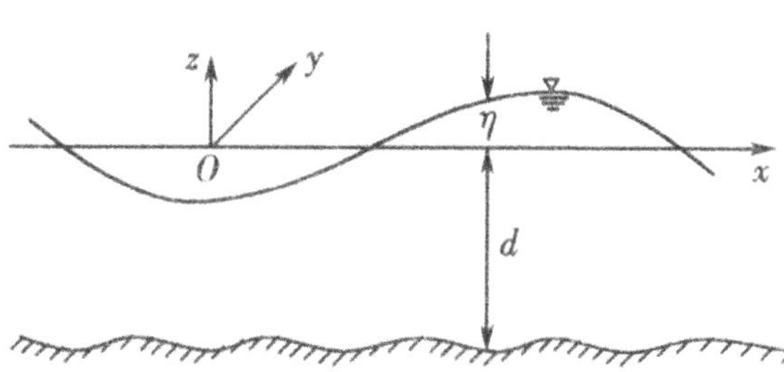

$$\frac{\partial \eta}{\partial t}+\frac{\partial}{\partial x}[Uh]+\frac{\partial}{\partial y}[Vh]=0 \tag{7.3.1}$$

$$\left.\begin{aligned}\frac{\partial U}{\partial t}+\frac{\partial UU}{\partial x}+\frac{\partial VU}{\partial y}&=-g\frac{\partial \eta}{\partial x}+T_x+M_x-\frac{\tau_{bx}}{\rho h}+\frac{\tau_{wx}}{\rho h}\\ \frac{\partial V}{\partial t}+\frac{\partial UV}{\partial x}+\frac{\partial VV}{\partial y}&=-g\frac{\partial \eta}{\partial y}+T_y+M_y-\frac{\tau_{by}}{\rho h}+\frac{\tau_{wy}}{\rho h}\end{aligned}\right\} \tag{7.3.2}$$

where U and V are the depth-averaged velocity components in x- and y-directions; T_x and T_y are the terms related to the radiation stress in x- and y-directions; M_x and M_y are the viscosity terms; τ_{bx} and τ_{by} are the bottom stress components in x- and y-directions; τ_{sx} and τ_{sy} are the surface shear stress components in x- and y-directions; $h=\eta+d$ is the total water depth.

When waves break, a part of water at wave front may roll. Here, the roller model proposed by Svendsen [7] is used. The roller energy per unit area can be expressed as Dally and Osiecki [2],

$$E_r=4\beta_r\frac{c}{gT}Q_bE \tag{7.3.3}$$

where $\beta_r\approx 0.9$; Q_b denotes the rate of breaking waves in the probability distribution of the wave height, see Eq. (5.5.12); c is the wave celerity. Considering the surface roller effect, the radiation stress components are expressed as,

$$T_x=-\frac{1}{\rho h}\left(\frac{\partial}{\partial x}(S_{xx}+R_{xx})+\frac{\partial}{\partial y}(S_{xy}+R_{xy})\right) \tag{7.3.4}$$

$$T_y=-\frac{1}{\rho h}\left(\frac{\partial}{\partial x}(S_{xy}+R_{xy})+\frac{\partial}{\partial y}(S_{yy}+R_{yy})\right) \tag{7.3.5}$$

in which R_{ij} is the stress induced by the surface roller of the breaking waves,

$$R_{xx}=2E_r\cos^2\alpha \tag{7.3.6}$$

$$R_{yy}=2E_r\sin^2\alpha \tag{7.3.7}$$

$$R_{xy}=R_{yx}=E_r\sin 2\alpha \tag{7.3.8}$$

$$S_{xx}=E\left[n\left(1+\cos^2\alpha\right)-\frac{1}{2}\right]=\frac{E}{2}\left[2n\left(1+\cos^2\alpha\right)-1\right] \tag{7.3.9}$$

$$S_{yy}=E\left[n\left(1+\sin^2\alpha\right)-\frac{1}{2}\right]=\frac{E}{2}\left[n\left(1+\sin^2\alpha\right)-1\right] \tag{7.3.10}$$

$$S_{xy}=S_{yx}=\frac{1}{2}En\sin 2\alpha=\frac{E}{2}n\sin 2\alpha; \tag{7.3.11}$$

where $n = \frac{1}{2}\left[1 + \frac{2kh}{\sinh(2kh)}\right]$; α is the angle between the wave propagation direction and the x-axis.

In the process of wave propagation, the radiation stress varies with the wave height.

The viscosity terms in the momentum equations read,

$$M_x = \frac{\partial}{\partial x}\left(\nu_b \frac{\partial U}{\partial x}\right) + \frac{\partial}{\partial y}\left(\nu_b \frac{\partial U}{\partial y}\right) \quad (7.3.12)$$

$$M_y = \frac{\partial}{\partial x}\left(\nu_b \frac{\partial V}{\partial x}\right) + \frac{\partial}{\partial y}\left(\nu_b \frac{\partial V}{\partial y}\right) \quad (7.3.13)$$

in which ν_b is the comprehensive viscosity coefficient expressed as,

$$\nu_b = \nu_{bw} + \nu_{bf} + \nu_{br} \quad (7.3.14)$$

where ν_{bw}, ν_{bf}, and ν_{br} are the viscosity coefficients related to wave motion, the bottom friction, and wave breaking, respectively. They can be calculated by,

$$\nu_{bw} = K_1 \frac{H^2 g T}{4\pi^2 d} \cos^2 \alpha \quad (7.3.15)$$

This relation proposed by Jonsson et al. [3] links the wave energy to the turbulence generation and takes into account of the wave effect in lateral mixing. In which K_1 is an empirical constant, H is the wave height, T is the wave period, and α is the wave direction angle.

Larson et al. [4] proposed an empirical formula for the viscosity coefficient related to the bottom friction,

$$\nu_{bf} = K_2 u_m H \quad (7.3.16)$$

in which u_m is the amplitude of the horizontal component of the wave particle velocity at seabed, which is given by,

$$u_m = \frac{gHT}{2L \cosh\left[\frac{2\pi(d+\eta)}{L}\right]}$$

where L is the wavelength; K_2 can be taken as 0.1.

Battjes [1] took turbulence as a function of local wave energy loss and proposed an expression of the viscosity coefficient related to the wave energy loss [6],

$$\nu_{br} = K_3(d + \eta)\left(\frac{\varepsilon_b}{\rho}\right)^{1/3} \quad (7.3.17)$$

where ε_b is the mean energy dissipation rate per unit area, as expressed in Eq. (5.5.2); K_3 can be taken as 0.025. By combining Eqs. (7.3.15), (7.3.16), and (7.3.17), the expression of viscosity coefficient Eq. (7.3.14) can be rewritten as,

$$\nu_b = K_1 \frac{H^2 gT}{4\pi^2 d} \cos^2\alpha + K_2 u_m H + K_3 (d+\eta) \left(\frac{\varepsilon_b}{\rho}\right)^{1/3} \tag{7.3.18}$$

In deep water, the wave particle velocity at the seabed u_m can be neglected. The lateral mixing of water masses is mainly caused by periodic wave oscillation. As the waves propagate from deep to nearshore shallow water, the wave particle velocity at the seabed increases gradually, and the effect of bottom friction on the lateral mixing of water masses increases gradually as well. The energy dissipation increases after the waves break. Therefore, the effects of wave motion, bottom friction, and wave breaking are included in the terms of lateral turbulent mixing in the surf zone.

The bottom shear stress components under the wave–current interaction can be given by,

$$\tau_{bx} = \frac{\rho g u\sqrt{u^2+v^2}}{C^2} + \frac{\pi\rho}{8} f_w \sqrt{u_b^2+v_b^2}\cdot u_b + \frac{B\rho}{\pi}\sqrt{2}\cdot\left(\frac{g}{C^2} f_w\right)^{\frac{1}{2}} \sqrt{u^2+v^2}\cdot u_b \tag{7.3.19}$$

$$\tau_{by} = \frac{\rho g v\sqrt{u^2+v^2}}{C^2} + \frac{\pi\rho}{8} f_w \sqrt{u_b^2+v_b^2}\cdot v_b + \frac{B\rho}{\pi}\sqrt{2}\cdot\left(\frac{g}{C^2} f_w\right)^{\frac{1}{2}} \sqrt{u^2+v^2}\cdot v_b \tag{7.3.20}$$

where u_b and v_b are the wave particle velocity components at seabed in x- and y-directions; C is the Chezy number; B is the wave—current interaction coefficient. According to Soulsby [6], B is related to the angle between the wave and current directions. If the waves propagate in the same direction as the current, B is taken as 0.917; and if waves propagate perpendicularly to the current, then B is taken as −0.1983. For the uncertain angle between the wave and current directions, B is taken as 0.359. f_w is the wave bottom friction coefficient, which can be obtained by [6],

$$f_w = 0.00251 \exp\left[5.21\cdot\left(\frac{A_L}{k_s}\right)^{-0.19}\right], \quad \frac{A_L}{k_s} > 1.57 \tag{7.3.21}$$

$$f_w = 0.3, \quad \frac{A_L}{k_s} \le 1.57 \tag{7.3.22}$$

in which k_s is the Nikuradse's equivalent sand roughness; A_L is the amplitude of the particle displacement just outside the wave boundary layer, which can be given by the small-amplitude wave theory,

$$A_L = \frac{H}{2\sinh kd} \tag{7.3.23}$$

where H is the wave height, and k is the wave number.

For the surface shear stress, the wind stress is the only one factor considered, and the components are given by,

$$\tau_{wx} = C_w \rho_a W_x \sqrt{W_x^2 + W_y^2},\ \tau_{wy} = C_w \rho_a W_y \sqrt{W_x^2 + W_y^2}$$

where C_w is the shear coefficient of the wind stress to the water surface; ρ_a is the air density; W_x and W_y are the wind velocity components 10 m above the water surface.

7.3.2 Nearshore Currents in Coastal Regions: Case Studies

1. Nearshore current in a rectangular coastal region [9]

The computational domain is a rectangular coastal region of 20 km × 20 km with a uniform slope as shown in Fig. 7.2. The open boundary is set at the slope toe with water depth of 10 m. The slope was set as 0.005. The lateral boundaries are taken as the solid boundaries. The coastline is treated as a solid boundary. A semidiurnal tide with a range of 2 m is prescribed. The wave height is 3 m with a period of 7 s and an incident wave angle of 30°. The dash line in Fig. 7.2 represents the breaker-line. Two cases are investigated; one is a pure tidal current case, and the other is a wave–current interaction case. A stationary state in the whole domain is the initial condition, i.e., $u = v = 0$ and $\eta = 0$ for $t = 0$. The Coriolis force is neglected, and the Manning number is set to be 0.02.

In the case of pure tidal current without wave, the wave action term T in Eq. (7.3.1) becomes zero. The grid spacing is $\Delta x = \Delta y = 500$ m, and the time step is $\Delta t = 90$ s. After 4–5 tidal periods, the current field tends to be steady. Figure 7.3 shows the steady current field at four different moments in a tidal period. The maximum current speed is up to 0.48 m/s.

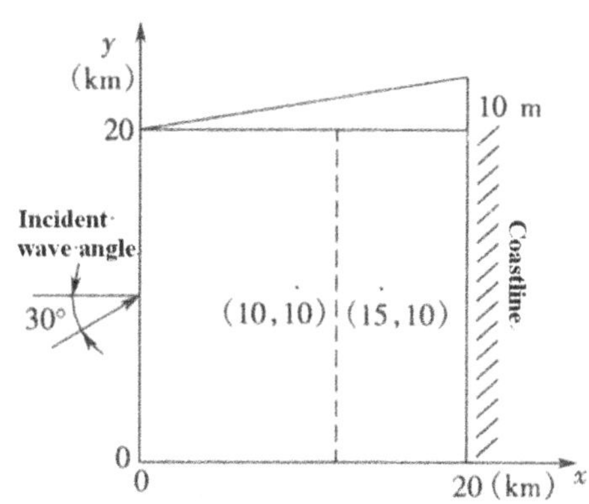

Fig. 7.2 Sketch of the model domain (the dash line represents the breaker-line)

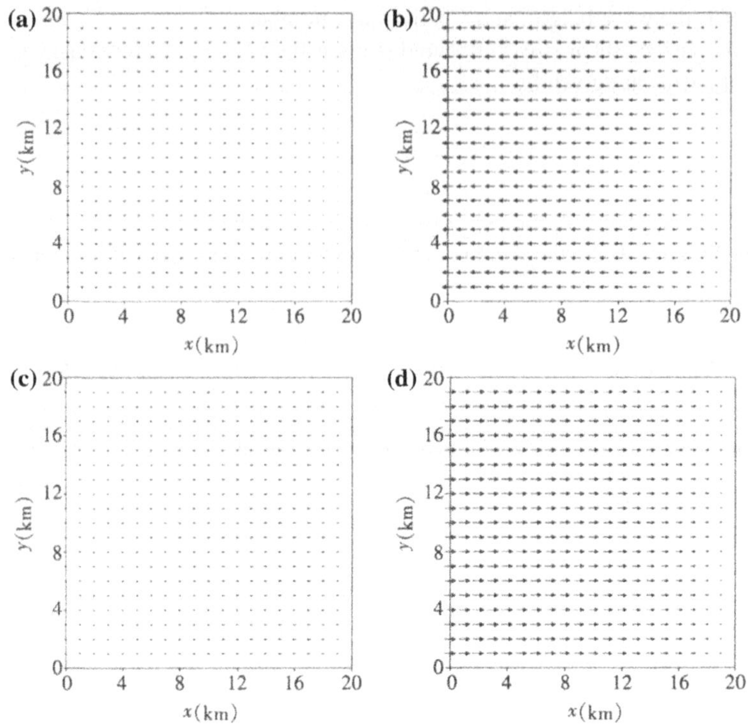

Fig. 7.3 Current field at four different moments in a tidal period T in the case of pure tidal current, **a** $t = 1/4\ T$; **b** $t = 2/4\ T$; **c** $t = 3/4\ T$; **d** $t = T$

In the case of wave–current interaction, the basic settings are the same as above for the case of pure tidal current, except for the grid spacing being 10.0 m when solving Eqs. (7.3.1) and (7.3.2). Figure 7.4 shows the steady current field. The maximum current speed is up to 0.81 m/s. Compared with the pure tidal current field, the current field under the wave–current interaction changes significantly due to the wave.

2. Nearshore currents over bathymetry varying in cosine-square expressions

The wave-induced currents over the protruding coastal bathymetry varying in cosine-square expressions are investigated here. The computational domain is 2100 m × 2800 m. The water depth h is expressed by,

$$x_c = 400\cos^2[\pi(y - 1400)/2000] \tag{7.3.24}$$

$$h = 20(2000 - x_c - x)/(2000 - x_c) \tag{7.3.25}$$

where the x-direction is perpendicular to the coastline, and the y-direction is parallel to the coastline. Figures 7.5 and 7.6 show the 3D bathymetry and the isobaths in the computational domain, respectively. There is a seaward hump on the seabed, and the

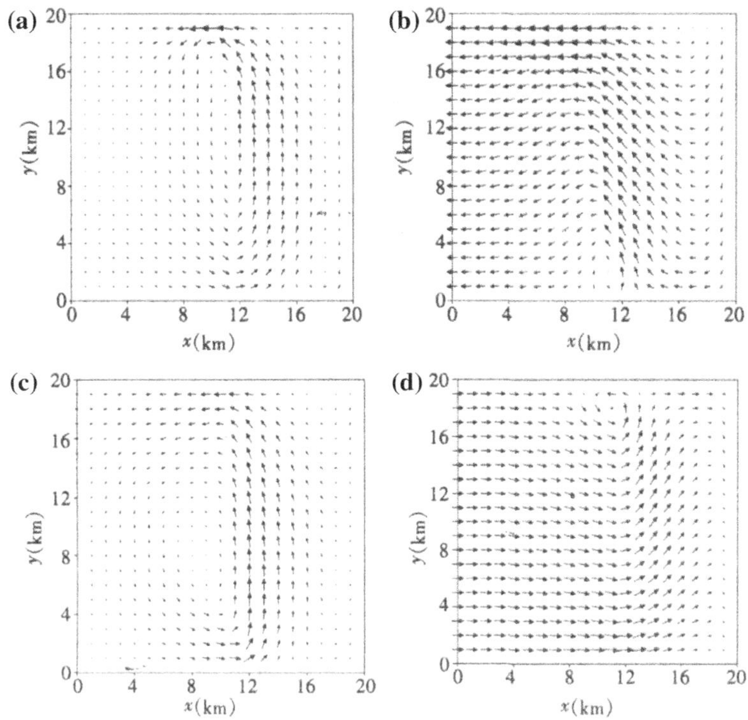

Fig. 7.4 Current field at four different moments in a tidal period T in the case of wave–current interaction, **a** $t = 1/4\ T$; **b** $t = 2/4\ T$; **c** $t = 3/4\ T$; **d** $t = T$

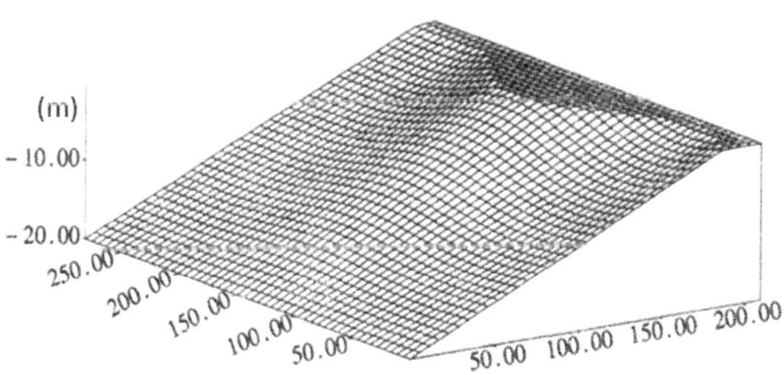

Fig. 7.5 Bathymetry in the computational domain

seabed is gentle on the both sides of the hump. This type of bathymetry is common in natural seas. Here, the bathymetry is regular for simplicity.

Over the bathymetry shown in Figs. 7.5 and 7.6, the incident wave propagates along the x-direction. There are certain angles between the wave propagation direction and the isobaths, i.e., the incident wave is oblique. Figure 7.7 shows the nearshore

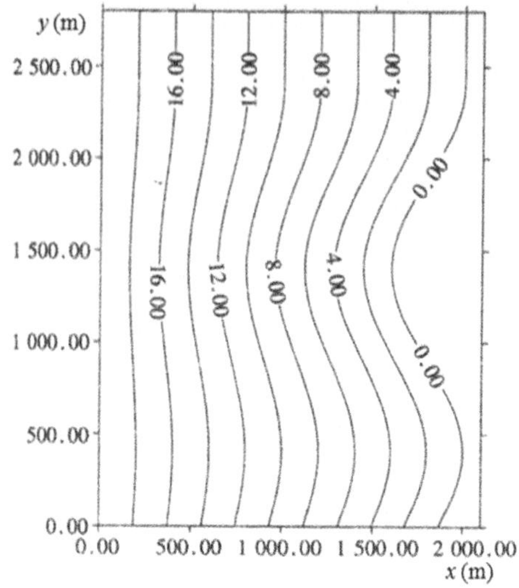

Fig. 7.6 Isobaths in the computational domain

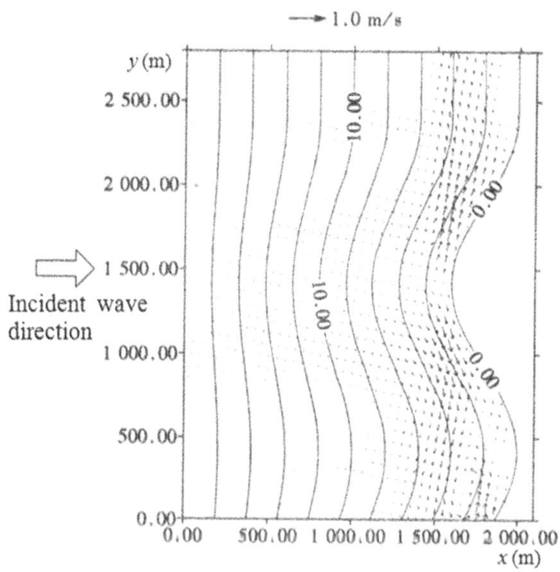

Fig. 7.7 Nearshore current field

current field induced by the incident wave propagating along the x-direction with wave height $H = 1.0$ m and wave period $T = 8.0$ s.

In the surf zone, the wave-induced currents flowed from the area with high wave height to the area with low wave height, i.e., the currents flowed from the shallow area to the deep areas because of the hump.

Now, we make a little change in Eqs. (7.3.24) and (7.3.25) to get a concave bathymetry in the center of the computational domain as shown in Figs. 7.8 and 7.9,

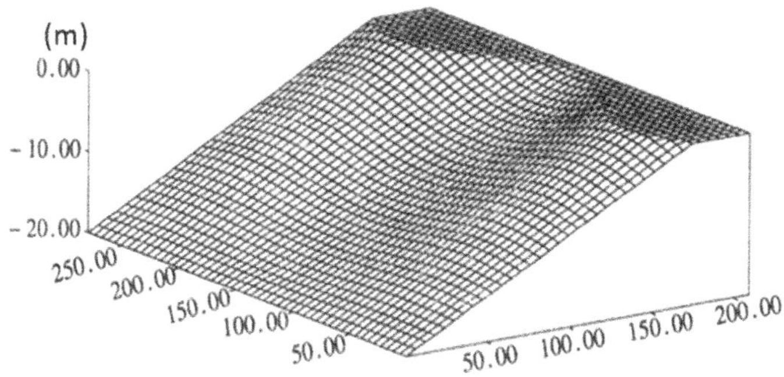

Fig. 7.8 Bathymetry in the computational domain

Fig. 7.9 Isobaths in the computational domain

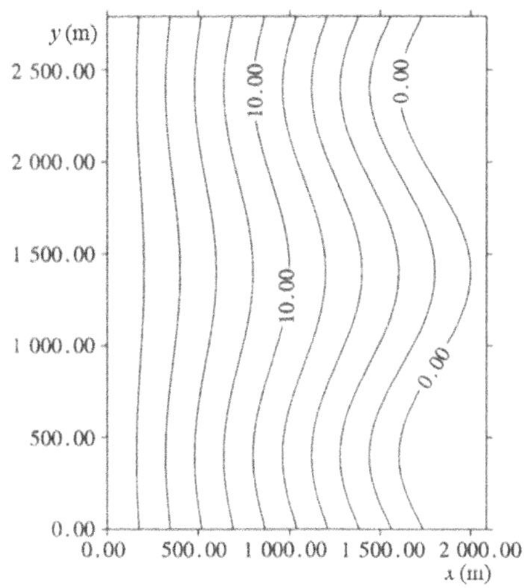

$$x_c = 400\cos^2[\pi(y - 400)/2000] \tag{7.3.26}$$

$$h = 20(2000 - x_c - x)/(2000 - x_c) \tag{7.3.27}$$

This type of bathymetry is also common in nature.

Similarly, the incident wave propagates along the x-direction with wave height $H = 1.0$ m and wave period $T = 8.0$ s. Figure 7.10 shows the corresponding nearshore current filed.

Again, the longshore currents flow from the shallow area with high wave height to the deep area with low wave height in the surf zone. When the currents on the two sides flow into the concave area in the center, the seaward current occurs for the continuity of water bodies. The seaward current is not limited in the surf zone.

Fig. 7.10 Nearshore current field

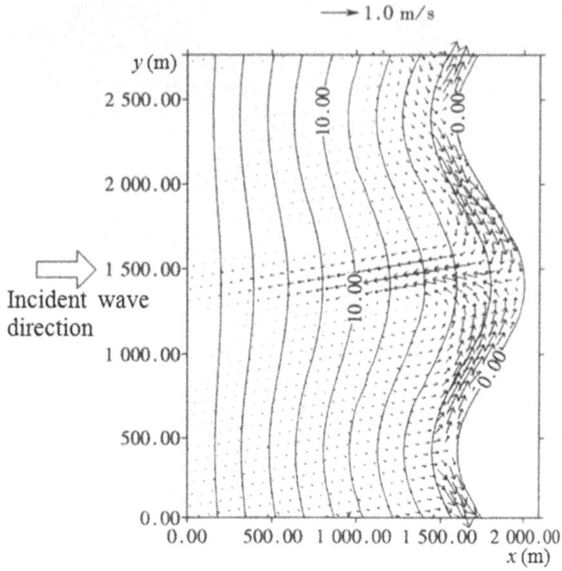

Although there is no particular driving force, the seaward current is still relatively strong outside of the surf zone.

7.3.3 Characteristics of the Longshore Current Induced by Waves Over Bathymetry with a Uniform Slope

When waves propagate obliquely into the coastal region with a uniform slope, the nearshore current induced by wave breaking is parallel to the shoreline. This current is called longshore current, and it is one of the main driving forces for longshore sediment transport and morphological evolution and has significant effect on the pollutant transport and dispersion in coastal regions.

Tao et al. [10] used physical and numerical models to study the longshore currents induced by the regular and irregular incident waves with different wave elements on various uniform slopes. The relations between the longshore current profiles and the impact factors such as the wave elements and seabed slope were analyzed [10].

Figure 7.11 shows comparisons of the numerical and experimental results for a regular wave. The depth in deep water before the slope is 18 cm, and the uniform slope was 1:100. The angle between the incident wave direction and the coastline is 30°. Figure 7.11a, b show the profiles of the wave height (H) and the corresponding longshore current speed (V) on a section perpendicular to the coastline, respectively.

For irregular waves, the profiles of the wave height (H) and the corresponding longshore current speed (V) on the section perpendicular to the coastline are shown in Fig. 7.12a, b, respectively.

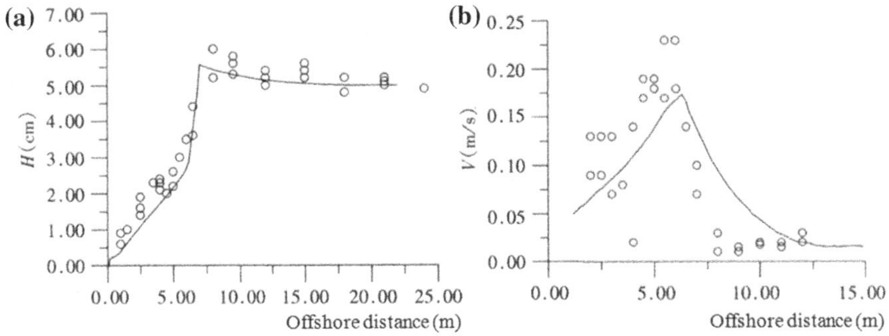

Fig. 7.11 Profiles of the wave height **a** and the longshore current speed **b** on a section perpendicular to the coastline for a regular wave. $T = 1.0$ s, $H = 5.0$ cm, slope 1:100, incident wave angle 30°; —: Numerical results; ○: Experimental results [10]

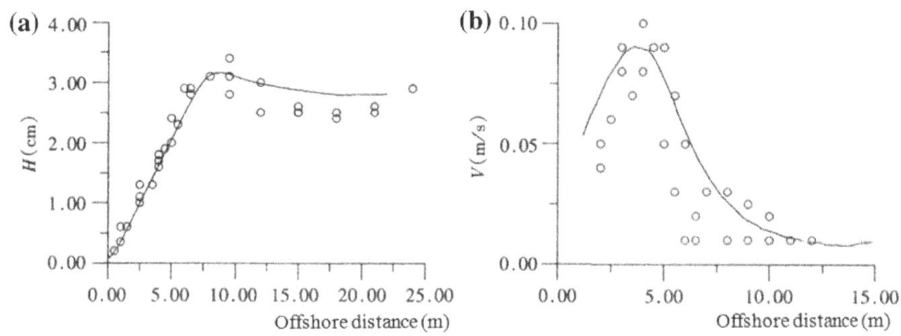

Fig. 7.12 Profiles of the wave height **a** and the longshore current speed **b** on a section perpendicular to the coastline for irregular waves. $\bar{T} = 2.0$s, $\bar{H} = 3.0$ cm, slope 1:100, incident wave angle 30°; —: Numerical results; ○: Experimental results [10]

The numerical results using the nearshore current model described above have a good agreement with the experimental results.

By comparing the numerical results for the regular and irregular waves, it could be seen that the breaker position is clear and the wave height decays abruptly after wave breaking in the regular wave case. The breaking process of irregular waves is relatively gradual. The longshore current profile on the section perpendicular to the coastline by the irregular waves is smoother than that by the regular wave, which reflects the difference of the wave breaking status between regular and irregular waves.

Figure 7.13 shows the profiles of the longshore current speed (V) on a section perpendicular to the coastline by regular waves with different wave heights over a uniform slope. The effect of wave height on the longshore current distribution can be seen from the figure. The offshore distance corresponding to the maximum longshore current speed increases with the increase of wave height, so that the offshore boundary

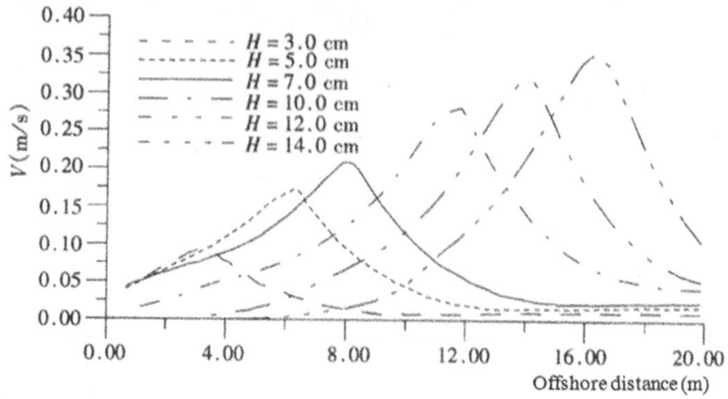

Fig. 7.13 Profiles of the longshore current speed on a section perpendicular to the coastline by the regular waves. $T = 1.0$ s, slope 1:100

of the longshore current moves further seaward, and the affected area of the longshore current is extended.

Figure 7.14 shows the profiles of the longshore current speed (V) on a section perpendicular to the coastline by irregular waves with different wave heights. Compared with the sharp profiles of longshore current speed induced by regular waves, the profiles of longshore current speed by irregular waves are relatively smooth. However, the common relation between the longshore current speed distribution and the incident wave height is not changed. As the incident wave height increases, the longshore current speed increases and the affected area of the longshore current is extended as well.

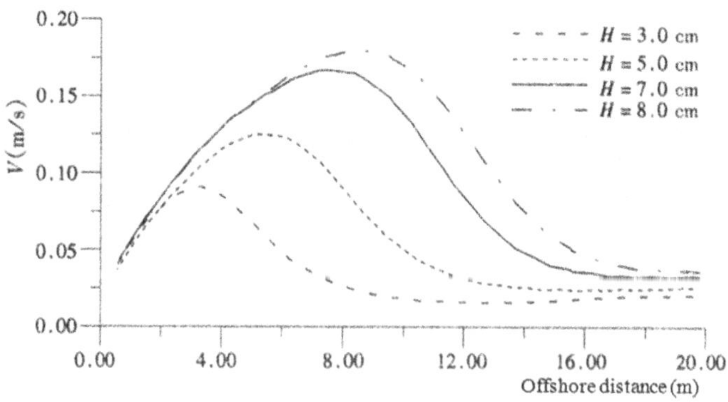

Fig. 7.14 Profiles of the longshore current speed on a section perpendicular to the coastline by the irregular waves. $T = 1.0$ s, slope 1:100

7.4 Mathematical Model of Pollutant Transport Under Interaction of Waves and Tidal Currents

7.4.1 *Governing Equation for Pollutant Transport*

The general three-dimensional (3D) advection-diffusion equation reads,

$$\frac{\partial c}{\partial t}+\frac{\partial(cu)}{\partial x}+\frac{\partial(cv)}{\partial y}+\frac{\partial(cw)}{\partial z}=\frac{\partial}{\partial x}\left(K_x\frac{\partial c}{\partial x}\right)+\frac{\partial}{\partial y}\left(K_y\frac{\partial c}{\partial y}\right)+\frac{\partial}{\partial z}\left(K_z\frac{\partial c}{\partial z}\right)+s_m \quad (7.4.1)$$

where c is the pollutant concentration; u, v, and w are the velocity components in the x-,y-, and z-directions; s_m is the pollutant change per unit volume in unit time; K_x, K_y, and K_z are the dispersion coefficients in the x-,y-, and z-directions.

In the right-hand coordinate system, by integrating Eq. (7.4.1) vertically from the sea surface $z=\eta$ to the seabed $z=-h$, using Leibnitz formula to change the order of integral and differential signs, and using the flux conditions at the sea surface and bottom, the depth-averaged two-dimensional (2D) advection-diffusion equation is obtained,

$$\frac{\partial hC}{\partial t}+\frac{\partial hUC}{\partial x}+\frac{\partial hVC}{\partial y}=\frac{\partial}{\partial x}\left(K_xh\frac{\partial C}{\partial x}\right)+\frac{\partial}{\partial y}\left(K_yh\frac{\partial C}{\partial y}\right)+hS_m \quad (7.4.2)$$

where $h=\eta+d$ is the total water depth; C, U, V, K_x, and K_y are the depth-averaged quantities.

7.4.2 *Simulations of Pollutant Transport Under Interaction of Waves and Tidal Currents in Coastal Regions*

1. Instantaneous point-source pollutant transport with pure tidal current over a uniform slope

The computational domain is identical to the region shown in Fig. 7.2 for the study of pollutant transport with the pure tidal current. The depth-averaged dispersion coefficients in Eq. (7.4.2) could be given by,

$$K_x=\frac{(\alpha U^2+\beta V^2)H\sqrt{g}}{C\sqrt{U^2+V^2}} \quad (7.4.3)$$

$$K_y=\frac{(\alpha V^2+\beta U^2)H\sqrt{g}}{C\sqrt{U^2+V^2}} \quad (7.4.4)$$

where $\alpha = 5.93$ and $\beta = 0.23$.

After getting the steady current field (with the same conditions as above for the case of pure tidal current in Sect. 7.3.2), a pollutant point-source with a unit concentration is set at the point with coordinates (15, 10) inside the surf zone in the computational domain.

In the pure tidal current case, after 14 tidal periods with the steady current field, the point-source pollutant with a unit concentration (1 g/mL) is discharged instantaneously at the point with coordinates (15, 10). Figure 7.15 shows the pollutant concentration distribution at the four different moments within the 15th tidal period. Because the current flows only in the *x*-direction, the pollutant is transported only in the *x*-direction, and only diffusion occurs in the *y*-direction. The pollutant moves seaward and shoreward along with the tidal current in the *x*-direction, while it is diffused within a tidal period. The phenomenon is similar in each tidal cycle, except that the diffusion area becomes larger with time.

2. Instantaneous point-source pollutant transport under the interaction of wave and tidal current over a uniform slope

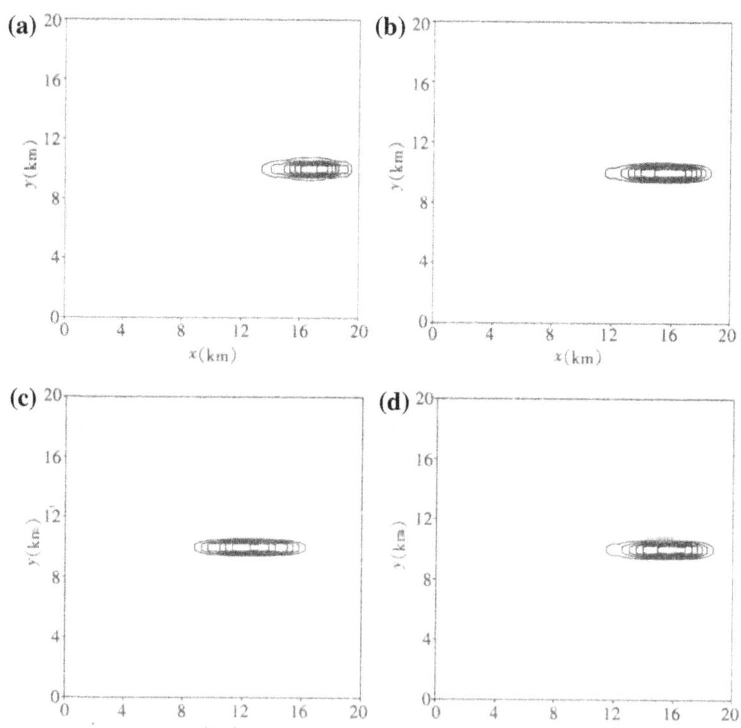

Fig. 7.15 Contours (with an interval of 0.02 g/mL) of pollutant concentration with the pure tidal current, the point-source at (15, 10); **a** $t = 1/4\ T$; **b** $t = 2/4\ T$; **c** $t = 3/4\ T$; **d** $t = T$

Under the interaction of wave and tidal current, the point-source pollutant with a unit concentration (1 g/mL) is discharged instantaneously at the point with coordinates (15, 10). Figure 7.16 shows the corresponding pollutant concentration distribution at the four different moments. Compared with Fig. 7.15 with the pure tidal current, the pollutant transport and dispersion changes significantly. The flow changes due to the wave action, which makes the pollutant move in the direction of longshore current. Meanwhile the pollutant moves seaward and shoreward along with the tidal current.

3. Continuous point-source pollutant transport under the interaction of wave and tidal current over a uniform slope

In the same computational domain over the uniform slope, the conditions are the same as above under the interaction of wave and tidal current, except that the pollutant is discharged continuously (1 mg/mL/s) at the point (17.5, 10) inside the surf zone. Figure 7.17 shows the calculated pollutant concentration distribution at the four different moments by the continuous point-source under the interaction of wave and tidal current. The pollutant moves along the coastline distinctly when the pollutant is discharged continuously at the location in the surf zone due to the interaction of

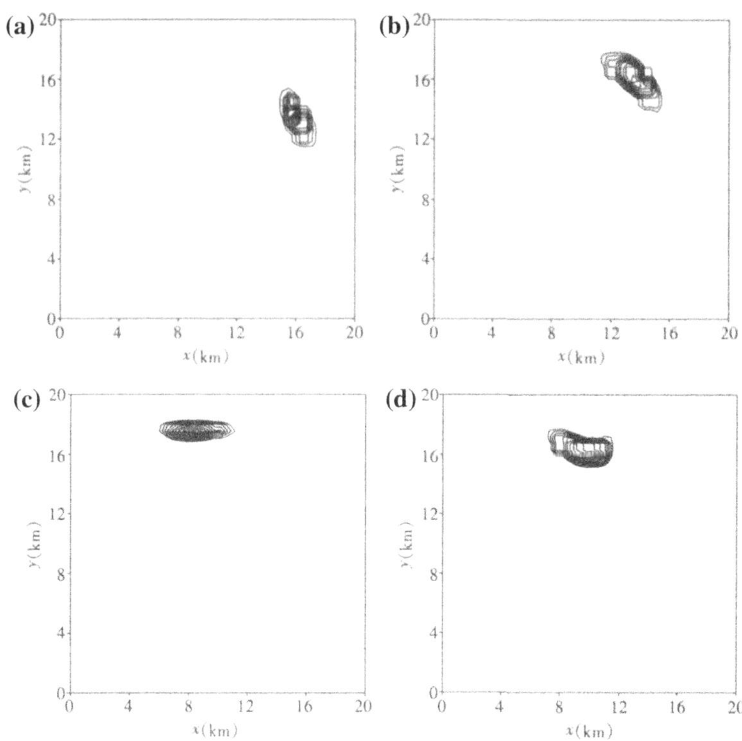

Fig. 7.16 Contours (with an interval of 0.02 g/mL) of pollutant concentration under the interaction of wave and tidal current, the point-source at (15, 10); **a** $t = 1/4\ T$; **b** $t = 2/4\ T$; **c** $t = 3/4\ T$; **d** $t = T$

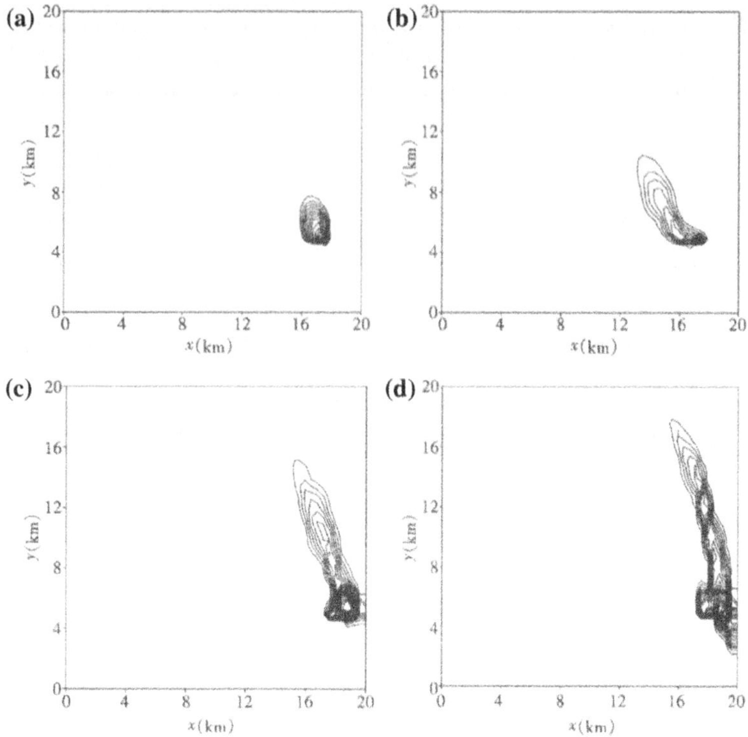

Fig. 7.17 Contours (with an interval of 2 g/mL and the minimum concentration of 2 g/mL) of pollutant concentration under the interaction of wave and tidal current, the continuous point-source at (17.5, 10); **a** $t = 1/4\ T$; **b** $t = 2/4\ T$; **c** $t = 3/4\ T$; **d** $t = T$

wave and tidal current, which is significantly different from the phenomenon when the pollutant source is located outside the surf zone.

Figure 7.18 shows the numerical pollutant distribution from different continuous point-sources under the wave action. When the pollutant source is located on the shoreward side of the breaker-line, the pollutant moves along the coastline due to the wave-induced nearshore current. When the pollutant source is located on the seaward side of the breaker line, the pollutant is diffused locally.

Figure 7.19 shows snapshots of experimental pollutant distribution in the physical wave basin. The pollutant is discharged continuously from different point-sources located inside and outside the surf zone under the wave action. It is proved by the experiments that the pollutant moves along the coastline when the source is on the shoreward side of the breaker-line, and the pollutant is diffused locally when the source is on the seaward side of the breaker-line.

The numerical simulations and the physical tests show that, in the coastal region with a mild slope, the pollutant transport and dispersion under pure tidal current action is significantly different from that under the wave action. With the existence of wave,

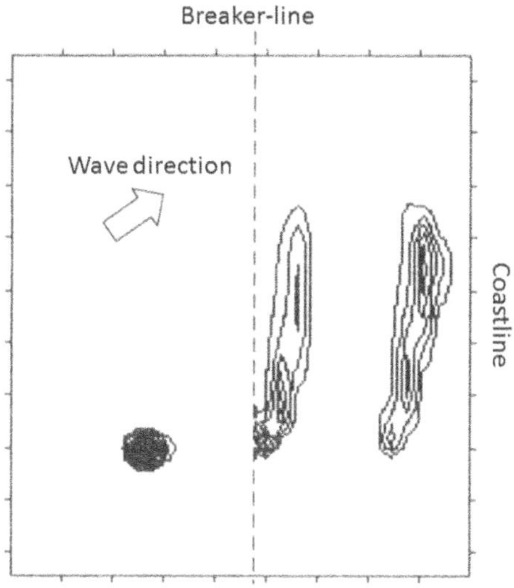

Fig. 7.18 Pollutant distribution from different continuous point-sources under the wave action

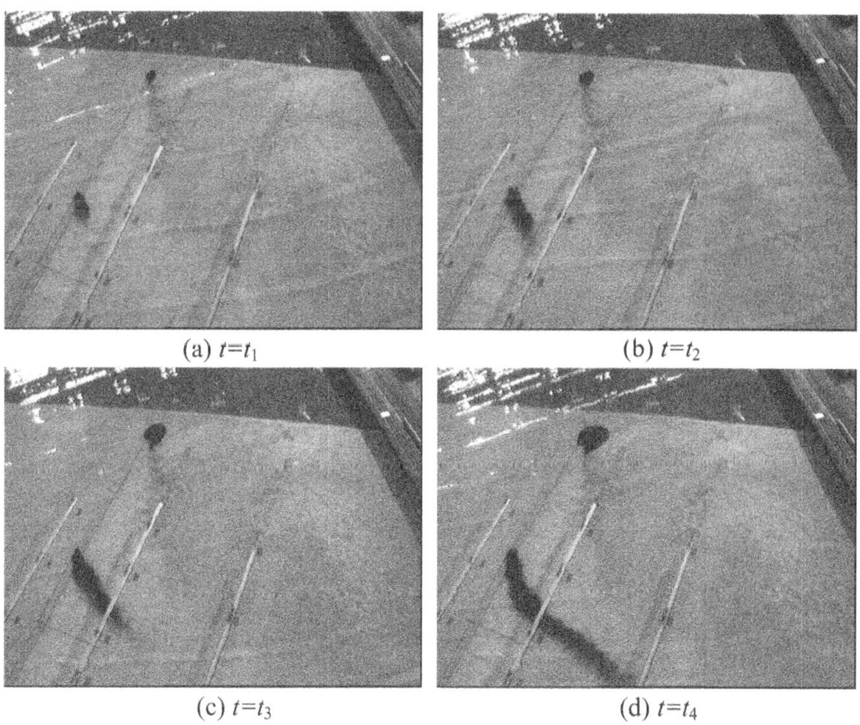

Fig. 7.19 Snapshots of pollutant distribution from different continuous point-sources under the wave action in the physical wave basin [10]; $T = 1.0$ s, $H = 3.0$ m, (slope 1:100)

the pollutant hugs the coastline due to the effect of the wave-induced nearshore current. As a result, it is difficult for the pollutant discharged nearshore to be transported into deep water rapidly.

7.5 Nearshore Currents and Pollutant Transport in Shallow Water with Mild Slope Under Wave Action and Interaction of Waves and Tidal Currents

The studies on wave breaking and the corresponding longshore currents over bathymetry with a uniform mild slope have indicated that the current field is affected by the wave action due to the wave-induced longshore current in the coastal region with a mild slope. In coastal regions, due to the varying bathymetry, the wave-induced nearshore currents include not only longshore current parallel to the shoreline, but also shoreward and seaward currents. The effects of nearshore currents are not only limited in the surf zone.

Tao et al. [12] studied the current fields in the coastal Bohai Bay under the action of waves from eastern (E) and east-south-eastern (ESE) directions by taking a typical muddy coastal region with mild slope in Bohai Bay as an example. Figures 7.20 and 7.21 show the bathymetries of the computational domains for the waves from E and ESE directions, respectively.

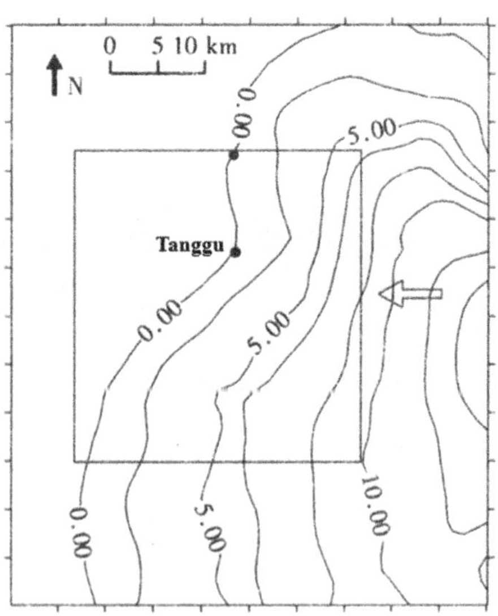

Fig. 7.20 Bathymetry of the computational domain for the wave from E direction

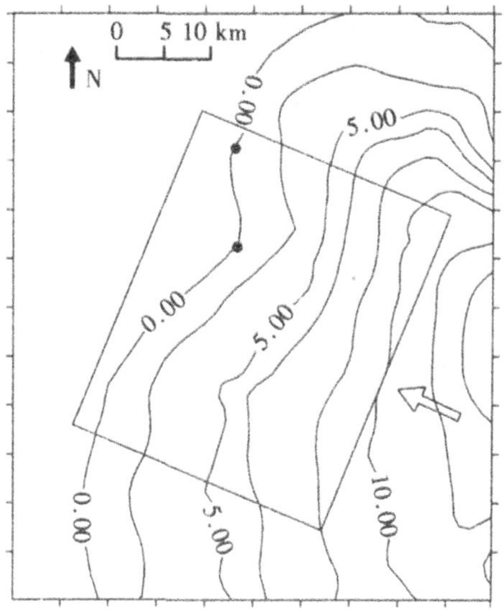

Fig. 7.21 Bathymetry of the computational domain for the wave from ESE direction

7.5.1 Nearshore Currents Under Wave Action and Interaction of Waves and Tidal Currents

1. Current fields in the coastal region under wave action

The determination of wave height distribution in the coastal region is the basis for the calculation of nearshore currents. The calculated wave height distributions for the waves from E and ESE directions are shown in Fig. 7.22.

Figure 7.23 shows the current fields under the wave action corresponding to the wave conditions in Fig. 7.22. The current fields are computed based on the waves and the subsequent radiation stress. The wave-induced nearshore currents are obtained using the mathematical model of nearshore currents under the wave action.

Due to the bathymetry with a mild slope, the nearshore currents mainly exist in the form of longshore current. It can be seen that the current speed is relatively small near the offshore boundary, the maximum current speed occurs in the middle of the surf zone, and the current becomes weak near the shore. This phenomenon is consistent with the characteristics of the nearshore currents described in the above section.

2. Current fields in the coastal region under interaction of waves and tidal currents

In the actual coastal region, wave-induced nearshore currents and tidal currents exist at same time. Figure 7.24 shows the current fields under the interaction of the tidal currents and the wave from E direction with a high wave height.

Figure 7.25 shows the current fields under the interaction of the tidal currents and the wave from E direction with a low wave height. It is clear that the higher the wave height is, the stronger the nearshore current is. The nearshore current zone mainly depends on the wave parameters and the tidal current with time.

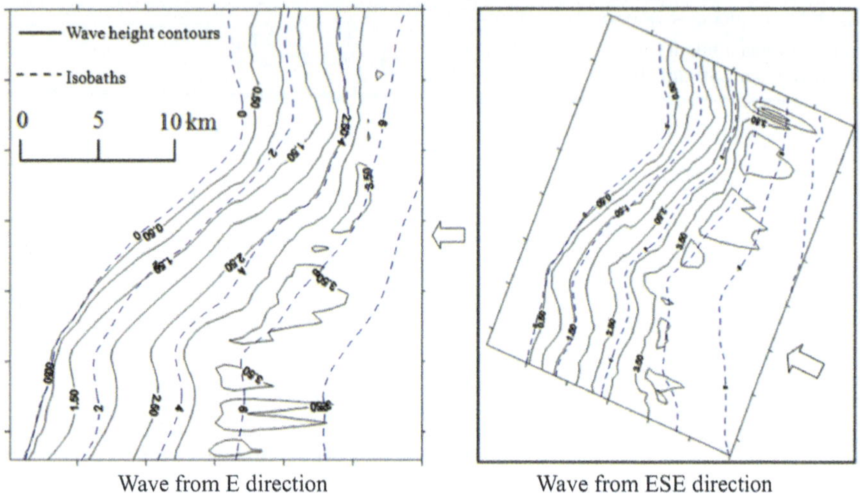

Fig. 7.22 Wave height distributions (return period of 50 years)

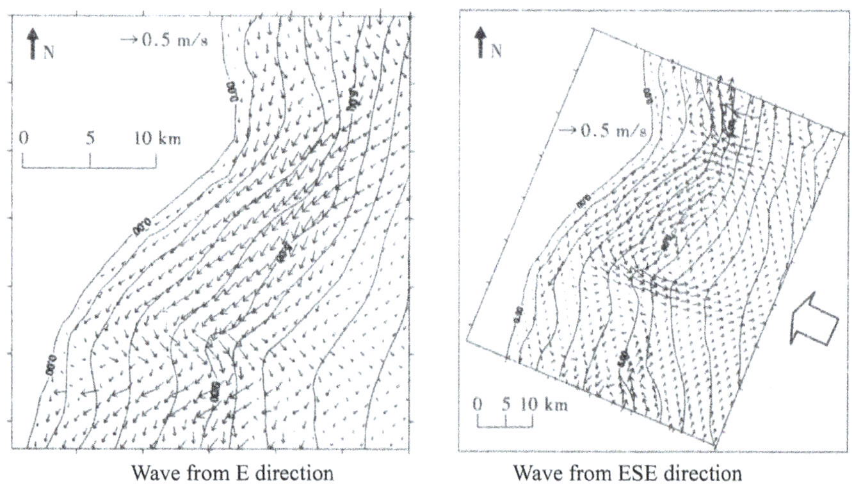

Fig. 7.23 Nearshore current fields under the wave action

It can be also seen from the figures that the nearshore currents in the shallow coastal region with mild slope can reach the same order of magnitude as the tidal current. The waves need to be considered in the simulation of the current fields in the coastal regions. Figure 7.26 shows the current fields under the interaction of the wave from ESE direction and tidal currents.

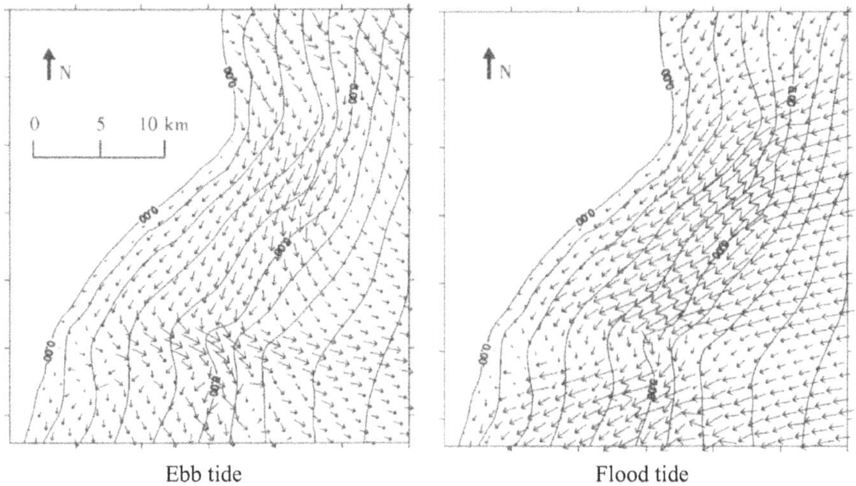

Fig. 7.24 Current fields under the interaction of the wave and tidal currents, incident wave from E direction with wave height of 2.9 m and period of 7.6 s

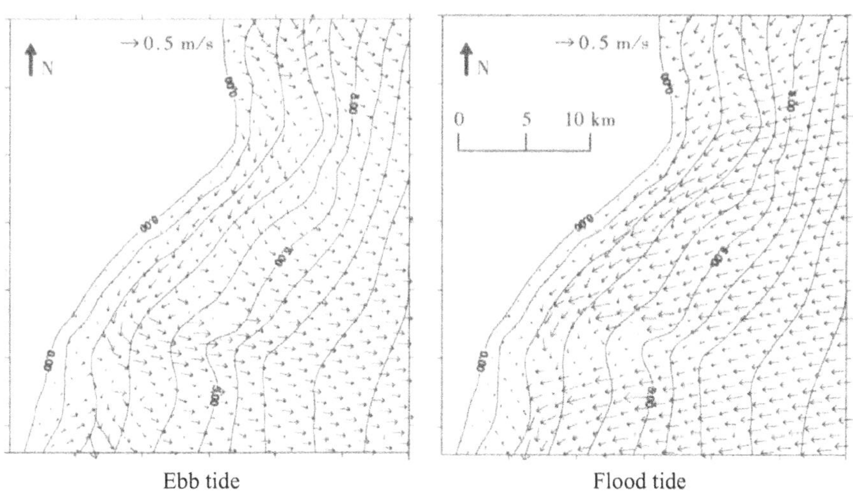

Fig. 7.25 Current fields under the interaction of the wave and tidal currents, incident wave from E direction with wave height of 1.52 m and period of 5.4 s

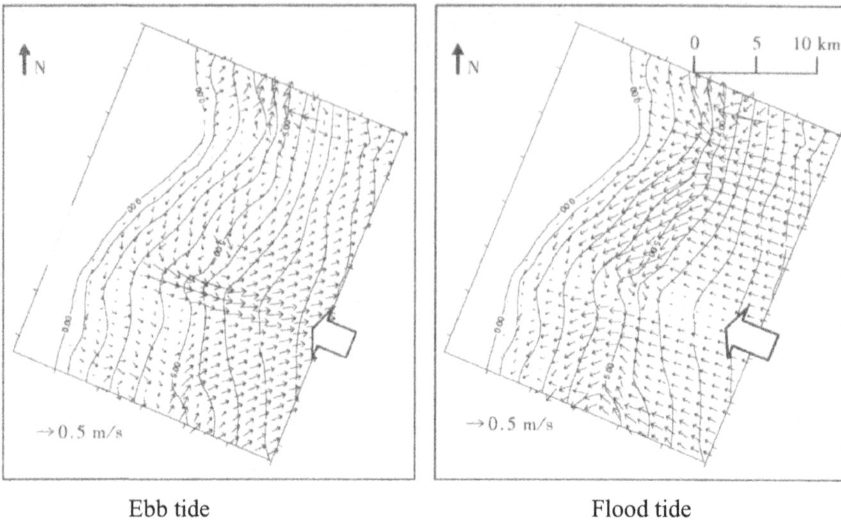

Fig. 7.26 Current fields under the interaction of the wave and tidal currents, incident wave from ESE direction with wave height of 3.31 m and period of 7.6 s

7.5.2 Pollutant Transport Under Interaction of Waves and Tidal Currents

Based on the current fields under the interaction of waves and tidal currents in the nearshore area of Bohai bay described above, a study on the pollutant transport is described in the following [11].

Here, a definition of contaminant exceedance percentage is used to describe the water pollution possibility at different locations. The contaminant exceedance percentage at each point is defined as the ratio of the cumulative time when the contaminant concentration is over a certain value to the total calculation time. It reflects the contaminant probability at each point. The contaminant exceedance percentage $P(x, y)$ is expressed as,

$$P(x, y) = \sum_{j=1}^{N} k_j \frac{\Delta t}{T} \tag{7.5.1}$$

where (x, y) is the coordinates of grid point; Δt is the time step; T is the total calculation time, $N = T/\Delta t$; $k_j = 1$ when the concentration at the grid point is over the certain value, otherwise $k_j = 0$.

The method to determine the distribution of contaminant exceedance percentage is that, by the concentration change with time at each point, the contaminant exceedance percentage at each point is calculated by counting the ratio of the cumulative time when the contaminant concentration is over the certain value to the total

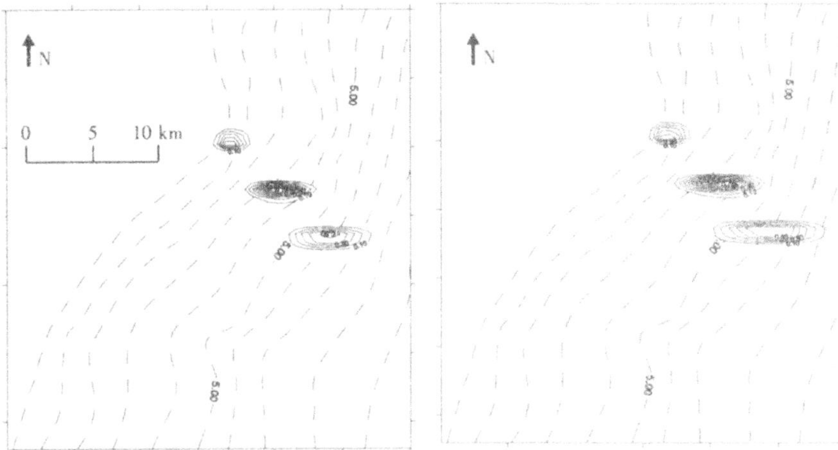

Instantaneous point-sources Continuous point-sources

Fig. 7.27 Distributions of contaminant exceedance percentage under pure tidal currents (24 h)

calculation time, and then, the contours of the contaminant exceedance percentage can be obtained by connecting the points with the same values.

The concept of the contaminant exceedance percentage can help depict the possibility of water pollution at different positions. In addition, the impact of contaminant discharge on the surrounding waters can be analyzed effectively.

Figure 7.27 shows the distributions of contaminant exceedance percentage by the point-source discharges at different locations after 24 h under the pure tidal currents. The offshore distances of the three point-sources are approximately 0 km, 5 km, and 10 km, respectively. With pure tidal currents, the affected areas by both instantaneous and continuous point-sources are relatively small. However, the water around the point-sources is contaminated heavily.

Figure 7.28 shows the distribution of contaminant exceedance percentage in the coastal region by the instantaneous point-source discharges after 24 h under the pure wave action. The corresponding distribution of contaminant exceedance percentage under the interaction of wave and tidal currents is shown in Fig. 7.29.

Comparing Figs. 7.28 and 7.29, it can be seen that the pollutant transport and dispersion is greatly affected by wave action in a certain coastal area. The affected areas by the pollutant discharges under the interaction of wave and tidal currents are similar to those under the pure wave action in general. The affected areas are relatively concentrated under the pure wave action. With the action of tidal currents, the affected areas stretch in the direction perpendicular to the coastline. Under the interaction of wave and tidal currents, the point-source pollutant outfall far from the shoreline has less impact on the surrounding environment, and the impact is mainly in the sea area far away from the outfall.

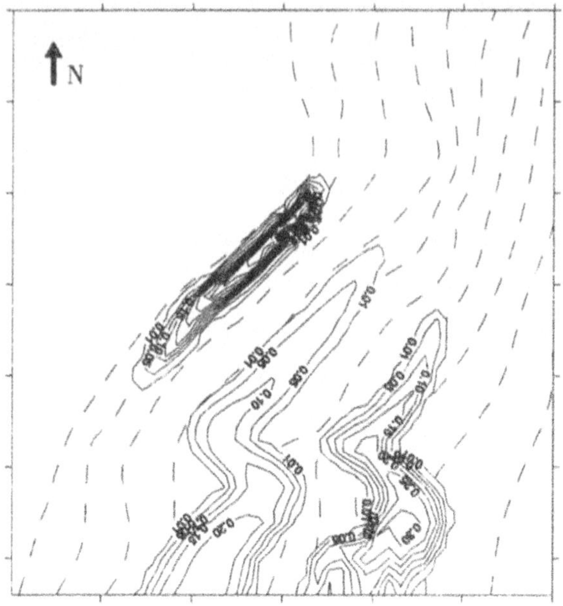

Fig. 7.28 Distribution of contaminant exceedance percentage under pure wave, (instantaneous point-sources, 24 h), incident wave from E direction with wave height of 2.9 m and period of 7.6 s

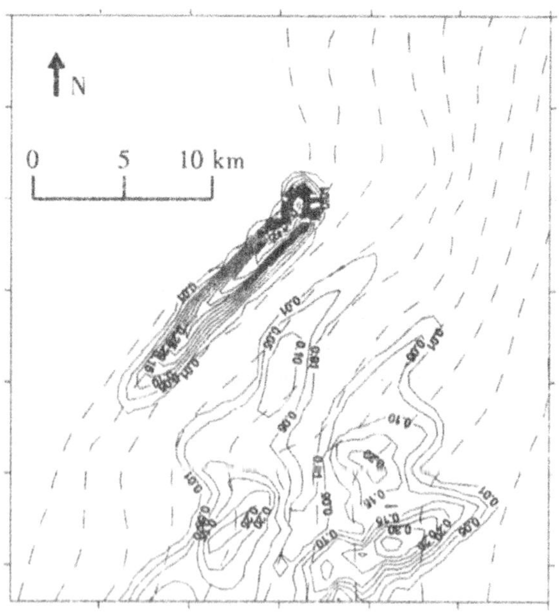

Fig. 7.29 Distribution of contaminant exceedance percentage under interaction of wave and tidal currents, (instantaneous point-sources, 24 h), incident wave from E direction with wave height of 2.9 m and period of 7.6 s

Figures 7.30 and 7.31 show the distributions of contaminant exceedance percentage by the continuous point-source discharges after 24 h under the pure wave action and under the interaction of wave and tidal currents, respectively.

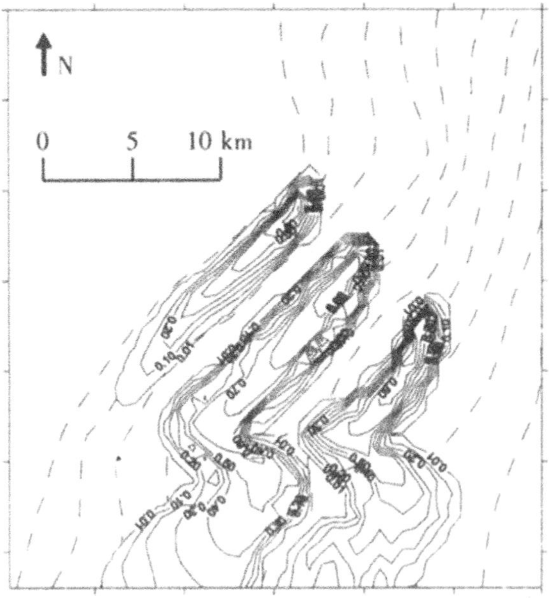

Fig. 7.30 Distribution of contaminant exceedance percentage under pure wave, (continuous point-sources, 24 h), incident wave from E direction with wave height of 2.9 m and period of 7.6 s

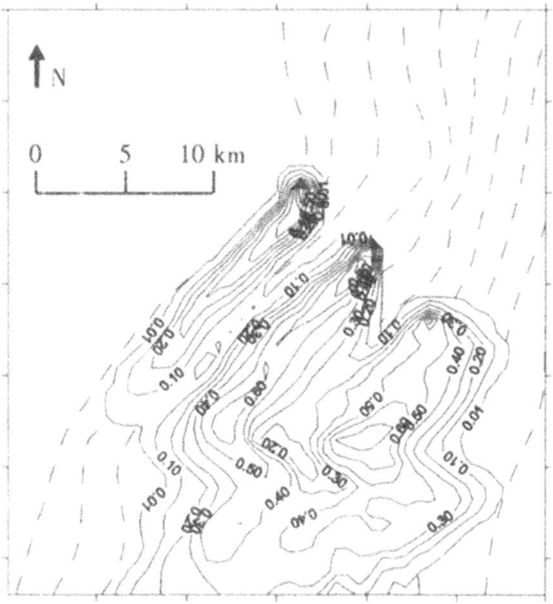

Fig. 7.31 Distribution of contaminant exceedance percentage under interaction of wave and tidal currents, (continuous point-sources, 24 h), incident wave from E direction with wave height of 2.9 m and period of 7.6 s

The contaminant exceedance percentage value at each point can clearly reflect the extent of pollution in different areas. Comparing the distributions of contaminant exceedance percentage by the instantaneous and continuous point-source discharges, it can be seen that the different locations of the outfalls have different impacts on the water environment. The outfall far from the shoreline has better diffusion condition with deeper water depth where the longshore current speed is relatively small. When the outfall is located where the longshore current is large, the pollutant will be transported and diffused along the shoreline.

Figures 7.32 and 7.33 show the distributions of contaminant exceedance percentage by the continuous point-source discharges after 24 h under the interaction of tidal currents and waves from ESE direction with different wave elements, respectively. The locations of the point-sources are identical to the above cases under the interaction of tidal currents and the wave from E direction.

Comparing Figs. 7.32 and 7.33, it can be seen that the affected area of pollutant transport and dispersion is obviously larger under the wave with higher incident wave height. In this typical coastal region, with the interaction of tidal currents and the waves from ESE direction, the effects of shoreward and seaward currents are significant as well. The pollutant discharged from the outfall far from the shoreline can still be transported shoreward along with the shoreward currents.

The above results indicate that, under the wave action in the coastal region with mild slope, the pollutant discharged near the shore might affect the coastal water far from the outfall. Knowledges on the pollutant transport and dispersion under the interaction of waves and tidal currents in the coastal region with mild slope can

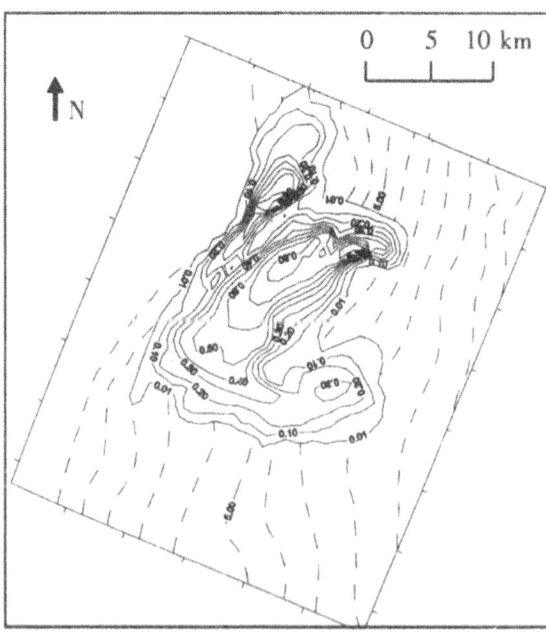

Fig. 7.32 Distribution of contaminant exceedance percentage under interaction of, wave and tidal currents, (continuous point-sources, 24 h), incident wave from ESE direction with wave height of 3.31 m and period of 7.6 s

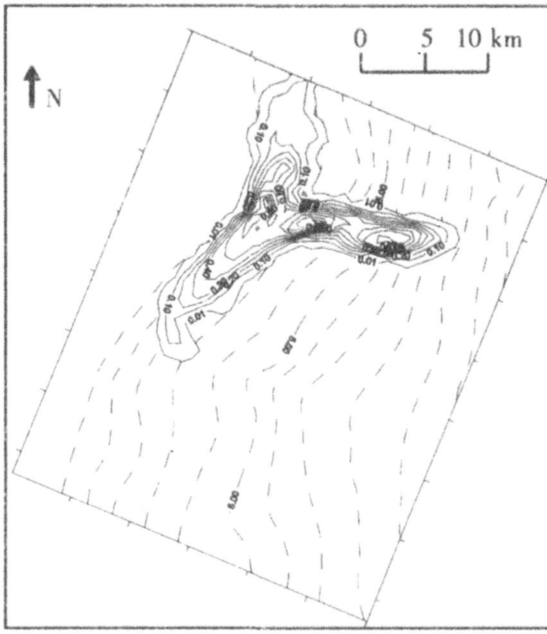

Fig. 7.33 Distribution of contaminant exceedance percentage under interaction of wave and tidal currents, (continuous point-sources, 24 h), incident wave from ESE direction with wave height of 1.4 m and period of 5.4 s

provide a scientific support to properly determine the scale of offshore sewage outfall project and the outfall locations. Meanwhile, it can also provide a scientific support in determining rational permit amounts of pollutant discharged to the sea from the continental sources and understanding the pollution status of the coastal waters.

References

1. Battjes J. Enengy loss and set-up due to breaking of random wave. In: Proceedings of 16th international conference on coastal engineering; New York: ASCE; 1978. p. 569–87.
2. Dally WR, Osiecki DA. The role of rollers in surf zone currents. In: 24th International Conference on Coastal engineering; Kobe; 1994. p. 1895–1905.
3. Jonsson et al. The wave friction factor revisited. Progress Report No. 37.
4. Larson M, Kraus NC. Numerical model of longshore currents for bar and trough beaches. J Waterw Port Coast Ocean Eng. 1991;117(4):326–47.
5. Longuet-Higgins MS. Longshore currents generated by obliquely incident sea waves, Parts 1 and 2. J Geophys Res. 1970;75:6778–789, 6790–801.
6. Soulsdy,R.L., Hamm,L., Klopman,G., et al. Wave-current interaction within and outside the bottom boundary layer. Coastal Eng., 21:41–69.
7. Svendsen IA, Madsen PA. A turbulent bore on a beach. J Fluid Mech. 1984;148:73–96.
8. Swart D H. Offshore sediment transport and equilibriam beach profiles. Publiation No.131, Delft Hydraulics Laboratory, Delft, The Netherlands, 1974.
9. Tao J, Han G. Effects of water wave motion on pollutant transport in shallow coastal water. Science in China Series E. 2002 Dec 1;45(6):593–605.

10. Tao S, Guang H, Jianhua Tao. Numerical simulation of wave-induced long-shore currents and experimental verification. J Hydraul Eng. 2001;11:1–7.
11. Tao S, Jianhua T. Numerical modelling and experimental verification of pollutant transport under waves in the nearshore zone. Acta Oceanol Sin. 2003;25(3):104–12.
12. Tao S, Jianhua T. The study of pollutant transport on the action of waves in the near-shore area of Bohai Bay. Oceanologia Et Limnologia Sinica. 2004;35(2):110–9.

Chapter 8
Numerical Simulation of Coastal Morphological Evolution

8.1 Introduction

Coastal areas as junctions of land and seas are the regions with most complex changes of various dynamic factors, especially the shallow coastal regions with mild slopes. The sediment transport in the coastal area is very complex due to the evolution of waves and currents and wave breaking. To establish a mathematical model for simulating the coastal morphological evolution, the study of coastal dynamic factors and sediment transport should be the starting point.

Under the long-term effects of coastal dynamic processes, the natural shorelines and coastal bathymetries tend to be in equilibrium states. However, the coastal dynamics are affected by changes due to human activities, river flow, coastal engineering constructions, and the sea level rise due to the climate change. These changes break the original equilibrium in coastal areas. Therefore, the prediction of coastal bathymetry changes under the new coastal dynamic conditions has become a necessary practice for coastal developments and coastal water utilizations.

8.1.1 Relations Among the Coastal Morphological Evolution, Coastal Dynamic Factors, and Sediment Transport

The direct causes of coastal morphological evolution are sediment erosion and deposition caused by the changes of coastal dynamic factors such as waves and tidal currents. The sediment erosion and deposition also depend on the kinematic characteristics of sediments. Therefore, the coastal morphological evolution, coastal dynamic factors, and sediment transport become an interdependent system as shown in Fig. 8.1.

To establish a mathematical model for the coastal morphological evolution, the first step is to study the mechanism of coastal dynamics and simulate its changes under

J. Tao, *Numerical Simulation of Water Waves*, Springer Tracts in Civil Engineering, https://doi.org/10.1007/978-981-15-2841-5_8

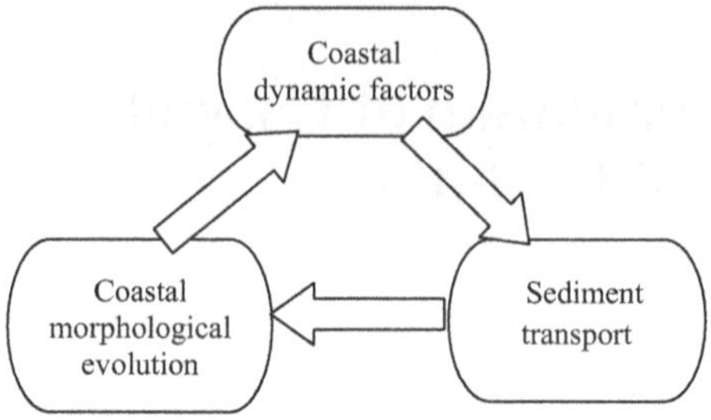

Fig. 8.1 Relations among the coastal morphological evolution, coastal dynamic factors, and sediment transport

the effects of human activities and natural conditions. These include wave refraction, diffraction, and deformation in coastal waters, effects of tidal currents and river flow on the waves, the mechanism and vertical structures of wave–current interaction, the mechanism of wave breaking and energy dissipation, and the complex nearshore current system. The second step is to study the mechanism of sediment motion and simulate its transport under complex dynamic conditions. These include the physical, hydraulic, and kinematic characteristics of cohesive and non-cohesive sediments, the sediment transport and the bottom shear stress under the wave action, the vertical distribution of suspended sediment in turbulence, etc. Due to the large scales in time and space in the mathematical model, it is impossible to track the movement of individual sediment particles. At present, the formulas of bed load, suspended load, or comprehensive sediment transport rate derived according to the kinematic characteristics are the main methods used. The third step is to establish a mathematical model of coastal morphological evolution. The reason for the change of erosion and deposition in the coastal area is that the in-equilibrium between the incoming and outgoing sediments (sediment-carrying capacity) along with the flow results in the erosion and deposition of the seabed at a certain spatial position. When the coastal morphology changes to a certain extent, it will also affect the waves, currents, and other dynamic factors. Therefore, in order to simulate the coastal morphological evolution, first the dynamic factors should be calculated based on the bathymetry condition, next the sediment transport such as sediment load is calculated, and then, the coastal morphological evolution is calculated. This procedure is repeated until the required calculation time is met.

8.1.2 Basics of Simulating Coastal Morphological Evolution

1. Coastal dynamic factors

(1) Shallow water waves and the corresponding mathematical models

Shallow water waves in coastal waters have high nonlinearity. The wave refraction, diffraction, and deformation under the effects of structures and bathymetry should be studied first. At present, the research in this field is relatively mature, and there have been many achievements. There are mainly two types of mathematical shallow water wave models for the simulation of coastal morphological evolution. One is the model based on the Boussinesq equations for shallow water dispersive waves. The other is the model based on the mild-slope equations which are also divided into the elliptic, parabolic, and time-dependent mild-slope equations. The key achievements in this field have been described in Chap. 4 of this book.

(2) Mechanism of wave breaking and its modeling

After wave breaking, the turbulence structure and the mechanism of energy dissipation in the surf zone are not yet fully understood. However, they are key issues affecting sediment transport in the coastal area. Regarding to the breaking wave models, some achievements have been described in Chap. 5 of this book.

(3) Vertical flow structure and bottom shear stress under wave–current interaction

A large number of studies have shown that the wave–current interaction affects not only the wave characteristics but also the vertical flow distribution and bottom shear stress which are different from those under the pure wave action. These will affect the sediment movement at the bottom and the carrying capacity of suspended sediment.

Klopman [1] gave the mean horizontal velocity profiles under the interaction of irregular waves and currents by experiments in a physical flume. It has been shown that the presence of waves affects the velocity profile significantly both inside and outside the bottom boundary layer. Pure waves (without current) can also cause flow near the bottom. The experiments in a physical flume carried out by Sun et al. [2] indicated the effect of wave–current interaction on the vertical flow structure as well.

Simons [3] measured the bottom shear stress under the interaction of irregular waves and the currents from different directions, which further proved that waves strongly affect the bottom shear stress. The bed shear stress measurements over a smooth bed using a high-resolution sensor in a basin by Arnskov et al. [4] indicated that waves suppress the turbulence in the flow, the turbulence transition is a complex function of waves and current velocity, and a sudden increase in shear stress results in turbulence.

(4) Nearshore current system under the wave–current interaction

The nearshore current system affects the sediment transport directly. The nearshore current system includes longshore currents, seaward currents, and shoreward currents. Goda [5], Zou et al. [6], and Sun et al. [7] carried out detailed experimental and numerical studies on the nearshore current system. The relations between the nearshore current system, nearshore bathymetry, and the wave–current interaction have been obtained. Some key achievements in this field have been described in Chap. 7 of this book

2. Sediment transport under waves

The early studies on sediment transport under waves took reference from the methods of river sediment. However, due to the complex hydrodynamic factors in the coastal waters, there are still lots of problems to be solved in the coastal sediment transport under waves.

(1) Dynamic characteristics of sediments

The dynamic characteristics of coastal sediments are closely related to the grain size. The coastal sediments with grain size of $0.3 < D < 2.0$ mm are called non-cohesive sediments, and the sediments with grain size of $D < 0.03$–0.05 mm are called cohesive sediments. Because the sediment particles are small, the molecular forces of the surrounding water affect the sediment transport. Macroscopically, the effect of molecular forces is called cohesion. For non-cohesive sediments under waves, the incipient motion of sediments, forming of sand ripples, and the lifting of sediment particles on the sand ripples have been studied adequately. However, for cohesive sediments under waves, further studies on flocculation, exchange with the seabed (such as lifting, settling, and consolidation), and liquefaction are needed. The domestic and foreign research achievements can be found in the Refs. [8, 9].

(2) Bed load and suspended load sediment movement under waves

The sediment transport under waves can be divided into two forms: bed load and suspended load.

Bed load movement consists of two kinds of movement forms, namely the movement of individual sediment particles and the "laminar flow" of particles in several layers. Suspended load movement consists of lifting of sediment particles off the sand ripples and suspending of sediments in the breaking waves. Suspended load movement is the main part of total sediment transport in sandy coastal areas. Liu [8] suggested that the bed load transport rate is approximately 26–43% of the total sediment transport rate in the coastal area with coarse sand having the grain diameter $D = 0.6$–1.0 mm, and the bed load transport rate is only approximately 0.2–5% of the total in the sandy coastal area with sand of $D = 0.1$–0.2 mm. Kana and Ward [10] measured the suspended sediment transport rate during storm periods. The site observations showed that the suspended sediment transport is dominant during the storm, and the bed load transport rate may reach less than or equal to the suspended load transport rate when the wave activity is weak.

Goda [11] reviewed the history and existing state of the studies on sediment movement in coastal morphology since the 1940s. He suggested that the studies on bed load were more adequate than the studies on suspended load. The inadequate understanding of sediment movement in the surf zone affected the studies on suspended load and the accurate prediction of the coastal morphology evolution.

To study the suspended sediment movement in the surf zone, it is necessary to study wave breaking, the three-phase interaction of gas, solid, and fluid in the surf zone, and the energy loss in the surf zone due to the sediments. It is very difficult to measure the bed load sediment on site. However, the suspended sediment transport rate can be calculated from the measured sediment concentration and the velocity distribution on the cross section of the surf zone. The bed load transport rate is generally derived from the estimated total sediment transport rate deducting the suspended sediment transport rate.

3. Longshore sediment transport rate and sediment-carrying capacity

In the establishment of the mathematical model of coastal morphological evolution, an important issue is the longshore sediment transport rate or sediment-carrying capacity. The longshore sediment transport rate comprehensively reflects the coastal dynamic factors and sediment kinematic characteristics under specific conditions. So far, the formulas for longshore sediment transport rate are semi-empirical formulas based on the theoretical analyses combined with a large amount of site measurements and experimental data. Several widely used formulas are as follows:

(1) CERC formula

On the basis of a large number of measurement data in various coastal areas, the U.S. Army Coastal Engineering Research Center (CERC) proposed a formula for the total longshore sediment transport rate, which is widely used and called CERC formula (USACE) [12]. This formula takes into account both bed load and suspended load. Since it was published, lots of studies have been carried out on it. Del Valle et al. [13] studied the empirical coefficient in the formula. Goda [11] compared the measured suspended sediment transport rates by Kana and Ward [11] with the calculated results using the CERC formula and pointed out that the measured and calculated suspended sediment transport rates matched well when the wave height was $H_{1/3} = 1.3$–2.4 m during the storm; however, when the wave height was down to $H_{1/3} = 0.8$–0.9 m after the storm, the measurements were only about 1/3 of the calculated results. Miller [14] presented the measured results during five storms and found that the measurements were in good agreements with the calculated results by using the CERC formula when the wave height was $H_{1/3} = 1.6$–3.5 m.

(2) Bailard formula

The energetic-based total load of sediment transport proposed by Bailard [15] is also a commonly used longshore sediment transport formula. This formula consists of two parts, bed load and suspended load, and includes the effect of bed slope.

Soulsby [16] made comparisons of the calculations by using the Bailard sediment transport formula with measurements. He pointed out the reasons of the Bailard formula being more common include that this energetic-based formula is different from the other sediment transport formulas starting from the shear stress, and this algebraic formula takes into account the combined effects of wave–current interaction. The studies showed that the formula is more suitable for the wave-dominant coastal areas, where the calculations agreed with the measurements better than the current-dominant coastal areas. Bayran et al. [17] compared the evaluations by six longshore sediment transport formulas including the CERC formula and the Bailard formula with the site data. They also suggested that the energetic-based formula is better than the formula starting from the shear stress.

(3) Dou Guoren's formulas

The formulas proposed by Dou Guoren include two separate types for bed load and suspended load, respectively. In the model of coastal morphological evolution, the morphology changes caused by the bed load and suspended load are calculated separately [18].

For the calculation of suspended load transport, Guoren [19] proposed a energetic-based formula of sediment-carrying capability under the interaction of waves and tidal currents and pointed out that the morphology changes due to the suspended load are caused by the in-equilibrium between the sediment-carrying capacity and the suspended sediment concentration.

(4) Liu Jiaju's formulas

Liu [8] proposed sediment transport formulas for bed load and suspended load. He carried out experiments in a flume and verified the proposed longshore sediment transport formula by using the observed data in Mauritania Port.

Regarding the mathematical models of coastal morphological evolution, lots of fundamental studies and some mathematical models for solving engineering problems have been carried out by Nanjing Hydraulic Research Institute, Tianjin Research Institute for Water Transport Engineering, Hohai University, Tianjin University, and some other institutes/universities in China. The European Union (EU) conducted a framework of Marine Science and Technology Programme (MAST) in 1990–1995. In this programme, a large-scale, well-organized coastal morphodynamic study including G6 and G8 projects [9] had been implemented. More than 100 scientists and researchers from 35 research organizations in EU participated in this work which cost 16 million Euro and lasted six years. The main purpose of the G6 and G8 projects for the coastal morphodynamic study was to provide the mechanism of physical phenomena for the development of mathematical models of coastal morphology dynamics. The study included the fundamental rules of coastal waves, currents, sediment transport, and their interactions, so as to establish mathematical shoreline models and coastal region models. The study of coastal dynamic models had also greatly promoted the study of the fundamental mechanisms of coastal waves, currents, and sediment transport. The study approaches included theoretical researches, experimental verifications, in situ observations, and numerical simulations. By means of this programme, some key hydraulic research institutes (universities) in EU countries have established unique mathematical models of coastal morphology dynamics. These models have been verified by a large amount of laboratory data and site data and already utilized for applications.

8.1.3 Classification of Mathematical Models of Coastal Morphological Evolution

Because of the complexity of the coastal waves, currents, and sediment transport, so far there is no mathematical model suitable for all situations. Presently, the mathematical models of coastal morphological evolution could only be established and classified according to the specific conditions [9]. Here are several means of classification.

1. Shoreline model and coastal region model

According to the morphological classification, the mathematical models of coastal morphological evolution can be divided into shoreline models and coastal region models.

(1) Shoreline model

Shoreline model is used for the situation with uniform longshore currents. It can be assumed that the shape of the cross-shore profile remains unchanged during the shoreline changes which just move seaward or shoreward in parallel. This model is also called "one-line theory model, and it is mainly used to calculate the development and change of shoreline during a period. The disadvantage is that it cannot calculate details of bathymetry changes.

(2) Coastal region model

A coastal region model is mainly used to calculate the regional seabed bathymetry changes. It is applicable for the situation when the longshore currents are not uniform due to the effect of the engineering structures. The bathymetry changes can be obtained through calculation of the distribution of regional sediment transport rates.

2. Mid-term model and long-term model

According to the temporal scales of morphological evolution, the mathematical models are divided into mid-term models and long-term models.

(1) Mid-term model

The relatively short to mid-term coastal evolution under the effect of the engineering structures is mainly studied. In general, the spatial scale is several hundred meters to several kilometers, and the time scale is several weeks to several years.

(2) Long-term model

It mainly focuses on the long-term and large-scale coastal evolution caused by climate change and sea level rise. The spatial scale is tens to hundreds of kilometers or more, and the time scale is several years to several decades.

3. Models of sandy coasts and muddy coasts

According to the properties of coastal sediment, the models can be divided into the ones for sandy coasts and muddy coasts. Different sediment sources and coastal dynamic conditions can form different coastal categories. There are many coastal categories, which are usually classified by sediment grain size in coastal engineering.

(1) Muddy coasts

The median grain size is $D_{50} < 0.05$ mm (or $D_{50} < 0.03$ mm), and the grains are cohesive. The beach slope is mild (1:500–1:2000). The sediment transport is dominated by suspended load, and the vertical distribution is relatively uniform. The settling velocity is represented by the equivalent grain size of floccules. This kind of coasts is often found near large river estuaries with fine sand delivered to the sea or coasts with weak ocean dynamics.

(2) Sandy coasts

The median grain size is 0.05 mm < D_{50} < 2.0 mm, and the beach slope is steep (1:5–1:500). For the sediments with median grain size 0.05 mm < D_{50} < 0.15 mm (or 0.03 mm < D_{50} < 0.11 mm), the grains are less cohesive. The phenomenon of flocculation no longer occurs, and the sediments are easy to be suspended. The sediment concentration distributed vertically is relatively small over the upper water column and large near the bottom, and the grains are easy to settle down. This kind of coasts is generally called silt-sandy coasts. For the sandy coast with median grain size 0.15 mm < D_{50} < 2.0 mm, the sediments are always in motion when the ocean dynamics are strong. From the sediment load point of view, it is mainly suspended load which is concentrated near the bottom. This kind of coasts is often found at coastal areas with strong coastal dynamics and without river flow into the sea or located near smaller river estuaries with limited sediment-carrying capacity. If the sediment source of a silt-sandy coast is terminated, under the sorting effect of coastal dynamics, the sediment grain size will be coarsened, and some coastal segments may be transited from silt-sandy category to sandy category.

(3) Gravel coasts

The median grain size is $D_{50} > 2.0$ mm. In the general coastal dynamic conditions, the sediment movement is slight. Sediment transport only occurs during storm waves. Because of the steep slope, the cross-shore extent of sediment transport is small. This kind of coasts is mostly caused locally by weathering and fragmenting of rocks or quarrying.

Among the above three types of coasts, the muddy and sandy coasts are the two most investigated types in engineering sediment studies.

In coastal engineering, there are many studies on mid-term shoreline models and coastal region models. At present, the studies of these two kinds of mathematical models for sandy coasts are relatively mature, and further studies on coastal morphological evolution of muddy beaches are needed. However, whether for sandy or muddy coasts, the sediment transport rates are mostly determined by semi-empirical formulas. The correct simulations of coastal morphological evolution are closely related to sediment transport rates. The models need to be calibrated and verified by a large amount of measurement data.

Several typical coastal morphological evolution models will be described in this chapter.

8.2 Mathematical Model of Shoreline Evolution

Near a sandy beach, the longshore current caused by wave breaking leads to sediment movement to form longshore sediment transport, which is equilibrium sediment transport in nature. The construction of coastal projects breaks the original equilibrium and causes deposition or erosion of the beach, which results in the change of shoreline morphology.

The sediment transport on a sandy beach consists of two parts: the onshore (or offshore) sediment transport which is perpendicular to the shoreline and the longshore sediment transport which is parallel to the shoreline. The effect of the former one on the beach is mainly seasonal change, and the effect of the latter one is long-term and important. The mathematical shoreline model is also called "one-line theory model." The basic assumption is that the shape of the cross-shore profile remains unchanged during the shoreline evolution and the shape is the same along shoreline. The direction of shoreline moving, seaward or shoreward, depends on the in-equilibrium of the local sediment transport rates. This model is mainly used to calculate the shoreline change during a certain period (such as one or several years).

8.2.1 Mathematical Shoreline Model

1. Governing equation

Let x be the main shoreline direction, and y be the offshore (or onshore) direction. Figure 8.2 shows a definition sketch. Considering the longshore sediment transport only, the continuity equation for sediment volumes reads

$$\frac{\partial y}{\partial t} + \frac{1}{D}\frac{\partial Q}{\partial x} = 0 \tag{8.2.1}$$

where $y = (x, t)$ is the position of the shoreline at t time; D is the active height of the coastal profile; Q is the longshore sediment transport rate which is mainly obtained by empirical formulas. When selecting the empirical formula, the application condition should be paid much attention, and the coefficients or parameters should be chosen properly in the formula.

Fig. 8.2 Definition sketch

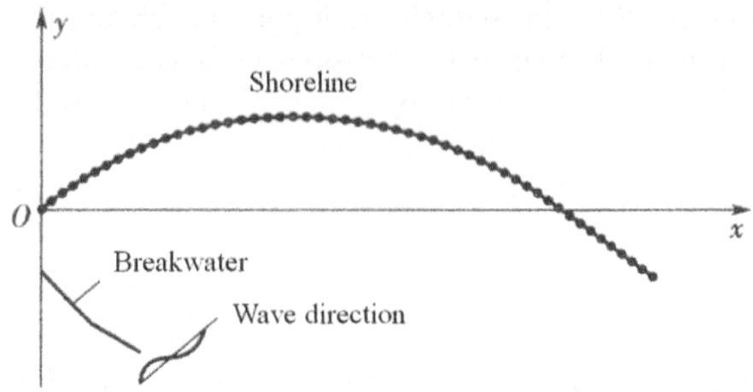

The CERC formula proposed in the Shore Protection Manual of the U.S. Army Coastal Engineering Research Center [12] is commonly used in shoreline models,

$$Q = AP_{\mathrm{lb}} \tag{8.2.2}$$

where A is a proportional factor. If the wave height is represented by the root-mean-square wave height, A can be taken as $A = 0.812 \times 10^{-4}$; P_{lb} is the longshore wave energy flux of breaking waves,

$$P_{\mathrm{lb}} = \left(E_f\right)_b \sin\alpha_b \cos\alpha_b \tag{8.2.3}$$

in which, $\left(E_f\right)_b$ is the wave energy flux per unit wave crest width; α_b is the wave angle at breaking, i.e., the angle between the wave direction and the shoreline at the breaking point (at t time). The linear wave theory gives

$$E_{\mathrm{fb}} = \frac{1}{8}\rho g H_b^2 c_{\mathrm{gb}} \tag{8.2.4}$$

where H_b is the wave height at breaking; c_{gb} is the group velocity at incipient breaking which is given by,

$$c_{\mathrm{gb}} = c_b\left[1 + \frac{2k_b d_b}{\sinh(2k_b d_b)}\right] \tag{8.2.5}$$

$$c_b = \left(\frac{g}{k_b}\tanh(k_b d_b)\right)^{1/2} \tag{8.2.6}$$

in which, T is the wave period, k_b is the wave number at the water depth of d_b at breaking. Let α_{b0} be the angle between the wave direction at breaking and the initial shoreline (at t_0 time), then as shown in Fig. 8.3, the relation between α_b and α_{b0} reads

$$\alpha_b = \alpha_{b0} - \arctan\left(\frac{\partial y}{\partial x}\right) \tag{8.2.7}$$

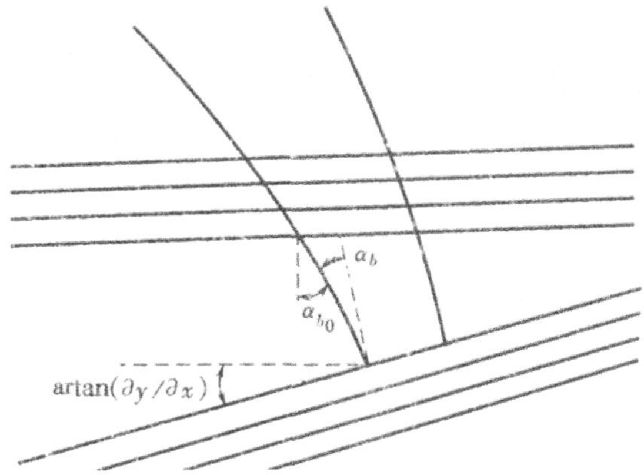

Fig. 8.3 Relation between α_b and α_{b0}

Overall, the sediment transport rate Q is a function of time t, spatial coordinates (x, y), and shoreline gradient $\frac{\partial y}{\partial x}$ (or α_b). Combining Eqs. (8.2.1), (8.2.2), (8.2.3), and (8.2.7) yields a parabolic formula for the longshore sediment transport rate Q,

$$\frac{\partial Q}{\partial t} = Z(x, t)\frac{\partial^2 Q}{\partial x^2} \tag{8.2.8}$$

$$Z(x, t) = \frac{\mathrm{AE_{fb}}\cos(2\alpha_b)}{D\left[1 + \left(\frac{\partial y}{\partial x}\right)^2\right]} \tag{8.2.9}$$

The above result is obtained by using the linear wave theory.

2. Determination of breaking wave parameters

In order to determine the breaker line and breaking wave parameters, the wave field in the near shore area after wave refraction and diffraction needs to be calculated first by using the known wave elements in the deep water. Then, the breaker line and the wave elements at breaking such as H_b, α_{b0}, α_b, k_b, and d_b are determined according to the wave breaking criteria.

There are different methods to determine the breaking wave parameters. Here, the empirical method for the calculation of breaking wave parameters proposed by Ozhan [20] is described. In this method, only wave refraction is taken into account for the wave deformation.

When the isobaths are almost parallel, the refraction coefficient from deep water to the breaking point reads

$$K_{\mathrm{rb}} = \left(\cos\alpha_0 / \cos\alpha_b\right)^{1/2}$$

The wave steepness after refraction is given by,

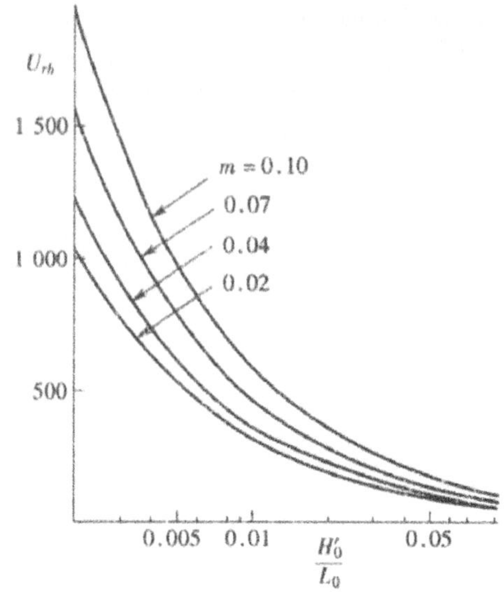

Fig. 8.4 Relations among the Ursell number, wave steepness H_0'/L_0, and beach slope m

$$\frac{H_0'}{L_0} = K_{\mathrm{rb}} \cdot \frac{H_0}{L_0} \tag{8.2.10}$$

Figure 8.4 shows relations among the Ursell number, the beach slope m, and the wave steepness H_0'/L_0. The Ursell number of breaking waves is defined as

$$Ur_b = H_b L_b^2 / d_b^3 \tag{8.2.11}$$

where H, L, d are the wave height, wave length, and water depth, respectively; the subscript "b" denotes the incipient breaking.

After finding the Ursell number from Fig. 8.4 based on the beach slope m and the wave steepness H_0'/L_0, the relative water depth and the water depth at breaking can be described by,

$$\frac{d_b}{L_b} = \left(\frac{H_b/d_b}{Ur_b}\right)^{1/2} \tag{8.2.12}$$

$$d_b = \frac{d_b}{L_b}\tanh\left(\frac{2\pi d_b}{L_b}\right) \Big/ \tanh\left(2\pi\frac{d_0}{L_0}\right) \cdot L_0 \tag{8.2.13}$$

The wave height and wave angle at breaking can be obtained by,

$$\frac{H_b}{d_b} = 0.72 + 5.6\,m, \tag{8.2.14}$$

where m is the beach slope.

According to the Snell's law, the wave angle at breaking is

$$\alpha_b = \arcsin\left(\sin\alpha_0 \cdot \tanh\frac{2\pi d_b}{L_b} / \tanh\frac{2\pi d_0}{L_0}\right) \tag{8.2.15}$$

By using Eqs. (8.2.10)–(8.2.15) and Fig. 8.4, the breaking wave elements can be obtained by means of the iterative method.

3. Initial and boundary conditions

Initial condition:

$$y(x, 0) = y_0(x) \tag{8.2.16}$$

where $y_0(x)$ is the initial position of the shoreline, which can be given by site measurement data.

Boundary conditions:

$$Q|_{x=0} = Q_0 \tag{8.2.17}$$

$$\frac{\partial Q}{\partial x}|_{x=+\infty} = 0 \quad or \quad \frac{\partial y}{\partial x}|_{x=+\infty} = 0 \tag{8.2.18}$$

in which, Q_0 is the sediment transport rate at the starting calculation point, which is determined by the specific condition.

4. Numerical scheme

Discretizing Eqs. (8.2.8) and (8.2.1) yields

$$\frac{Q_j^{n+1} - Q_j^n}{\Delta t} = \psi Z_j^{n+1}\frac{Q_{j+1}^{n+1} - 2Q_j^{n+1} + Q_{j-1}^{n+1}}{\Delta x^2} + (1-\psi)Z_j^n\frac{Q_{j+1}^n - 2Q_j^n + Q_{j-1}^n}{\Delta x^2} \tag{8.2.19}$$

$$\frac{y_j^{n+1} - y_j^n}{\Delta t} + \frac{1}{D}\left[\theta\frac{Q_{j+1}^{n+1} - Q_{j-1}^{n+1}}{2\Delta x} + (1-\theta)\frac{Q_{j+1}^n - Q_{j-1}^n}{2\Delta x}\right] = 0 \tag{8.2.20}$$

$$Q_j^n = Q(jx, nt), \quad y_j^n = y(jx, nt)$$

in which, ψ and θ are the weighting factors. When $\psi = \theta = 1/2$, the accuracy of the above difference equations is up to $O(\Delta t^2, \Delta x^2)$.

5. Procedure of calculating shoreline evolution

(1) Calculating the wave height distribution in the nearshore area with the initial shoreline using a numerical method. Generally, the wave height, wave direction, and wave period in deep water are the calculation bases. Because of the coastal

bathymetry changes and the sheltering effect of breakwaters, the distributions of coastal wave height and wave direction should be calculated by using a numerical method for combined effects of wave refraction, diffraction, and shoaling.

(2) Determining the location and parameters of wave breaking and providing the breaking wave height H_b, the water depth d_b, and the wave angle α_b at the incipient breaking in the nearshore area with the initial shoreline.

(3) Calculating the longshore sediment transport rate of the initial shoreline by solving Eq. (8.2.2).

(4) Calculating the longshore sediment transport rate and the position of shoreline at the next time step by solving Eqs. (8.2.19) and (8.2.20).

The longshore sediment transport rate Q_j^{n+1} at $t + \Delta t$ time is obtained by Eq. (8.2.19) first. Then, the shoreline change y_j^{n+1} is solved by Eq. (8.2.20).

(5) When the shoreline has changed significantly, going back to (1) and re-calculating the wave height followed by steps (2), (3), and (4), repeating these steps until the shoreline change at the required calculation time is obtained.

A flowchart for calculating shoreline evolution is shown in Fig. 8.5.

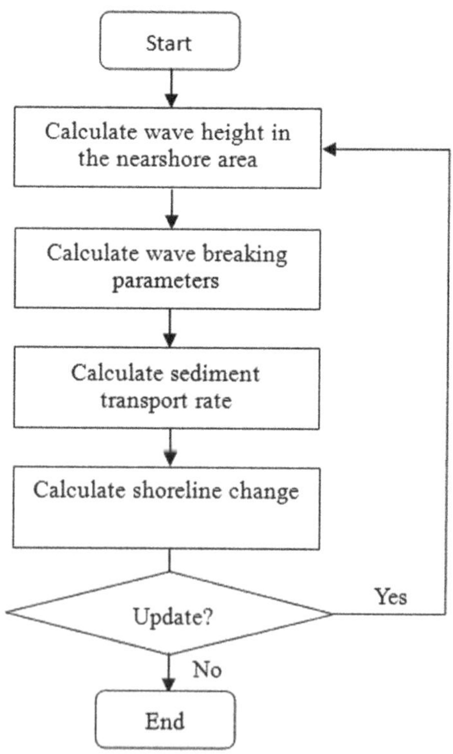

Fig. 8.5 Flowchart for calculating shoreline evolution

8.2.2 Study on Shoreline Evolution and Protection Works in the Downstream of the Breakwater at Friendship Port in Mauritania

The Friendship Port (Port of Nouakchott) in Mauritania is located on the east coast of the Atlantic Ocean with typical sandy beach. The shoreline is in the north–south direction. The breakwater at the port is a single breakwater. The dominant wave directions are from WN and WNW directions. The breakwater was built in 1980s and broke the equilibrium of longshore sediment transport, which caused deposition in the upstream of the breakwater and erosion in the downstream of the breakwater. The downstream erosion caused danger to the port facilities. In order to predict the development of downstream erosion and the effect of remedial measures, numerical simulations on the development of downstream shoreline erosion have been carried out [21].

Because the main purpose was to study the shoreline erosion in the downstream of the breakwater, the shoreline evolution model described in this section could be used for the simulations.

1. Wave refraction and diffraction

Because the local shoreline is relatively straight, the wave refraction and diffraction could be calculated separately. The refraction theory was used to calculate the wave height change from deep to shallow water in the far field. The diffraction theory was used to calculate the wave diffraction due to the breakwater in the near field.

(1) Wave refraction and shoaling

After wave refraction and shoaling, the wave height becomes

$$H = K_s \cdot K_r \cdot H_0 \tag{8.2.21}$$

where H_0 is the incident wave height in deep water, i.e., the wave height before refraction and shoaling; K_s is the shoaling coefficient, and K_r is the refraction coefficient. The Snell's law gives

$$\frac{\sin\alpha}{c} = \text{const} \tag{8.2.22}$$

where α is the wave angle, c is the wave celerity. When the isobaths are parallel straight lines, the refraction coefficient K_r can be given by,

$$K_r = \left(\cos\alpha_0 / \cos\alpha\right)^{\frac{1}{2}} \tag{8.2.23}$$

The shoaling coefficient K_s is determined by,

$$K_s = \left(\frac{2kd_0 + \sinh 2kd_0}{2kd + \sinh 2kd}\right)^{\frac{1}{2}} \cdot \frac{\cosh kd}{\cosh kd_0} \tag{8.2.24}$$

in which, k is the wave number which satisfies the dispersion relation,

$$\frac{\omega^2}{g} = k \tanh kd$$

where ω is the angular frequency.

For irregular waves, in the refraction theory, it is assumed that each individual wave component deforms independently, and the deepwater wave spectrum develops into a new shallow water spectrum in shallow water. Meanwhile, it is assumed that the isobaths are parallel straight lines, and wave diffraction and the energy dissipation due to bottom friction are neglected.

According to the law of energy conversation, the wave spectrum after wave refraction reads

$$S(\omega, \theta) = K_s^2 S(\omega, \theta_0) \tanh kd_0 / \tanh kd \tag{8.2.25}$$

where the wave angle θ can be derived by the wave angle in deep water θ_0 as

$$\theta = \sin^{-1}(\sin\theta_0 \cdot \tanh kd / \tanh kd_0)$$

Thus, in the case of isobaths being approximately parallel straight lines, when the wave spectrum $S(\omega, \theta_0)$ at the water depth of d_0 is known, the wave spectrum in shallow water can be obtained by Eq. (8.2.25).

If the wave height obeys the Rayleigh distribution, the root-mean-square wave height reads

$$H_{rms}^2 = 8 \iint S(\omega, \theta) d\omega d\theta \tag{8.2.26}$$

Then, the wave height ratio (ratio of root-mean-square wave height to incident root-mean-square wave height) at a certain point is given by,

$$K_H = \left[\int_0^{\infty} \int_{-\pi/2}^{\pi/2} K_r^2 K_s^2 \cdot S(\omega, \theta_0) d\theta_0 d\omega \Big/ \int_0^{\infty} \int_{-\pi/2}^{\pi/2} S(\omega, \theta_0) d\theta_0 d\omega \right]^{\frac{1}{2}} \tag{8.2.27}$$

(2) Wave diffraction

The ratio of the diffracted wave height due to the breakwater H' to the incident wave height H is denoted by the diffraction coefficient K_d. According to the diffraction theory (See Chap. 4), K_d is expressed as

$$K_d = \frac{H'}{H} = \frac{2\omega}{gH}\sqrt{\Phi_{\mathrm{Re}}^2 + \Phi_{\mathrm{Im}}^2} \tag{8.2.28}$$

where Φ is the wave potential, and ω is the angular frequency.

For irregular waves, the diffracted wave height can be superposed by the diffracted wave height of each individual regular wave component.

According to the random wave theory, the diffraction coefficient of irregular waves reads

$$K_d = \left[\frac{\sum\limits_{i=1}^{m}\sum\limits_{j=1}^{n}(K_d)_{ij}^2 S(\omega_i, \theta_j)\delta\omega_i\delta\theta_j}{\sum\limits_{i=1}^{m}\sum\limits_{j=1}^{n} S(\omega_i, \theta_j)\delta\omega_i\delta\theta_j}\right]^{\frac{1}{2}} \tag{8.2.29}$$

where $(K_d)_{ij}$ is the diffraction coefficient of the wave component at the ith frequency in the jth direction; $\delta\omega_i$ and $\delta\theta_j$ denote the frequency interval and the direction interval, respectively. The diffraction coefficient of the wave component $(K_d)_{ij}$ can be obtained by using the regular wave theory.

3. Mean wave energy flux and longshore sediment transport rate

The group velocity and dispersion relation for the calculation of longshore sediment transport rate could be obtained by using the linear wave theory and the nonlinear cnoidal wave theory, respectively.

The calculation of the mean wave energy flux and the longshore sediment transport rate by using the linear wave theory has been described above. For the calculation using the nonlinear cnoidal wave theory, see Ref. [21] for the information.

4. Results and discussion

Figure 8.6 shows a comparison of the numerically predicted shoreline and the observed shoreline on site in half a year on the basis of the initial shoreline. It is seen that the numerical prediction and the observation matched well.

Figure 8.7 shows numerical predictions of shoreline evolution in 30 years from July 1986 to July 2016 by using different wave theories for the wave field. The

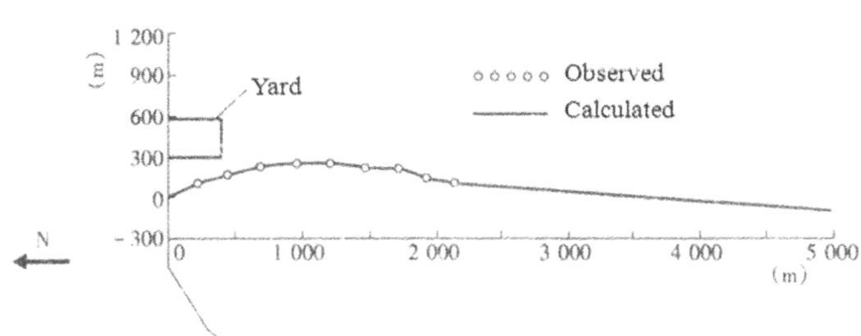

Fig. 8.6 Comparison of calculated and observed shoreline in half a year

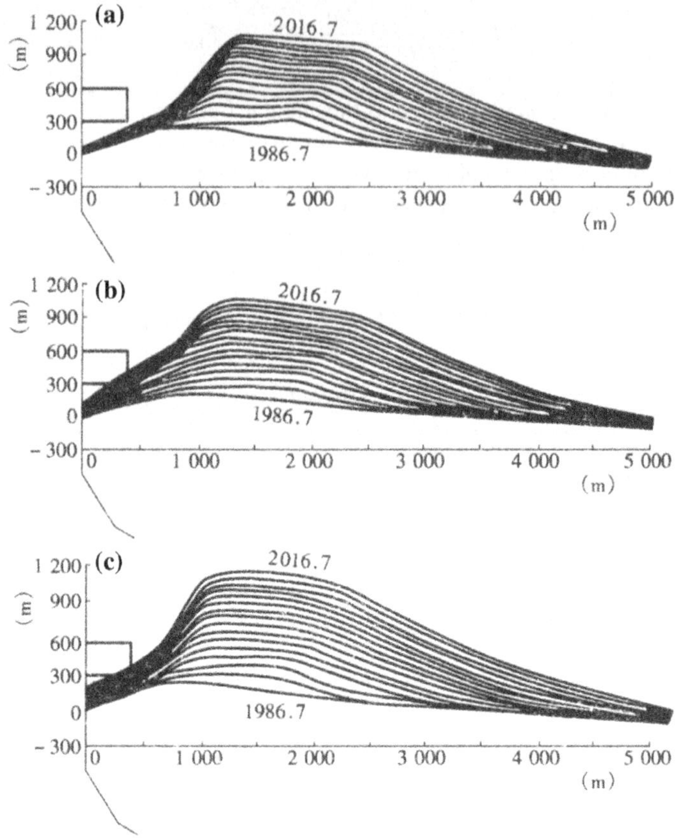

Fig. 8.7 Predictions of shoreline evolution in the downstream of the breakwater at the friendship port in 30 years. **a** Linear wave theory; **b** irregular wave theory; **c** cnoidal wave theory

results by using the linear regular wave theory, the irregular wave theory, and the nonlinear cnoidal wave theory are shown in Fig. 8.7a–c, respectively. It is seen by comparing the three different results that the erosion would harm the port yard in both Fig. 8.7b, c. This indicates that the prediction by using the linear wave theory was not conservative enough. It was better to use irregular waves or cnoidal waves than linear waves for the wave calculation in practical engineering.

Figure 8.8a, b shows the numerically predicted results of shoreline evolution with a groin located at 670 m and 800 m downstream of the breakwater, respectively. It is seen that the groin would make the erosion area move to the downstream of the groin, which could protect the facilities in the port effectively.

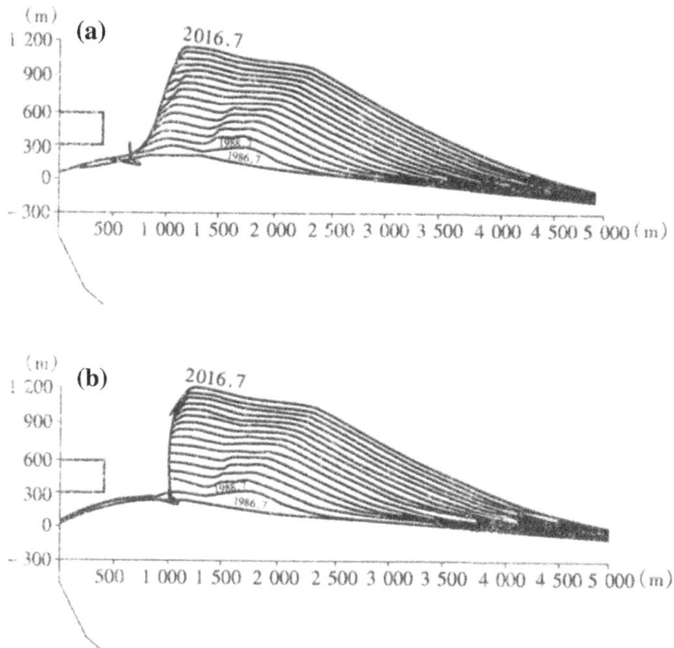

Fig. 8.8 Predictions of shoreline evolution in 30 years with a groin. **a** Groin located at 670 m downstream of the breakwater; **b** groin located at 800 m downstream of the breakwater

8.3 Region Model of Sandy Beach Evolution Under Wave Action

When the longshore current is not uniform due to the interference of the engineering structures, the sediment transport rate in the coastal area is unevenly distributed, which leads to uneven changes of seabed bathymetry. The mathematical model for simulating this bathymetric evolution is called a coastal region model. This kind of model, mostly being a medium term model, mainly studies the seabed evolution under the effects of engineering structures.

A coastal region model mainly consists of four sub-models: the wave model, the model of longshore current under wave action, the sediment transport model, and the seabed evolution model. Applying different theories and methods in the four sub-models may make the region model different. However, it can meet the engineering requirements after model calibrations and verifications by means of the laboratory data or in situ observation data.

8.3.1 *Sub-models of the Region Model of Sandy Beach Evolution*

1. Nearshore wave model

The nearshore seabed evolution is directly related to the variation of sediment transport rate. The sediment transport mainly occurs in the surf zone caused by wave breaking. Therefore, the calculation of the nearshore wave field and the determination of the breaker line are key steps in the coastal region model.

The mathematical models describing the nearshore wave field mainly include the ray method model based on the energy conversation in wave propagation, the mild-slope equation model, and the Boussinesq model which both incorporate combined effects of wave refraction and diffraction. When waves propagate from deep to shallow water over a seabed with approximately uniform slope, the ray method model could be used to calculate the wave deformation due to the wave shoaling and refraction in the area far from the engineering structures, and then, the wave field near the structures could be calculated by a wave diffraction model. When the bathymetry in the computational domain is complex, wave refraction and diffraction should be taken into account simultaneously. Then, the mild-slope equation model or the Boussinesq model could be applied. It is better to use the Boussinesq equation model in the region near the structures due to the importance of wave reflection. It is noted that when the computational domain is very large, the elliptic or time-dependent mild-slope equation model and the Boussinesq equation model might be limited due to the excessive workload. Then, the parabolic mild-slope equation model could be used in the far field, and the Boussinesq equation model can be used in the near field of structures.

2. Determination of the breaker line

When waves propagate from deep to shallow water and reach a critical state, waves will break. Sediment transport and coastal morphological change are very active in the surf zone. Therefore, the correct calculations of the wave field and seabed change in the surf zone are required. Firstly, the breaker line should be determined.

Goda [22] proposed a wave breaker index using a critical relative water depth. The formula for the inception of wave breaking reads

$$\left.\begin{aligned} \eta_b &= \left(b - \tfrac{a}{T^*}\right)^{-1} \\ a &= 1.36g\left(1 - \mathrm{e}^{-19n}\right) \\ b &= 1.56\left(1 + \mathrm{e}^{-19.5n}\right)^{-1} \\ T^* &= gT^2/H_b \end{aligned}\right\} \tag{8.3.1}$$

where η_b is the critical relative water depth, $\eta_b = \frac{h_b}{H_b}$; h_b is the water depth at the breaking point; H_b is the breaker height; T is the wave period; n is the bottom slope. By means of the empirical formula for the determination of the breaker height H_b

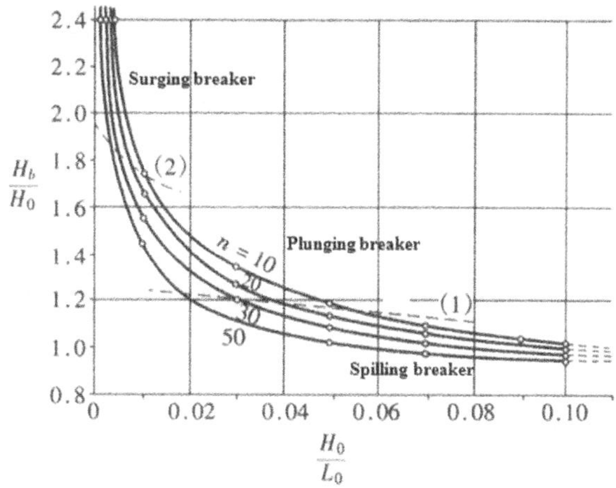

Fig. 8.9 Determination for the breaker height H_b (Goda)

proposed by Goda, η_b can be obtained as shown in Fig. 8.9. In the figure, H_0 and L_0 are the wave height and the wavelength in deep water.

3. Calculation of sediment transport rate

There are two forms of sediment transport in water flow: suspended load and bedload. Therefore, the sediment transport rate also consists of suspended load transport rate and bedload transport rate. Many scholars have explored the sediment transport rate for many years. However, most of the formulas achieved are empirical. Here, the empirical formulas proposed by Bailard [15] are given, in which the effects of waves and currents are taken into account. As mentioned above, the sediment transport rate consists of two parts, suspended load transport rate and bedload transport rate.

$$\boldsymbol{q} = \boldsymbol{q}_B + \boldsymbol{q}_s \tag{8.3.2}$$

$$\boldsymbol{q}_B = \frac{C_f \varepsilon_B}{(\rho_s/\rho - 1) g \tan\phi} \cdot \left[\langle |\boldsymbol{u}_t|^2 \cdot \boldsymbol{u}_t \rangle - \frac{\tan\beta}{\tan\phi} \cdot \langle |\boldsymbol{u}_t|^3 \rangle \boldsymbol{i} \right] \tag{8.3.3}$$

$$\boldsymbol{q}_S = \frac{C_f \varepsilon_S}{(\rho_s/\rho - 1) g \omega_s} \cdot \left[\langle |\boldsymbol{u}_t|^3 \boldsymbol{u}_t \rangle - \frac{\varepsilon_S}{\omega_s} \tan\beta \cdot \langle |\boldsymbol{u}_t|^5 \rangle \boldsymbol{i} \right] \tag{8.3.4}$$

where $\boldsymbol{q}_B$ and $\boldsymbol{q}_S$ are the time-averaged volumetric sediment transport rate over one wave period $q = (q_x, q_y)$ for the bedload and suspended load, respectively; ρ_s is the density of sand, and ρ is the density of water; $\boldsymbol{u}_t$ is the total velocity vector in the wave–current field; ϕ is the angle of internal friction of the sand grains; β is the inclined angle of seabed; ε_B and ε_S are empirical factors for the bedload and suspended load transport, which could be taken as $\varepsilon_B = 0.21$ and $\varepsilon_S = 0.025$; C_f is the bottom friction coefficient; $\boldsymbol{i}$ is the unit vector in the downslope direction of the local seabed; $< > = \frac{1}{T}\int_0^T (\ )\mathrm{d}t$ represents the averaged value over a wave period; ω_s is the settling velocity of sand grains in water, $\omega_s = \omega_{s0} + R_* \frac{\partial \varepsilon_z}{\partial z}$, in which ω_{s0} is the settling velocity of the suspended sediment in still water, $\varepsilon_z = \frac{b_* g T H}{4\pi} \cdot \frac{\cosh[k(h-z)]}{\cosh(kh)}$, and for shallow water waves, $R_* = A\left(\frac{H}{b_* \pi h} + 1\right)$, $A \approx 1$, b_* is an empirical factor with $b_* = (0.001 \sim 0.003)$; h is the water depth; T is the wave period; k is the wave number; and H is the wave height. The sediment transport rate is directly related to the combined wave–current field in the nearshore area.

4. Sediment equilibrium equation for seabed evolution

If the spatial sediment transport rate distribution of is given, the bed level z_b at each spatial point can be obtained by solving the sediment balance equation which reads

$$\frac{\partial z_b}{\partial t} + \frac{\partial q_x}{\partial x} + \frac{\partial q_y}{\partial y} = 0 \tag{8.3.5}$$

where $q_x = q_x(t, x, y)$ and $q_y = q_y(t, x, y)$ are the components of sediment transport rate per unit width in the x- and y-directions, respectively, i.e., the effective volume ($\mathrm{m}^3/\mathrm{s} \cdot \mathrm{m}$) through the vertical section per unit width in unit time at a certain spatial point and a certain moment. Obviously, the determination of the sediment transport rate distribution is the most critical issue for the correct calculation of seabed bathymetry changes.

When the steepening bottom slope is taken into account, the sediment balance equation under gravity can be modified as

$$\frac{\partial z_b}{\partial t} = \frac{1}{1-\lambda_0}\left[-\frac{\partial}{\partial x}(q_x - \varepsilon_s |q_x| \frac{\partial z_b}{\partial x}) - \frac{\partial}{\partial y}(q_y - \varepsilon_s |q_y| \frac{\partial z_b}{\partial y})\right] \tag{8.3.6}$$

where ε_s is an empirical coefficient; λ_0 is the bed porosity. Here, q_x and q_y are the components of net sediment transport rate per unit width in the x- and y-directions, respectively, i.e., the volume of compact imporous sand through the vertical section per unit width in unit time.

8.3.2 Calculation Case of Bathymetry Evolution Around a Breakwater

A mathematical region model of coastal morphology was established by Zhang and Tao [23]. The model had been verified by the experimental data of the bathymetry evolution around an offshore breakwater.

The experimental data of the bathymetry evolution around an offshore breakwater were cited from the reference by Watanabe et al. [24]. The bathymetry and wave conditions had been chosen the same as the experimental data for the calculation. Figure 8.10 shows a bathymetry sketch. The initially uniform bed slope was 1/20, and the median diameter of grains was 0.2 mm. The offshore breakwater was set in

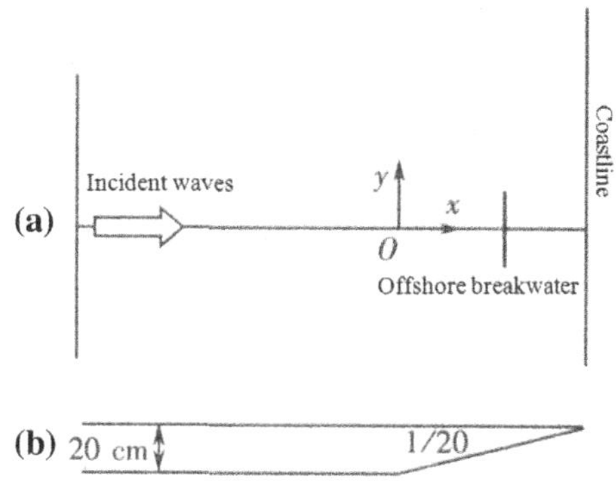

Fig. 8.10 Bathymetry sketch. **a** Plan view; **b** elevation view

parallel with the initial shoreline and located at 1.8 m away from the shoreline. The incident wave height was 4.5 cm, and wave period was 0.87 s.

Due to the symmetry of the results, the calculation was carried out in half of the region ($0 < x \leq 4$ and $0 < y \leq 3.6$ m).

1. Calculation of the wave field

The time-dependent mild-slope equation had been used for calculation of the wave field. The wave breaker index proposed by Watanabe had been applied [24]. In the numerical model, the grid spacing was 5.0 cm, and time step was set to be $T/30$. The deepwater open boundary was set at the water depth of 20 cm. The two lateral boundaries of $y = 3.6$ m and $y = 0$ (the symmetrical center line) were taken as solid boundaries with full reflection condition. The time-dependent mild-slope equation was solved numerically with boundary conditions at $t = 0$. Figures 8.11 and 8.12 show the position of breaker line and the wave height distribution at $t = 0$, respectively.

2. Calculation of the nearshore current field

The nearshore current field was obtained by solving Eqs. (7.3.1) and (7.3.2) in Chap. 7. Figure 8.13 shows the current velocity distribution at $t = 0$. In the figure, a typical circulation cell is seen clearly behind the breaker.

3. Calculation of sediment transport and bathymetry changes

Using the formulas of sediment transport in Eqs. (8.3.2)–(8.3.4) and the sediment balance Eq. (8.3.6), the new bathymetry at the next time $t = t_1$ had been calculated on the basis of the initial bathymetry condition at $t = 0$. After that by taking the calculated bathymetry at $t = t_1$ as the updated initial condition, the wave field, current field, sediment transport, and the bathymetry changes had been re-calculated. Then, the bathymetry after a certain time period could be obtained by repeating this iteration.

Figures 8.14 and 8.15 show the bathymetry at $t =$ 02:37 (h:min) and $t =$ 05:05, respectively. In the two figures, the calculated results and the experimental results are shown in (a) and (b), respectively, for the comparison.

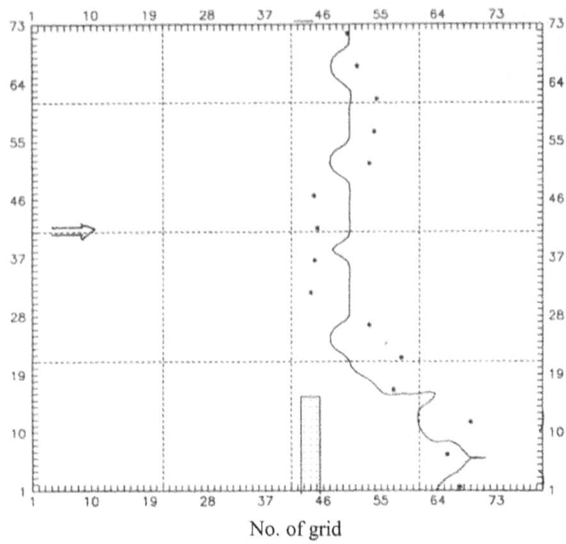

Fig. 8.11 Position of breaker line at $t = 0$

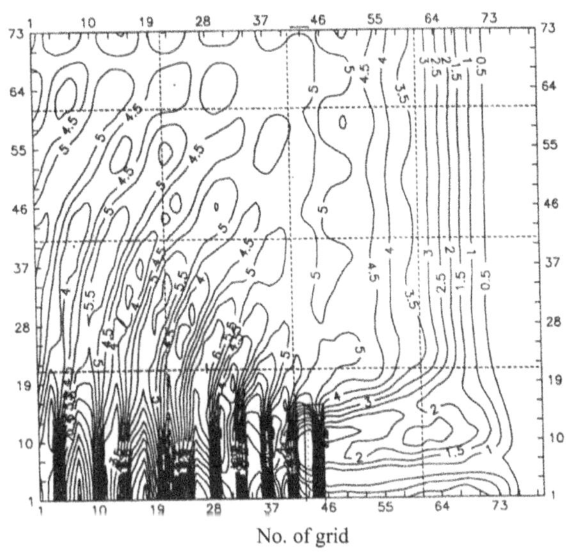

Fig. 8.12 Wave height distribution at $t = 0$

Figure 8.16 shows comparisons of the calculated and experimental bathymetry results on the cross-shore sections at $t = 05{:}05$. It is seen from the comparisons that the calculated results of the coastal region model were in good agreement with the experimental results.

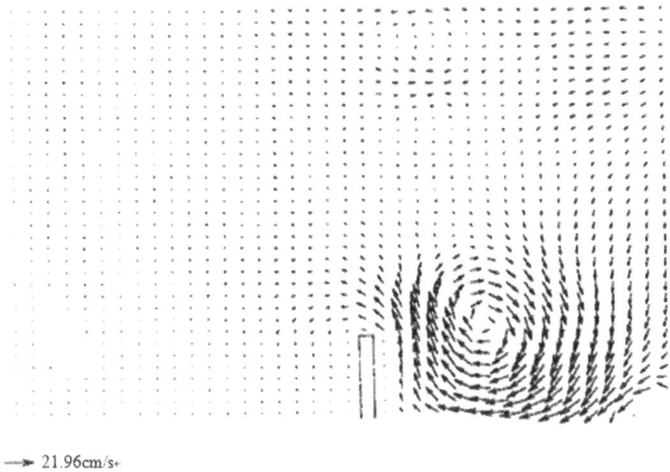

Fig. 8.13 Current velocity distribution at $t = 0$

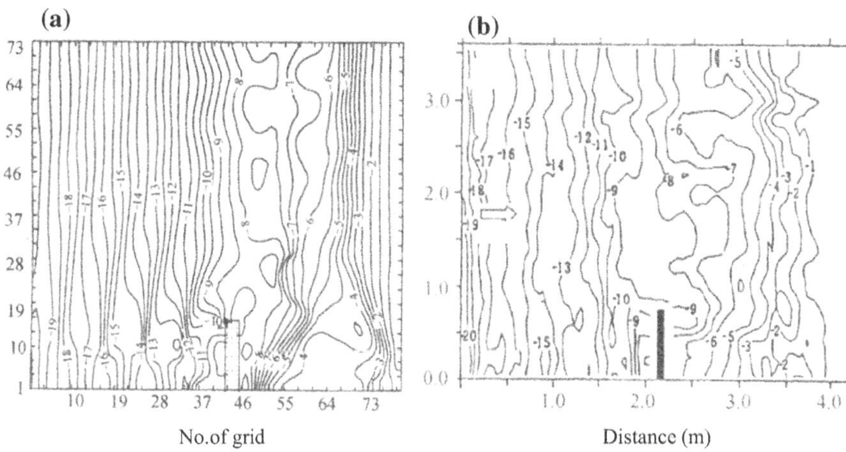

Fig. 8.14 Bathymetry at $t = 02{:}37$ (h:min). **a** Calculated result; **b** experimental result [24]

8.3.3 *Bathymetry Evolution Around a Sunken Ship Near Shore*

The bathymetry evolution caused by a sunken ship near the shore at Nouakchott in Mauritania was simulated [23] using the established coastal region model. The initial bathymetry was given by the bathymetric survey in July 1986. The bathymetry changes around the sunken ship after 30 days and 60 days were calculated. The dominant wave direction was from the west-northwest, the wave period was 8 s, the wave height was 0.67 m, and the design water level was 1.69 m. In the calculation,

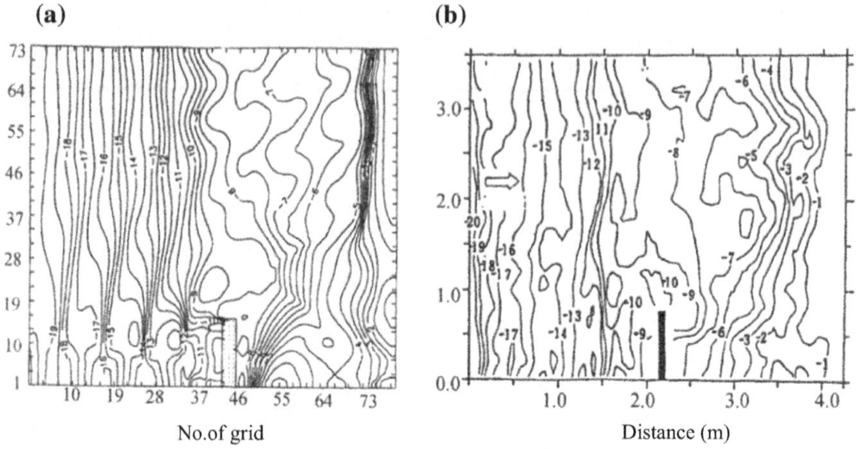

Fig. 8.15 Bathymetry at t = 05:05 (h:min). **a** Calculated result; **b** experimental result [24]

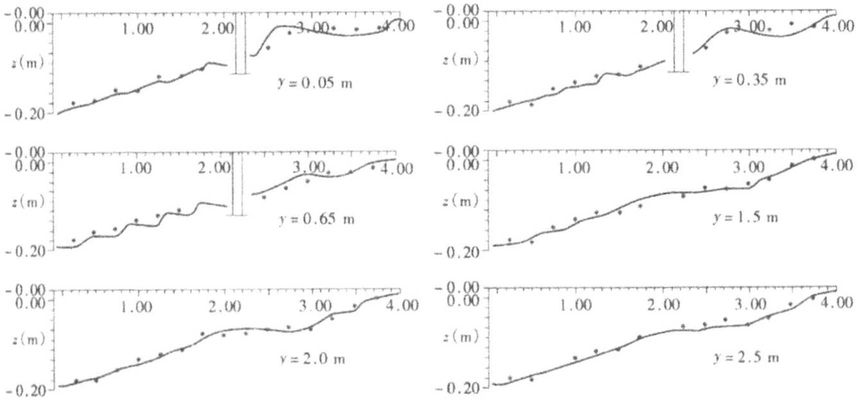

Fig. 8.16 Comparisons of the calculated (—) and experimental (*) bathymetry results on the cross-shore sections at t = 05:05

the dominant wave direction was selected as the x-direction, the grid spacing was 5 m, and the calculation domain was about 700 m × 800 m around the sunken ship. Figure 8.17a shows the initial bathymetry near the sunken ship locally (where the bathymetry changes were significant). Figure 8.17b shows the changed bathymetry after 60 days. It is seen from the comparison that the sediment deposition occurred between the sunken ship and the shoreline has a trend to form a tombolo.

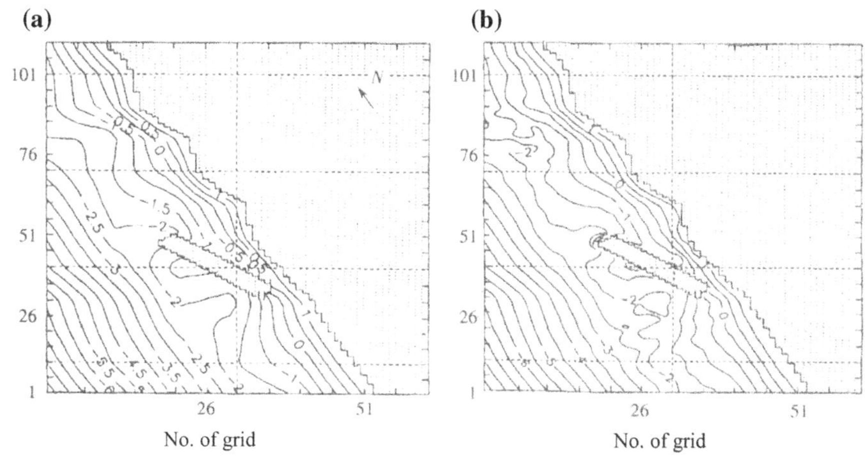

Fig. 8.17 Calculated bathymetry changes around a sunken ship. **a** Initial bathymetry, **b** changed bathymetry after 60 days

8.3.4 Comparison of Sub-models in Several Coastal Region Models

In the past two to three decades, region models of coastal morphological evolution have been developed rapidly. The studies mostly focus on the two-dimensional depth-averaged models, also contain some three-dimensional or quasi three-dimensional models. A great amount of achievements have been made in the sub-models of waves, currents, sediment transport, and seabed evolution. Through the G6 and G8 projects of coastal morphodynamic study in the MAST program of EU, the key hydraulic research institutes and universities in Europe have developed several coastal region models. Intercomparison of the coastal region morphodynamic models at that time developed by five different institutes had been presented by Nicholson et al. [25] by comparing the numerical results of the different models with the same experimental data.

The five models developed by Danish Hydraulic Institute (DHI), Delft Hydraulics (DH) in the Netherlands, HR Wallingford (HR) in UK, Service Technique Central de Ports Maritimes (STC) in France, and University of Liverpool (UL) in UK have similar structures, which include sub-models of waves, currents, sediment transport, and seabed evolution.

Table 8.1 presents the approaches applied in the sub-models of the five coastal region models at that time [25].

It has been indicated by comparing the results of the different models, and the selection of sediment transport formula has a great influence on the coastal bathymetry results; the achieved or approaching equilibrium bathymetry calculated in each model is due to the interaction of hydrodynamics and morphodynamics. The influence factors of hydrodynamics include (1) wave type (regular or irregular

Table 8.1 Approaches applied in the sub-models in the five coastal region models

Models	Wave model			Current model			Wave-current interaction	Sediment transport model	Remarks
	Governing equations	Wave breaking index	Friction resistance	Governing equations	Viscosity coefficient	Friction resistance			
DHI	Parabolic approximate mild-slope equation	Battjes and Janssen (1978)	None	Shallow water equations	Constant	Friction affected by waves (FredsØe 1984)	Partly, current affected by waves through friction	Steady advection–diffusion equation (Deigaard et al. 1986)	2D depth-averaged form
DH	Wave action density balance equation	Battjes and Janssen (1978)	Modified friction coefficient (Holthuijsen et al. 1989)	Shallow water equations	Constant	Shear stress under wave–current interaction (Soulsby et al. 1993)	Partly	Bijker formula (1968)	2D depth-averaged form
HR	Mild-slope equation	Battjes and Janssen (1978)	Friction coefficient under wave–current action (Yoo and O'Connor 1987)	Shallow water equations	Energy dissipation (De Vriend and Stive 1987)	Modified channel friction coefficient (Chesher et al. 1993)	Partly	Chesher and Miles formula (1992)	2D depth-averaged form

(continued)

Table 8.1 (continued)

Models	Wave model			Current model			Wave-current interaction	Sediment transport model	Remarks
	Governing equations	Wave breaking index	Friction resistance	Governing equations	Viscosity coefficient	Friction resistance			
STC	Time-dependent mild-slope equation	Munk (1949)	None	Shallow water equations	Constant	Approach of Nishimura (1982)	None	Bijker formula (1968)	2D depth-averaged form; reflection waves
UL	Modified momentum and energy conservation equations	Nairn (1990)	Friction coefficient under wave–current action (Yoo and O'Connor 1987)	Shallow water equations	Energy dissipation (De Vriend and Stive 1987)	Friction coefficient under wave–current action (Yoo and O'Connor 1988)	Fully	O'Connor and Nicholson formula (1995)	2D depth-averaged form

waves); (2) friction coefficient; (3) viscosity coefficient; (4) wave–current interaction; (5) wave refraction, diffraction, and reflection.

Through the studies in the G8 project, it has been particularly pointed out that 3D or quasi-3D models should be developed to simulate the regional bathymetry change for the further study, and it is difficult to use a 2D depth-averaged model for simulating scour.

8.4 Mathematical Model of Estuarine Morphological Evolution

In estuary areas, river flow, tidal currents, and waves are all driving forces of sediment transport. They have significant effects on sediment movement. The sediment transport also consists of bedload and suspended load.

The key to establish a model for estuarine morphological evolution is taking into account the interaction of river flow, tidal currents, and waves. In general, the river flow discharge could be taken as an input of boundary condition or treated as a source term at estuary. Regarding to the wave–current interaction, different models use different methods. A common method is to take into account the wave-induced radiation stress and the bottom friction under the wave–current interaction in the hydrodynamic equations and consider the interaction of waves and tidal currents in the sediment transport formulas, such as the method used by Xin [26]. Dou [27] developed a mathematical model of estuarial morphological evolution on the basis of the sediment-carrying capacity under tidal currents and waves proposed by himself [19]. Based on the principle of energy superposition, he derived the sediment-carrying capacity under the combined action of tidal currents and waves by superposing the tidal current energy and wave energy for carrying sediment, which had been verified by a large amount of measurement data. Then, the radiation stress for the momentum exchange between waves and currents is no longer introduced in the mathematical current model. This method is suitable for the estuaries with strong river flow, such as the Yangtze Estuary. Here, Dou Guoren's method is described in the following.

8.4.1 2D Hydrodynamic Equations

The governing equations of 2D hydrodynamics are

$$\frac{\partial \eta}{\partial t} + \frac{\partial (hU)}{\partial x} + \frac{\partial (hV)}{\partial y} = 0 \tag{8.4.1}$$

$$\frac{\partial (hU)}{\partial t} + \frac{\partial \left(hU^2\right)}{\partial x} + \frac{\partial (hUV)}{\partial y} + fV + gh\frac{\partial \eta}{\partial x} + \frac{U\sqrt{U^2+V^2}}{C^2 h}$$

$$= \frac{\partial}{\partial x}\left(\nu_w \frac{\partial (hU)}{\partial x}\right) + \frac{\partial}{\partial y}\left(\nu_w \frac{\partial (hV)}{\partial y}\right) \tag{8.4.2}$$

$$\frac{\partial (hV)}{\partial t} + \frac{\partial (hUV)}{\partial x} + \frac{\partial \left(hV^2\right)}{\partial y} - fU + gh\frac{\partial \eta}{\partial y} + \frac{V\sqrt{U^2+V^2}}{C^2 h}$$
$$= \frac{\partial}{\partial x}\left(\nu_w \frac{\partial (hV)}{\partial x}\right) + \frac{\partial}{\partial y}\left(\nu_w \frac{\partial (hV)}{\partial y}\right) \tag{8.4.3}$$

where η is the surface elevation; h is the total water depth; C is the Chezy number; U and V are the depth-averaged velocity components in the x- and y-directions; f is the Coriolis parameter; ν_w is the viscosity coefficient.

8.4.2 2D Mathematical Model of Sediment Transport

1. Suspended load transport equation

In Cartesian coordinate system, the horizontal 2D transport equation of suspended sediment reads

$$\frac{\partial (hS)}{\partial t} + \left[\frac{\partial (hUS)}{\partial x} + \frac{\partial (hVS)}{\partial y}\right] + \alpha\omega(S - S_*) = 0 \tag{8.4.4}$$

where S is the depth-averaged suspended sediment concentration; α is a tuning coefficient to be determined by verification calculation; ω is the settling velocity of suspended sediment, which is the flocculation settling velocity under the flocculation condition; S_* is the sediment-carrying capacity, which can be expressed as the following under the combined action of tidal currents and waves,

$$S_* = \alpha_0 \frac{\gamma\gamma_s}{\gamma_s - \gamma}\left[\frac{(U^2+V^2)^{3/2}}{C^2 h\omega} + \beta_0 \frac{H_w^2}{HT\omega}\right] \tag{8.4.5}$$

where γ and γ_s are the water density and the sediment density, respectively; H_w and T are the mean wave height and wave period; the coefficients α_0 and β_0 are determined by measurement data, which could be taken as $\alpha_0 = 0.023$, $\beta_0 = 0.04 f_w$, and f_w is the friction coefficient of waves; $C = \frac{1}{n}h^{1/6}$ is the Chezy number, and n is the Manning number for bed roughness; U and V are the depth-averaged velocity components in the x- and y-directions, which can be solved by Eqs. (8.4.1)–(8.4.3).

In the absence of wave, the sediment-carrying capacity in pure current S_{*F} could be determined by,

$$S_{*F} = \alpha_0 \frac{\gamma\gamma_s}{\gamma_s - \gamma} \cdot \frac{v^3}{C^2 h\omega} \tag{8.4.6}$$

where v is the current speed.

When the Chezy number C is given by the Manning number, Eq. (8.4.6) can be rewritten as

$$S_{*F} = \alpha_0 \frac{\gamma \gamma_s}{\gamma_s - \gamma} \cdot \frac{n^2 v^2}{h^{4/3} \omega} \tag{8.4.7}$$

When only waves exist in the absence of current, the sediment-carrying capacity of waves is

$$S_{*w} = \alpha_0 \beta_0 \frac{\gamma \gamma_s}{\gamma_s - \gamma} \cdot \frac{H_w^2}{hT\omega} \tag{8.4.8}$$

2. Bedload transport equation

The bedload transport equation reads

$$\frac{\partial ((hN)}{\partial t} + \left[\frac{\partial (hUN)}{\partial x} + \frac{\partial (hVN)}{\partial y} \right] + \frac{\alpha_b}{\beta} \omega_b (N - N^*) = 0 \tag{8.4.9}$$

$$N^* = \frac{q_b^*}{\beta_H \sqrt{U^2 + V^2}} \tag{8.4.10}$$

where β_H is the thickness of the bedload layer; N is the bedload sediment mass per unit volume; α_b is the bedload settling coefficient; ω_b is the settling velocity of bedload grains; q_b^* is the bedload transport capacity per unit width in unit time, which could be given by the Dou Guoren's formula,

$$q_b^* = \frac{k_2}{C_0^2} \frac{\gamma \gamma_s}{\gamma_s - \gamma} m \frac{(U^2 + V^2)^{3/2}}{\omega_b} \tag{8.4.11}$$

in which, $m = \begin{cases} 0, & \text{for } V_k > \sqrt{U^2 + V^2} \\ \sqrt{U^2 + V^2} - V_k, & \text{for } V_k \le \sqrt{U^2 + V^2} \end{cases}$, V_k is the critical current speed for incipient motion of bedload grains and expressed by the Dou Guoren's formula as

$$V_k = 0.265 \ln\left(11 \frac{h}{\Delta}\right) \sqrt{\frac{\gamma_s - \gamma}{\gamma} g d_{50} + 0.19 \left(\frac{\gamma_0}{\gamma_0^*}\right)^{2.5} \frac{\varepsilon_k + gh\delta}{d_{50}}} \tag{8.4.12}$$

in which, γ_0 is the dry bulk density of the sediment on bed; γ_0^* is the steady dry bulk density; d_{50} is the median grain size of bedload grains; ε_k is a viscous parameter ($\varepsilon_k = 2.56\ \text{cm}^3/\text{s}^2$ for natural sand); δ is the thickness of water film, $\delta = 0.21 \times 10^{-4}$ cm; Δ is the bed roughness height given by,

$$\Delta = \begin{cases} 0.5\,\text{mm}, & d_{50} \le 0.5\,\text{mm} \\ d_{50}, & d_{50} > 0.5\,\text{mm} \\ h/100, & h/\Delta \le 100 \end{cases}$$

k_2 is a factor; c_0 is the dimensionless Chezy number. For fine sediment, $\gamma_0 = \gamma_0^*$ can be used in Eq. (8.4.12).

3. Bed evolution equation

The equation of bed deformation caused by suspended load reads

$$\gamma_0 \frac{\partial Z_s}{\partial t} = \alpha\omega(S - S_*) \tag{8.4.13}$$

where Z_s is the thickness of erosion-deposition caused by suspended load; γ_0 is the dry bulk density of the sediment on bed; ω is the settling velocity; S is the suspended sediment concentration; S_* is the suspended sediment-carrying capacity.

The equation of bed deformation caused by bedload reads

$$\gamma_0 \frac{\partial \eta_b}{\partial t} = \alpha_b\omega_b(N - N^*) \tag{8.4.14}$$

where η_b is the thickness of erosion–deposition caused by bedload.

The total thickness of erosion-deposition caused by suspended load and bedload is

$$Z = Z_s + Z_b \tag{8.4.15}$$

8.4.3 Procedure of Calculating Estuarial Morphological Evolution

(1) Calculating the current field

The current field is calculated by using Eqs. (8.4.1)–(8.4.3) to obtain the velocity components $U(x, y)$ and $V(x, y)$ in the computational domain.

(2) Calculating the sediment transport capacities of suspended load and bedload

The sediment transport capacities of suspended load and bedload are solved by using Eqs. (8.4.5)–(8.4.8) and (8.4.11), respectively.

(3) Calculating the suspended sediment concentration and the bedload mass

The depth-averaged suspended sediment concentration S at each point in the domain is obtained by solving the non-equilibrium suspended load transport Eq. (8.4.4). The bedload sediment mass per unit volume N is obtained by solving the non-equilibrium bedload transport Eq. (8.4.9).

(4) Calculating the bed level changes

The bed level changes caused by the non-equilibrium suspended load and bedload are calculated by using Eqs. (8.4.13)–(8.4.14) and (8.4.15), respectively.

(5) Iteration

When the bed level has changed significantly, one should re-calculate the tidal current field, the suspended sediment concentration, the bedload mass, and the bed level changes caused by the non-equilibrium suspended load and bedload. Repeating these steps until the calculation terminates at the required time.

A flowchart of calculating estuarial morphological evolution is shown in Fig. 8.18.

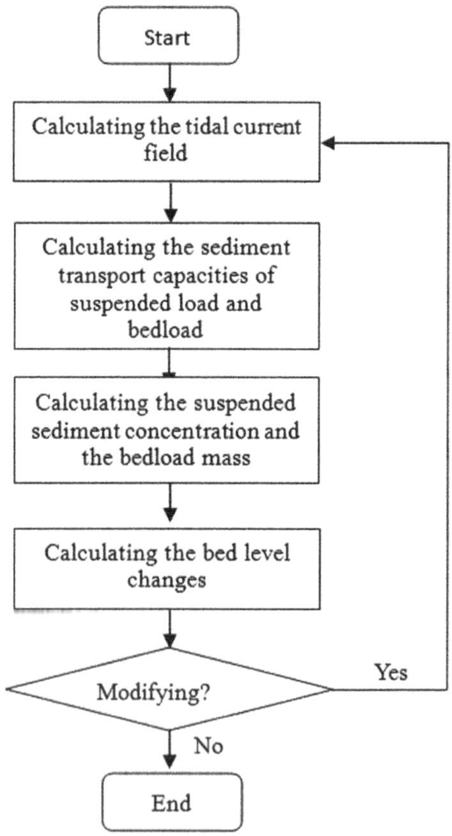

Fig. 8.18 Flowchart of calculating estuarine morphological evolution

8.5 Long-Term Model of Coastal Morphological Evolution

Long-term prediction of coastal sediment movement and morphological evolution is a very important issue because it is closely related to human activities and environmental changes. Long-term model mainly focuses on large spatial scales and long-time processes such as the effects of climate change and sea level rise on the coastal morphology, i.e., the prediction of coastal morphology resulted by the long-term interactions of hydrodynamic factors, sediment and seabed. In recent years, with the demands of sustainable economic development and the increasing threat of sea level rise and climate change, this issue has attracted more and more attention.

Various mid-term small-scale models of coastal morphological evolution have been described above. Although long-term operations of the small-scale models are allowed by the present computer's capabilities, they are not successfully applied to the long-term coastal morphological evolution. Because systems in nature are nonlinear, there are many inherent un-predictabilities during the long-term evolution. Some factors neglected in a small-scale model may play important roles in a large-scale model. There are non-unique solutions to a nonlinear multi-scale problem, which has been verified in many fields, such as meteorology, oceanography, and structural dynamics. Despite of some pioneering works, the studies on long-term model of coastal morphological evolution are still at an early stage.

The long-term coastal morphological evolution is not only related to the shoreline and the surrounding environment but also related to evolutions of some more complex systems, such as reefs, bays, and estuaries.

In the review of approaches to long-term modeling of coastal morphology in the G6 and G8 projects (coastal morphodynamics) of the MAST program, De Vriend et al. [28] pointed out that the main bases of the long-term model are the physical principles of the short-term coastal morphological model and field measurements. At present, the studies are at an early stage of development, and the achievements are relatively scattered. There are mainly two types of long-term models on coastal morphological evolution presently. A mostly common type is shoreline model (Hanson [29]), which is utilized in a deterministic way without bothering about the statistics and reliability of the prediction. In fact, this type should be restricted to short-term applications. Another type concerns the equilibrium state of the system under the given external conditions, such as the method of equilibrium beach profile proposed by Dean [30].

Reduction of the data in a wide range of space and time scales for the description of the physical process is a general character in various models of long-term coastal morphological evolution. The key is the reduction of the input information, the model, the output, and the measured data, such as selecting the representative tides and the representative wave conditions. Florence [31] simulated the morphological evolution of a tidal inlet, the Arcachon Bay on the French Atlantic coast in 82 years from 1905 to 1987 by using a long-term coastal morphological model, which is an example of long-term model that could be referred to.

References

1. Klopman G. Vertical structure of the flow due to wave and currents, part 2: Laser-dopple flow measurement for wave following and opposing a current. Delft Hydraulics Rep. 1994;H840:330.
2. Sun H, Han G, Jianhua Tao. The flow structure along the vertical line under the interaction of current and wave. J Hydraulic Eng. 2001;32(7):63–8.
3. Simons RR. Bottom stresses under random wave with a current superimposed. In: Proceedings of the 24th international conference on coastal engineering. 1995; 565–78.
4. Anskov MM, Fredso J, Sumer BM. Bed shear stress measurements over a smooth bed in three-dimensional wave-current motion. Coast Eng. 1993;20:277–316.
5. Goda Y, Watanabe N. A longshore current formula for random breaking waves. Coast Eng Japan. 1991; 34(2).
6. Zou Z, Chang M, Qiu D, Wang F, Dong G. Experimental study of longshore currents. J Hydrodyn. 2002;17(2):174–80.
7. Sun T, Han G, Tao J. Numerical simulation of wave-induced long-shore currents and experimental verification. J Hydraulic Eng. 2002;1:1–7.
8. Liu J. Study of sediment movement under wave action. In: Proceedings of the symposium on basic theoretical research of sediment in China; 1992.
9. Tao J. Research progress on coastal waves, currents, sediment and coastal morphodynamics. In: Proceedings of the 11st national symposium on hydrodynamics; 1996.
10. Kana TW, Ward LG. Nearshore and suspended sediment load during storm and post storm condition. In: Proceedings of the 17th international conference on coastal engineering. Sydney; 1980.
11. Goda Y. A new approach to beach morphology with the focus on suspended sediment transport. In: Proceedings of the first conference on Asian and Pacific coastal engineering. 2001; 1–24.
12. Coastal Engineering Research Center. Shoreline protection manual. U.S. Army. Corps of Engineers; 1984.
13. Del Valle R, Medina R, Losada MA. Dependence of coefficient K on grain size. J Waterw Port Coast Ocean Eng. 1993;119(5):568–74.
14. Miller HC. Field measurement of longshore sediment transport during storm. Coast Eng. 1999;36(4):301–21.
15. Bailard JA. An energetic total load sediment transport model for a plane sloping beach. J Geophys Res. 1981;86(c11):10938–54.
16. Soulshy RL. The "Bailard" sediment transport formula: comparisons with data and models. In: Report HR Wallingford; 1996.
17. Bayram A, Larson M, Miller HC, Kraus NC. Performance of longshore sediment transport formulas evaluate with field data. In: Proceedings of the 27th international conference on coastal engineering. Sydney; 2000. p. 3114–27.
18. Dou G, Dong F, Dou X, Li T. Study on mathematical model of estuarial and coastal sediment. Sci China Ser A. 1995;25(9):995–1001.
19. Dou D, Dong F, Dou X. Sediment-carrying capacity of tidal currents and waves. Chin Sci Bull. 1995;40(5):443–6.
20. Ozhan E. Wave energy flux computations for estimating longshore sands transport rate. In: Proceedings of the second international conference on coastal and port engineering in developing countries; 1987.
21. Jianhua Tao, Zhenfu Tian. Numerical modeling of nearshore wave field under combined regular waves and irregular waves. China Ocean Eng. 1991;5(1):75–84.
22. Goda Y. Irregular wave deformation in the surf zone. Coast Eng Jpn. 1975;18:13–26.
23. Haiwen Zhang, Jianhua Tao. Numerical modelling of sand beach evolution around coastal structures. Acta Oceanol Sin. 2000;19(2):127–36.
24. Watanabe A, Maruyama K, Shimizu T, et al. Numerical prediction model of three-dimensional beach deformation around a structure. Coast Eng Jpn. 1986;29:179–94.

25. Nicholson J, Broker I, Roelvink JA, Price D, et al. Intercomparison of coastal area morphodynamic models. Coast Eng. 1997;31(1):97–123.
26. Xin W. Numerical model of 2D estuarial suspended sediment motion under the interaction of tidal flow and waves. Ocean Eng. 1997;15(2):30–47.
27. Dou X, Li T, Dou G. Numerical model study on total sediment transport in the Yangtze River estuary. J Nanjing Hydraulic Res Inst. 1999;2:136–45.
28. De Vriend HJ, Capobianco M, Chesher TJ, et al. Approaches to long-term modelling of coastal morphology: a review. Coast Eng. 1993;21(1–3):225–69.
29. Hanson H. Genesis: a generalized shoreline change numerical model for engineering use. University of Lund, Department of water Resources Engineering, Report 1007; 1987.
30. Dean RG. Equilibrium beach profiles: characteristics and applications. J Coast Res. 1991;7(1):53–84.
31. Florence C. Long-term morphological modeling of a tidal inlet: the Arcachon Basin, France. Coast Eng. 2001;42(2):115–42.

Chapter 9
Incompressible Viscous Fluid Model for Simulating Water Waves

9.1 Introduction

The numerical simulation methods for solving water wave problems introduced in the above chapters are not directly from the Navier–Stokes (N-S) equations or the Reynolds equations which describe the viscous fluid motion. In these methods, the viscosity is taken into account in the form of energy dissipation or neglected. However, the viscosity cannot be neglected for some complex flows, especially for the interaction between water waves and medium-scale or small-scale structures. It is effective but difficult to simulate water wave motion directly using the N-S equations or the Reynolds equations. The main difficulties of using the N-S equations or the Reynolds equations for the viscous fluid flow to solve water wave problems are as follows.

(1) Treatment of fast varying free surface. The water wave free surface varies with the wave motion, so that the computational domain varies with time. In general, the water wave free surface is nonlinear. Therefore, how to treat the nonlinear free surface is one of the biggest challenges for using an incompressible viscous fluid model to simulate water waves. (2) Method of solving incompressible fluid flow problems. It is generally known that, there is no pressure term in the continuity equation of the N-S equations for the incompressible fluid flow so that the pressure field needs to be guessed or obtained by iteration in the solving process. Regarding the solving method, it is more difficult to solve the incompressible N-S equations than the compressible N-S equations. (3) The flow in which water waves interact with structures is generally turbulent. In general, water waves need to be simulated in a large area in engineering. But when using the N-S equations directly for the simulation, the spatial and temporal scales in the model should be smaller than the scales of the small and medium vortexes in turbulence. Therefore, it is quite difficult to use the N-S equations directly for the simulation. It is necessary to select an appropriate turbulence model and use it to close the Reynolds equations, i.e., the time-averaged N-S equations. (4) The simulation of

J. Tao, *Numerical Simulation of Water Waves*, Springer Tracts in Civil Engineering, https://doi.org/10.1007/978-981-15-2841-5_9

water waves requires a numerical scheme with high accuracy. In the interaction of nonlinear water waves and structures, the wave surface near the structures deforms greatly and non-uniformly. The vortex system behind the structures is also very complex. Therefore, a high resolution scheme is required for the simulation. Because of the large computational domain, quick calculation method and fast computer system are also required.

9.2 Mathematical Model of Incompressible Viscous Fluid Flow

9.2.1 Navier–Stokes Equations

The three dimensional (3D) N-S equations have been given in Eqs. (2.1.7)–(2.1.11) in Chap. 2. The 3D N-S equations in conservation form read,

$$\frac{\partial u}{\partial x}+\frac{\partial v}{\partial y}+\frac{\partial w}{\partial z}=0 \tag{9.2.1}$$

$$\frac{\partial u}{\partial t}+\frac{\partial uu}{\partial x}+\frac{\partial uv}{\partial y}+\frac{\partial uw}{\partial z}=-\frac{1}{\rho}\frac{\partial p}{\partial x}+f_x+\frac{1}{\rho}\left(\frac{\partial \tau_{xx}}{\partial x}+\frac{\partial \tau_{xy}}{\partial y}+\frac{\partial \tau_{xz}}{\partial z}\right) \tag{9.2.2}$$

$$\frac{\partial v}{\partial t}+\frac{\partial uv}{\partial x}+\frac{\partial vv}{\partial y}+\frac{\partial vw}{\partial z}=-\frac{1}{\rho}\frac{\partial p}{\partial y}+f_y+\frac{1}{\rho}\left(\frac{\partial \tau_{yx}}{\partial x}+\frac{\partial \tau_{yy}}{\partial y}+\frac{\partial \tau_{yz}}{\partial y}\right) \tag{9.2.3}$$

$$\frac{\partial w}{\partial t}+\frac{\partial uw}{\partial x}+\frac{\partial vw}{\partial y}+\frac{\partial ww}{\partial z}=-\frac{1}{\rho}\frac{\partial p}{\partial z}+f_z+\frac{1}{\rho}\left(\frac{\partial \tau_{zx}}{\partial x}+\frac{\partial \tau_{zy}}{\partial y}+\frac{\partial \tau_{zz}}{\partial z}\right) \tag{9.2.4}$$

where

$$\tau_{xx}=2\mu\frac{\partial u}{\partial x},\quad \tau_{xy}=\tau_{yx}=\mu\left(\frac{\partial u}{\partial y}+\frac{\partial v}{\partial x}\right),\quad \tau_{yy}=2\mu\frac{\partial v}{\partial y}$$

$$\tau_{xz}=\tau_{zx}=\mu\left(\frac{\partial u}{\partial z}+\frac{\partial w}{\partial x}\right),\quad \tau_{yz}=\tau_{zy}=\mu\left(\frac{\partial v}{\partial z}+\frac{\partial w}{\partial y}\right),\quad \tau_{zz}=2\mu\frac{\partial w}{\partial z}$$

in which, u, v, and w are the velocity components in the x, y, and z directions; f_x, f_y, and f_z are the components of unit mass force in the x, y, and z directions; ρ is fluid density; $\nu=\frac{\mu}{\rho}$ is kinematic viscosity coefficient; t is time; x, y, and z are the Cartesian coordinate components.

9.2.2 Reynolds Equations

When the Reynolds number in a flow is high or a water wave problem in engineering is studied, the flow is generally at the state of turbulence. Very fine spatial grid and very small time step are required to solve the N-S equations directly in order to simulate the random vortexes in turbulence. It means that the spatial and temporal scales in the model should be smaller than the scales of the vortexes in turbulence. Although the computer technology has developed to a very high level, it is still very difficult to solve engineering problems by directly solving the N-S equations. In fact, when solving engineering problems, the physical quantities of interest are the averages over a certain time period or a certain area. Therefore, the time-averaged values of the physical quantities in turbulence are mainly studied in practice. The equations to describe these time-averaged quantities are called the Reynolds equations. It is defined in time-average method that,

$$u = \bar{u} + u' \tag{9.2.5}$$

$$v = \bar{v} + v' \tag{9.2.6}$$

$$w = \bar{w} + w' \tag{9.2.7}$$

$$p = \bar{p} + p' \tag{9.2.8}$$

where $\bar{u} = \frac{1}{\Delta t}\int_0^{\Delta t} u\mathrm{d}t$, $\bar{v} = \frac{1}{\Delta t}\int_0^{\Delta t} v\mathrm{d}t$, and $\bar{w} = \frac{1}{\Delta t}\int_0^{\Delta t} w\mathrm{d}t$ are the time-averaged velocity components in the x, y, and z directions; $\bar{p} = \frac{1}{\Delta t}\int_0^{\Delta t} p\mathrm{d}t$ is the time-averaged pressure; u', v', and w' are the pulsating quantities of the velocity components in the horizontal and vertical directions; p' is the pulsating quantity of the pressure. Time-averaging the N-S equations yields the 3D Reynolds equations,

$$\frac{\partial \bar{u}}{\partial x} + \frac{\partial \bar{v}}{\partial y} + \frac{\partial \bar{w}}{\partial z} = 0 \tag{9.2.9}$$

$$\begin{aligned}\frac{\partial \bar{u}}{\partial t} + \bar{u}\frac{\partial \bar{u}}{\partial x} + \bar{v}\frac{\partial \bar{u}}{\partial y} + \bar{w}\frac{\partial \bar{u}}{\partial z} &= f_x - \frac{1}{\rho}\frac{\partial \bar{p}}{\partial x} + (\nu + \nu_t)\frac{\partial^2 u}{\partial x^2} + (\nu + \nu_t)\frac{\partial^2 u}{\partial y^2} \\ &+ (\nu + \nu_t)\frac{\partial^2 u}{\partial z^2} + 2.0\frac{\partial \nu_t}{\partial x}\frac{\partial \bar{u}}{\partial x} - \frac{2}{3}\frac{\partial k}{\partial x} + \frac{\partial \nu_t}{\partial y}\left[\frac{\partial \bar{u}}{\partial y} + \frac{\partial \bar{v}}{\partial x}\right] + \frac{\partial \nu_t}{\partial z}\left[\frac{\partial \bar{u}}{\partial z} + \frac{\partial \bar{w}}{\partial x}\right]\end{aligned} \tag{9.2.10}$$

$$\begin{aligned}\frac{\partial \bar{v}}{\partial t} + \bar{u}\frac{\partial \bar{v}}{\partial x} + \bar{v}\frac{\partial \bar{v}}{\partial y} + \bar{w}\frac{\partial \bar{v}}{\partial z} &= f_y - \frac{1}{\rho}\frac{\partial \bar{p}}{\partial y} + (\nu + \nu_t)\frac{\partial^2 v}{\partial x^2} + (\nu + \nu_t)\frac{\partial^2 v}{\partial y^2} \\ &+ (\nu + \nu_t)\frac{\partial^2 v}{\partial z^2} + 2.0\frac{\partial \nu_t}{\partial y}\frac{\partial \bar{v}}{\partial y} - \frac{2}{3}\frac{\partial k}{\partial y} + \frac{\partial \nu_t}{\partial x}\left[\frac{\partial \bar{u}}{\partial y} + \frac{\partial \bar{v}}{\partial x}\right] + \frac{\partial \nu_t}{\partial z}\left[\frac{\partial \bar{v}}{\partial z} + \frac{\partial \bar{w}}{\partial x}\right]\end{aligned} \tag{9.2.11}$$

$$\frac{\partial \bar{w}}{\partial t}+\bar{u}\frac{\partial \bar{w}}{\partial x}+\bar{v}\frac{\partial \bar{w}}{\partial y}+\bar{w}\frac{\partial \bar{w}}{\partial z}=f_z-\frac{1}{\rho}\frac{\partial \bar{p}}{\partial z}+(\nu+\nu_t)\frac{\partial^2 w}{\partial x^2}+(\nu+\nu_t)\frac{\partial^2 w}{\partial y^2}$$
$$+(\nu+\nu_t)\frac{\partial^2 w}{\partial z^2}+2.0\frac{\partial \nu_t}{\partial z}\frac{\partial \bar{w}}{\partial z}-\frac{2}{3}\frac{\partial k}{\partial z}+\frac{\partial \nu_t}{\partial x}\left[\frac{\partial \bar{u}}{\partial z}+\frac{\partial \bar{w}}{\partial x}\right]+\frac{\partial \nu_t}{\partial y}\left[\frac{\partial \bar{v}}{\partial z}+\frac{\partial \bar{w}}{\partial y}\right] \tag{9.2.12}$$

where ν_t is the eddy viscosity; k is the turbulent kinetic energy.

The two-dimensional (2D) incompressible Reynolds equations read,

$$\frac{\partial \bar{u}}{\partial x}+\frac{\partial \bar{v}}{\partial y}=0 \tag{9.2.13}$$

$$\frac{\partial \bar{u}}{\partial t}+\frac{\partial \bar{u}\bar{u}}{\partial x}+\frac{\partial \bar{u}\bar{v}}{\partial y}=f_x-\frac{1}{\rho}\frac{\partial \bar{p}}{\partial x}-\frac{2}{3}\frac{\partial k}{\partial x}+\frac{\partial}{\partial x}\left[2(\nu+\nu_t)\frac{\partial \bar{u}}{\partial x}\right]$$
$$+\frac{\partial}{\partial y}\left[(\nu+\nu_t)\left(\frac{\partial \bar{u}}{\partial y}+\frac{\partial \bar{v}}{\partial x}\right)\right] \tag{9.2.14}$$

$$\frac{\partial \bar{v}}{\partial t}+\frac{\partial \bar{v}\bar{u}}{\partial x}+\frac{\partial \bar{v}\bar{v}}{\partial y}=f_y-\frac{1}{\rho}\frac{\partial \bar{p}}{\partial y}-\frac{2}{3}\frac{\partial k}{\partial y}+\frac{\partial}{\partial x}\left[(\nu+\nu_t)\left(\frac{\partial \bar{u}}{\partial y}+\frac{\partial \bar{v}}{\partial x}\right)\right]$$
$$+\frac{\partial}{\partial y}\left[\left(2(\nu+\nu_t)\frac{\partial \bar{v}}{\partial y}\right)\right] \tag{9.2.15}$$

9.2.3 Turbulence Model

In the Reynolds equations, ν_t and k are unknown so that the number of the equations is smaller than the number of the unknowns and the equations are not closed. In order to close the Reynolds equations, according to the number of equations needed to be solved, turbulence models can be classified into zero-equation models, one-equation models, two-equation models, Reynolds stress models, and so on.

In a zero-equation model, instead of using a differential equation to describe the turbulence terms, the eddy viscosity is connected to the time-averaged velocity directly. Since Prandtl proposed the first model describing the distribution of ν_t in 1925, this kind of model has been proved to be quite applicable to a flow with a thin shear layer. The eddy viscosity ν_t is proportional to a typical velocity scale V and a typical length scale l_m which characterize the large-scale turbulence,

$$\nu_t \propto V l_m \tag{9.2.16}$$

Prandtl proposed an assumption that the velocity scale V of a pulsating motion is equal to the product of the velocity gradient and the square of the length scale l_m, then the eddy viscosity ν_t can be obtained by,

$$\nu_t = l_m^2 \left(\frac{\partial \bar{u}}{\partial y} \right) \tag{9.2.17}$$

where l_m is called the mixing-length and normally determined by the empirical method. The effect of the pulsation terms is neglected in the mixing-length model proposed by Prandtl, which hints that the turbulence is in equilibrium locally. When the turbulent convection cannot be neglected, the mixing-length model is not appropriate. The mixing-length model is widely used only in the flow with a free shear layer presently. It is difficult to determine the distribution of the mixing-length l_m in a multi-dimensional complex flow filed so that the mixing-length model is rarely used in this case.

In a one-equation model, a differential equation is used to describe the convection and diffusion of the turbulent kinetic energy k which is the most important scale in turbulence. By taking $\sqrt{k}$ as the velocity scale to characterize the pulsating motion, and according to the following Kolmogorov–Prandtl expression,

$$\nu_t = C'_\mu \sqrt{k} L \tag{9.2.18}$$

in which, C'_μ is an empirical constant; L is a characteristic length scale of the turbulent pulsation which is not equal to the mixing-length l_m in general; then the differential equation for the turbulent kinetic energy k reads,

$$\begin{aligned} \frac{\partial k}{\partial t} + \frac{\partial \bar{u} k}{\partial x} + \frac{\partial \bar{v} k}{\partial y} &= \frac{\partial}{\partial x}\left[\left(\nu + \frac{\nu_t}{\sigma_k}\right)\frac{\partial k}{\partial x}\right] + \frac{\partial}{\partial y}\left[\left(\nu + \frac{\nu_t}{\sigma_k}\right)\frac{\partial k}{\partial y}\right] + \nu_t\left(\frac{\partial \bar{u}}{\partial y} + \frac{\partial \bar{v}}{\partial x}\right)^2 \\ &\quad + 2\nu_t\left(\frac{\partial \bar{u}}{\partial x}\right)^2 + 2\nu_t\left(\frac{\partial \bar{v}}{\partial y}\right)^2 - C_d \frac{k^{3/2}}{L} \end{aligned} \tag{9.2.19}$$

where C'_μ and $C_d \approx 0.09$; σ_k is the Prandtl number of the pulsating kinetic energy which is generally taken as $\sigma_k = 1$. Compared with the zero-equation model, the one-equation model is improved. However, it still needs to select the characteristic length scale by experience. It is difficult to determine the specific distribution of L in a complex flow, which restricts the model application.

In a one-equation model, the characteristic length scale of the turbulence L is given by experience. In fact, the length scale is also a variable which can be obtained by a differential equation. It means that the characteristic length scale of the turbulence can be given by means of the relation $\varepsilon = \nu\overline{\left(\frac{\partial u'_i}{\partial x_k}\right)^2} = c_D \frac{k^{3/2}}{L}$. Thus, a turbulence model with two equations is formed. The 2D $k - \varepsilon$ model reads,

$$\frac{\partial k}{\partial t} + \frac{\partial \bar{u} k}{\partial x} + \frac{\partial \bar{v} k}{\partial y} = \frac{\partial}{\partial x}\left[\left(\nu + \frac{\nu_t}{\sigma_k}\right)\frac{\partial k}{\partial x}\right] + \frac{\partial}{\partial y}\left[\left(\nu + \frac{\nu_t}{\sigma_k}\right)\frac{\partial k}{\partial y}\right] + G - \varepsilon \tag{9.2.20}$$

$$\frac{\partial \varepsilon}{\partial t} + \frac{\partial \bar{u} \varepsilon}{\partial x} + \frac{\partial \bar{v} \varepsilon}{\partial y} = \frac{\partial}{\partial x}\left[\left(\nu + \frac{\nu_t}{\sigma_\varepsilon}\right)\frac{\partial \varepsilon}{\partial x}\right] + \frac{\partial}{\partial y}\left[\left(\nu + \frac{\nu_t}{\sigma_\varepsilon}\right)\frac{\partial \varepsilon}{\partial y}\right]$$

$$+ C_{\varepsilon 1}\frac{\varepsilon}{k}G - C_{\varepsilon 2}\frac{\varepsilon^2}{k} \tag{9.2.21}$$

where $G = 2\nu_t S_{ij}\frac{\partial \bar{u}_i}{\partial x_j} = \nu_t\left\{\left(\frac{\partial \bar{u}}{\partial y} + \frac{\partial \bar{v}}{\partial x}\right)^2 + 2\left[\left(\frac{\partial \bar{u}}{\partial x}\right)^2 + \left(\frac{\partial \bar{v}}{\partial y}\right)^2\right]\right\}$, $C_{\varepsilon 1} = 1.43$, $C_{\varepsilon 2} = 1.92$, $\sigma_k = 1.00$ and $\sigma_\varepsilon = 1.30$.

Here, k is the turbulent kinetic energy; ε is the dissipation rate of the turbulent kinetic energy; S_{ij} is the strain rate tensor.

The spatial distributions of k and ε at a certain moment are obtained by solving the governing equations of k and ε. Then, the spatial distribution of the eddy viscosity can be obtained by,

$$\nu_t = C_\mu \frac{k^2}{\varepsilon} \tag{9.2.22}$$

where $C_\mu = 0.09$.

The $k - \varepsilon$ turbulence model is widely used in practical applications and has strong adaptability.

For convenience, the symbol of time-average (-) is omitted in the following sections of this chapter.

9.2.4 Boundary and Initial Conditions

1. Free surface boundary conditions

For small-scale flow problems with interface, such as bubble motion in water, it is necessary to take surface tension into account. But when simulating the wave-structure interaction, the free surface tension and the tangential wind stress can be neglected because the wave length is long and the computational domain is not very large. Meanwhile, the tangential stress on the free surface is continuous, which gives that,

$$\text{in the normal direction: } p - \mu\left(\frac{\partial u_i}{\partial x_j} + \frac{\partial u_j}{\partial x_i}\right)n_i n_j = p_a \tag{9.2.23}$$

$$\text{in the tangential direction: } \mu\left(\frac{\partial u_i}{\partial x_j} + \frac{\partial u_j}{\partial x_i}\right)n_i t_j = 0 \tag{9.2.24}$$

Regarding the free surface kinematic boundary condition, the free surface is a material surface when the waves are non-breaking. It is assumed that the equation for the free surface reads,

$$F(x, y, \eta(x, y, t), t) = 0 \tag{9.2.25}$$

Then, the free surface kinematic boundary condition reads,

$$\frac{D}{Dt}[F] = \frac{\partial \eta}{\partial t} + u\frac{\partial \eta}{\partial x} + v\frac{\partial \eta}{\partial y} = 0 \tag{9.2.26}$$

When the waves break at the moving free surface, the kinematic boundary condition at the moving free surface can be obtained by using the following volume of fluid (VOF) method.

With the $k - \varepsilon$ turbulence model used for the governing equations, it can be assumed that the turbulent kinetic energy k and its dissipation rate ε are not diffused in the direction normal to the free surface, which gives,

$$\frac{\partial k}{\partial x_i} n_i = 0, \quad \frac{\partial \varepsilon}{\partial x_i} n_i = 0 \tag{9.2.27}$$

where n_i is the normal vector. These provides boundary conditions for k and ε.

2. Inflow boundary conditions

When solving a flow problem with free surface, it is needed to give inflow boundary conditions. For a fluctuation problem or the boundary condition at the wave paddle of wavemaker, the time series of velocity and surface elevation at the inlet can be given, i.e.,

$$\eta = \eta(t), \quad u = u(t), \quad v = v(t) \tag{9.2.28}$$

Since the turbulence model equations need to be solved, the boundary conditions of k and ε at the inlet should be also given. In a flow problem with free surface, they can be given according to the measurement data at the inlet or on the basis of the average flow velocity and the calculated characteristic length. In fact, there are few measurement data on turbulence parameters at the inlet in practice. Therefore, in practical calculation, they can be estimated by,

$$k = \frac{3}{2}(\bar{u}I)^2 \tag{9.2.29}$$

$$\varepsilon = C_\mu^{\frac{3}{4}} \frac{k^{\frac{3}{2}}}{\ell} \tag{9.2.30}$$

where $I = \frac{u'}{\bar{u}} \approx 0.16(\mathrm{Re})^{-\frac{1}{8}}$ is the turbulence intensity defined as the ratio of the pulsating velocity to the average flow velocity, which is generally small; ℓ is the characteristic turbulence length scale, which is related to the scale of large eddies in the turbulence; C_μ is a characteristic constant in the turbulence model.

3. Outflow boundary conditions

Flow problems are normally located in the project areas where flow has been fully developed. It is generally deemed that the turbulence has reached self-equilibrium.

It can be assumed for the outflow boundary conditions that the gradients of the flow parameters in the outlet direction are zero at the outlet,

$$\frac{\partial Q}{\partial t} + U_n \frac{\partial Q}{\partial n} = 0 \text{ or } \frac{\partial Q}{\partial n} = 0 \tag{9.2.31}$$

where Q denotes u, v, p, k, ε and η.

4. Solid boundary conditions

The fluid is viscous in turbulent flow problems. The non-slip boundary condition should be satisfied at the solid wall, i.e., $u_i = 0$. Moreover, the turbulent kinetic energy k and its dissipation rate ε are zero at the solid wall. However, the grid spacing is usually larger than the boundary layer thickness in the practical calculation, so that the non-slip boundary condition cannot be directly used. Therefore, an appropriate treatment in the near-wall region is needed.

A wall function method proposed by Launder and Spalding [1] is utilized to treat the solid boundary conditions.

According to the log-law of the wall, it is assumed that the velocity at the wall boundary is zero. Figure 9.1 shows a sketch for the wall function. The velocity in the near-wall region satisfies,

$$\frac{U_P}{U^*} = \frac{1}{\kappa} \ln(f Y_P) \tag{9.2.32}$$

where U_p is the resultant velocity at the point P near the wall, which is parallel to the wall boundary; f is the surface roughness of the wall, which can be taken as $f = 9.0$ for the horizontal smooth wall; U^* is the bottom friction velocity, also called the resistant velocity; κ is the Von Karman constant, which is generally taken as $\kappa = 0.41$; $Y_p = \frac{dU^*}{\upsilon}$ is the dimensionless friction length, in which d is the dimensional length of the node P to the solid wall. In the calculation, the P point should be properly set to ensure that Y_p is within 30–100.

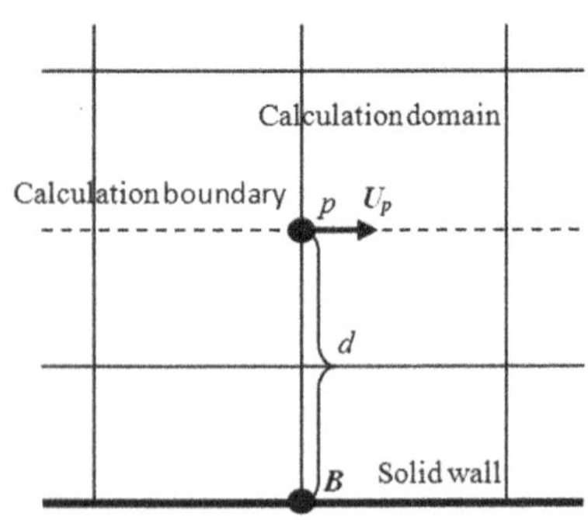

Fig. 9.1 Sketch for the wall function

For the calculation of the turbulent kinetic energy k and its dissipation rate ε in the near-wall region, it is not necessary to solve the governing equations of k and ε in the near-wall region. They can be obtained directly by,

$$k = \frac{U^*}{\sqrt{C_\mu}}, \tag{9.2.33a}$$

$$\varepsilon = \frac{U^{*3}}{\sqrt{\kappa Y_p}} \tag{9.2.33b}$$

5. Initial conditions

At the initial time of simulating water waves, the pressure can be taken as the hydrostatic pressure and the surface elevation is taken as the still water level. The initial velocity in the domain is zero. The turbulent kinetic energy k and its dissipation rate ε in the domain can be taken as the values with small perturbation.

9.3 Free Surface Treatments for Simulating Water Waves by Reynolds Equation Model

Most of the water flow problems in nature are free surface flows, such as flows in rivers, lakes, and oceans. One of the important characteristics of them is the water–gas interface varying with time. They are the problems of two phases which are impermeable to each other. Because the densities of the two flow media are greatly different (the density ratio of water to air is 1000:1), when the water flow is studied, the impact of air on water can be neglected so that the problem is simplified to a single-phase flow problem with moving free surface. The computational domain changes at each moment along with the free surface varying with time, so that the flow problem with free surface is a difficult one in fluid mechanics. The key to solve the free surface problem is the treatment of free surface.

Some difficulties in numerically solving unsteady viscous flow problems with free surface are as follows. (1) The free surface conditions are nonlinear, and the equations of the boundary conditions are hyperbolic in general. However, the governing equations of flow are parabolic. Thus, the whole problem is mixed. (2) The governing equations of flow are nonlinear. (3) The free surface boundary in the computational domain varies with time. (4) The free surface boundary sometimes changes greatly and cusps occur, which leads to singular solution. (5) The free surface boundary may merge or roll such as when wave breaking occurs. All these difficulties come down to one point of how to capture the shape and position of the time-varying free surface. Presently, there are several methods to determine the shape and position of the free surface.

(1) Rigid-lid hypothesis method. It is assumed that the free surface is fixed, and the normal velocity is zero.
(2) Water-level function method. The depth-averaged continuity equation, which is obtained by using the free surface kinematic boundary condition and integrating the continuity equation over the water depth, is taken as the governing equation of water level. By solving this equation, the position of the free surface can be determined. This method is applicable to the flow problems with free surface varying not dramatically and having only one single position without overlap (see Chap. 3).
(3) Lagrangian and ALE method [2]. The governing equations are solved directly in the moving computational domain, and the computational grid moves along with the fluid. Then, the unsteady variations in the fluid domain can be simulated in detail. The main weakness of this method is that the free surface with discontinuity cannot be simulated.
(4) Coordinate transformation method. The region of the fluid in physical space is transformed into a rectangular or cuboid region in a fixed coordinate system for the computational domain by means of a certain coordinate transformation, such as sigma coordinate transformation method (see Chap. 3).
(5) MAC method. An Eulerian grid system which does not move with the fluid is combined with Lagrangian marker particles moving with the fluid. The marker particles set in the computational grid with fluid always move along with the fluid at the local velocity. Therefore, the free surface movement can be described by the movement of the marker particles. The weaknesses of this method are that, more storage space and computation time are needed to track the movement of marker particles, and the adaptability to irregular boundaries is poor.
(6) Level-set method. The level-set function is also called the distance function which is defined as a signed distance from each computational point in the computational domain to the fluid interface. The fluid interface is determined by solving the distance function. The level-set method is usually used to solve two-phase flow problems. Because the flow problems of the two phases should be solved at the same time, the calculation for the gas phase is added even though only the solution to the water flow problem is of interest.
(7) VOF method. The moving free surface is captured by solving a transport equation on a function of fluid volume. The method is quite simple and convenient. It also has good stability. It has some advantages like the MAC method, but also saves time and storage space. Meanwhile, it overcomes the shortcoming of the water-level function method which cannot deal with the multi-valued free surface problems, and the weakness of the Lagrangian and ALE method which cannot simulate the free surfaces with discontinuity.

Brief introductions of the level-set method and VOF method are given in the following.

9.3.1 Level-Set Method

The level-set method was proposed by Osher and Sethain [3] for solving the two-phase fluid motion. It is mainly to implicitly track the interface of the two-phase flow using a level-set function by solving the N-S equations which should be satisfied by the fluid in the whole computational domain.

The level-set function denoted by $\phi(x, y, t)$ is defined as the signed normal distance from each computational point in the computational domain to the interface of the two fluids. When the point (x, y) is above the interface at the t time, $\phi(x, y, t)$ is positive. When the point (x, y) is below the interface, $\phi(x, y, t)$ is negative. The interface of the two fluids is made up by the points with $\phi(x, y, t) = 0$.

Considering a curve AB in a 2D region denoted by $\gamma(0)$, as shown in Fig. 9.2, it is assumed that the velocity of the curve moving along the normal direction is a function of the curvature κ at the point on the curve and denoted by $F(\kappa)$, and the curve moves to $\gamma(t)$ at the t time. Let the point on $\gamma(t)$ be located at $X(x, t) = (x(s, t), y(s, t))$, $s \in [0, S]$. The point $\boldsymbol{A}(t)$ is located at $(x(0, t), y(0, t))$, and the point $\boldsymbol{B}(t)$ located at $(x(S, t), y(S, t))$.

It is assumed that the curve $\gamma(t)$ is the position of the free surface when solving the fluid flow problem. Then, the distance function $\phi(x, y, t)$ given by the level-set method is the signed normal distance from the point (x, y) to the free surface $\gamma(t)$. When the point (x, y) is located above (below) the interface, $\phi(x, y, t)$ is positive (negative). The interface of the two fluids formed by the points with $\phi(x, y, t) = 0$ gives,

$$\phi(x, y, t) = 0, \quad (x, y) \in \gamma(t) \tag{9.3.1}$$

If the particle velocity at the free surface $\gamma(t)$ is $\boldsymbol{u}$, then,

$$F = \boldsymbol{u} \cdot \boldsymbol{n} \tag{9.3.2}$$

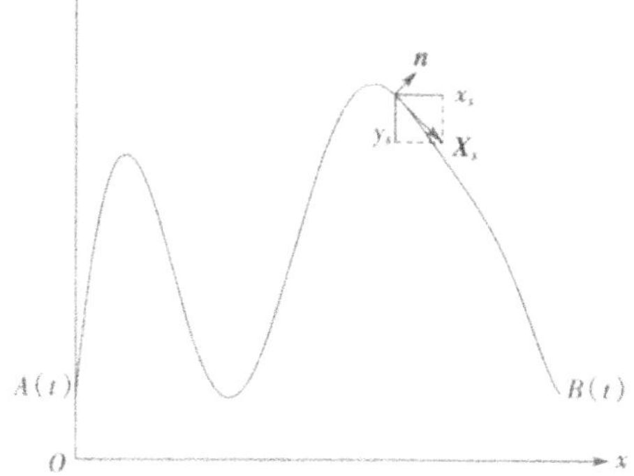

Fig. 9.2 Sketch of a curve in a 2D region

$$n = -\nabla\phi/|\nabla\phi| \tag{9.3.3}$$

Through derivation [3], it yields

$$\phi_t + (\boldsymbol{u} \cdot \nabla)\phi = 0 \tag{9.3.4}$$

which is the motion equation satisfied by the level-set function $\phi(x, y, t)$ in the whole flow field. The free surface can be implicitly captured by solving this equation.

However, when calculating relatively intense motion such as the diffusion of bubbles, $\phi(x, y, t)$ will no longer be a distance function after a time period if it is re-calculated by Eq. (9.3.4) in the flow filed at each time step. Moreover, the distance function $\phi(x, y, 0)$ at the initial time can be given by the definition in principle. But this is not accurate, also the workload is heavy. In order to overcome the above weaknesses, Sussman et al. [4] proposed an iterative method to keep $\phi(x, y, t)$ as a distance function. However, this method is effective only for some simple interfaces. When the interface shape is complex, it may lead to a large amount of computation, very slow convergence or even no convergence. Therefore, a more appropriate and effective method is needed.

Chen et al. (1995) [5] proposed a distance function $\phi(x, y, t)$ by solving the following bi-harmonic equation,

$$\begin{cases} \Delta^2\phi = 0 \\ \left.\frac{\partial\phi}{\partial n}\right|_{\Gamma} = 1 \\ \phi|_{\Gamma} = 0 \end{cases} \tag{9.3.5}$$

Some good results have been achieved. By this method, the solution is analytic everywhere in the domain. Moreover, this method can be applied to some complex boundaries and interfaces.

The solution of the bi-harmonic equation (9.3.5) is analytic everywhere in the domain and maintained as a signed distance function to Γ. Therefore, if Γ is taken as the interface with $\phi = 0$, this solution satisfies the definition of the function. For the above bi-harmonic equation, Zhu [6] proposed a boundary integral method, which has advantages of fast convergence and high accuracy. This method is described briefly as follows.

By introducing intermediate variables φ and q,

$$\begin{cases} \varphi = [\Delta\phi] = \Delta\phi|_{\Gamma_{in}} - \Delta\phi|_{\Gamma_{out}} \\ q = \left[\frac{\partial\Delta\phi}{\partial n}\right] = \left.\frac{\partial\Delta\phi}{\partial n}\right|_{\Gamma_{in}} - \left.\frac{\partial\Delta\phi}{\partial n}\right|_{\Gamma_{out}} \end{cases} \tag{9.3.6}$$

the integral expression of ϕ is obtained,

$$\phi(y) = -\int_{\Gamma} q(x)E(|x - y|)\mathrm{d}\gamma_x + \int_{\Gamma} \varphi(x)\frac{\partial E(|x - y|)}{\partial n_x}\mathrm{d}\gamma_x + p(y) \tag{9.3.7}$$

where $E(|x-y|)=\frac{1}{8\pi}|x-y|^2\ln|x-y|$ is the fundamental solution of the biharmonic operator in 2D cases. The determination method of q, φ and p, and solving Eq. (9.3.7) can be found in Refs. [7, 6].

As the boundary integral method with advantages of fast convergence and high accuracy is used to determinate ϕ, it overcomes the difficulties encountered when setting the level-set function.

In the calculation, the distance function in the whole field is obtained by solving Eq. (9.3.5) at the initial time. At the intermediate moments, Eq. (9.3.4) is used to get $\phi(x, y, t)$. The semi-discrete scheme reads,

$$\frac{\phi^{n+1}-\phi^n}{\Delta t}=-\frac{1}{2}\left\{[(\boldsymbol{u}\cdot\nabla)\phi]^n+[(\boldsymbol{u}\cdot\nabla)\phi]^{n+1}\right\} \tag{9.3.8}$$

The second-order upwind scheme can be used for the difference discretization of the terms on the right side.

A projection method [8] can be used to solve the N-S equations in the computational domain including air and water. If the time step is small, and the shape of water waves does not change dramatically, $\phi(x, y, t)$ can still remain as a distance function after a certain time period.

It is noted that, the two-phase flow of air and water is in the computational domain when solving the problem of free water surface. The density and the kinematic viscosity coefficient in the domain can be expressed respectively as,

$$\rho=\begin{cases}\rho_a, & \phi>0\\ \frac{1}{2}(\rho_a+\rho_w), & \phi=0\\ \rho_w, & \phi<0\end{cases} \tag{9.3.9}$$

$$\nu=\begin{cases}\nu_a, & \phi>0\\ \frac{1}{2}(\nu_a+\nu_w), & \phi=0\\ \nu_w, & \phi<0\end{cases} \tag{9.3.10}$$

Because of the great density ratio of water to air, directly using Eq. (9.3.9) to determine $\rho(\phi)$ may result in strong discontinuity at the interface, which will lead to unexpected instability of the solution. Therefore, the density near the interface is smoothed by,

$$\rho=\begin{cases}\rho_a, & \phi>\alpha\\ \rho_a+\frac{\rho_w-\rho_a}{2\alpha}(\alpha-\phi), & -\alpha\le\phi\le\alpha\\ \rho_w, & \phi<-\alpha\end{cases} \tag{9.3.11}$$

This ensures the stability of the difference scheme (here, α is a very small distance being taken as 1.5–2.5 times of grid spacing).

9.3.2 Volume of Fluid (VOF) Method

In order to track the movement of free surface, a function f is defined at each point in the computational domain, which is a function of time and space, $f = f(x, y, t)$. When the point is occupied by a fluid particle, $f = 1$ at this point; otherwise, $f = 0$. In 2D cases, after the computational domain is discretized into computational grid in space, let F be the ratio of the area occupied by fluid in a grid cell to the area of this cell and expressed as,

$$F(x, y, t) = \frac{1}{\Delta A} \iint\limits_{A} f(\varsigma, \eta, t)\mathrm{d}\varsigma\mathrm{d}\eta \tag{9.3.12}$$

When $F = 1$, the cell is full of fluid, and it is an internal cell in fluid. When $F = 0$, the cell is empty, and it is an external cell. When F is in between 0 and 1, there is fluid interface in the cell, which either intersects the free surface or contains bubbles smaller than the cell size.

F is a function of time and space, $F = F(x, y, t)$. It can be explained as the movement of markers without mass and viscosity, which are fixed with the fluid particles and moving along with them. The motion equation reads,

$$\frac{DF}{Dt} = 0, \text{ i.e., } \frac{\partial F}{\partial t} + \frac{\partial uF}{\partial x} + \frac{\partial vF}{\partial y} = 0 \tag{9.3.13}$$

By solving F, it can be determined which cells the free surface is located in. Furthermore, F varies most rapidly in the normal direction of the free surface so that the direction of the free surface can be determined by the gradient of F. The basic idea of the VOF method is to construct a function of fluid volume to track the fluid flow quantity in each grid cell, and construct the shape of the free surface according to this function and its gradient. By using the VOF method, the problem of calculating the moving free surface is turned into a problem of calculating F in the whole computational domain. For the wave problems discussed in this chapter, when the wave surface rolls, merges or breaks, there might be multi values for the free surface elevation sometimes. Because the VOF method is to calculate the function of fluid volume at each grid cell, the location and shape of the free surface is constructed by the values of the fluid volume function F, so that it is relatively easy to handle the wave rolling, merging and breaking.

The computational domain for using the VOF method must be larger than where water can arrive. See Fig. 9.3 for a sketch. F is 1 in the shaded region, and 0 in the blank region. After the computational grid is generated, F is in between 0 and 1 for the grid cells at free surface.

According to the definition of F, it is known that the F function is not a continuous function but a step function. Therefore, the equations cannot be discretized by an ordinary difference scheme. If so, the discontinuities of F might be smoothed out or numerical oscillations might be generated at the discontinuity points of F, so that

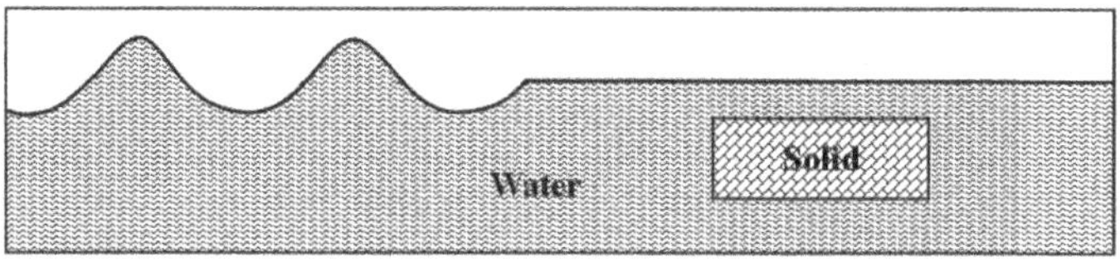

Fig. 9.3 Sketch of computational domain

the original meaning of the F function is lost. In order to overcome this difficulty, a method of free surface reconstruction is needed to treat the free surface. Presently, the methods of free surface reconstruction mainly include the piecewise constant and piecewise linear methods. A brief introduction to the methods of free surface reconstruction is given in the following.

The different methods of free surface reconstruction can be classified into piecewise constant volume tracking methods, piecewise linear volume tracking methods, the methods of implicit tracking free surface based on high resolution difference scheme, and so on.

1. Piecewise constant volume tracking methods

In 1970s and 1980s, treatments on the free surface were at the exploratory stage of development. At that time, an approach of piecewise constant approximation was utilized for the free surface. By using this approach the free surface in a grid cell is represented by a horizontal or vertical line segment with low accuracy. Typical methods include, the donor-acceptor method proposed by Hirt and Nichols [9], the Simple Line Interface Calculation (SLIC) method proposed by Noh and Woodward [10]. The difference of them is selecting different geometries to approximate the free surface. In the donor-acceptor method by Hirt and Nichols, the staircase geometry is used to approximately represent the free surface. The free surface in one grid cell is parallel to the straight line of the cell boundary. Which side of the cell boundary being parallel to the free surface depends on the distribution of the fluid volume function in this cell and the adjacent cells. But in the SLIC method, the piecewise constant is used to represent the free surface. The free surface in one grid cell is a straight line segment only parallel to the cell boundary in a certain direction. This simple treatment on the free surface has a certain advantage when simulating multi-phase flow with free surface. In the design of an earlier volume tracking method, a combined upwind and downwind scheme was usually used to calculate the change of the fluid volume in one grid cell, and only the fluid volume in each grid cell was needed instead of the exact position of free surface. Subsequently, some improvements have been made by many of scholars. For instance, Chorin [11] and Barr and Ashurst [12] have developed the free surface tracking method to be multi-dimensional using direction splitting schemes.

In the SLIC method proposed by Noh and Woodward, the free surface is represented by straight line segments parallel to one of the coordinate axes. When calculating the flow quantity in the x or the horizontal direction, all the free surface segments are perpendicular to the horizontal direction, as shown in Fig. 9.4b. When

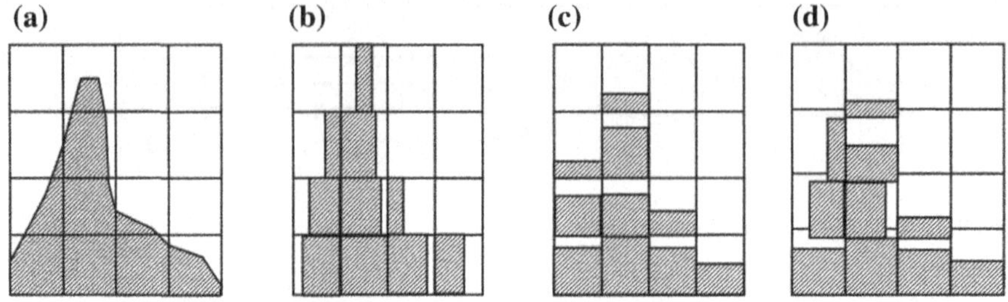

Fig. 9.4 Piecewise constant approximations of the free interface

calculating the flow quantity in the y or the vertical direction, all the free surface segments are parallel to the horizontal direction, as shown in Fig. 9.4c. This is a direction splitting algorithm. When calculating the flow quantity in one direction, only F in the unidirectionally adjacent cells are needed. Once the shape of the free surface is determined in a certain cell and the unidirectionally adjacent cells, the flow quantity of F that flows out of or into this cell can be determined by the local flow state.

The donor-acceptor method proposed by Hirt and Nichols is similar to the SLIC method by using the straight lines to represent the free surface segments. But it is not like the SLIC in which calculation of the flow quantity in one direction is only related to the unidirectionally adjacent cells. In this method, the flow quantity flowing out of or into a cell is related to all the adjacent cells, as shown in Fig. 9.4d. The direction of the free surface being horizontal or vertical depends on the magnitude of the component of the normal vector of free surface in this direction.

2. Piecewise linear volume tracking methods

In 1980s, piecewise linear volume tracking methods have been gradually utilized to reconstruct the free surface, as shown in Fig. 9.5. Oblique lines are used to represent the free surface segment in one grid cell or with the cell boundary as the center. The accuracy has been greatly improved by these methods. The typical ones include a method proposed by Youngs [13] and the Flux Line-segment model for Advection

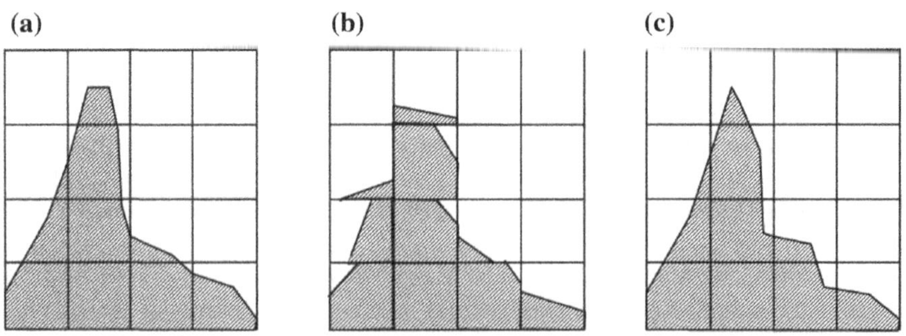

Fig. 9.5 Piecewise linear approximations of the free interface

and Interface Reconstruction (FLAIR) method [14]. Afterwards, Kim and No [15] even used a secondary order curve $y = ax^2 + bx + c$ to approximate the free surface, which was not widely applied due to the complexity of the implementation.

In the method proposed by Youngs, oblique lines are used to represent the free surface segment in one grid cell, as shown in Fig. 9.5b. The specific approach is to determine the direction of the free surface first based on the gradient of a fluid volume function F in the control volume. Then the position of the free surface in the control volume is determined on the basis of the fluid volume function in the control volume. The volume flux through one side of the control volume within a time step is determined by a method similar to the Lagrangian method, instead of directly using the difference method to solve the governing equation of F. Based on the volume of the fluid flowing through the boundaries into the adjacent cells within a time step, the volume function values in the cell and the adjacent cells are modified. Then the shape of the free surface at the next time step is determined.

According to the design criteria proposed by Pilliod and Puckeet [16], the accuracy of the algorithms is second-order in 2D cases, and less than second-order in 3D cases. Since Youngs' algorithm was published, it has been further developed by many of scholars. Colella et al. [17] combined the adaptive mesh refinement and VOF to track the free surface. Kothe et al. [18] applied the 3D algorithm of Youngs to an unstructured grid, and used the second-order Runge-Kutta method for time integral terms.

The FLAIR method [14] was proposed by Ashgitz and Poo. The FLAIR technique is to represent the free surface by using oblique lines. However, the surface is approximated by a set of line segments fitted at the boundary of every two neighboring computational cells, as shown in Fig. 9.5c.

9.4 Discretization and Solution of the Incompressible Reynolds Equations

9.4.1 Solution of the Incompressible Reynolds Equations

Since there is no pressure term in the continuity equation of the incompressible Reynolds equations, the pressure field needs to be guessed or obtained by iteration in the solving process. In order to reduce the computational workload, a two-step projection method proposed by Chorin [11, 19] for solving the Reynolds equations is introduced here.

In the first step, the contribution of the pressure gradient to the velocity field is not taken into account, i.e., neglecting the pressure gradient terms in the momentum equations. Using forward difference scheme for the time terms yields,

$$\frac{\hat{u}_i^{n+1} - u_i^n}{\Delta t} = -u_j^n \frac{\partial u_i^n}{\partial x_j} + g_i + \frac{\partial \tau_{i,j}^n}{\partial x_j} \tag{9.4.1}$$

where, $\tau_{i,j}^n$ consists of the viscous stress and Reynolds stress terms. Thus an intermediate velocity $\hat{u}_i^{n+1}$ can be obtained at each time step. However, this velocity field obtained here does not necessarily satisfy the continuity equation. Therefore, projecting the intermediate velocity onto a space of divergence-free velocity field is needed in the second step,

$$\frac{u_i^{n+1} - \hat{u}_i^{n+1}}{\Delta t} = -\frac{1}{\rho^n}\frac{\partial p^{n+1}}{\partial x_i} \tag{9.4.2}$$

$$\frac{\partial u_i^{n+1}}{\partial x_i} = 0 \tag{9.4.3}$$

By taking the divergence of Eq. (9.4.2), the Poisson equation on the pressure p^{n+1} is obtained and expressed as,

$$\frac{\partial}{\partial x_i}\left(\frac{1}{\rho^n}\frac{\partial p^{n+1}}{\partial x_i}\right) = \frac{1}{\Delta t}\frac{\partial \hat{u}_i^{n+1}}{\partial x_i} \tag{9.4.4}$$

In the projection method, an intermediate velocity is calculated by Eq. (9.4.1) first. Then the pressure filed is obtained by solving Eq. (9.4.4). Finally, the velocity field is corrected by using Eq. (9.4.2).

9.4.2 Discretization of the Computational Domain

When solving water waves, the water depth is generally shallow and the horizontal scale is much larger than the vertical scale, especially in the nearshore regions. Taking an equidistant square grid will result in a very large number of meshes in the horizontal directions. Therefore, a non-equidistant rectangular grid is typically used for spatial discretization.

A discretization method for the computational domain is introduced in the following.

A staggered grid is used for the variables in the computational domain, as shown in Fig. 9.6. The scalar variables such as the pressure p, the turbulent kinetic energy k, the turbulent kinetic energy dissipation rate ε, the eddy viscosity ν_t and the fluid volume function F, are stored in the cell centers of the control volumes. The velocity components are located at the centers of the cell boundaries of the control volumes. The horizontal velocity component is located at the centers of the two perpendicular cell boundaries, and the vertical velocity component is at the centers of the two horizontal cell boundaries. The control volume for solving the equations on scalar variables is shown in Fig. 9.6a. Meanwhile, the control volumes for discretizing the horizontal and perpendicular momentum equations are shown in Fig. 9.6b, c respectively.

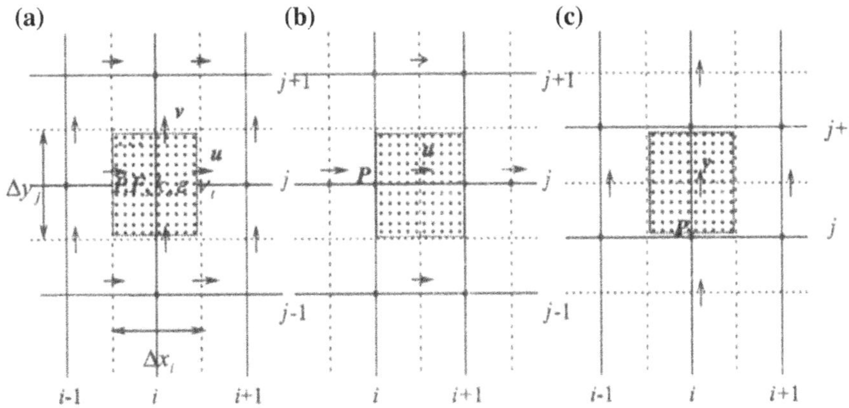

Fig. 9.6 Sketch of staggered grid. **a** Computational cell for p, k, ε, ν_t, F; **b** cell for the velocity component u; **c** cell for the velocity component v

Let the cell spacings in the x and y directions for discretizing the governing equations of scalar variables (the pressure p and $k - \varepsilon$ equations, etc.) be Δx_i and Δy_j. Then $\Delta x_{i+\frac{1}{2}}$ and Δy_j are the cell spacings for the momentum equation in the x direction, Δx_i and $\Delta y_{j+\frac{1}{2}}$ for the momentum equation in the y direction. Where,

$$\Delta x_{i+\frac{1}{2}} = \frac{1}{2}(\Delta x_i + \Delta x_{i+1}), \tag{9.4.5a}$$

$$\Delta y_{j+\frac{1}{2}} = \frac{1}{2}(\Delta y_j + \Delta y_{j+1}) \tag{9.4.5b}$$

The corresponding spacings of the cells upstream read,

$$\Delta x_{i-\frac{1}{2}} = \frac{1}{2}(\Delta x_i + \Delta x_{i-1}), \tag{9.4.6a}$$

$$\Delta y_{j-\frac{1}{2}} = \frac{1}{2}(\Delta y_j + \Delta y_{j-1}) \tag{9.4.6b}$$

9.4.3 Discretization of the Equations

A principle to establish the discretized incompressible Reynolds equations is presented in the following. The detailed procedure can be found in Ref. [20].

When solving the horizontal momentum equation, the control volume shown in Fig. 9.6b is used. Because of the importance of the convection term in the momentum equation, the discretization of it should be paid much attention. The convection term in the momentum equation is discretized at the point $\left(i+\frac{1}{2}, j\right)$, i.e.,

$$\left(u\frac{\partial u}{\partial x}+v\frac{\partial u}{\partial y}\right)_{i+\frac{1}{2},j}=\left(u\frac{\partial u}{\partial x}\right)_{i+\frac{1}{2},j}+\left(v\frac{\partial u}{\partial y}\right)_{i+\frac{1}{2},j} \tag{9.4.7}$$

where, $u_{i+\frac{1}{2},j}$ is located at the storage node for the horizontal velocity component; $v_{i+\frac{1}{2},j}$ is not located at the storage node for the perpendicular velocity component, which should be obtained by interpolation of the values at the adjacent storage nodes.

For the velocity gradient $\left(\frac{\partial u}{\partial x}\right)_{i+\frac{1}{2},j}$ in the convection term, numerical dissipation will be introduced in the calculation if the upwind scheme is employed. Also it is easy to cause numerical instability when using the central scheme. Therefore, it is better to combine the upwind scheme with the central scheme to obtain better accuracy and stability.

$$\begin{aligned}\left(\frac{\partial u}{\partial x}\right)_{i+\frac{1}{2},j}=&\left\{\left[1+\alpha\operatorname{sgn}\left(u_{i+\frac{1}{2},j}\right)\right]\Delta x_{i+1}\left(\frac{\partial u}{\partial x}\right)_{i,j}\right\}/\Delta x_\alpha\\&+\left\{\left[1-\alpha\operatorname{sgn}\left(u_{i+\frac{1}{2},j}\right)\right]\Delta x_{i}\left(\frac{\partial u}{\partial x}\right)_{i+1,j}\right\}/\Delta x_\alpha\end{aligned} \tag{9.4.8}$$

where,

$$\Delta x_\alpha=\Delta x_{i+1}+\Delta x_i+\alpha\operatorname{sgn}\left(u_{i+\frac{1}{2},j}\right)(\Delta x_{i+1}-\Delta x_i) \tag{9.4.9}$$

α is a weight factor of the upwind scheme and the central scheme, $\alpha=0$ for the central scheme and $\alpha=1$ for the upwind scheme; sgn is the sign function.

Similarly, the spatial discretization for the gradient of u in the y direction $\left(\frac{\partial u}{\partial y}\right)_{i+\frac{1}{2},j}$ can be obtained by using the same method.

The viscous stress terms in the momentum equation in the x direction are also discretized at the point $\left(i+\frac{1}{2}, j\right)$, for which the central scheme can be employed.

After the discretized equation of Eq. (9.4.1) is established, the intermediate velocity $\hat{u}_i^{n+1}$ can be obtained. The same discretization method is applied for the discretization of the momentum equation in the y direction. The control volume shown in Fig. 9.6c is used. The discretization center is located at the point $\left(i, j+\frac{1}{2}\right)$. After having the discretized momentum equation in the y direction, the intermediate velocity $\hat{v}_i^{n+1}$ can be obtained.

When solving the Poisson equation of the pressure in the second step, the following equation needs to be solved,

$$\frac{\partial}{\partial x}\left(\frac{1}{\rho^n}\frac{\partial p^{n+1}}{\partial x}\right)+\frac{\partial}{\partial y}\left(\frac{1}{\rho^n}\frac{\partial p^{n+1}}{\partial y}\right)=\frac{1}{\Delta t}\left(\frac{\partial \hat{u}}{\partial x}+\frac{\partial \hat{v}}{\partial y}\right)^{n+1} \tag{9.4.10}$$

The control volume shown in Fig. 9.6a is used. Because the pressure is stored at the center of the grid cell, the central difference scheme is employed for the terms on the left side of Eq. (9.4.10). As the fluid density is stored at the node (i, j), the fluid density at the points $\left(i \pm \frac{1}{2}, j\right)$ and $\left(i, j \pm \frac{1}{2}\right)$ can be obtained by interpolation of the values at the adjacent storage nodes. For instance, the density at $\left(i + \frac{1}{2}, j\right)$ reads,

$$\rho^n_{i+\frac{1}{2},j} = \frac{\rho^n_{i,j}\Delta x_{i+1} + \rho^n_{i+1,j}\Delta x_i}{\Delta x_i + \Delta x_{i+1}} \tag{9.4.11}$$

After getting the pressure at the central nodes at the time layer $(n + 1)$, the velocity field satisfying the continuity equation at the time layer $(n + 1)$ can be obtained by correcting the intermediate velocity $\hat{u}_i^{n+1}$ using Eq. (9.4.2).

When discretizing the k equation in the turbulence model, the control volume is the same as for solving the Poisson equation of the pressure. The turbulent kinetic energy k is stored at the central node (i, j). Combination of the upwind scheme and the central scheme can also be employed for the gradient of k in the x direction.

For discretizing the governing equation of the turbulent kinetic energy dissipation rate ε, the control volume shown in Fig. 9.6a is used. The same difference scheme as that for the k equation is employed.

After solving the governing equations of k and ε, the spatial distribution of the turbulent viscosity at the time layer $(n + 1)$ can be obtained by

$$(\nu_t)_{i,j} = C_\mu \frac{k^2_{i,j}}{\varepsilon_{i,j}} \tag{9.4.12}$$

9.4.4 Procedure of Solving the Discretized Equations

When the governing equations are discretized into the algebraic equations, the unknown variables need to be solved in a certain order.

(1) The initial conditions are given. The initial velocity is zero. The turbulent kinetic energy k and its dissipation rate ε at the initial moment are the values of k and ε at the wave-maker boundary. The initial water surface is still water surface. The initial fluid volume function F is obtained according to the still water surface. The initial pressure is hydrostatic.
(2) Based on the initial velocity or the velocity at the time step n, the intermediate velocity components $\hat{u}^{n+1}_{i+\frac{1}{2},j}$ and $\hat{v}^{n+1}_{i,j+\frac{1}{2}}$ at the time step $n + 1$ are obtained by solving the discretized equation of Eq. (9.4.1).
(3) The pressure at the time step $n + 1$ is obtained by solving the discretized equation of the Poisson equation (9.4.4), which satisfies the vertical dynamic boundary condition at the free surface.

(4) The velocity components $u^{n+1}_{i+\frac{1}{2},j}$ and $v^{n+1}_{i,j+\frac{1}{2}}$ at the time step $n+1$ are solved by the discretized equation of Eq. (9.4.2).
(5) Calculating ε and k at the time step $n+1$, respectively. Then, the turbulent viscosity is solved by Eq. (9.4.12).
(6) The fluid volume function in each grid cell at the time step $n+1$ F^{n+1}_{ij} is obtained by using the Youngs' method. Then the shape and the position of the free surface at this new time step are re-constructed.
(7) Judging whether the calculated time has reached the given computation time. If so, the calculation is completed. Otherwise, going back to the step (2).

In the above calculation procedure, the boundary conditions and the required stability conditions should be considered at the same time.

9.5 Verification and Application of the Reynolds Equation Model for Water Waves

9.5.1 Transformation of Wave Passing Over a Step

To study the local wave transformation with wave-structure interaction, Seabra-Santos et al. [21] and Maxworthy [22] carried out experimental studies on the interaction between a solitary wave and a step, and also numerical simulations by using the CEILW (curvature effects including long-wave) equations [21]. Xie and Tao [8] studied on this problem using the Navier-Stokes equations for the simulations and applying the level-set method for tracking the free surface.

Figure 9.7 shows a sketch of the experiment setup from Seabra-Santos et al. [21]. The incident solitary wave height is $H = 3.65$ cm in front of the step. The still water depth is $d = 20$ cm, and height of the step is $d_1 = 10$ cm. The length of the wave flume is $OB = 30$ m, and the length of the step is $AB = 15$ m. The computation timer started ($t = 0$) when the crest of the incident wave arrived at 3 m away from the front of the step. Six wave gauges were located at $x = 6, 9, 12, 15, 18, 21$ m, respectively in the flume in order to measure the time series of surface elevation.

To verify the Reynolds equation model by using the VOF method for the treatment of free surface, Liu [20] carried out numerical simulations with the same conditions as the experiments, and made comparisons of time series of surface elevation at

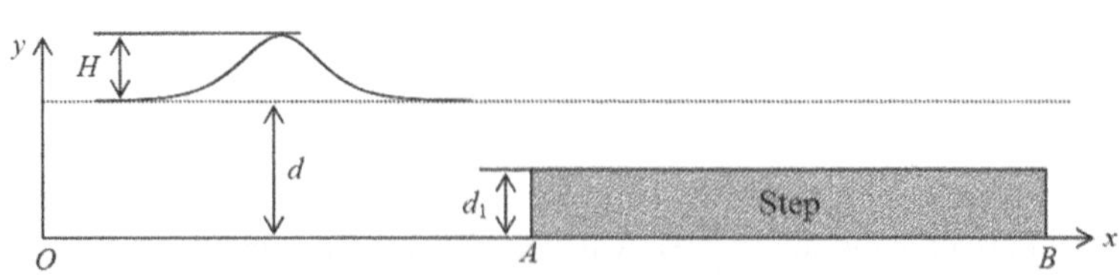

Fig. 9.7 Sketch of the experiment setup for interaction of a solitary wave with a step

the wave gauges. The coordinate system was set as shown in Fig. 9.7 with the coordinate origin located at the point O. Figure 9.8 shows numerical time series of surface elevation at different locations before and after the step compared with the experimental results. It can be seen that the numerical results of the Reynolds equation model are in good agreement with the experimental results. Figure 9.8a–c show the reflected wave propagation. The reflected wave is a dispersive wave with two small wave peaks. The wave trough between the two peaks is below the still water level somewhere. The reflected wave height is not large. Figure 9.8d–f show

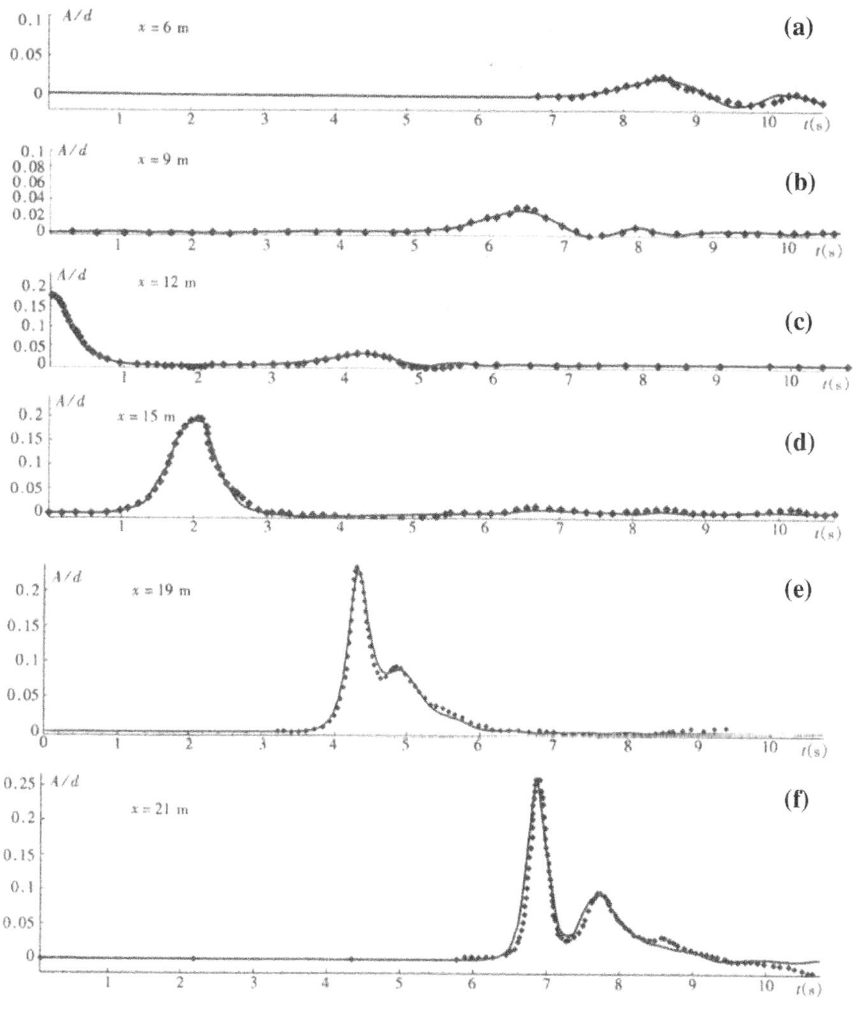

Fig. 9.8 Time series of surface elevation at different locations before and after the step (A is the surface elevation above the still water level; d is the still water dept; t is the time). ◆◆◆ Experimental results [21]; ___ Numerical results from Ref. [20]

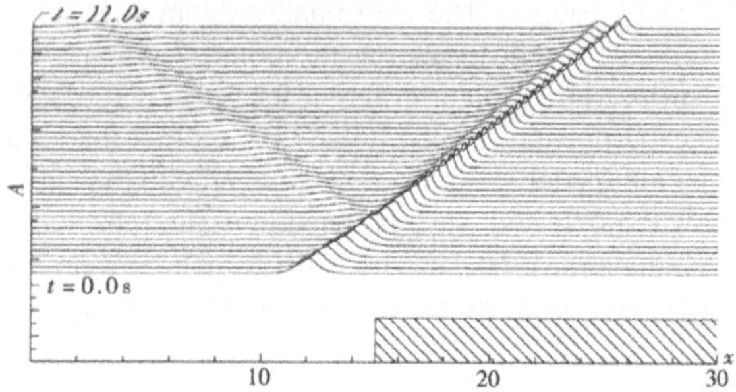

Fig. 9.9 Free surface shapes of the solitary wave passing over a step at different time

the transmitted wave propagation. Due to the effect of the step, a part of the water body behind the solitary wave is separated from the main wave peak because of the flow structure change, which causes the dispersive wave with two peaks.

Figure 9.9 shows the calculated free surface shapes of the solitary wave passing over a step at different time using the Reynolds equation model. The whole process of deformation and propagation of the transmitted and reflected waves for a solitary wave interacted with a step can be seen from the figure. The results agree with those obtained by Seabra-Santos et al. using the CEILW equations [21] and by Xie and Tao [8] using the level-set method.

9.5.2 *Simulation and Verification of Velocity Field Around a Submerged Rectangular Breakwater*

When waves interact with structures, vortices may occur near the structures sometimes. The generation and development of these vortices play an important role in scour around the structures, sediment transport, and pollutant diffusion and dispersion in the adjacent water areas. To verify the correctness of the Reynolds equation model simulating the flow field around a structure, Liu [20] used a Reynolds equation model to simulate the interaction between a solitary wave and a submerged rectangular breakwater, which had been experimentally studied in a wave flume by Fei and Lee [23].

Figure 9.10 shows a sketch of the experiment setup in a physical flume for studying the interaction between a solitary wave and a submerged rectangular breakwater by Zhuang et al. [21]. The rectangular breakwater is 15 in. (0.381 m) long and 4.5 in. (0.1143 m) high. The still water depth is $d = 9$ in. (0.2286 m). Apparently, the structure occupies 50% of the still water depth in the vertical direction. The incident solitary wave height is $H = 2.7$ in. (0.06858 m) so that the ratio of wave height to water depth is $H/d = 0.3$. Titanium dioxide was used as a tracer and sprayed out from a nozzle behind the structure in the experiments, as shown in Fig. 9.10, to track the vortex generated behind the structure when a solitary wave passed over the structure. Meanwhile, the time series of the horizontal and vertical velocity components at the

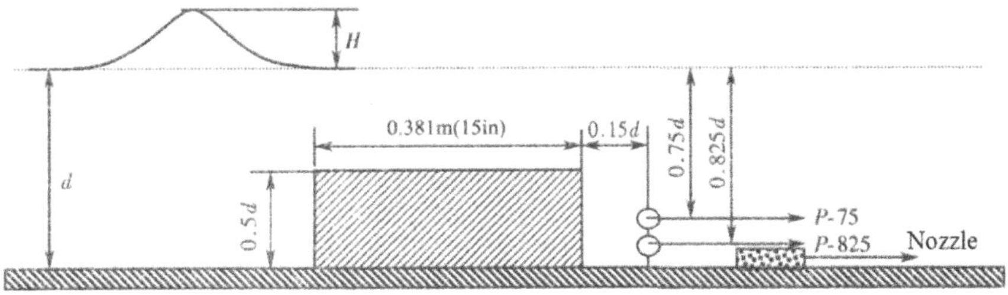

Fig. 9.10 Sketch of the experiment setup for interaction of a solitary wave and a submerged rectangular breakwater

points P-75 and P-825 were measured by the two-component Doppler velocimeter with four light beams. The points P-75 and P-825 are located at $0.15d$ away from the vertical edge of the structure, and $0.75d$ and $0.825d$ below the still water level respectively.

Figure 9.11 shows the comparison of dimensionless velocity at P-825. The horizontal axis for time is normalized by $\sqrt{g/d}$. The vertical axis for the horizontal and vertical velocity components respectively is normalized by $\sqrt{gd}$. In Fig. 9.11, "◆◆◆" represents for the experimental results, the solid line for the numerical results of the Reynolds equation model, the dash-dotted line for the numerical results by Fei and Lee [23] using the combined method (the vortex function method in the local area behind the structure, and the potential flow method in the other area), and the dotted line for the numerical results by Fei and Lee [23] using the potential flow method in the whole domain. It can be seen that, there is a big difference between the numerical results of the potential flow method and the experimental results. There

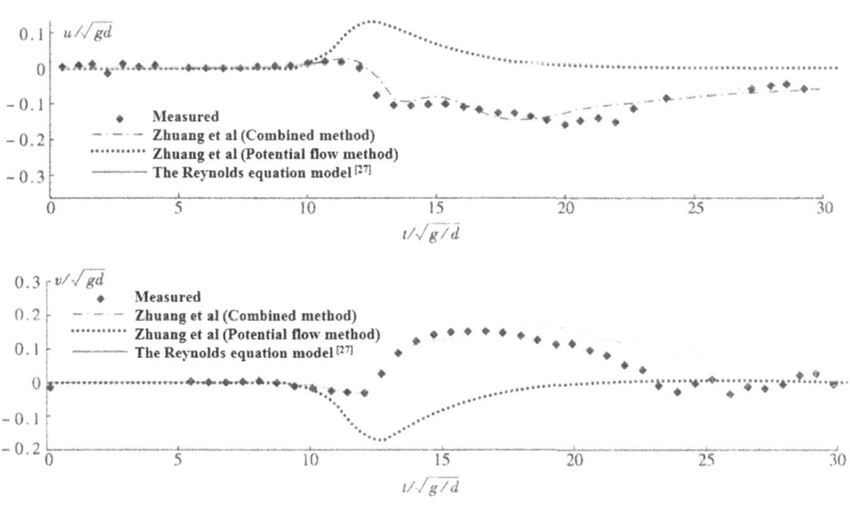

Fig. 9.11 Dimensionless time series of horizontal and vertical velocity components at P-825

is no vortex behind the breakwater using the potential flow method to simulate the interaction between the solitary wave and the rectangular breakwater. The viscous flow model described in this chapter can produce results that agree well with the actual situation, in which both the horizontal and vertical velocity components turn directions in the process of interaction between the solitary wave and the structure (the horizontal velocity component turns from the positive to the negative, the vertical velocity component turns from the negative to the positive). This indicates that vortex occurs behind the breakwater. Figure 9.12 shows the comparison of the velocity at *P*-75. The trend of the velocity change at *P*-75 is almost the same as that at *P*-825.

Figure 9.13 shows a comparison of simulated flow field at a certain time by the Reynolds equation model with the experimental result of vortex displayed by the transport of titanium dioxide from the nozzle. It shows clearly that the Reynolds

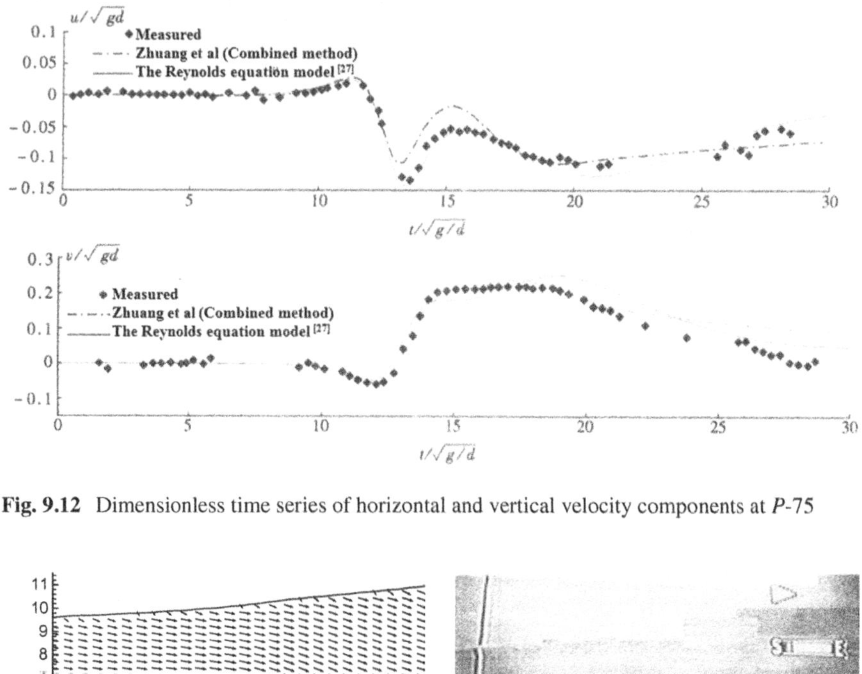

Fig. 9.12 Dimensionless time series of horizontal and vertical velocity components at *P*-75

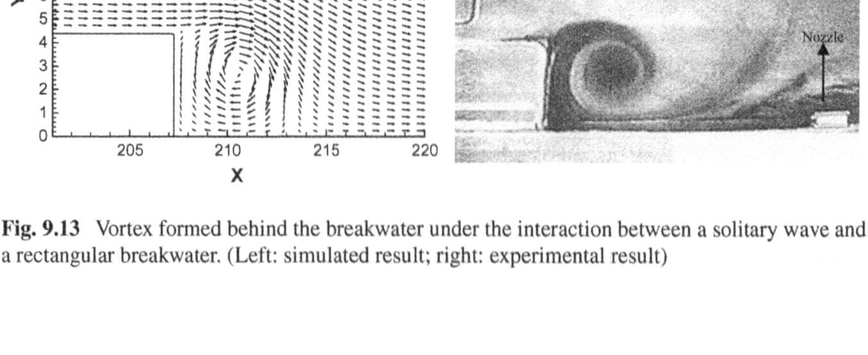

Fig. 9.13 Vortex formed behind the breakwater under the interaction between a solitary wave and a rectangular breakwater. (Left: simulated result; right: experimental result)

equation model can simulate the development of the detached vortex behind the structure under the wave action well.

9.5.3 Simulation and Verification of Wave Uplift Forces on the Wharf Upper-Structure

In the study of wave-structure interaction, wave force is one of the concerns for engineering designs. For the wharf upper-structure, the lower support columns are omitted under a certain water level, and the top of the structure is above the still water level. When the wave crest arrives, the wave is transmitted below the structure so that the structure is subjected to the wave impact. French [24] studied solitary wave forces on horizontal platforms experimentally. Lai et al. [25] calculated the wave forces on the above structure by using the finite element method based on a potential flow model. Liu [20] used the Reynolds equation model to simulate the same cases, and compared the simulated wave forces with the results in Refs. [24, 25].

Figure 9.14 shows a sketch of the experiment setup. The still water depth is $d =$ 15 in. (0.381 m). A rectangular block with length of $L = 4d$ (1.524 m) is located at $S = 0.2d$ (0.0762 m) above the still water level. The wave forces on the block have been calculated for the solitary wave cases with the incident wave height of $H =$ $0.24d$, $0.32d$ and $0.4d$.

The calculation conditions are the same as the experimental conditions. The computational domain is 14 m long in the horizontal direction, and 0.6 m high in the vertical direction. The coordinate system and definitions are shown in Fig. 9.14. The lower left corner of the block is (7.229, 0.4572 m). The computation of the velocity and pressure starts when the crest of the solitary wave arrives at $x = 3.5$ m. In the following figures of wave forces, time starts from when the solitary wave reaches the block. When calculating the wave forces, the thickness of the block is set to unit length.

Figures 9.15, 9.16 and 9.17 show the calculated vertical wave forces using the Reynolds equation model [20] compared with the experimental results [24] and the calculated results in Ref. [25] for the three cases of $H/d = 0.24$, 0.32 and 0.40,

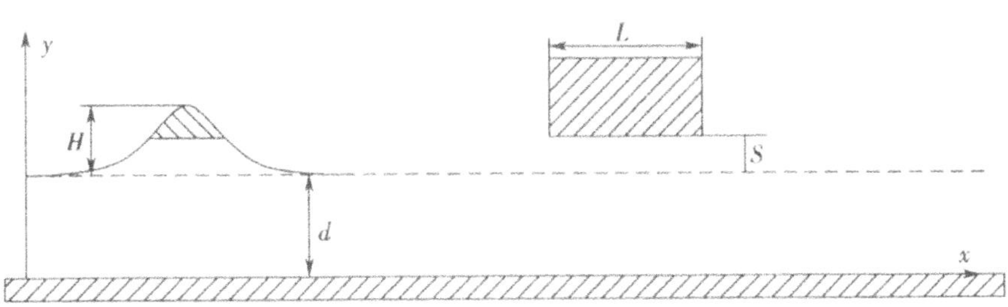

Fig. 9.14 Sketch of the experiment setup and definitions

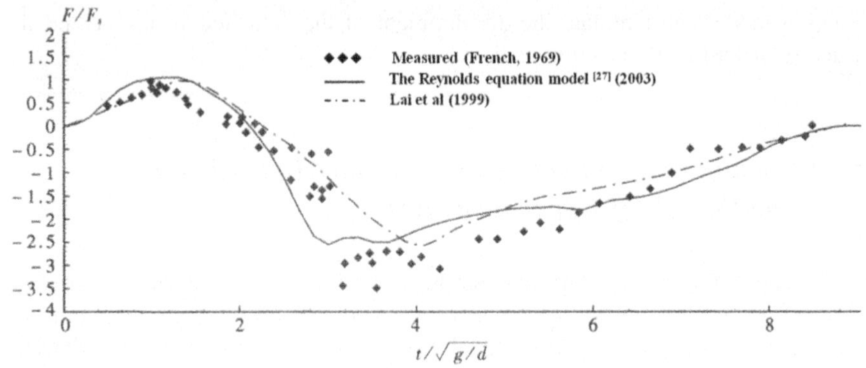

Fig. 9.15 Calculated uplift force using the Reynolds equation model compared with the measurements and the other calculated results ($H/d = 0.24$)

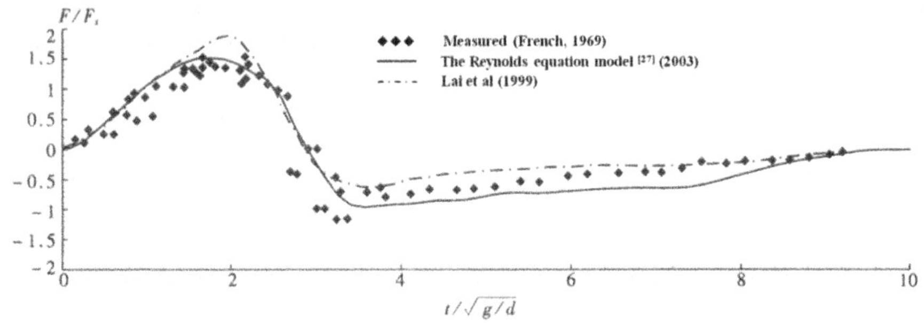

Fig. 9.16 Calculated uplift force using the Reynolds equation model compared with the measurements and the other calculated results ($H/d = 0.32$)

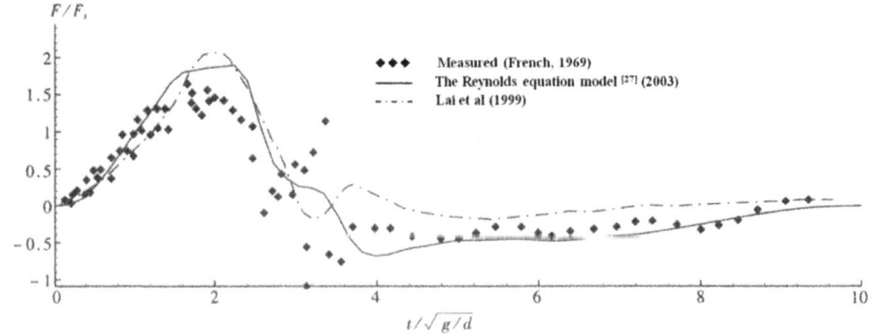

Fig. 9.17 Calculated uplift force using the Reynolds equation model compared with the measurements and the other calculated results ($H/d = 0.40$)

respectively. The magnitudes of the forces are normalized by the weight of the shaded water body in Fig. 9.14 (the water body above the bottom of the rectangular block with a thickness of 1 m) F_s. The horizontal time axis is normalized by $\sqrt{g/d}$.

From the three figures of the uplift forces on the structure by solitary waves, it can be seen that the numerical results of the Reynolds equation model are in agreement with the experimental results and the calculated results from the reference. Due to the consideration of the fluid viscosity, the numerical results of the Reynolds equation model are relatively smaller than the calculated results of the potential flow model, but closer to the experimental results. By comparing the three figures, it is found that, the time of the wave force direction changing from upward to downward is associated with the incident wave height under the condition of constant still water depth. When the incident solitary wave height is relatively small, the water body above the bottom of the structure is less. It takes shorter time for the water body at the wave crest to enter below the structure, so that the period of the upward vertical force is shorter. When the incident solitary wave height is relatively high, there is much more water body above the bottom of the structure, so that it takes longer time for the water body at the wave crest to get below the structure, which causes a longer period of the upward vertical force. Meanwhile, the magnitude of the wave force on the structure is greatly associated with the incident wave height. When the incident solitary wave height is smaller, the uplift force is lower but the downward force is higher.

9.6 Numerical Simulation of Wave-Structure Interaction Using a 3D Reynolds Equation Model

The flow velocity at the head of the breakwater or pierhead varies greatly and unevenly under wave action. Vortices are formed in general around the head of the breakwater, where the flow structure is complex, leading to local scour easily. Even more, engineering accidents can happen due to the instability of the breakwater head caused by local scour. Therefore, the wave field and flow field around the head of breakwater are very important not only in design phase, but also in construction phase. After the breakwater is built, the pollutant transport in the project area is of great significance to project management as well.

The local wave and flow around the head of breakwater are typical 3D problems, which cannot be simulated by a 2D numerical model. The numerical simulation of local wave and flow near the head of breakwater under wave action by using a 3D Reynolds equation model is described in this section.

Liu [20] used a Reynolds equation model to simulate wave diffraction caused by a solitary wave passing through a gate between two piles for the study of variation of wave surface elevation and flow field around breakwater gap.

Figure 9.18 shows a sketch of the numerical model setup [20]. The still water depth in the computational domain is $d = 5.0$ m. The incident solitary wave height

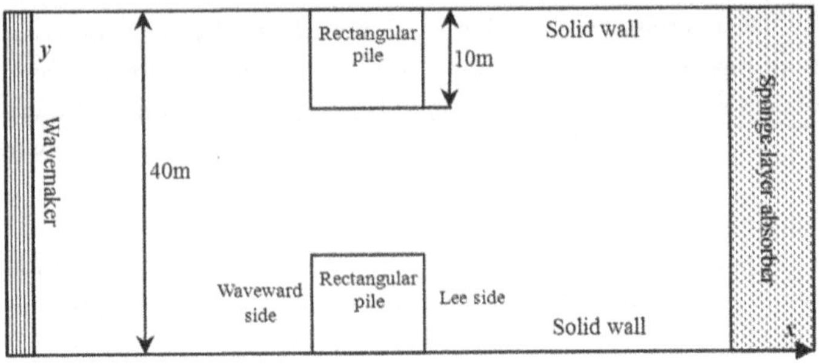

Fig. 9.18 Sketch of the numerical model setup

is $H = 1.0$ m. The numerical wave basin is 40 m wide and 300 m long. The two rectangular piles are located at $x = 150.0$ m. Each of them is 10 m wide and 10 m long. The elevation of the top of each pile is 7.5 m. The boundaries of the two sides of the basin and the piles are set to be solid walls. The right side of the wave basin is open boundary with sponge layers. The width of the sponge layers is approximately the effective wavelength of the solitary wave. The whole wave basin is taken as the computational domain instead of setting a symmetrical boundary at the central plane to test the symmetric merit of the algorithm. The solitary wave is generated by a wavemaker on the left side of the basin. The time axis is set to start from when the wave crest arrives at $x = 170.0$ m.

9.6.1 Surface Elevation of a Solitary Wave Passing Through a Gate Between Two Piles

Figure 9.19 shows the variation of wave surface elevation around the gate between the two piles with time interval of 1 s. The simulated wave surface is generally symmetrical to the central axis. Apart from the wave diffraction on the leeside, complex wave deformation occurs at the gate between the two piles. In the incident wave side, wave run-up occurs due to the interaction between the solitary wave and the two piles. Due to the wave diffraction around the pile head, the maximum run-up height near the solid boundary of the basin is higher than that at the pile head. It is clear to see the wave diffraction and the oscillatory deformation behind the piles in the leeside. Around the corners of the gate, the wave surface changes significantly. The complex oscillatory deformation still occurs around the piles when the reflected and transmitted waves propagate away from the piles.

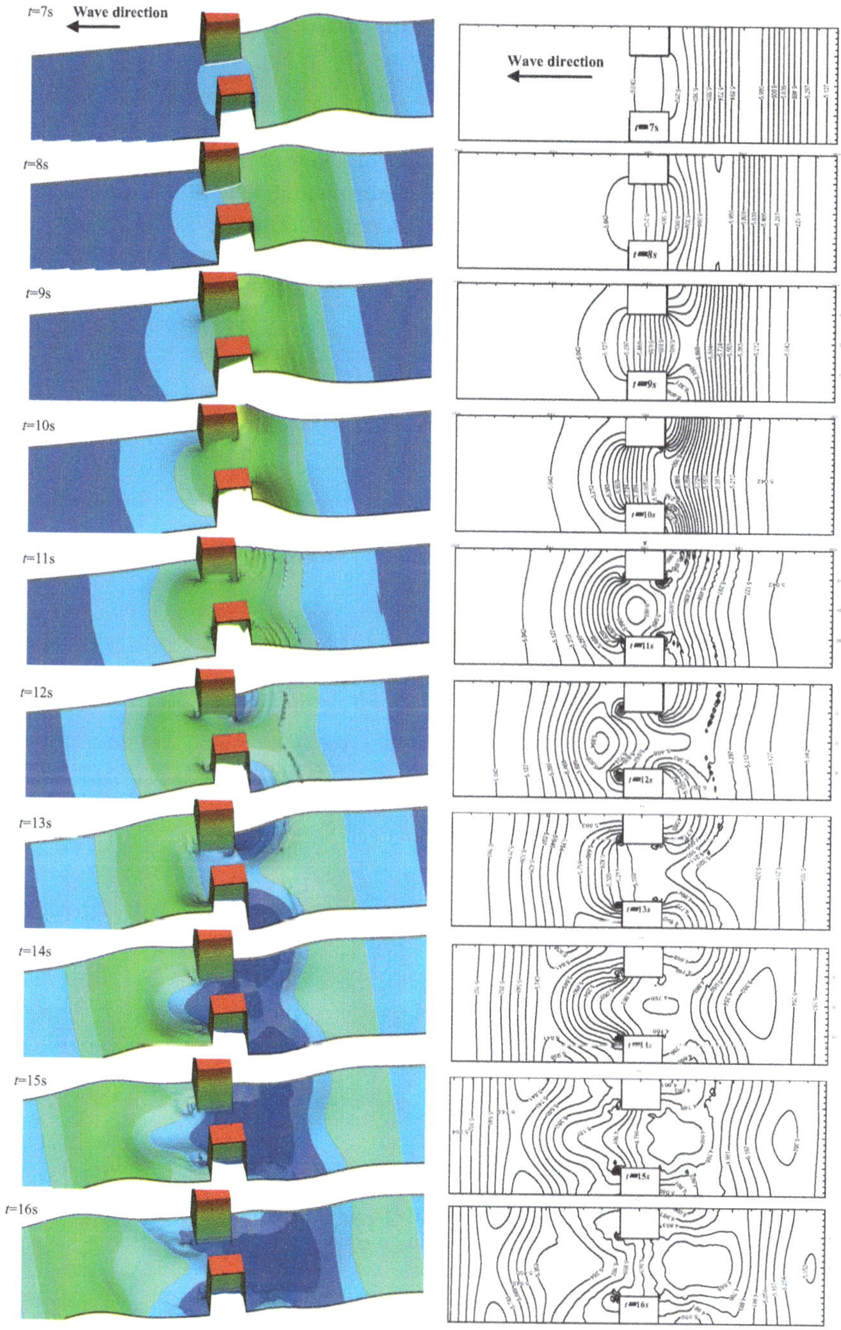

Fig. 9.19 Variation of the wave surface elevation around the two rectangular piles. (Left: wave surface; right: surface elevation contour lines)

9.6.2 Pressure and Velocity on a Horizontal Section

Figure 9.20 shows the variation of pressure and velocity around the two piles on a horizontal section close to the still water level ($z = 4$ m) with time interval of 1 s (corresponding to the time steps in Fig. 9.19). The shape of the pressure distribution at each time step is generally similar to the corresponding one of the wave surface elevation. Regarding the velocity on the horizontal section of $z = 4$ m, small-scale vortices occur in the incident wave side in front of the piles due to the diffraction.

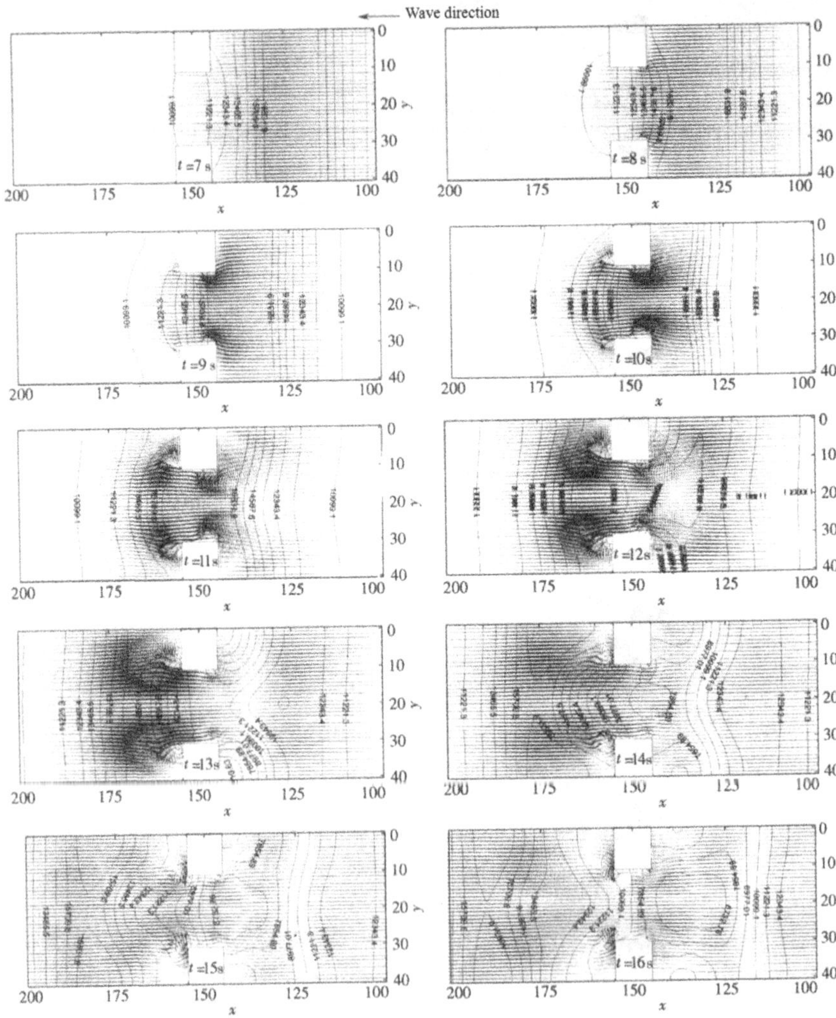

Fig. 9.20 Variation of the velocity and pressure (contour lines) around the two rectangular piles on the horizontal section of $z = 4$ m

But in the leeside behind the piles, relatively large-scale vortices are formed, and their decay is relatively slow.

9.6.3 Velocity Variation on Vertical Transects

Figure 9.21 shows the velocity variation around the two piles on a vertical transect ($x = 142$ m) in the incident wave side. When the solitary wave-front starts to interact with the two piles, the velocity becomes upward and the surface rises up. Subsequently, because of the wave run-up in the incident wave side and the wave diffraction around

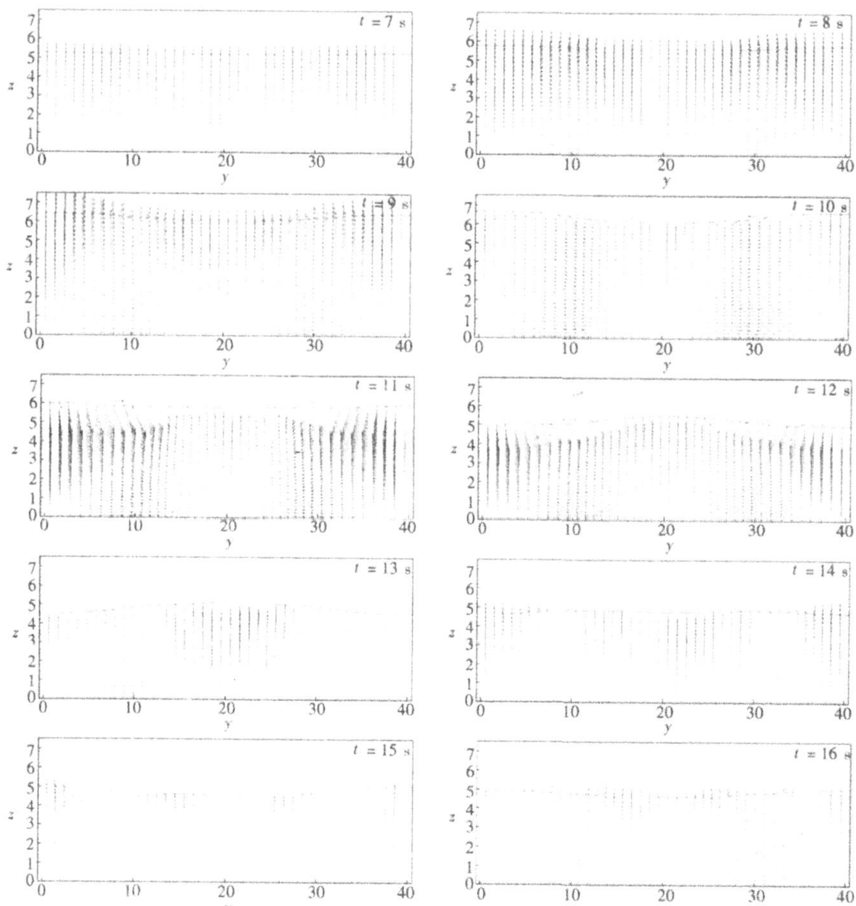

Fig. 9.21 Variation of the surface and velocity around the two piles on the vertical transect of $x = 142$ m. →: 1 m/s

the pile heads, the velocity direction becomes obliquely upward pointing to the central axis of the gate. After the wave run-up reaches the highest point in the incident wave side, the water body flows downward gradually under the action of gravity. Meanwhile, because of the diffraction, the velocity direction is obliquely downward pointing to the central axis of the gate. Afterwards, the speed decreases rapidly on the vertical $(y - z)$ transect. The wave surface on the section oscillates due to the gravity, and the velocity varies in both speed and direction with the wave surface.

Figure 9.22 shows the variation of the surface and velocity around the two piles on a vertical transect ($x = 158$ m) in the lee side. The process of surface variation is opposite to that in the incident wave side. At the beginning, the surface elevation rises rapidly at the gate without blocking effect of the piles. Because of the diffraction in

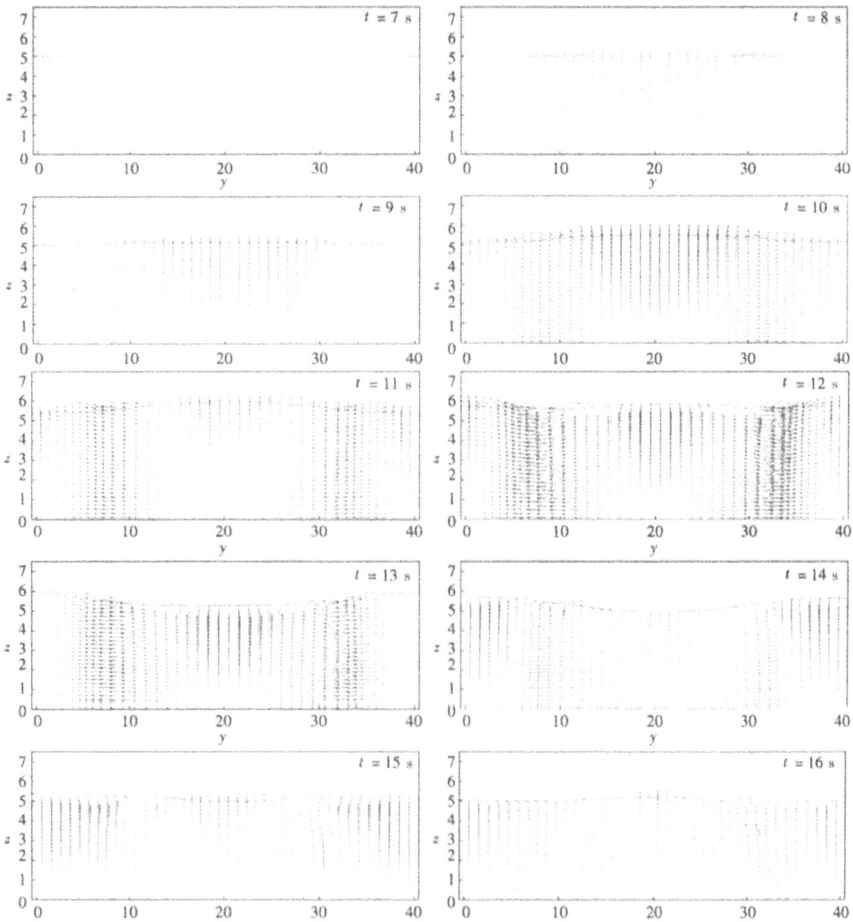

Fig. 9.22 Variation of the surface and velocity field around the two piles on the vertical transect of $x = 158$ m. →: 1 m/s

the lee side, the velocity direction is obliquely upward departing from the gate, and the surface rises up as well. Subsequently, the surface falls down and the velocity direction is obliquely downward away from the gate because of the wave reflection from the solid boundaries of the basin. As the two sides of the basin are taken as solid boundaries, the transmitted wave surface away from the gate in the leeside oscillates and the velocity oscillates along with the surface as well.

Figure 9.23 shows the variation of the surface and velocity on a horizontal transect close to the gate ($y = 7$ m) in the process of the interaction between the solitary wave and the two piles. The diffracted wave height is relatively high in the leeside because of the diffraction near the gate, and the section being close to the gate. Also because

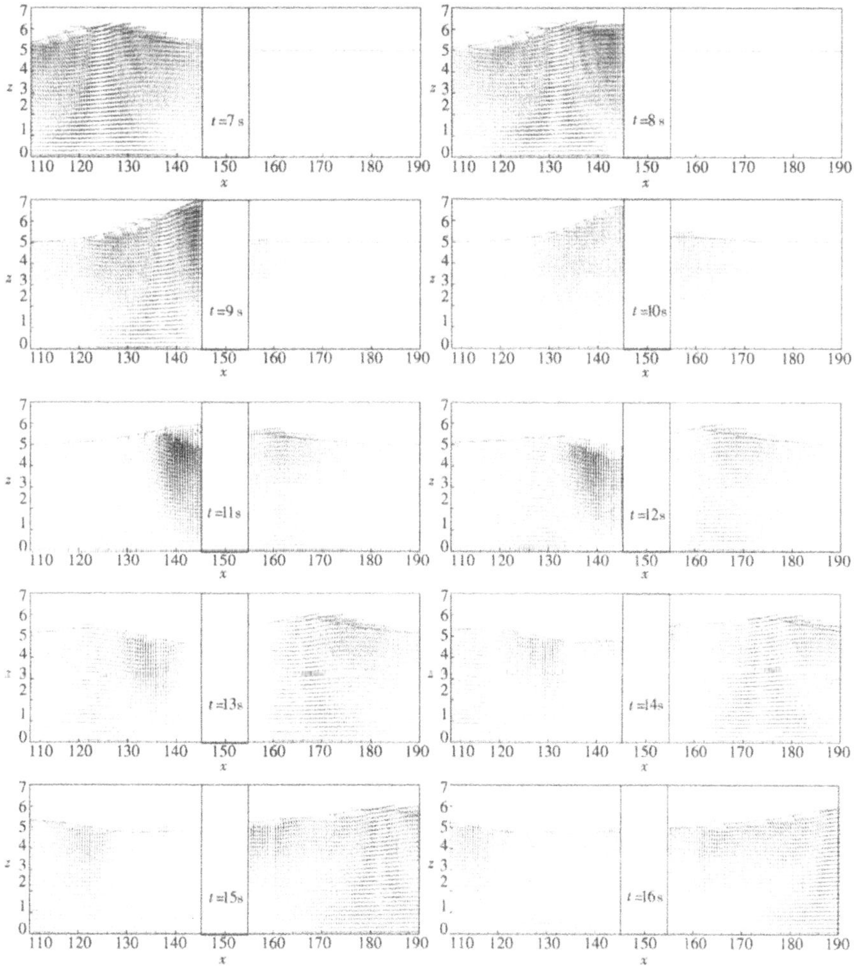

Fig. 9.23 Variation of the surface and velocity around the two piles on the horizontal transect of $y = 7$ m. →: 1 m/s

of the diffraction near the gate, the oscillation amplitude of the reflected wave is high in the leeside, and sometimes the surface elevation is even lower than the still water level.

References

1. Launder BE, Spalding DB. The numerical computation of turbulent flows. Comput Methods Appl Mech Eng. 1974;3(2):269–89.
2. Lin P, Liu PL-F. Numerical study of breaking waves in the surf zone. J Fluid Mech. 1998;359:239–64.
3. Osher SJ, Sethain JA. Fronts propagating with curvature-dependent speed: algorithms based on Hamilton-Jacobi formulations. J Comput Phys. 1988;79(1):12–49.
4. Sussman M, Smereka P, Osher SJ. A level set approach for computing solutions to incompressible two-phase flow. J Comput Phys. 1994;114(1):146–59.
5. Chen Y, Bi Y. Spatial spline and body-fitted coordinates. In: Proceedings of the 1995 national conference on hydrodynamics; 1995.
6. Zhu J. The boundary integral equation method for solving the Dirichlet problem of a biharmonic equation. Math Numer Sin. 1984;3:278–88.
7. Yuan D, Tao J. Simulation of the flow with free surface by level set method. Acta Mech Sin. 2000;32(3):264–71.
8. Xie W, Tao J. Interaction of a solitary wave and a front step simulated by level set method. Appl Math Mech. 2000;21(7):686–92.
9. Hirt CW, Nichols BD. Volume of fluid (VOF) method for the dynamics of free boundaries. J Comput Phys. 1981;39:201–25.
10. Noh WF, Woodward PR. SLIC (simple line interface method). Lect Notes Phys. 1976;59:273–85.
11. Chorin AJ. Flame advection and propagation algorithms. J Comput Phys. 1980;35(1):1–11.
12. Barr PK, Ashurst WT. An interface scheme for turbulent flame propagation. Technical Report SAND 82-8773, Sandia National Laboratories; 1984.
13. Youngs DL. Time-dependent multi-material flow with large fluid distortion. Numerical methods in fluid dynamics. New York: Academic Press; 1982. p. 273–85.
14. Ashgriz N, Poo JY. FLAIR: flux line-segment model for advection and interface reconstruction. J Comput Phys. 1991;93(2):449–68.
15. Kim SO, No CH. Second-order model for free surface convection and interface reconstruction. Int J Numer Methods Fluids. 1998;26:79–100.
16. Pilliod JE, Puckett EG. Second-order accurate volume of fluid algorithms for tracking material interfaces. J Comput Phys. 1997;199(2):465–502.
17. Colella P, et al. A numerical study of shock wave refractions at a gas interface. In: Pulliam T, editors. Proceedings of the AIAA ninth computational fluid dynamics conference; 1989. p. 426–39.
18. Kothe DB, et al. Volume tracking of interfaces having surface tension in two and three dimensions. Technical Report AIAA. 96-0859;1996.
19. Chorin AJ. Numerical solution of the Navier-Stokes equations. Math Comput. 1968;22.
20. Liu C. Using unsteady Reynolds Navier-Stokes equations to simulate the water wave interaction with coastal structures. Dissertation for Ph.D. degree, Tianjin: Tianjin University; 2003.
21. Seabra-Santos FJ, Renouard DP, Temperville AM. Numerical and experimental study of the transformation of a solitary wave over a shelf or isolated obstacle. J Fluid Mech. 1987;176:117–34.
22. Maxworthy T. Experiments on collisions between solitary waves. J Fluid Mech. 1976;76(1):177–85.

23. Fei Z, Lee J-J. A viscous rotational model for wave overtopping over marine structure. In: Proceedings of 25th international conference on coastal engineering, ASCE, 1996:2178–91.
24. French JA. Wave uplift pressure on horizontal platforms, Report No. KH_R_19, W. M. Keck Laboratory of Hydraulics and Water Resources, Pasadena: California Institute of Technology; 1969.
25. Lai CP, Lee J-J. Interaction of finite amplitude waves with platforms or docks. J Waterw, Port, Coast, Ocean Eng. 1989;115(1):19–39.

Chapter 10
Numerical Wave Flume and Numerical Wave Basin

10.1 Introduction

10.1.1 Numerical Experiments of Water Waves on Computer

In the design of coastal and marine engineering, physical model tests in laboratory are usually carried out for the study of wave action on structures. For instance, the planar, sectional, or whole physical model tests are needed for cases such as the harbor layout, artificial island shape and revetment designs, breakwaters, design and construction of offshore oil platforms, pipelines, and also the flow field, vortex field, and movements of sediment and contaminants near structures under wave action.

Wave flumes and wave basins are the most fundamental facilities for physical model tests in laboratory. The performance of the wave flume/basin mainly depends on the performance of the wavemaker, the absorber, the extent of the test section, and the measuring instruments. It is very expensive to build a wave flume or basin with good performance in laboratory. Furthermore, physical model tests are subject to scale limitations. For instance, the gravitational acceleration and density are consistent in prototype and model. When water is used for the experiments in the gravitational field, only one force similarity, either for the gravity or for the viscous force, can be achieved according to the similitude theory. Then physical model tests will encounter great difficulties when solving the problems of which both the gravity and the viscous force are important, or with very large spatial extent.

With the development of computer and computing technologies, great progresses have been made in theories, methods, and applications of numerical models to simulate the interaction of water waves with structures, as described in the previous chapters of this book. Using modern numerical calculation theories, methods, and techniques, the establishment of a dedicated software system, including a numerical wave flume for sectional wave experiments, a horizontal two-dimensional (2D) numerical wave basin for planar experiments, and a three-dimensional (3D)

J. Tao, *Numerical Simulation of Water Waves*, Springer Tracts in Civil Engineering, https://doi.org/10.1007/978-981-15-2841-5_10

numerical wave basin for the whole experiments on the wave-structure interaction, will play a significant role in engineering design and scientific research. Numerical experiments on the wave-structure interaction in the numerical flume or basin do not suffer from scaling problems and are free of the spatial extent limitation. In numerical models, the numerical simulations can be carried out in a large region with the wave direction easily changed. Also, the prototype is simulated directly without the problem of scale. The greatest advantages of numerical wave flumes and basins are short experimental time and easily comparing various layouts for design optimization.

The integrated use of physical and numerical wave flumes/basins, which utilizes a numerical wave flume/basin for comparing various layouts and a local physical model for the optimized layout, can deliver even better results.

10.1.2 Key Components of a Numerical Wave Flume/Basin

1. Mathematical model and software system

A mathematical model is the core of a numerical wave flume/basin, which can be vertically 2D, horizontally 2D or 3D depending on the requirement of the wave flume/basin. The mathematical model consists of the governing equations and well-posed initial and boundary conditions. The governing equations may be the N–S equations, the Reynolds equations, the mild-slope equation, the Boussinesq equations, the Laplace equation, or others. The mathematical model needs to use a high-accuracy numerical method to discretize the governing equations. The commonly used methods include finite difference methods, finite element methods, boundary element methods, and finite volume methods. The computational domain can be discretized with structured or unstructured grids. The fundamental requirement of establishing a numerical wave flume or basin is to choose a mathematical model which describes the physical problem correctly and meets the requirements in mathematics, and the corresponding numerical method which requires high accuracy, low numerical dissipation and dispersion, and fast calculation speed.

2. Numerical wavemaker system

Whether the numerical wavemaker system can produce the waves with expected wave characteristics or spectrum in prescribed location is one of the key issues of the numerical wave flume/basin. At present, there are mainly three types of wave generation methods: (1) giving the wave surface or velocity at the incident boundary; (2) a piston-type or flap-type wave generation method; (3) adding a source/sink term to the governing equations. The strengths and weaknesses of various methods and the details will be described below.

3. Non-reflecting open boundary system

In either physical or numerical wave flumes/basins, the experiment extents or the computational domains are fixed. All the open boundaries need non-reflecting condition. Similar to placing passive absorbers at the open boundaries in a physical model, the non-reflecting condition should be treated at the open boundaries in a numerical wave flume/basin. For linear waves, the radiation condition (the Sommerfeld condition) can be used as the non-reflecting condition. However, the radiation condition for nonlinear waves is not mature. Further studies are needed for the non-reflecting boundary condition, which are mostly treated by numerical techniques at present.

4. Pre- and post-processing system

The pre-processing is to generate the bathymetry and structures in the computational domain and input initial conditions and the related parameters. A good numerical wave flume/basin should have a sound pre-processing system and a friendly human-computer interface. Users can take advantages of this system to modify the bathymetry, structures, and various parameters conveniently.

The post-processing is mainly to extract the numerical simulation results according to the users' requirements and represent a large amount of data by various graphics or dynamic demonstration. The post-processing also needs a friendly user interface.

5. Graphic display system

Outputs of the numerical simulation are generally time series of the output data on spatial grid nodes in tabular form, which is difficult to understand for users. In order to visually export the output results, the graphic display can be contour lines or a 3D dynamic demonstration for a certain variable. According to the characteristics of great spatial and temporal variations of wave-structure interaction, and by means of the image display technology, selecting a suitable tool to build a graphic display software is an important part of a numerical wave flume/basin. The grid spacing and time step for plot data collection are usually different from those for the simulations. They should be determined carefully by comparison.

10.2 Numerical Wave Flume Based on the Reynolds Equations

The numerical wave flume is mainly used for vertical 2D (sectional) problems which can be solved by a physical model in a wave flume. At present, there are mainly two kinds of numerical wave flumes including the ones based on the potential flow theory and the ones based on viscous flow theory. The flumes based on potential flow theory can simulate the wave deformation process and the wave forces on the structures in some cases, but they cannot simulate the wave breaking near structures or the development of the vortex system. Using the flumes based on viscous flow theory, not only the wave forces on the structures, the nonlinear wave deformation around the structures, but also the complex flow filed near the structures can be obtained. With developments of computer and computing technologies, the numerical wave flumes based on the viscous flow theory have been more widely used.

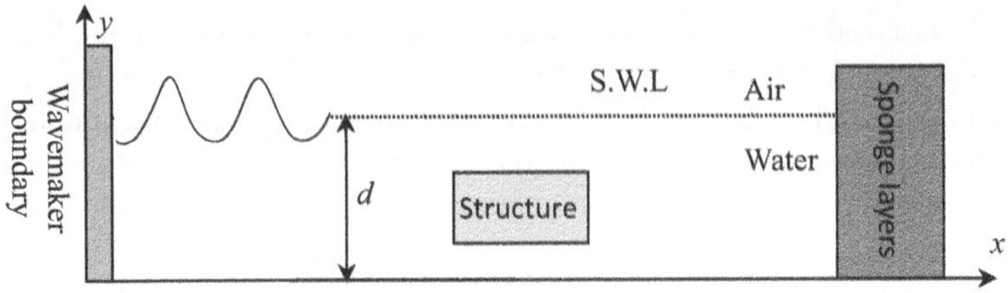

Fig. 10.1 Sketch of the numerical wave flume

A numerical wave flume based on the viscous incompressible flow theory is introduced in the following. There are several key issues on the numerical wave flume including free surface tracking, solving the equations for incompressible viscous flow, and a high-accuracy numerical method. In the numerical wave flume, waves may break near the structure due to the wave-structure interaction, so that the flow is turbulent. Therefore, a proper turbulence model is also needed in the mathematical model for the numerical wave flume.

In order to generate given incident waves at the pre-arranged location in the numerical wave flume, a numerical wave generation zone is needed in the numerical wave flume. An open boundary processing zone is also needed for the treatment of the non-reflecting open boundary condition. Figure 10.1 shows a sketch of the numerical wave flume.

10.2.1 Governing Equations of the Mathematical Model

The 2D Reynolds equations are used for the governing equations because of the turbulent flow due to the wave-structure interaction.

The 2D Reynolds equations have been given in Eqs. (9.2.12)–(9.2.14). ν_t or μ_t can be obtained by using a certain turbulence model, which makes it possible to determine the final form of the governing equations.

$$\frac{\partial \bar{u}}{\partial x} + \frac{\partial \bar{v}}{\partial y} = 0$$

$$\frac{\partial \bar{u}}{\partial t} + \frac{\partial \bar{u}\bar{u}}{\partial x} + \frac{\partial \bar{u}\bar{v}}{\partial y} = f_x - \frac{1}{\rho}\frac{\partial \bar{p}}{\partial x} - \frac{2}{3}\frac{\partial k}{\partial x} + \frac{\partial}{\partial x}\left[2(\nu + \nu_t)\frac{\partial u}{\partial x}\right] + \frac{\partial}{\partial y}\left[(\nu + \nu_t)\left(\frac{\partial \bar{u}}{\partial y} + \frac{\partial \bar{v}}{\partial x}\right)\right]$$

$$\frac{\partial \bar{v}}{\partial t} + \frac{\partial \bar{v}\bar{u}}{\partial x} + \frac{\partial \bar{v}\bar{v}}{\partial y} = f_x - \frac{1}{\rho}\frac{\partial \bar{p}}{\partial y} - \frac{2}{3}\frac{\partial k}{\partial y} + \frac{\partial}{\partial x}\left[(\nu + \nu_t)\left(\frac{\partial \bar{u}}{\partial y} + \frac{\partial \bar{v}}{\partial x}\right)\right] + \frac{\partial}{\partial y}\left[\left(2(\nu + \nu_t)\frac{\partial \bar{v}}{\partial y}\right)\right]$$

The $k-\varepsilon$ turbulence model is widely used, which can close the Reynolds equations. The $k-\varepsilon$ model equations have been given in Eqs. (9.2.19) and (9.2.20).

$$\frac{\partial k}{\partial t}+\frac{\partial \bar{u}k}{\partial x}+\frac{\partial \bar{v}k}{\partial y}=\frac{\partial}{\partial x}\left[\left(\nu+\frac{\nu_t}{\sigma_k}\right)\frac{\partial k}{\partial x}\right]+\frac{\partial}{\partial y}\left[\left(\nu+\frac{\nu_t}{\sigma_k}\right)\frac{\partial k}{\partial y}\right]+G-\varepsilon$$

$$\frac{\partial \varepsilon}{\partial t}+\frac{\partial \bar{u}\varepsilon}{\partial x}+\frac{\partial \bar{v}\varepsilon}{\partial y}=\frac{\partial}{\partial x}\left[\left(\nu+\frac{\nu_t}{\sigma_\varepsilon}\right)\frac{\partial \varepsilon}{\partial x}\right]+\frac{\partial}{\partial y}\left[\left(\nu+\frac{\nu_t}{\sigma_\varepsilon}\right)\frac{\partial \varepsilon}{\partial y}\right]$$
$$+C_{\varepsilon 1}\frac{\varepsilon}{k}G-C_{\varepsilon 2}\frac{\varepsilon^2}{k}$$

in which

$$G=2\nu_t S_{ij}\frac{\partial u_i}{\partial x_j}=\nu_t\left\{\left(\frac{\partial \bar{u}}{\partial y}+\frac{\partial \bar{v}}{\partial x}\right)^2+2\left[\left(\frac{\partial \bar{u}}{\partial x}\right)^2+\left(\frac{\partial \bar{v}}{\partial y}\right)^2\right]\right\},$$

$$C_{\varepsilon 1}=1.43,\ C_{\varepsilon 2}=1.92,\ \sigma_k=1.00,\ \sigma_\varepsilon=1.30.$$

Here, k is the turbulent kinetic energy; ε is the dissipation rate of the turbulent kinetic energy; S_{ij} is the strain rate tensor. The spatial distributions of k and ε at a certain moment are obtained by solving the differential equations of k and ε. Then, the spatial distribution of the eddy viscosity can be obtained by Eq. (9.2.21),

$$\nu_t=C_\mu\frac{k^2}{\varepsilon}$$

where $C_\mu=0.09$.

10.2.2 Treatment of Free Surface

For the numerical wave flumes based on the viscous flow theory, the main free surface treatment methods include the MAC method, the level-set method and the VOF method. In the numerical wave flumes by using VOF for the treatment of free surface, the donor-acceptor method proposed by Hirt and Nichols is mostly utilized.

Presently, the Youngs method for free surface tracking is fairly good. The detailed description of the VOF method has been given in Chap. 9 for the reference.

10.2.3 Numerical Wave Generation

Numerical wave generation method is a key technique in a mathematical wave model. Among the methods of wave generation mentioned above in Sect. 10.1.2, giving the wave surface or velocity at the incident boundary is relatively easier, but it is hard to remove re-reflected waves from the incident boundary. The wave generation method by adding a source/sink term to the governing equations can avoid the interference of reflection from the boundary to the wave field in the computational domain, which is widely used in the models based on the potential flow theory, especially in the depth-averaged Boussinesq equation model. There are also successful examples in the models based on the viscous flow theory without turbulence model. However, when the turbulence model is included, it is difficult to determine the turbulent kinetic energy and the corresponding dissipation rate at the source/sink in the numerical wave flume, so that the applications are limited.

The numerical wavemaker with active absorption is introduced in the following. Ursell and Dean [7] proposed a wave generation theory for piston-type or flap-type wavemakers for small-amplitude waves. Below a description of the numerical piston-type wavemaker is given.

The periodic horizontal displacement of the moving wave paddle reads

$$x(t) = \frac{S_0}{2}\sin\omega t \tag{10.2.1}$$

Then, the horizontal velocity of the wave paddle is $u(t) = \frac{S_0}{2}\cos\omega t$ where S_0 is the stroke of the wave paddle; ω is the wavemaker frequency. Based on the linear wavemaker theory, the wave surface elevation in the basin with constant water depth of d reads

$$\eta = \frac{S_0}{2}\left\{\frac{4\sinh^2(kd)}{2kd+\sinh(2kd)}\cos(kx-\omega t) + \sum_{n=1}^{\infty}\frac{4\sin^2(\mu_n d)}{2\mu_n d+\sin(2\mu_n d)}e^{-\mu_n x}\sin\omega t\right\} \tag{10.2.2}$$

in which

$$gk\tanh(kd) - \omega^2 = 0, \tag{10.2.3a}$$

$$\mu_n g\tan(\mu_n d) - \omega^2 = 0 \tag{10.2.3b}$$

The second term in Eq. (10.2.2) $\sum_{n=1}^{\infty}\frac{4\sin^2(\mu_n d)}{2\mu_n d+\sin(2\mu_n d)}e^{-\mu_n x}\sin\omega t$ is evanescent and can be neglected where it is located more than three times of water depth far away from the paddle. Therefore, the surface elevation at places far away from the paddle reads

$$\eta = \frac{S_0}{2}\left\{\frac{4\sinh^2(kd)}{2kd+\sinh(2kd)}\cos(kx-\omega t)\right\} = \frac{H}{2}\cos(kx-\omega t) \tag{10.2.4}$$

in which η is the expected incident wave surface elevation; H is the expected incident wave height.

The ratio of the wave height to the stroke for the small-amplitude waves at places more than 3 times of water depth far away from the paddle reads

$$\frac{H}{S_0} = \frac{4\sinh^2 kd}{2kd + \sinh 2kd} = \frac{2(\cosh 2kd - 1)}{2kd + \sinh 2kd} = W \tag{10.2.5}$$

Let the wavemaker be located at $x = 0$. Then, the surface elevation in Eq. (10.2.2) at the wavemaker yields

$$\eta_0 = W\frac{S_0}{2}\cos\omega t + L\frac{S_0}{2}\sin\omega t = \frac{W}{\omega}u(t) + \frac{L}{\omega}u\left(t - \frac{T}{4}\right) \tag{10.2.6}$$

in which

$$L = \sum_{n=1}^{\infty}\frac{4\sin^2\mu_n d}{2\mu_n d + \sin 2\mu_n d} \tag{10.2.7a}$$

$$u(t) = \frac{\eta(t)\omega}{W} \tag{10.2.7b}$$

In the above, the re-reflection caused by the reflected waves from the structures or the outer boundary impinging on the wavemaker is not taken into account. However, the re-reflection problem must be solved in order to generate steady incident waves in the wave flume. As we know that, the second term of the surface elevation in Eq. (10.2.2) is evanescent at places far from the periodically moving paddle. By means of this characteristic, the re-reflected waves can be prevented by generating waves opposite to the reflected waves in the computational domain.

The principle of the numerical wavemaker with active absorption is as follows: When the horizontal velocity of the paddle is $u_0 = \frac{\eta(t)\omega}{W}$, the surface elevation generated at places far from the paddle is $\eta(t)$ which is the expected incident surface elevation. When the reflected waves impinge on the wavemaker, the surface elevation at the paddle is changed. In order to remove the effect of the reflected waves, it is assumed that the surface elevation at the moving paddle is η_1 which includes the surface elevation of the reflected waves. When the paddle moves with a horizontal velocity $u_s = \frac{(\eta_1 - \eta)\omega}{L}$, the reflected waves at the paddle are removed, and their effect on the incident waves at places far from the paddle is little. Then, the horizontal velocity of the paddle at $x = 0$ becomes

$$u_m = \frac{\eta\omega}{W} - \frac{(\eta_1 - \eta)\omega}{L} \tag{10.2.8}$$

The flux inflowing from the wavemaker boundary within one wave period is not zero. For the cnoidal wave generation, the net flux caused by the moving paddle per one wave period is $\Delta Q = \frac{\pi H_0^2}{4W}$, in which H_0 is the surface elevation at the paddle. In order to maintain the water quantity balance in the computational domain, the velocity is modified by

$$u'_m = \begin{cases} u_m\left(1 - \frac{\pi H_0}{8d}\right), & \text{for } u_m \geq 0 \\ u_m\left(1 + \frac{\pi H_0}{8d}\right), & \text{for } u_m < 0 \end{cases} \tag{10.2.9}$$

The wave paddle can be taken as a slip boundary.

To gradually generate incident waves at the paddle, let

$$H_0 = H(1 - \mathrm{e}^{-5\frac{t}{2T}}) \tag{10.2.10}$$

where T is the incident wave period.

The turbulence intensity at the entrance is so small that it can be assumed to be zero. However, when the turbulent kinetic energy and the corresponding dissipation rate are zero, the eddy viscosity ν_t is infinity in the $k-\varepsilon$ turbulence model. Therefore, it is assumed that the turbulent kinetic energy and the corresponding dissipation rate are small and given by

$$k = \frac{1}{2}U_t^2 \tag{10.2.11}$$

in which $U_t = \delta C$; C is the phase velocity of the incident waves at the paddle; δ is a constant.

$$\varepsilon = C_\mu \frac{k^2}{\nu_t} \tag{10.2.12}$$

in which $\nu_t = \varsigma\nu$; ς is a constant; ν is the viscosity of water; C_μ is a constant in the turbulence model.

10.2.4 Non-reflective Open Boundary

The open boundary problems have been studied by many scholars in order to make waves travel through the open boundary without reflection. The Sommerfeld radiation conditions can be taken as the open boundary conditions for the small-amplitude wave problems. $f_t + c_x f_x = 0$ and $f_t - c_y f_y = 0$ in which f is any variable of u, v, η, k, and ε; c_x and c_y are the phase velocities in the x and y directions. For nonlinear wave problems, there is not a theoretically mature radiation condition at present. In the numerical wave flume, the sponge-layer absorption boundary condition proposed by Larsen and Dancy [3] can be used to treat the open boundaries, so that the waves decay in the sponge layers to eliminate the reflected waves.

In the experience of using sponge-layer open boundary condition, mostly it is aimed to solve the Boussinesq equations for which the discretized grid can be relatively coarse. To solve the N–S equations, the discretized grid is much finer and this sponge-layer open boundary condition is not suitable. Liu [9] carried out a detailed study on the thickness of the sponge layers and the parameter selection in the sponge layers. He proposed a method of the sponge-layer parameter selection independent of grid scale. It has been proved by the numerical simulations that the selected parameters using this method gave good results of wave absorption for calculating various waves with different grid sizes.

The thickness of the sponge layers is kept to be one wavelength λ. When the incident wave is a solitary wave, the thickness of the sponge layers is approximately the effective length of the solitary wave. Assume the grid in the horizontal direction is uniform with a grid spacing of Δx in the sponge layers, then the number of the calculation nodes is $\lambda/\Delta x$ in the sponge layers. Then, the decaying function $\mu(x)$ is expressed by

$$\mu(x) = \begin{cases} e\left[\left(2^{-\frac{10L}{\lambda}} - 2^{-10}\right)\ln(\alpha)\right], & 0 \le L \le \lambda \\ 1, & L > \lambda \end{cases} \tag{10.2.13}$$

It is found from the numerical simulations that, when $\alpha = 1.1$ the sponge layers give good results of wave absorption, and there is almost no effect of wave reflection on the waves in the computational domain.

In the wave model by using the Reynolds equations, the velocity and surface elevation are decayed in the sponge layers, but the pressure is not treated in the sponge layers, which gives

$$u = u/\mu(x), v = v/\mu(x), \eta = \eta/\mu(x), k = k/\mu(x), \varepsilon - \varepsilon/\mu(x), p = p$$

The decaying function in the sponge layers can be chosen as Eq. (10.2.13). It has been proved by the numerical simulations that the wave energy is extremely small when the waves arrive at the right boundary behind the sponge layers, so that the right boundary can be taken as a solid boundary.

10.2.5 Verification of the Numerical Wave Flume

In order to verify the numerical wave flume based on the Reynolds equations described in this section, Liu [9] established a numerical wave flume by using the VOF method for the free surface treatment, numerical wavemaker with active absorption for the wave generation system, and the sponge layers for the non-reflective open boundary condition. He carried out the numerical experiments using the established numerical wave flume. The water depth of the wave flume is $d = 10$ m, the incident wave height is 1.8 m, and the wave period is 8 s. Based on the small-amplitude wave theory, the incident wavelength is $\lambda = 70.9$ m.

Figure 10.2 shows the calculated surface elevation using the numerical wave flume compared with the analytic solution. Figure 10.3 shows the wave envelope in the wave flume when the wave is stable, for which the time interval is 1 s. The wave is basically stable in locations far more than 1–2 times of wavelength from

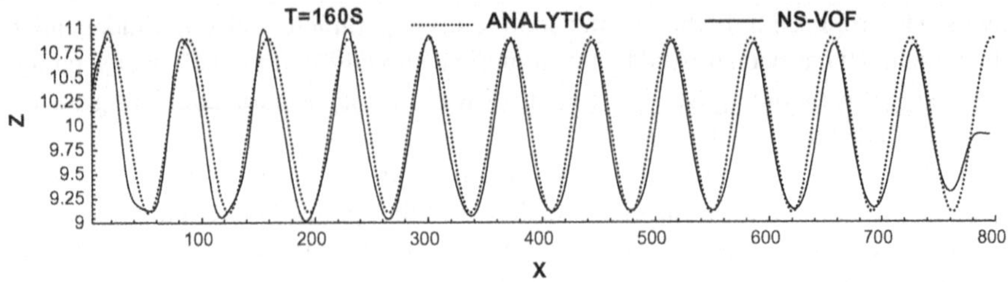

Fig. 10.2 Calculated small-amplitude wave surface elevation (NS-VOF) compared with the analytic solution

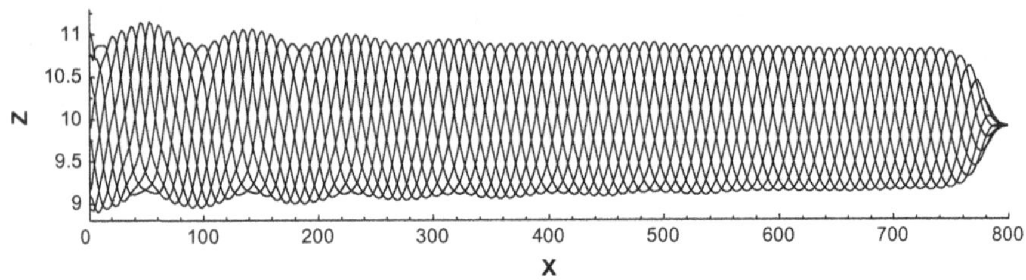

Fig. 10.3 Wave envelope in the wave flume

the wavemaker of the left boundary. The effect of wave absorption by the sponge layers in front of the right boundary is apparent. Figure 10.4 shows the velocity field in the flume when the wave is stable. It can be seen that the wave is steady in the experimental zone of the flume (far more than 1–3 times of wavelength from the wavemaker).

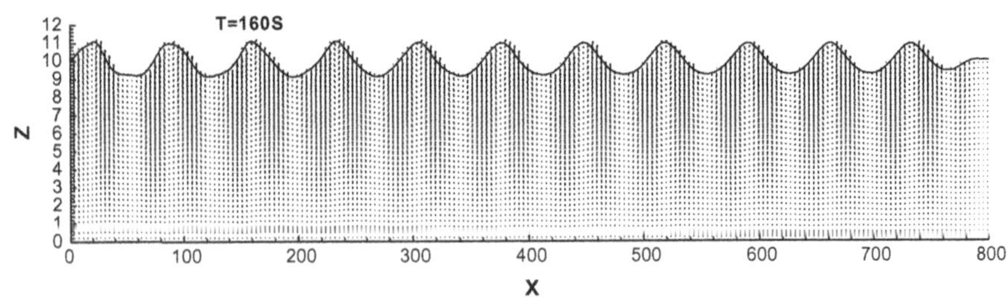

Fig. 10.4 Stabled velocity field in the flume at a certain moment

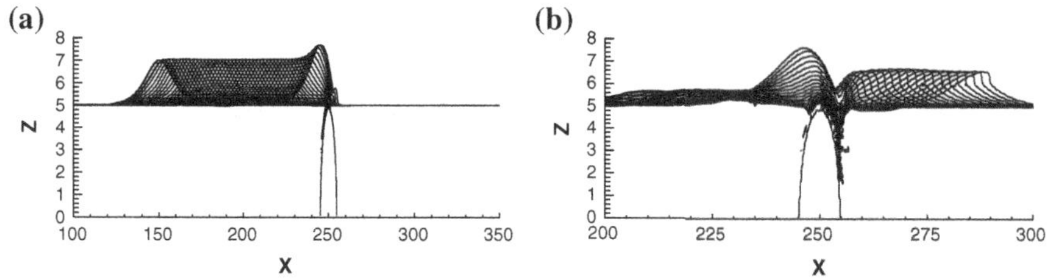

Fig. 10.5 Uplift and deformation of the solitary wave surface near the crest of the semicircular breakwater. **a** Uplift of the wave surface near the breakwater crest; **b** Reflected and transmitted solitary wave deformation

10.2.6 Numerical Experiments on the Interaction of a Solitary Wave and a Semicircle Breakwater in the Numerical Wave Flume

Liu [9] simulated the interaction of a solitary wave with a semicircular breakwater in the numerical wave flume. The breakwater crest is at the still water level. Figure 10.5a shows a process of the solitary wave surface being steep and uplifting near the breakwater crest. Figure 10.5b shows the wave deformation when the solitary wave passes over the semicircular breakwater. It is seen that wave breaking occurs in the leeside of the breakwater during overtopping. The velocity fields in the process of the interaction of the solitary wave with the semicircle breakwater have been shown in Figs. 10.6a–f and 10.7a, b show the vortices at certain moments during and after overtopping, respectively. A pair of large vortices is generated in the leeside of the semicircular breakwater, and the vortex intensity near the wave surface is larger. As the overtopping declines, the vortices spread to the wave surface gradually. Figure 10.8a–f show the pressure contours at certain moments.

10.3 Numerical Wave Basin Based on the Boussinesq Equations

The numerical wave basin is mainly used to simulate the horizontal problems which normally need to be solved by physical model tests in a shallow water wave basin, such as the layout design of harbors, breakwaters and navigation channels, and the wave forecast around artificial islands. With developments of computer and computing technologies, especially the further studies on the Boussinesq equations to describe the nonlinear shallow water waves in the recent years, the conditions for establishing a numerical wave flume which replaces the physical model to simulate nearshore waves in a certain extent have been more mature. A numerical wave basin "NWB" [6] based on the Boussinesq equations was developed successfully by the Computational Fluid Dynamics Laboratory at Tianjin University, China, in the early

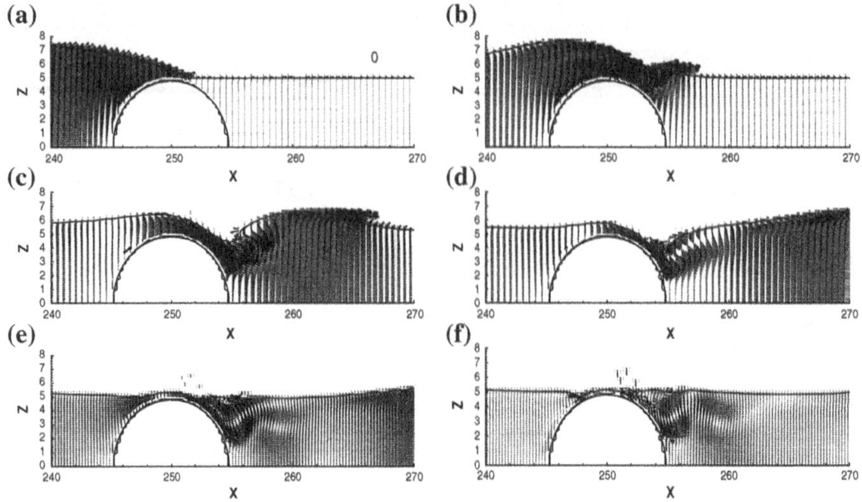

Fig. 10.6 Velocity fields in the process of the interaction of the solitary wave with the semicircular breakwater. **a** At the beginning of overtopping; **b** At a certain moment when the solitary wave is passing over the semicircle; **c** At a certain moment when the deformed wave travels backward; **d** At the moment when the solitary wave is about to pass over the semicircle; **e** At the moment when the solitary wave just completely passes over the semicircle; **f** At a certain moment when the wave passes over the semicircle and travels forward for a while

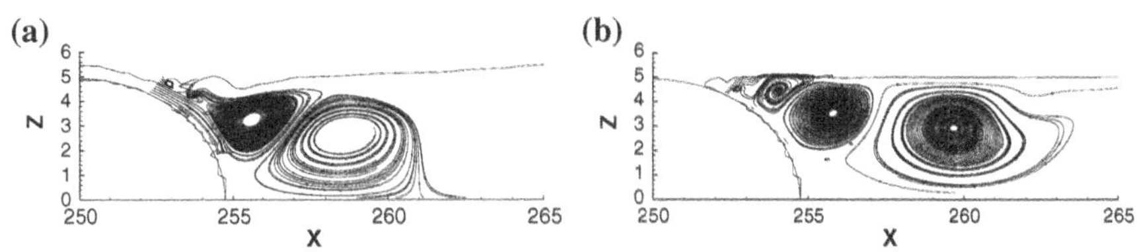

Fig. 10.7 Vortices when the solitary wave acts on the semicircular breakwater. **a** At a certain moment during overtopping **b** At a certain moment after overtopping

1990s. A convenient and practical software system was integrated with a graphical user interface coded by Turbo C 2.0, and the dynamic display of wave motion by means of moving the computational grid. The core and the pre- and post-processing systems of this software have been improved and expended unceasingly in the past years. Nowadays, the core computing program and the pre- and post-processing systems have been moved to the Windows operating system. Meanwhile, a friendly human-computer interface and a dynamic display module with good visual effects of wave motion have been developed with the tools of Visual C++ and Open GL. The improved numerical wave basin "NWB-Boussinesq" ("NWB-B" for short) is a visualization system of general mathematical model to simulate shallow water waves, which has similar functions of a physical wave basin. Numerical experiments for

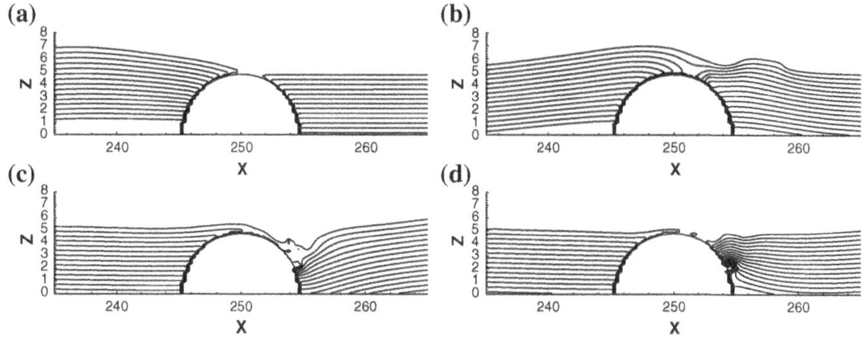

Fig. 10.8 Pressure variation when the solitary wave acts on the semicircular breakwater. **a** Pressure at the beginning of overtopping; **b** pressure at a certain moment during overtopping; **c** pressure at the moment when overtopping is about to finish; **d** pressure at a certain moment when overtopping is over

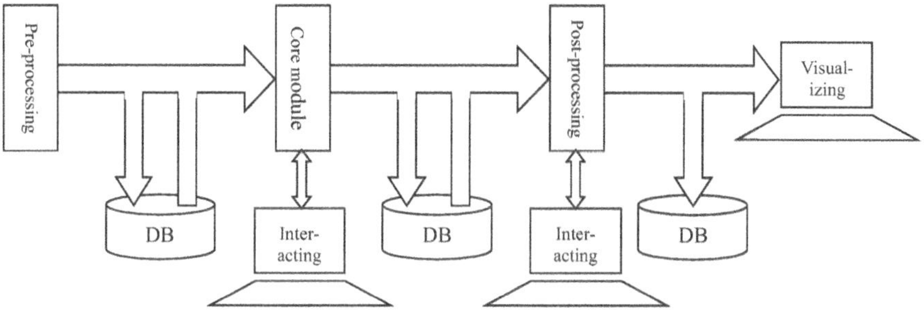

Fig. 10.9 Structural diagram of the NWB software system

various water wave problems in engineering can be carried out using this numerical wave basin.

The numerical wave basin has three important components (see Fig. 10.9): the pre-processing module, the core computing module, and the post-processing module. These three modules are introduced, respectively, in the following.

10.3.1 Core Computing Module

The core computing module of "NWB" is the mathematical model based on the Boussinesq equations. It consists of the governing equations and treatments of the boundary conditions.

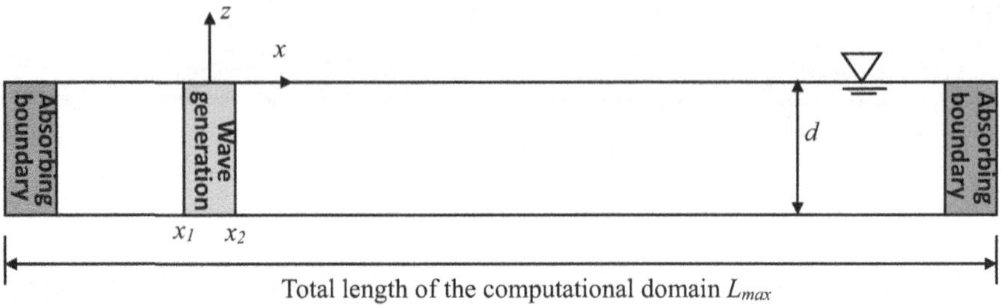

Fig. 10.10 Sketch of 1D computational domain with a wave generation zone

1. Governing equations

The improved Boussinesq equations proposed by Madsen (in Chap. 4) with a suitable range from relatively deep to shallow water have been applied in NWB-B. The numerical solving method for the Boussinesq equations has been described in Chap. 4, which is not repeated here.

2. Numerical wave generation system

The original wave generation method by adding a source term to the governing equations is a "line source" wave generation. The source term is added along a line (for 2D model) or at a point (for 1D model) in the computational domain [1]. When the nonlinearity of the target wave is high, it is prone to cause discontinuities in the vicinity of the wave generation region which can result in numerical instability. For this reason, Wei and Kirby [2] improved the "line source" method and proposed a "source function" wave generation method based on a spatially distributed source. In this method, a smoothly distributed source term is added in a limited area to avoid the numerical instability caused by the "line source" method.

"NWB-B" employed a source term added to the continuity equation using the "source function" wave generation method. Figure 10.10 shows a sketch of 1D computational domain with a wave generation zone [8].

The linearized improved Boussinesq equations read

$$\frac{\partial \eta}{\partial t} + \frac{\partial P}{\partial x} = 0 \tag{10.3.1a}$$

$$\frac{\partial P}{\partial t} + gd\frac{\partial \eta}{\partial x} + \alpha d^2 \frac{\partial^3 P}{\partial x^2 \partial t} = 0 \tag{10.3.1b}$$

in which P is the discharge per unit width (flux); η is the surface elevation; d is the still water depth; $\alpha = \frac{1}{3} + B$ and $B = \frac{1}{21}$.

Adding a source function $f(x, t)$ to the continuity Eq. (10.3.1a) gives

$$\frac{\partial \eta}{\partial t} + \frac{\partial P}{\partial x} = f(x, t) \tag{10.3.2}$$

Eliminating η from Eqs. (10.3.2) to (10.3.1b) yields

$$gd\frac{\partial^2 P}{\partial x^2} - \frac{\partial^2 P}{\partial t^2} - \alpha d\frac{\partial^4 P}{\partial x^2 \partial t^2} = gd\frac{\partial f}{\partial x} \tag{10.3.3}$$

The velocity potential ϕ is introduced which satisfies $u = \nabla\phi$, where u is the velocity. The relation between the velocity u and the flux P is $P = ud$, which gives

$$P = d\nabla\phi \tag{10.3.4}$$

By substituting Eqs. (10.3.4) into (10.3.3), and integrating it over x, we have

$$gd\frac{\partial^2 \phi}{\partial x^2} - \frac{\partial^2 \phi}{\partial t^2} - \alpha d^2\frac{\partial^4 \phi}{\partial x^2 \partial t^2} = gf(x,t) \tag{10.3.5}$$

The velocity potential $\phi(x, t)$ and the source function $f(x, t)$ are Fourier transformed by

$$\phi(x,t) = \frac{1}{2\pi}\int_{-\infty}^{\infty} \hat{\phi}(x,\omega)\exp(-i\omega t)d\omega \tag{10.3.6}$$

$$f(x,t) = \frac{1}{2\pi}\int_{-\infty}^{\infty} \hat{f}(x,\omega)\exp(-i\omega t)d\omega \tag{10.3.7}$$

in which ω is the angular frequency of waves. Substituting Eqs. (10.3.6) and (10.3.7) into Eq. (10.3.5) yields a second-order ordinary differential equation for $\hat{\phi}$ with respect to x. $\hat{\phi}$ and $\hat{f}$ can be solved by using Green's function method. Then, the source function $f(x, t)$ can be obtained.

For given wave parameters ω, η_0 and the water depth d, the wave number k is determined by the dispersion relation. Then, the source function is expressed as

$$f(x,t) = \frac{1}{2\pi}\int D(\omega)\exp(-\beta x^2)\exp(-i\omega t)d\omega = \exp(-\beta x^2)F(t) \tag{10.3.8}$$

in which, $D(\omega)$ is the amplitude of the source term.

$$D(\omega) = \frac{2gdk\eta_0}{\omega I[1-\alpha(kd)^2]} = \frac{2\omega\eta_0}{kI[1-\alpha(kd)^2]} \tag{10.3.9}$$

$$I = \sqrt{\frac{\pi}{\beta}}\exp\left(-\frac{k^2}{4\beta}\right) \tag{10.3.10}$$

$$\left.\begin{aligned} \alpha &= -1/3 \\ \beta &= \frac{80}{\delta_w^2 L^2} \end{aligned}\right\} \tag{10.3.11}$$

The source function width W ($W = |x_2 - x_1|$) is related to the wavelength L as

$$W = \delta_w(L/2)$$

in which δ_w is a factor of the source function width.

It is suggested that the typical value of δ_w is in the range of 0.3–0.5 and the corresponding source function width is about 0.15–0.25 times of the wavelength in reference [2]. It is found through the trial calculation of typical cases using "NWB-B" by Qin [8] that: the wider source function region, the much higher target wave height than the input wave height; the narrower source function width, the closer to the "line source" wave generation, which is prone to sudden change so that the generated waves are unstable near the wave generation zone. Therefore, comprehensive consideration should be taken for the selection of the source function width.

3. Non-reflective open boundary.

The open boundary refers to the boundary between the computational domain and the external water. It is required for the open boundary that there is no wave reflected from the external water back to the computational domain. The open boundary in the Boussinesq equation model can be treated by the radiation boundary condition or the sponge-layer absorbing boundary condition.

(1) Linear radiation open boundary condition

In a linear wave diffraction problem, the scattered waves passing through the open boundary of a finite domain can be expressed as $f(x - ct) + f(x + ct)$ in general, which consist of the waves propagating both outward and inward. If the open boundary is at infinite distance apart from the wave source, there is only the wave propagating outward $f(x - ct)$. Then the solution at the open boundary satisfies

$$\frac{\partial \phi_S}{\partial t} + c\frac{\partial \phi_S}{\partial \boldsymbol{n}} = 0 \tag{10.3.12}$$

in which $\boldsymbol{n}$ is the exterior normal vector at the boundary; ϕ_S is the scattered wave potential function.

On the basis that there is only wave energy propagating outward without any energy propagating inward at infinity, Sommerfeld [4] proposed the well-known radiation condition expressed in Eq. (4.3.12) as

$$\lim_{r\to\infty} \sqrt{r}\left(\frac{\partial \phi_S}{\partial t} + c\frac{\partial \phi_S}{\partial r}\right) = 0$$

in which ϕ_S is the scattered wave potential function; c is the phase velocity; $r = \sqrt{x^2 + y^2}$.

When the wave amplitude is small, the radiation condition in Eq. (4.3.12) can be applied for the open boundary condition in the Boussinesq equation model. Although this condition requires that the open boundary of the computational domain is far enough from the wave source (i.e., at infinity in mathematics), it can be utilized in practical calculation.

$$\frac{\partial f}{\partial t} + c_x \frac{\partial f}{\partial x} = 0 \tag{10.3.13}$$

in which f can be any variable such as P or η; c_x is the phase velocity in the x direction. A sided difference scheme can be used to solve this equation. c_x can be obtained by

$$c_x = \sqrt{gd(1 + \eta/d)} \tag{10.3.14}$$

(2) Sponge-layer absorbing boundary condition

For nonlinear waves with large wave height, the linear radiation open boundary condition is not valid. The sponge-layer absorbing boundary condition proposed by Larsen and Dancy [3] can be applied to decay waves and reduce the reflected wave energy passing to the computational domain. If the above linear open boundary method is also applied to release the waves, the wave reflection can be effectively reduced to achieve the non-reflective boundary. In the mathematical model, the following decaying function can be used

$$\mu(x) = \begin{cases} \exp\left[\left(2^{-x/\Delta x} - 2^{-x_a/\Delta x}\right)\ln a\right], & 0 \le x \le x_a \\ 1, & x > x_a \end{cases} \tag{10.3.15}$$

in which x_a is the width of the absorbing boundary; x is the distance from the grid point to the boundary; Δx is the grid spacing; $a = x_a/\Delta x$ is generally taken as 10–20.

Then, the decaying surface elevation and flux in the absorbing zone read

$$\eta = \eta/\mu(x),\ P = P/\mu(x) \tag{10.3.16}$$

In "NWB-B," the method of combing the sponge-layer absorbing boundary condition and the radiation open boundary condition is applied to absorb waves.

10.3.2 Pre-processing System

When using a numerical wave basin to solve practical engineering problems, a large number of parameters such as the bathymetry, wave parameters, and grid parameters need to be input. Furthermore, the bathymetry and the other parameters may also be altered when it is needed to study the wave conditions in different layouts. It is tedious

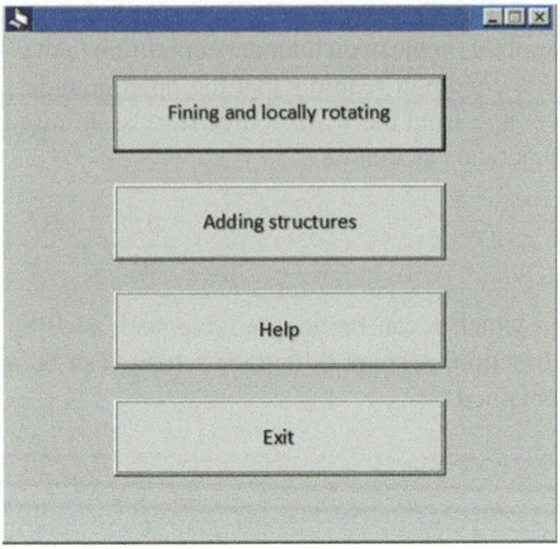

Fig. 10.11 Home page for bathymetry-processing

and error-prone to manually input these data. For this reason, a friendly user interface and a powerful pre-processing system should be provided in the numerical wave basin software system to reduce the users' workload and the risk of errors as much as possible. Then users can use this pre-processing system to easily complete the input and modification of various parameters, and also work on the original bathymetry according to the various design layouts such as dredging navigation channels or basins, adding breakwaters or artificial islands, rotating and cutting the bathymetry. There are four key interfaces of the bathymetry-processing in the pre-processing system in "NWB-B". At the home page for bathymetry-processing, functions of Fining and locally rotating', 'Adding structures', 'Help' and 'Exit' are included. At the page for the input of bathymetry parameters, the parameters for the original bathymetry and processed bathymetry can be given. At the page for adding navigation channels and basins, as well as the page for adding breakwaters and jetties, the parameters for original bathymetry, processing, and bathymetry output can be set. Figure 10.11 shows the home page for bathymetry-processing as an example.

10.3.3 Post-processing System

The original numerical simulation results are large amounts of data, which are tedious and difficult to understand so that it is difficult to find some valuable insights directly. It is very helpful to solve the problems by extracting a variety of intuitive charts from these data according to the requirements of engineers, technicians and decision-makers. The post-processing system of the numerical wave basin should be designed based on this demand. By means of this post-processing system, the original calculation results can be conveniently processed to output the charts of wave surface and wave height contours at various time, time series of surface elevation at each measurement points, and other charts or data that users are interested in. Furthermore, the post-processing system should also provide a dynamic display function, which can display the results in an animation format on the computer screen. Users can see the whole process of wave propagation, evolution, and deformation clearly from the animated display, which is of great help to evaluate or modify the project layouts. It is better to package the core computing module with the post-processing system together through the main page of "NWB-B," so that users can easily modify the calculation parameters and check the calculation results. At present, a friendly main page of "NWB-B" in Windows has been developed by using Visual C++. Meanwhile, functions of graphic plotting and dynamic display of wave field have been accomplished on the basis of Open GL which represents the industrial graphics standard. Figure 10.12a shows a snapshot of the main page of "NWB-B" for an example. The bathymetry corresponding to the calculation case in Fig. 10.12a is shown in Fig. 10.12b. In this representative calculation case, there are one navigation channel, one submerged breakwater, one rectangular pile, and two breakwaters. A variety of wave phenomena including refraction, diffraction, reflection, and nonlinear shoaling near the structures can be displayed in this case. It can be seen clearly in Fig. 10.12a that (1) wave reflection in front of the breakwaters and the pile is obvious, and the water behind the breakwaters and the pile is sheltered significantly but with certain wave diffraction; (2) as the water depth becomes shallow, the waves over the submerged breakwater deform significantly with violent rolling; (3) wave refraction occurs significantly due to the existence of the navigation channel. These phenomena in the calculation results are consistent with the well-known rules of waves impinging or passing over structures in coastal engineering.

10.3.4 Application of the Numerical Wave Basin

1. Numerical simulation of typhoon-generated waves in Zhuhai Port

In this example, there are two basins in the north and south of Zhuhai Port as shown in Fig. 10.13 for the sketch of this layout. The main channels, the north basin and the south basin, are all in deep water. The bathymetry in this port is complex.

Under the impact of typhoon-generated waves on November 25, 1998, the revetment on the side of Pingpai Hill of the north basin being almost parallel with the incident wave direction was destroyed. According to the observation of witnesses, the wave height increased gradually along the main channel after the typhoon-generated waves in the SSE direction entered into the harbor region from the sea; when the waves reached the opposite to Pingpai Hill, they changed direction suddenly and impinged to the direction of Pingpai hill to result in the revetment damage.

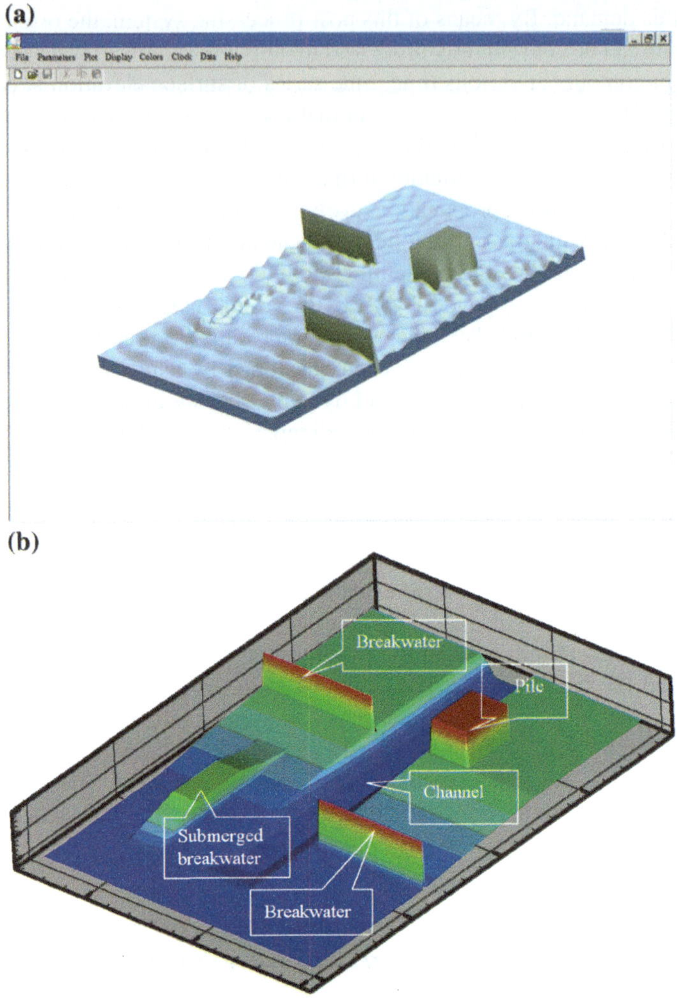

Fig. 10.12 **a** "NWB-B" main page; **b** Bathymetry of the calculation case

Was the revetment destruction accidental or inevitable? In order to further investigate the cause of the disaster, a numerical wave basin (NWB-B) based on the nonlinear shallow water wave theory was used to simulate and dynamically demonstrate the typhoon-generated waves in various directions [5, 10, 11]. Meanwhile, the mechanism of wave refraction and wave amplitude converge in channel and basin was analyzed by using the linear water wave refraction theory. Figure 10.14 shows the wave height ratio in the port for the south-south-eastern (SSE) incident typhoon-generated waves simulated by the numerical wave basin. In this case, the angle between the incident wave and the main channel is 12.5°, which is less than the critical angle of 27.53° according to the linear wave theory. Because the wave energy

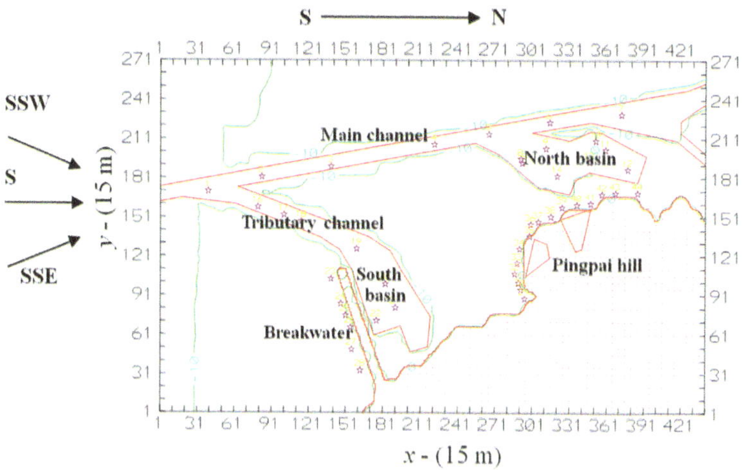

Fig. 10.13 Sketch of the layout of Zhuhai Port

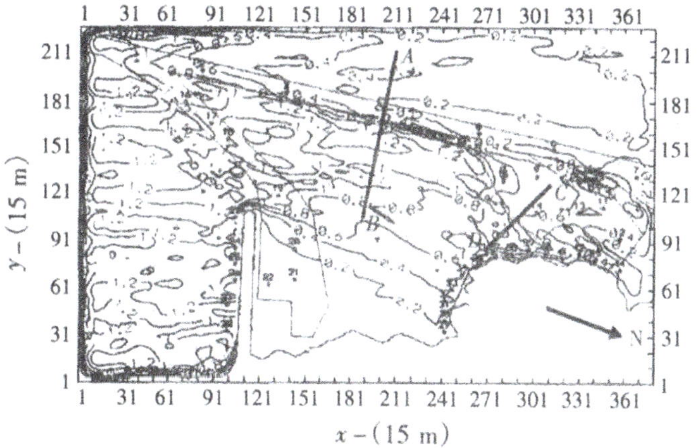

Fig. 10.14 Wave height ratio for SSE typhoon-generated waves

cannot cross the main channel, most of the wave energy is gathered on the east side of the main channel (the side having the basins) with the wave height ratio up to 1.2–1.4. The wave height ratio on the west side of the main channel is only 0.2–0.4. These indicate that the typhoon-generated wave energy fails to cross the main channel and gathers in the port region. The gathered waves travel along the main channel on the east side. When the waves reach the north port, they fail to enter into the basin because the angle between the wave direction and the north port is less than the critical angle. Then, the waves turned the direction, and the wave energy converged at the south-eastern side of the north basin and propagated to Pingpai hill, which results in the wave height in front of the revetment of Pingpai Hill being too high so as to cause the revetment damage. Therefore, the numerical results of the typhoon-generated waves are in agreement with the observation of witnesses. Figure 10.15 shows the wave height profiles at the cross section *AB* of the main channel and the cross section *CD* of the north basin. It is clear to see the significant wave refraction and amplitude converge due to the main channel and the north basin. Figure 10.16 shows the surface elevation of SSE typhoon-generated waves at a typical moment. The surface elevation of south-south-western (SSW) typhoon-generated waves at a typical moment is shown in Fig. 10.17. On the basis of the study described above, the reason of the revetment damage due to the typhoon-generated waves and the repair proposal had been put forward in the Ref. [11].

2. Yangtze River Estuary Deepwater Channel Regulation Project

Yangtze River Estuary Deepwater Channel Regulation Project was a major project of the China Ministry of Transportation, Shanghai, and Jiangsu province with a large scale and great cost. In order to ensure the safety of the project and save cost, numerical simulations of the wave fields before and after the project in various layouts in the whole project area were needed [12].

In this example, due to the complex bathymetry in the project area, waves propagating from deep water to the project area can cause refraction, diffraction, and complex deformation. In order to determine the design wave parameters of the jetties, numerical simulations on the waves propagating from deep water to the project area in various directions, various return periods, and various water levels in the

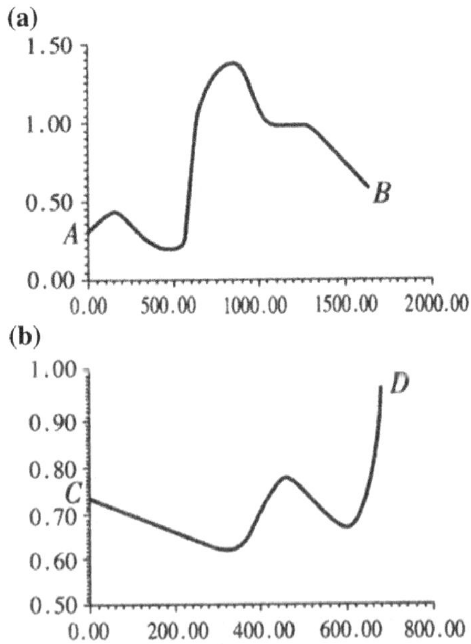

Fig. 10.15 Wave height ratio **a** along the section *AB*; **b** along the section *CD*

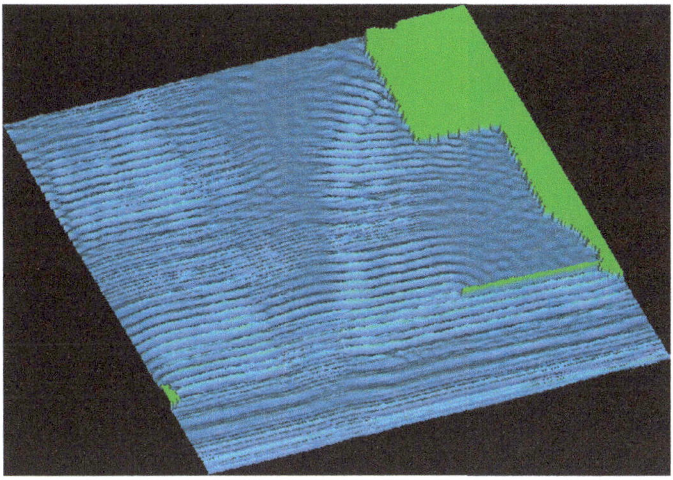

Fig. 10.16 Surface elevation of *SSE* typhoon-generated waves at a typical moment

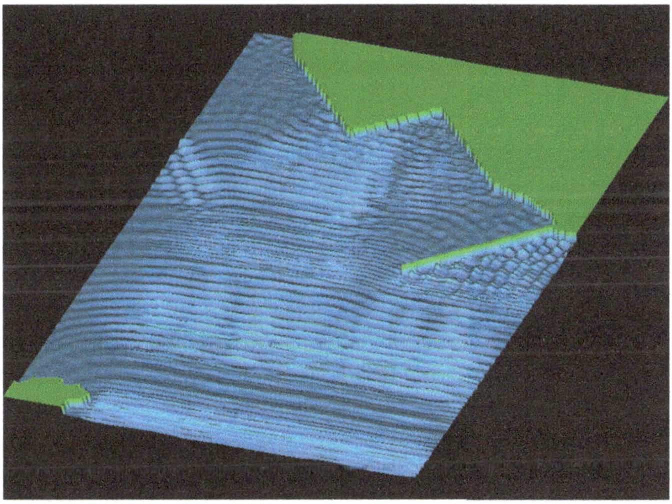

Fig. 10.17 Surface elevation of *SSW* typhoon-generated waves at a typical moment

existing bathymetry have been carried out first by using the numerical wave basin "NWB-P" [13] based on the higher-order approximate parabolic mild-slope equation. To determine the design wave parameters of a certain groin, it should be considered that the structures including the north jetty, the south jetty, the channel and the group of groins on the east side of the studied groin have been built. Due to the existence of the structures, the wave refraction, diffraction, and reflection must be taken into account. In this case, the numerical wave basin "NWB-B" based on the

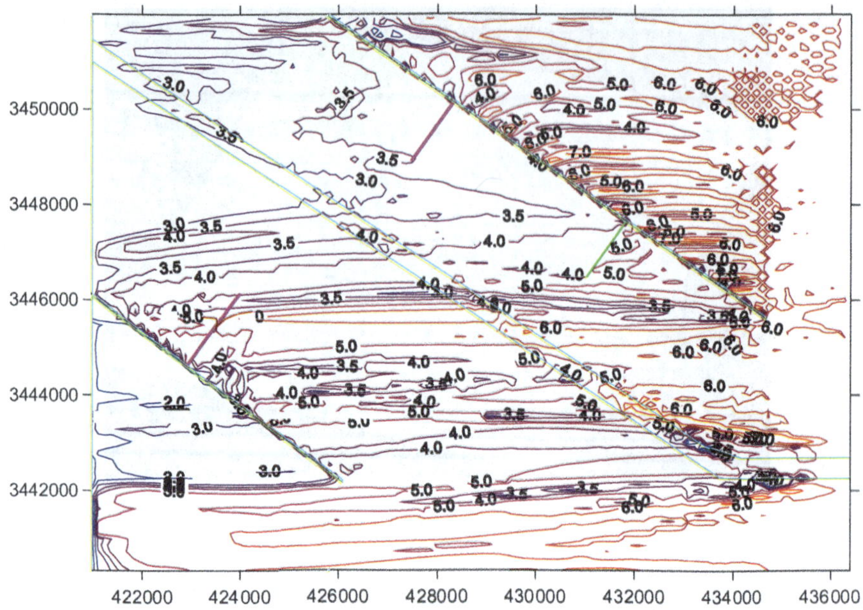

Fig. 10.18 Contours of wave height in the sub-region with jetties and the channel

Boussinesq equation model have been used for the simulations. Because the whole computational domain is huge, multi-block computational method has been applied. The whole computational domain has been divided into several connected small sub-regions which have been solved from the east to the west, respectively. Figure 10.18 shows the contours of wave height in the first sub-region. It is clear to see the wave reflection and diffraction under the impact of the jetties, and the wave refraction due to the channel (in the channel section being parallel to the incident wave direction, the waves are dispersed to the two sides of the channel). Figure 10.19 shows the contours of wave height around the head of the north jetty. Due to the wave reflection, the waves similar to standing waves are formed around the head of the jetty.

Using "NWB-P" and "NWB-B," the detailed wave height distributions in the vicinity of the jetties, around the head of jetties and at each groin have been obtained by the accurate numerical simulations of the waves in the whole computational domain and the corresponding analyses, which can provide the basis for the engineering design and construction.

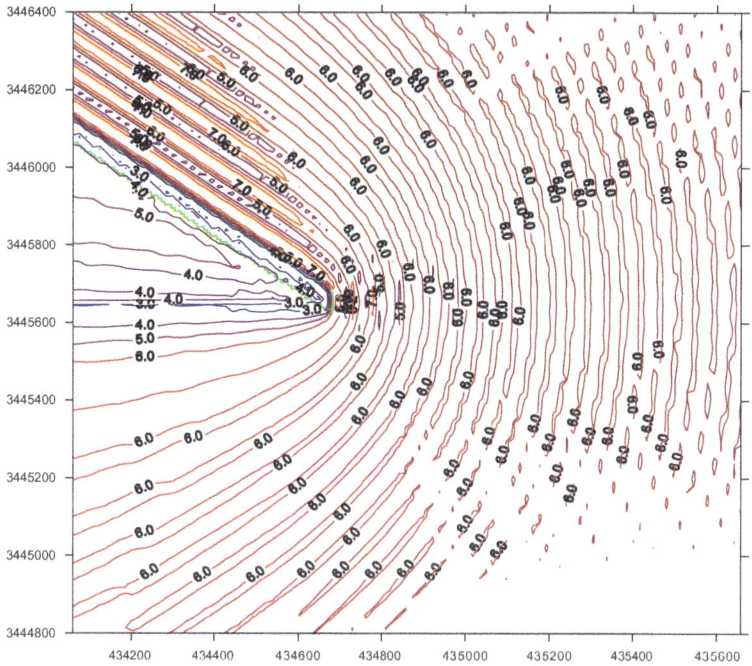

Fig. 10.19 Contours of wave height around the head of the north jetty

References

1. Wei G, Kirby JT. Time-dependent numerical code for extended Boussinesq equations. J Waterw Port Coast Ocean Eng. 1995;121(5):251–61.
2. W G, Kirby J, Sinha Amar. Generational of waves in Boussinesq models using a source function method. Coast Eng. 1999;36(4):271–99.
3. Larsen J, Dancy H. Open boundaries in short wave simulations—a new approach. Coast Eng. 1983;7(3):285–97.
4. Sommerfeld A. Partial differential equation in physics. New york: Academic Press; 1949.
5. Tao J, Long W. The critical angle of wave refraction and multiple refraction by harbor approach channel and basin. In: Proceedings of the 29th IAHR World Congress; September 2001; Beijing, p. 472–78.
6. Tao J, Wu Y. Numerical wave basin and applications. In: Hydrodynamics: theory and applications: proceedings of the second international conference on hydrodynamics; 1996; Hong Kong, 1996.
7. Ursell F, Dean RG, Yu YS. Forced small-amplitude waters: a comparison of theory and experiment. J Fluid Mech. 1960;7(1):33–52.
8. Qin W. Improvement of Boussinesq-equation mathematical model and the engineering applications. Dissertation for Master Degree, Tianjin" Tianjin University; 2000.
9. Liu C, Using unsteady Reynolds Navier-Stokes equations to simulate the water wave interaction with coastal structures. Dissertation for Ph.D. degree, Tianjin: Tianjin University; 2003.
10. Wen L, Tao J. Influence of navigational channels and basins upon wave field in harbor area and analysis of damage by typhoon waves to a revetment in Zhuhai Port. China Harbour Eng. 2000;4:39–41.

11. Tao J, et al. The report on numerical simulations of wave field in Zhuhai Port. Technical Report in Department of Mechanics, Tianjin: Tianjin University; 1998.
12. Tao J, et al. Estimation of design wave parameters of structures in Yangtze Estuary Deepwater Channel Regulation Project in Phase II. Technical Report in Department of Mechanics, Tianjin: Tianjin University; 2000.
13. Tao J, Han G. A random wave model based on higher-order approximation parabolic mild slope equation. China Harbour Eng. 2001;6:20–5.

Chapter 11
Applications of Numerical Simulation of Water Waves in Coastal Waters and Coastal Engineering

11.1 Study on Water Exchange Characteristics of Bohai Sea

Water exchange represents important transport characteristics of water bodies, playing key roles in understanding the transportation and diffusion of materials such as nutrients, contaminants, planktons, and fine sediments. It is one of the most important mechanisms giving rise to the self-cleaning capacities of water bodies, thus directly affecting the health of marine environment and ecosystem of a sea area. The assessment of water exchange characteristics is a highly desired procedure in the planning of any coastal or ocean engineering projects.

Bohai Sea is an inner sea of China, bordering coastal regions of fast-growing economies featuring dense ports and large populations. Its limited water body bears tremendous environmental stresses placed upon it by the out-proportioned social and economic developments. The strait connecting the Bohai Sea to the outer Yellow Sea is a relatively narrow water channel compared against the dimensions of the inner sea, restricting the capacity for water exchange between the inner and outer seas. Investigations of the time scale, circulation characteristics, and transport path of water exchange among the bays along the Bohai coast, and ultimately to the Yellow Sea, help to make policies in the ecological environment protection and to promote coordinated developments among all regions around the Bohai Sea.

Previous studies have demonstrated that tidal current is the most important dynamic factor for water exchange in the Bohai Sea. The sea is geologically characterized by its large lateral area, shallow water, and gradually varied bed slope, which leads to the negligible vertical water movements in comparison with the horizontal currents. As good approximations, two-dimensional (2D) horizontal long wave models will be sufficient for the simulation of the tidal water motions in the Bohai Sea.

J. Tao, *Numerical Simulation of Water Waves*, Springer Tracts in Civil Engineering, https://doi.org/10.1007/978-981-15-2841-5_11

A 2D convection–diffusion model and an average age model for conservative materials will be formulated in this section. From the distributions of average ages of water bodies, circulations, hydrodynamic characteristics, and time scales of water exchange in the Bohai Sea are investigated. The water bodies of the bay areas in the Bohai Sea are labeled, and the source and destination of each water body are traced. Finally, by comparing the analysis results to the conclusions obtained from the average age distribution of the seawater, the overall circulation in the Bohai Sea and the water exchange capacity of each bay are discussed.

11.1.1 Validation of the 2D Long Wave Model of Bohai Sea

Over many years of research efforts by the Computational Fluid Dynamics group at Tianjin University [1, 2], the 2D long wave models of the Bohai Sea described in this chapter have been systematically validated by comparing the simulation results against data from observations and measurements.

1. **Validation against data of continuous observations**

Continuous hydrological data are observed at three stations marked as V1, V2, V3, and V5 in Fig. 11.1. Data in July 2003 over a 72-h duration at stations V1, V2, V3, as well as data in November 2012 over a 30-h duration at station V5, are used for

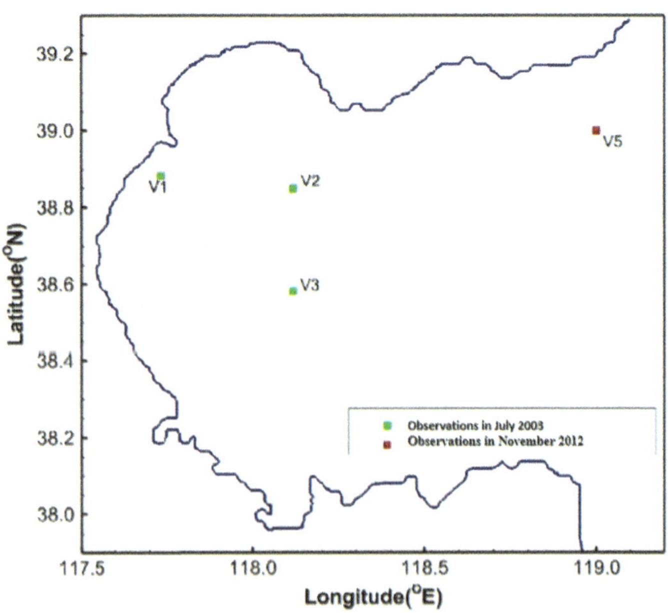

Fig. 11.1 Locations of observation stations

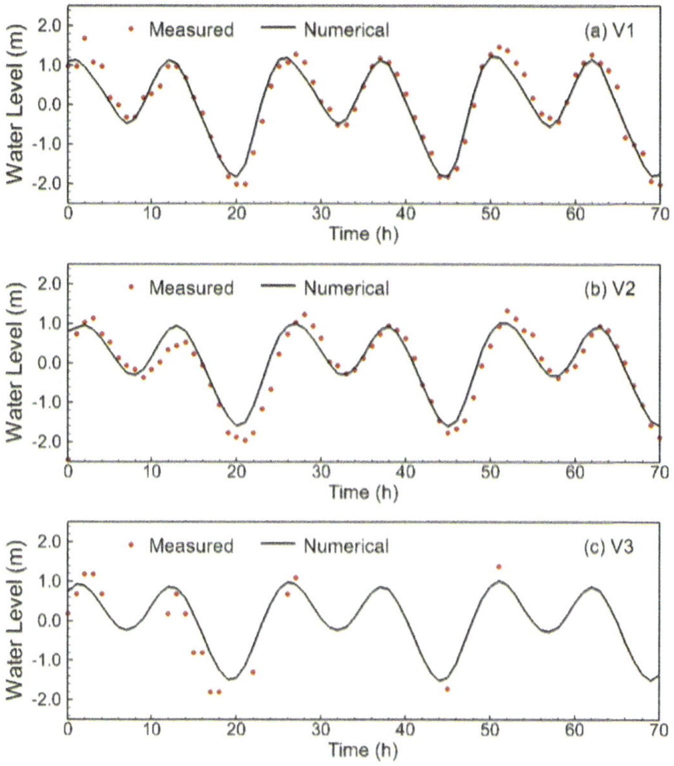

Fig. 11.2 Validation on water level in 2003 at stations V1, V2, and V3

validation that compares water level, current speed, and direction to the simulation results from the established flow models for the Bohai Sea.

Figures 11.2, 11.3 and 11.4 show the comparisons of the simulated and measured results of water level, current direction, and speed in July 2003 at stations V1, V2, and V3, which have demonstrated the general agreements between the calculated results and the observations. The correlation coefficients between the calculated and the measured data for water level, as well as that for current speed, are both better than 0.89. The root-mean-square errors of the calculated water-level results against the observations, normalized by tidal range, are within 20%. The root-mean-square errors of the calculated results for current speed against the observations, normalized by maximum speed in observations, are less than 10%.

Figure 11.5 shows the comparisons of the calculated water level, current direction, and speed against the observations in November 2012 at station V5, with similar observed.

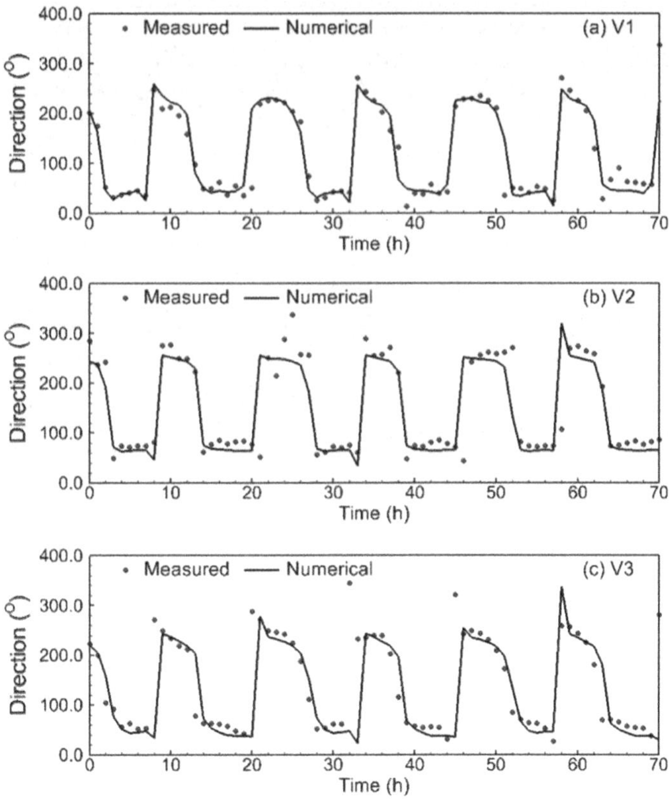

Fig. 11.3 Validation on current direction in 2003 at stations V1, V2, and V3

2. **Model validation against harmonic constants**

The tidal current in the Bohai Sea was simulated for an entire year of 2003. By means of the harmonic analysis on the modeled water level, the harmonic constants of the four main semidiurnal constituents (M2, S2, K2, and O2) of astronomical tides at twelve observation stations were obtained. The locations of these stations for validating harmonic constants are shown in Fig. 11.6.

Figure 11.7 shows the comparison of the calculated harmonic constants of M2 constituent with the ones from the Admiralty Tide Tables [3]. It can be seen that they are in generally good agreements with exceptions of a few cases. The average error of the calculated results of the dominant M2 constituent is around 12%.

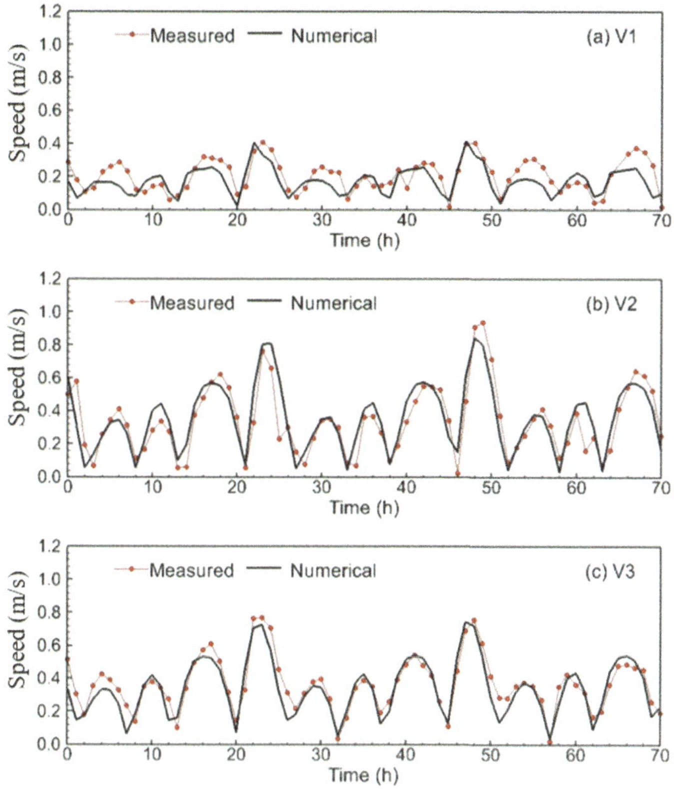

Fig. 11.4 Validation on current speed in 2003 at stations V1, V2, and V3

11.1.2 Convection–Diffusion Model and Age Model

The distribution of the concentration of a tracer material can be calculated using a convection–diffusion model based on the flow field in the whole Bohai Sea area obtained from the 2D long wave model. Then from the flow field and the concentration distribution of the tracer, a model of time scales enables simulations and analyzes the characteristics of water exchange. Here, for the study of water exchange, a popular concept of "age" is adopted as the time scale and the corresponding governing relations are called "age model."

1. **Convection–diffusion model**

The distribution of tracer concentration needed for calculating average age is governed by a convection–diffusion process. In ways similar to the 2D long wave model, a depth-integrated 2D convection–diffusion model is used in this section due to the relatively shallow water depth and the nearly uniform vertical mixing in the Bohai Sea. The governing equation, i.e., the depth-integrated convection–diffusion equation

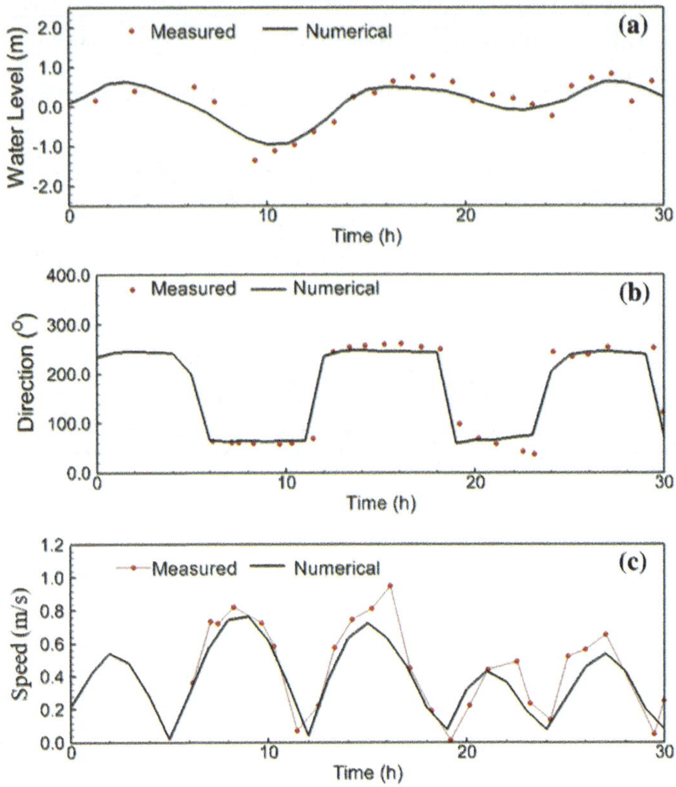

Fig. 11.5 Validation on water level, current speed, and direction in November 2012

Fig. 11.6 Stations for validating harmonic constants, S1: Nanhuangcheng Dao; S2: Qimujiao; S3: Dagu; S4: Caofeidian; S5: Laomigou Kou; S6: Dapuhe Kou; S7: Qinhuang Dao; S8: Hulu Dao; S9: Liaohe Kou; S10: Changxing Dao; S11: Xizhong Dao; S12: Lvshun Gang

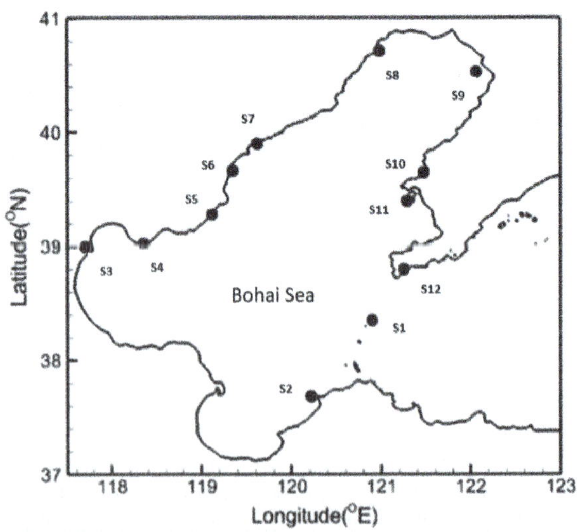

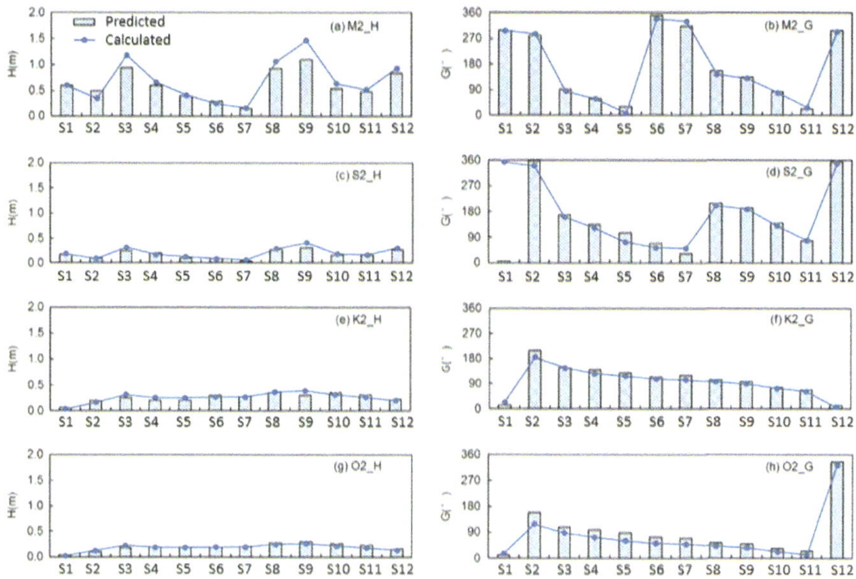

Fig. 11.7 Validation on harmonic constants, solid lines: calculated height (H) or phase lag (G), columns: harmonic constants from Admiralty Tide Tables

is as follows

$$\frac{\partial hC_{\mathrm{c}}}{\partial t}+\frac{\partial pC_{\mathrm{c}}}{\partial x}+\frac{\partial qC_{\mathrm{c}}}{\partial y}=\nabla_{\mathrm{h}}(\mathbf{D}_{\mathrm{h}}\cdot\nabla_{\mathrm{h}}C_{\mathrm{c}})+S_{\mathrm{c}} \tag{11.1.1}$$

where h is the total water depth; C_{c} is the depth-averaged concentration of some conservative material; p and q are the flow rates per unit width in the x- and y-directions, respectively; S_{c} is the rate of generation (source or sink) of the conservative material per horizontal unit area; D_{h} is a matrix of diffusion coefficients; ∇_{h} is the horizontal gradient operator which can be written as

$$\nabla_{\mathrm{h}}=\left(\frac{\partial}{\partial x},\frac{\partial}{\partial y}\right) \tag{11.1.2}$$

Similar to the methods used in solving the 2D long wave model, Eq. (11.1.1) can be solved using the finite difference method with alternating direction implicit (ADI) approach, the details of which are not described here.

Non-diffusive flux boundary condition is imposed on closed boundaries

$$\begin{cases}\left.\frac{\partial C_{\mathrm{c}}}{\partial \mathbf{n}}\right|_{\Gamma_{\mathrm{c}}}=0\\ \left.\frac{\partial^2 C_{\mathrm{c}}}{\partial \mathbf{n}^2}\right|_{\Gamma_{\mathrm{c}}}=0\end{cases} \tag{11.1.3}$$

$$\begin{cases} \left.\frac{\partial \alpha_a}{\partial \mathbf{n}}\right|_{\Gamma_c} = 0 \\ \left.\frac{\partial^2 \alpha_a}{\partial \mathbf{n}^2}\right|_{\Gamma_c} = 0 \end{cases} \tag{11.1.4}$$

where α is the average age of the tracer, which is to be later defined by Eq. (11.1.7).

On open boundaries, the concentration and age concentration of the conservative material are specified

$$c_c(x, y; t) = \breve{c}_c(x, y; t) \tag{11.1.5}$$

$$\alpha_a(x, y; t) = \breve{\alpha}_a(x, y; t) \tag{11.1.6}$$

Here, the tracer material is defined to be the water from the Yellow Sea flowing into the Bohai Sea to study the movement of this part of water after entering the Bohai Sea, thus yielding the understanding of the hydrodynamic condition in the Bohai Sea. Therefore, the tracer concentration of the water in the Yellow Sea at the open boundary (see Fig. 11.9) is set to 100%. As the age of the water in the Yellow Sea is zero before it enters the Bohai Sea, the age of the tracer on these open boundaries is set to 0.

2. **Age model**

Bolin and Rodhe [4] proposed the concept of water age and defined it as the elapsed time of a water particle entering into a certain area, as illustrated in Fig. 11.8. This definition of water age is based on the Lagrangian description of motion of continuum and utilizes the concept of water particles. Delhez et al. [5] developed a general age theory, defined the age of a parcel of a constituent to be the time elapsed since the parcel left the region where its age is prescribed to be zero, and proposed a method to calculate the mean age of any constituent in the water using the mass transport equation. This method has since been widely adopted to characterize water exchange in many sea areas around the world [6, 7] and is well suited to establish the age model in this chapter (Fig. 11.9).

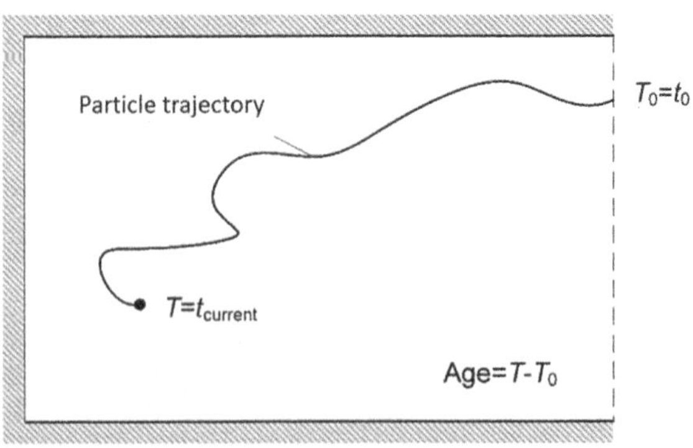

Fig. 11.8 Sketch for age concept

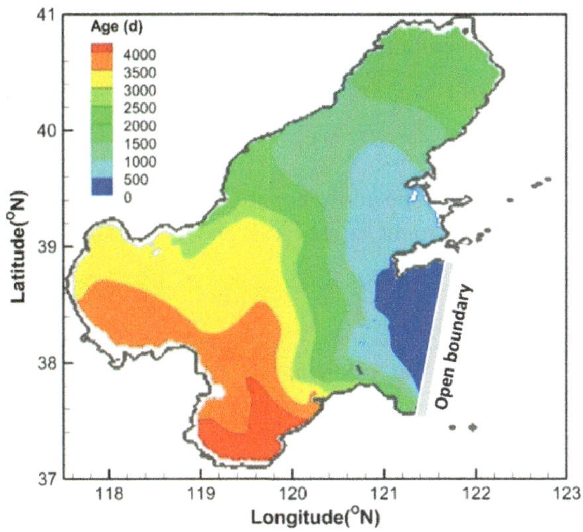

Fig. 11.9 Mean age distribution of water flowing from Yellow Sea into Bohai Sea

According to Delhez et al., the age of a parcel of a constituent in water is defined as

$$a_{\mathrm{a}}(t, \mathbf{x}) = \frac{\alpha_{\mathrm{a}}(t, \mathbf{x})}{c_{\mathrm{c}}(t, \mathbf{x})} \tag{11.1.7}$$

where $c_{\mathrm{c}}(t, \mathbf{x})$ is the concentration of the constituent, $\alpha_{\mathrm{a}}(\mathrm{t}, \mathbf{x})$ is the age concentration function of the constituent which is defined as

$$\alpha_{\mathrm{a}}(t, \mathbf{x}) = \int_0^\infty \tau \cdot c(t, \mathbf{x}, \tau)\mathrm{d}\tau \tag{11.1.8}$$

where τ is the age of a certain parcel of the constituent, $c(t, \mathbf{x}, \tau)$ the concentration with the age τ at position $\mathbf{x} = (x, y, z)$ at time t, $c(t, \mathbf{x}, \tau)$ and $c_{\mathrm{c}}(t, \mathbf{x})$ satisfy the following relation

$$c_c(t, \mathbf{x}) = \int_0^\infty c(t, \mathbf{x}, \tau)\mathrm{d}\tau \tag{11.1.9}$$

It can be seen that Eq. (11.1.9) represents the expected age of a constituent in the water at a certain position and a certain time, i.e., the mean age of the constituent.

The concentration of the conservative material follows the convection–diffusion equation,

$$\frac{\partial c_{\mathrm{c}}}{\partial t} + \nabla \cdot (\mathbf{u}c_{\mathrm{c}}) = \nabla(\mathbf{D} \cdot \nabla c_{\mathrm{c}}) \tag{11.1.10}$$

where **D** is the matrix of diffusion coefficients.

The concentration $c(t, \mathbf{x}, \tau)$ satisfies the following equation,

$$\frac{\partial c}{\partial t} + \frac{\partial c}{\partial \tau} + \nabla \cdot (\mathbf{u}c) = \nabla(\mathbf{D} \cdot \nabla c) \tag{11.1.11}$$

Multiplying the above equation by τ, integrating over τ in the range $(0, \infty)$, and taking into account Eq. (11.1.8) and a commonsense condition

$$\lim_{\tau \to \infty} c(t, \mathbf{x}, \tau) = 0 \tag{11.1.12}$$

we will obtain the following equation,

$$\frac{\partial \alpha_a}{\partial t} + \nabla \cdot (\mathbf{u}\alpha_a) = \nabla(\mathbf{D} \cdot \nabla \alpha_a) + c_c \tag{11.1.13}$$

It can be seen that the age concentration function obeys the mass transport equation and has the same expression with the governing equation of the concentration of conservative material Eq. (11.1.10). Thus, the mean age of a constituent in the water can be obtained by solving Eqs. (11.1.8), (11.1.10), and (11.1.13).

11.1.3 Water Exchange Characteristics in Bohai Sea

1. **Age distribution of the outer water and the overall transportation characteristics in Bohai Sea**

Figure 11.9 shows the calculated mean age distribution of the water from the Yellow Sea flowing into the Bohai Sea under the action of tide. The division of the Bohai Sea area is shown in Fig. 11.10. It shows that the water age is at its lowest values in the Bohai strait, higher in the Liaodong Bay, and the highest in the Laizhou Bay, with typical ranges being about 2–6 years in the Liaodong Bay, 10 years in the Bohai Bay, and about 11 years in the Laizhou Bay. This is consistent with the observations that the water from the Yellow Sea first reaches the Liaodong Bay, and the Laizhou Bay is located in the lower reaches of the circulation in the Bohai Sea.

From the results depicted in Fig. 11.9, the mean age of the water from the Yellow Sea ranges from 0 to 11 years, making the time scale of water exchange between the Yellow Sea and the Bohai Sea to be in the order of 10^0–10^1 years. There exists a counterclockwise circulation of current in the Bohai Sea. The Yellow Sea water enters into the Central Bohai Sea from the north of Bohai Strait. One branch flows northward to the Liaodong Bay. Another branch flows further eastward to the Luan River Estuary near the mouth of the Bohai Bay, and enters into the Bohai Bay from the north bank, and then enters into the Laizhou Bay from the Yellow River Estuary in the south of Bohai Bay, and finally flows out of the Bohai Sea from the south bank of the Bohai Strait. Zhang et al. [8] concluded on the circulation in the Bohai Sea that

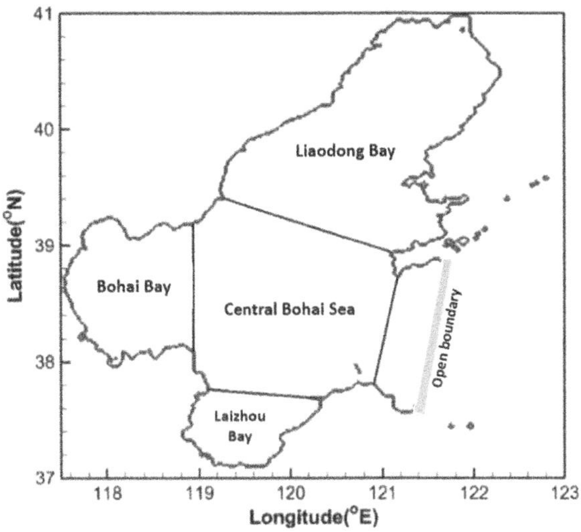

Fig. 11.10 Divided regions of Bohai Sea

there is an anticlockwise circulation in the Central Bohai Sea in both summer and winter; the Liaodong Bay is in the upper reaches of the circulation, and the Laizhou Bay is in the lower reaches. This conclusion is basically consistent with the result obtained here based on the mean age distribution.

2. **Characteristics of water exchange among regions of Bohai Sea**

In order to further understand the characteristics of water exchanges in various regions of the Bohai Sea and the mutual water exchanges among regions, the Bohai Sea is divided into four areas as the Liaodong Bay, Bohai Bay, Laizhou Bay, and the Central Bohai Sea, according to their geographical locations (as shown in Fig. 11.10). The water in each region is labeled and the whereabouts of the each labeled water is tracked as it flows around in the Bohai Sea.

Figure 11.11 shows time histories of remnant functions of labeled waters in each region. The remnant function is defined as the volumetric proportion of the labeled water retained in the source region or flowed into the other regions. First, it is seen that the water from the Central Bohai Sea region starts to appear in all other regions in somewhat immediate terms. By contrast, Fig. 11.11d shows that when the water in the Central Bohai Sea flows out, the water supplied from the surrounding bays to this region is significantly less than what is needed to make up the loss due to outflow. Thus, the loss of water in this region must be supplemented by the water from the Yellow Sea. As a result, the Central Bohai Sea functions as a hub of water exchange between bay regions and the Yellow Sea.

Figure 11.11a shows that when the water in Liaodong Bay flows out, the water supply from the Central Bohai Sea is at a relatively low level. Similar to the Central Bohai Sea, the principle of water conservation dictates that the water in this bay region be supplemented by the water from the outer Yellow Sea through the Central Bohai Sea. It is found by examining Fig. 11.11 that, after water from the Liaodong

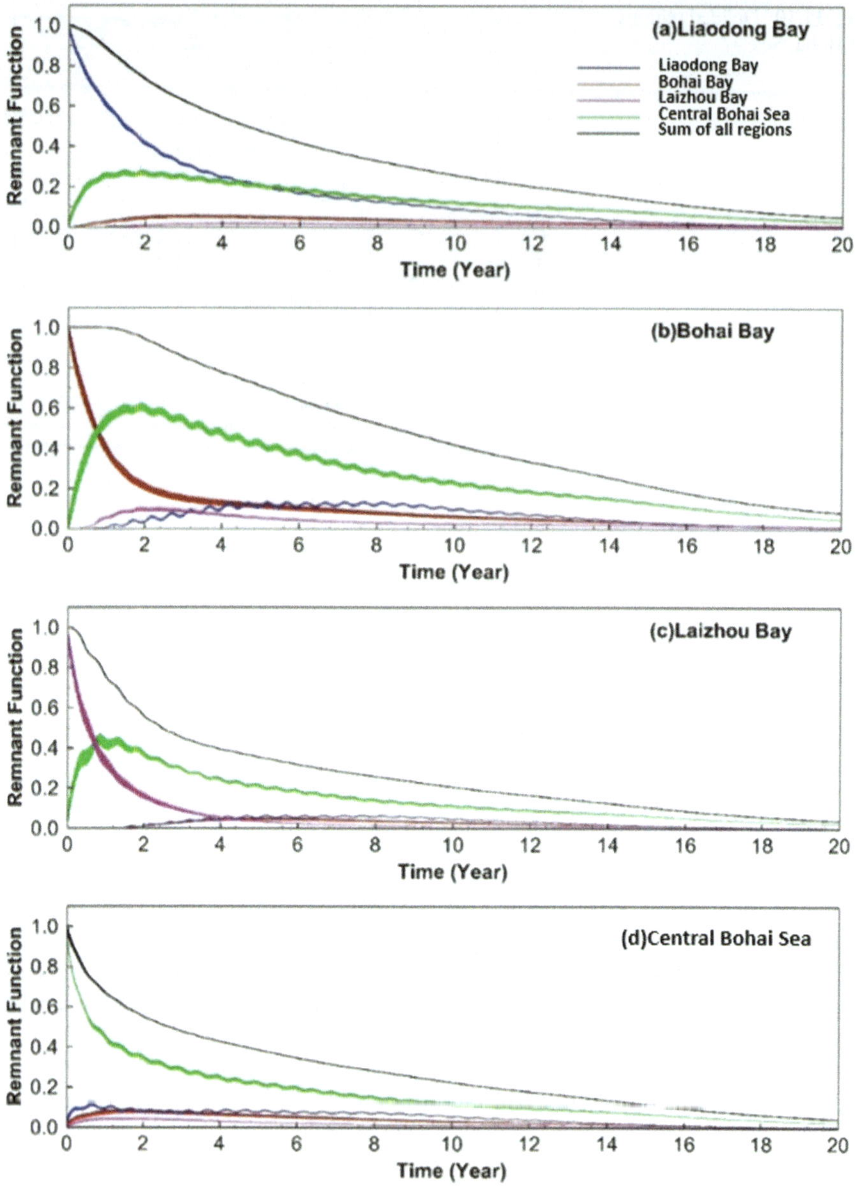

Fig. 11.11 Remnant function of marked water in each region of Bohai Sea

Bay enters the Central Bohai Sea, it takes additional two to three months for it to appear in the Bohai Bay. The amount of the water from the Liaodong Bay in the Bohai Bay peaks in about two years. Only at this instant, the water from the Liaodong Bay starts to appear in the Laizhou Bay, which indicates that the water from the Liaodong Bay enters into the Laizhou Bay mostly through Bohai Bay in a counterclockwise circulating pattern.

Figure 11.11b shows that the water from the Bohai Bay enters the Laizhou Bay shortly after flowing into the Central Bohai Sea and appears in the Liaodong Bay later. As the volume of the Bohai Bay is much smaller than that of the Liaodong Bay, the amount of the water flowing into the Liaodong Bay from the Bohai Bay is small compared with the large amount of the water flowing into it from the Yellow Sea. Therefore, the water from the Bohai Bay does not significantly affect the mean water age of the Liaodong Bay.

The outflows of water from the Laizhou Bay are shown in Fig. 11.11c. Again, most of the water in the Laizhou Bay flows into the Central Bohai Sea. By comparing the time instants when the water amounts from each of the three bays reach their peaks in the Central Bohai Sea and the declines that follow, it can be seen that the water of the Laizhou Bay peaks early in the Central Bohai Sea and then declines quickly. Furthermore, the proportions of the water from the Laizhou Bay flowing into the Bohai Bay and the Liaodong Bay are small, indicating that the water from the Laizhou Bay mostly flows out of the Bohai Sea through the near-by Bohai Strait.

Overall, the analyses on the water exchange among the various regions in the Bohai Sea show that the age values of the water from the Yellow Sea through the Bohai Strait are the smallest in the Liaodong Bay, larger in the Bohai Bay, and the largest in the Laizhou Bay. In a long time scale, the water of the Yellow Sea flows into the Bohai Sea through the north of Bohai Strait and out through the south of Bohai Strait, confirming the so-called North In, South Out phenomenon along the strait.

11.2 Numerical Simulation of Water Quality for Pearl River Estuary and Adjacent Coastal Areas in South China Sea

11.2.1 Introduction

1. **Objectives**

Pearl River Estuary (PRE) is located in the center of China's most economically developed region, highlighted by metropolitan cities such as Shenzhen, Hang Kong, and Macao. The rapid economic growth since the early 1980s has converted the region into a factory for the world. The out-proportioned activities of manufacturing and consumptions by the associated human population constantly generate massive-scale wastes, of which a large portion in liquid form is eventually dumped into rivers and seas in the region. According to China's Seawater Quality Standard (GB 3097-1997), water quality in the whole PRE and some adjacent sea areas reaches a concerning level of WQL 5.

The objective of this study was to gain an improved understanding of the distributions of environmental indicators and their seasonal variations in the PRE and assess the effects of pollution reductions that would be implemented. Based on the in situ measured data and the numerical simulation results, the hydrodynamics, pollutant transport, as well as the environmental carrying capacity of the PRE and adjacent sea areas were studied. The ultimate goal was to device a tool that predicts the pollutant response concentrations to various pollutant sources so that the overall pollution level distributions can be generated quantitatively and graphically for any given pollutant sources to aid the waste control and management.

Scientific references were provided by this study to the local environment protection agency for developing effective water management measures to reduce pollution in the PRE,which laid the foundation for achieving the goal of "Land–Sea Coordination for Total Pollutant Control."

2. **Model domain**

This study focused on the water quality in the Pearl River Estuary (PRE) and adjacent coastal areas, within a square area defined by 21°30′–23°00′ N and 112°30′–115°00′ E, as shown in Fig. 11.12 [9, 10]. The coastal line stretches from the northeast to the southwest with the South China Sea located at the south side and the Pearl River Delta at the north side. Hong Kong is located at the east bank of the PRE. The water depth is less than 20 m for the most part of the estuary and bays. The total area of water in this domain is about 30,000 km^2.

The PRE is the biggest estuary of the Pearl River, which is the second-largest river in China and the thirteenth-largest river in the world in terms of average annual water discharge. Located at about 22° N latitude, the Pearl River is a typical subtropical river. The total annual discharge is about 330×10^9 m^3 according to the historic data. Passing through a complex river network in the Pearl River Delta, the freshwater discharge flows into the PRE and coastal areas mainly through eight major entrances, namely Hu-men, Jiao-men, Hongqi-men, Heng-men, Medao-men, Jiti-men, Hutiao-men, and Ya-men from the northeast to the southwest, respectively (see Fig. 11.12), where the word "men" means a major entrance in Chinese. Apart from the major entrances, three rivers controlled by sluices, namely Dalongdong, Qianshan, and Shenzhen, are located near the Guanghai Bay, Medao-men, and PRE, respectively.

Diamonds: eight major entrances; solid circles: sampling stations; squares: three sluice controlled rivers (1: Dalongdong, 2: Qianshan, and 3: Shenzhen).

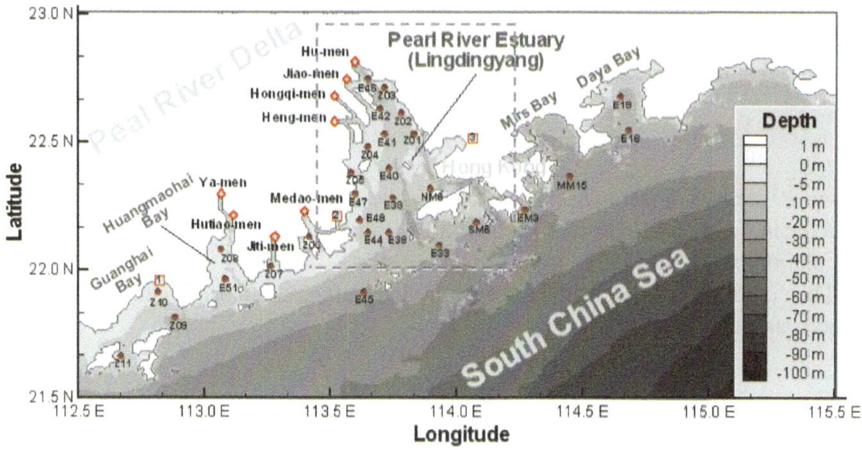

Fig. 11.12 Bathymetry of the PRE and adjacent sea areas

11.2.2 Water Quality Model

1. **Governing equation of the water quality model**

The governing equation of depth-integrated 2D water quality model (see Eq. (7.4.2) in Chap. 7) can be written as

$$\frac{\partial(H\phi)}{\partial t}+\frac{\partial(p\phi)}{\partial x}+\frac{\partial(q\phi)}{\partial y}$$
$$=\frac{\partial}{\partial x}\left(HD_{xx}\frac{\partial\phi}{\partial x}+HD_{xy}\frac{\partial\phi}{\partial y}\right)+\frac{\partial}{\partial y}\left(HD_{yx}\frac{\partial\phi}{\partial x}+HD_{yy}\frac{\partial\phi}{\partial y}\right)+S_\phi \quad (11.2.1)$$

where ϕ is the depth-averaged mass concentration of solute in unit/m^3, with the "unit" being mass such as kg, g, or mol; x and y are the Cartesian coordinates in the horizontal directions, (m); t is the time, (s); H is the total water depth; $p = HU$ and $q = HV$ are the discharge per unit width in the x- and y-directions, respectively, (m^3/s/m); U and V are the depth-averaged velocity components in the x- and y-directions, respectively, (m/s), which can be obtained by the hydrodynamic model; D_{xx}, D_{xy}, D_{yx}, and D_{yy} are the depth-averaged comprehensive diffusion coefficients, (m^2/s), which include the turbulent diffusion effect and the effect from the non-uniform flow velocity distribution over the depth, and can be determined by experience or calculated by empirical formula; S_ϕ is a source term representing the magnitude of solute mass per unit area due to sources, (unit/s/m^2), taking into account the input from rivers, the pollutant decay, and the atmospheric deposition.

2. **Initial and boundary conditions**

The initial solute concentration at $t = 0$ is

$$\phi(x, y, 0) = \phi^0(x, y) \tag{11.2.2}$$

where ϕ^0 denotes the known concentration in the whole domain at the initial moment, which can be set to a constant in the whole domain (a cold start) or given by trial calculation or other calculation results (a hot start).

The conditions of zero diffusion flux are used along the closed boundaries (Γ_0), which can be written as

$$\left.\frac{\partial \phi}{\partial n}\right|_{\Gamma_0} = 0 \text{ and } \left.\frac{\partial^2 \phi}{\partial n^2}\right|_{\Gamma_0} = 0 \tag{11.2.3}$$

At the open boundary, the situation is more complex. If water flows out of the domain, the concentration at the boundary can be non-specified and it is considered that the concentration values outside and inside the domain are the same. On the contrary, if water flows into the domain, the concentration at the boundary should be specified with known values

$$\phi(x, y, t)|_{\Gamma_1} = \phi^*(x, y, t)\big|_{\Gamma_1} \tag{11.2.4}$$

3. **Boundary conditions and model parameters of the water quality model for the PRE and adjacent coastal areas**

(1) Design hydrological conditions of the rivers flowing into the sea

The Pearl River, as an extensive river system, is one of the seven major rivers in China. The Pearl River is also used as a catch-all for the watersheds of four river systems including the Xi ("west"), Bei ("north"), and Dong ("east") rivers, and the other rivers flowing into the Pearl River Delta. Table 11.1 presents the multiyear average annual runoff discharge of the major rivers flowing into the Pearl River Delta at the control stations. The multiyear average annual discharge of the Xi and Bei rivers accounts for 84% of the total discharge. Therefore, the runoff for the calculation of pollutant concentration and capacity of the PRE is mainly controlled by the discharge conditions at the hydrological stations of Xi and Bei rivers.

Table 11.1 Average annual runoff discharge of the major rivers

Major rivers	Xi River (Makou)	Bei River (Sanshui)	Dong River (Boluo)	Zeng River (Qilinzui)	Tan River (Shizui)	Liuxi River (Niuxinling)	Local production	Total
Annual discharge ($\times 10^9$ m^3/a)	238	39.5	23.6	4	3.2	1.5	20.2	330
Ratio (%)	72.0	12.0	7.2	1.2	1.0	0.5	6.1	100.0

Table 11.2 Annual discharge flowing into the PRE at the eight major entrances

No.	Major entrances	Annual discharge ($\times 10^9$ m^3/a)	Ratio (%)
1	Hu-men	60.31	18.50
2	Jiao-men	56.50	17.33
3	Hongqi-men	20.88	6.41
4	Heng-men	36.58	11.22
5	Medao-men	92.24	28.30
6	Jiti-men	19.69	6.04
7	Hutiao-men	20.21	6.20
8	Ya-men	19.57	6.00
Total		325.98	100.00

The runoff in the Pearl River Delta flows into the PRE and coastal areas mainly through eight major entrances (Hu-men, Jiao-men, Hongqi-men, Heng-men, Medao-men, Jiti-men, Hutiao-men, and Ya-men). The multiyear average annual net volume flowing into the sea is about 326×10^9 m^3. Table 11.2 presents the multiyear average annual discharge flowing into the sea at the eight major entrances.

The multiyear average monthly discharge was used in the water quality model for the PRE and adjacent sea areas. The frequency statistical discharge was used to calculate the response concentration distribution of oxygen consumption (chemical oxygen demand, CODMn) and toxic pollutant (petroleum, oil).

(2) Characteristics of offshore tidal waves

The main tidal characteristics in the PRE and adjacent sea areas are mixed semidiurnal tide containing both semidiurnal and diurnal constituents. In this study, eight tidal constituents, including four main constituents (M2, S2, K1, and O1) and four minor constituents (N2, P2, K2, and Q2), were taken into account in the hydrodynamic model to calculate water level and flow field as the basis of the water quality model.

(3) Flux of pollutants entering into the sea

Pollutants entering into the PRE and adjacent sea areas mainly come from land, sea, and air sources. The land sources can be divided into two types: from rivers to the sea and direct discharge into the sea. The sea sources include marine cultivation, port activities, illegal dumping, and maritime accident. The air sources mainly include two processes of atmospheric deposition (wet and dry) with vertical inflow flux, and they are major sources of nutrients such as nitrogen and phosphorus.

Therefore, the pollutant flux into the PRE and adjacent sea areas can be divided into four types: the flux from rivers, from land directly, from the marine pollutant sources, and from atmospheric wet and dry deposition. The pollutant flux of these four types into the model domain and the corresponding ratio of pollutant flux contribution are presented in Table 11.3.

Table 11.3 Pollutant flux into the sea from different sources

Types	Chemical oxygen demand		Permanganate index		Total nitrogen		Total phosphorus		Oil	
	Flux (ton/a)	Ratio (%)	Flux (ton/a)	Ratio (%)	Flux (ton/a)	Ratio (%)	Flux (ton/a)	Ratio (%)	Flux (ton/a)	Ratio (%)
From rivers	2,856,688.0	89.72	865,663.0	89.73	587,163.0	81.49	59,138.0	92.58	15,845.0	95.12
From land directly	323,690.0	10.17	98,088.0	10.17	39,695.0	5.51	3051.0	4.78	255.6	1.53
Marine pollutant sources	3602.7	0.11	1000.8	0.10	2582.8	0.36	760.6	1.19	557.4	3.35
Atmospheric wet and dry deposition	–	–	–	–	91,085.2	12.64	930.8	1.46	–	–
Total	3,183,980.7	100	964,751.8	100	720,526.0	100	63,880.4	100	16,658.0	100

It can be seen from the table that the pollutant flux from rivers into the sea accounts for the vast majority of the total flux for all concerned substances. The flux from land directly into the sea occupies certain ratios. The atmospheric wet and dry deposition has a certain effect on total nitrogen, accounting for 12.64% of the total flux. The ratios of the flux from marine pollutant sources to the total flux are small for all substances.

In this study, the pollutant flux from rivers into the sea (including the major eight entrances and Zhaxi), as the main pollutant source, was taken into account in the water quality model. Meanwhile, the flux from land directly into the sea and the effect of atmospheric wet and dry deposition were also considered. The marine pollutant sources were neglected in the model due to the small contributions to the total pollutant flux.

(4) Boundary conditions of the water quality model

Boundary conditions of the water quality model mainly refer to the concentration conditions of water quality indicators at the open boundaries of the domain. Based on the distribution characteristics of the water quality survey, the boundary conditions of the water quality model were specified as follows. The salinity boundary condition was set to 35‰, and the concentrations of CODMn, inorganic nitrogen (including ammonium-nitrogen, nitrite-nitrogen, and nitrate-nitrogen), phosphate, and oil were set to zero along the boundaries with the water depth being deeper than 50 m.

(5) Decay coefficient of water quality indicator

In the process of pollutant transport in water, the amount of pollutants will be reduced to some extent due to sedimentation and decomposition. The first-order decay coefficient (or called integrated attenuation coefficient) K_ϕ (per day) was used to represent integrated effects of sedimentation and decomposition in the water quality model for each water quality indicator in this study.

11.2.3 Verifications of Hydrodynamic Model and Water Quality Model

1. **Verification of hydrodynamic model**

The hydrodynamic model was verified against the data from the survey and previous studies. Figure 11.13 shows comparisons of time series of water depth, current speed, and current direction between the model results and the measurements, at station E38 in the period from 12:00 pm on October 19, 2006, to 12:00 pm on October 20, 2006. It can be seen that the modeled water depth, current speed, and direction agree well with the measurements. Figure 11.14 shows comparisons between the modeled and observed tidal harmonic constants, in which the data are obtained from: (a) the Admiralty Tidal Table [3], and (b) a survey by Mao et al. [12] in both dry and wet

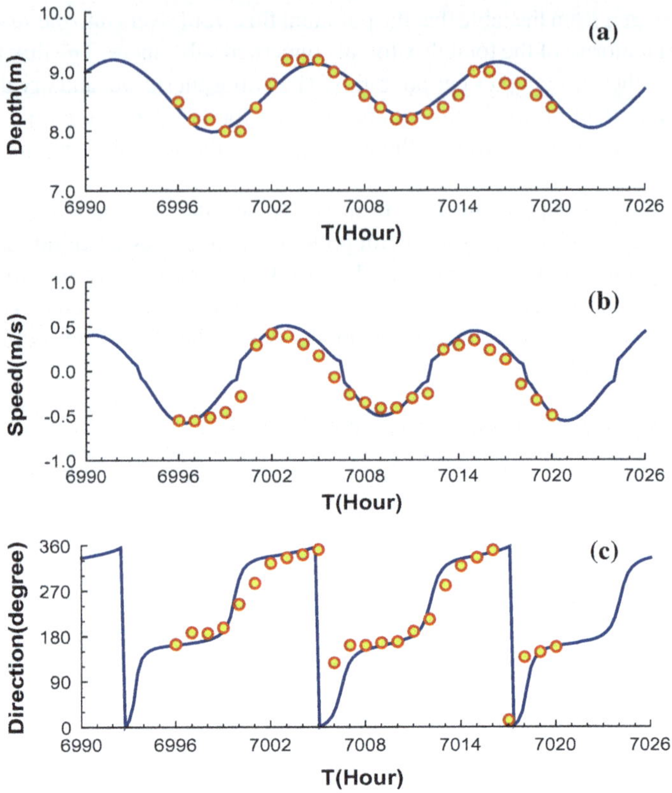

Fig. 11.13 Comparison between modeled (line) and measured (points) water depth (**a**) current speed (**b**) and current direction (**c**) at Station E38 during 12:00 pm on October 19, 2006, to 12:00 pm on October 20, 2006

Fig. 11.14 Comparison between modeled and measured tide amplitudes; measured data obtained from HoN [11] and Mao et al. [12]

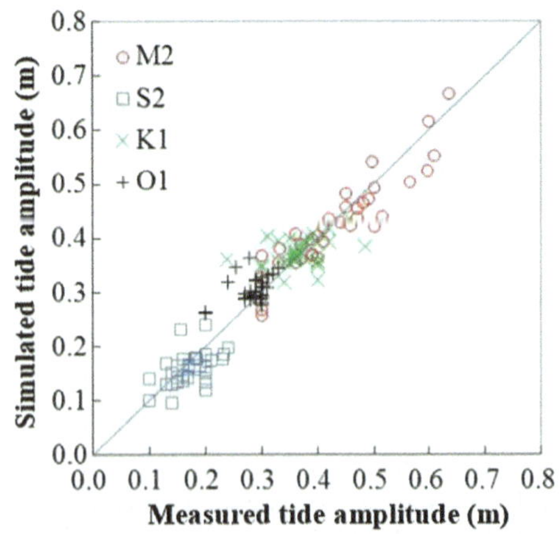

seasons in 1998. The good agreement of harmonic constants indicates the reliability of the model simulation for long-term hydrodynamic processes.

2. **Verification of water quality model**

The water quality model consists of two sub-models: the advection–diffusion model and the source model. The advection–diffusion model was verified against salinity distribution first, for which the source term was set to zero. Then two environmental indicators including dissolved inorganic nitrogen (DIN) and chemical oxygen demand (COD) were simulated. The simulated distributions of salinity level, DIN, and COD concentrations in three typical seasons are shown in Fig. 11.15. The salinity level shows a decreasing trend from the outer estuary to the inner estuary, while a generally increasing trend can be found for DIN and COD concentrations. These trends show that the DIN and COD carried by freshwater from rivers are diluted by seawater. Comparisons of the simulated results and the measurements are shown in Fig. 11.16. The simulated salinity level, DIN, and COD concentrations generally

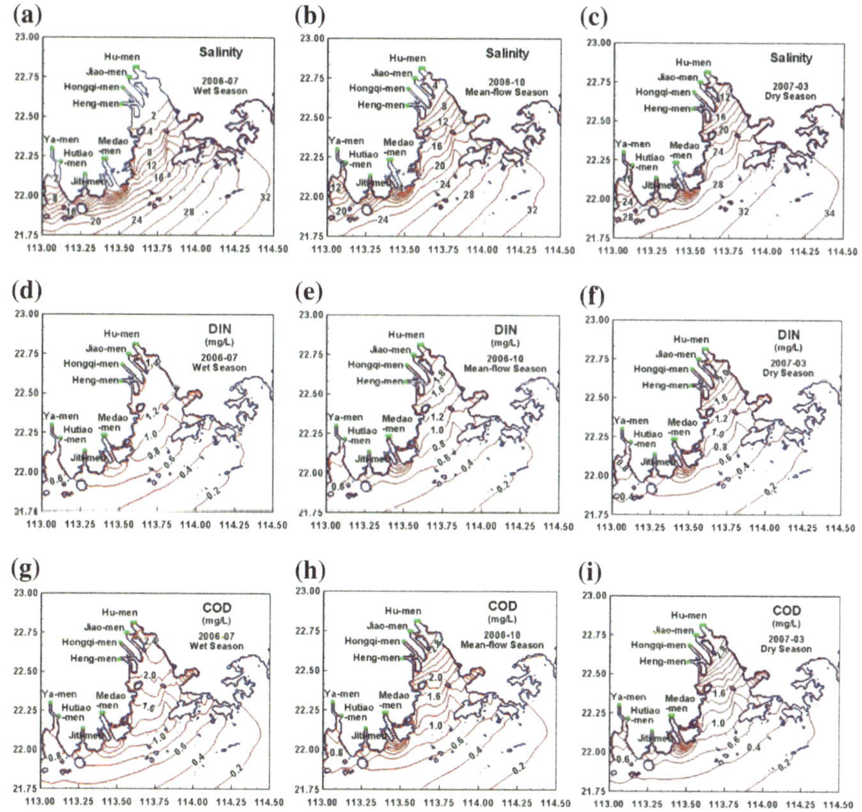

Fig. 11.15 Distributions of simulated salinity level, DIN, and COD concentrations. The contour is plotted by the monthly average values

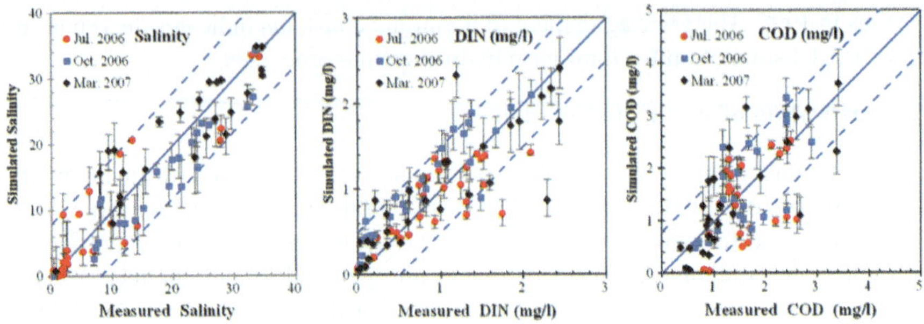

Fig. 11.16 Comparison of simulated and measured salinity level, DIN, and COD concentrations. Error bars are used to denote the variation range due to tidal fluctuation

agree well with the measurements, which is also supported by the previous in situ observations.

The deviation of the simulated COD concentration from the measurements is larger than that of DIN. This is considered to be related to the low COD concentration in this area. Under such a condition, complex chemical and biological reactions also contribute to the variation of COD concentration. This is also supported by the statistical results that DIN has a closer relationship with salinity than COD. In addition, the major environmental factor is DIN rather than COD because of high level of DIN and low level of COD. The accuracy of the model for simulating COD concentration is sufficient for the purpose of providing references for planning of integrated pollution reduction.

11.2.4 Distribution of Pollutant Response Concentration

The pollutant response concentration or called response coefficient refers to the pollutant concentration under the effect of a single unit external source (such as input from rivers or atmospheric deposition). When the discharge amounts of certain pollution sources are known, the overall pollutant concentration can be obtained by superposition of the response concentrations. Thus, the allowable discharge amounts under different conditions can be estimated according to the criteria of the seawater quality standard.

The external source terms include the terms of input from rivers and atmospheric deposition. They can be written as a superposition of single unit external source terms

$$\underbrace{S_{\mathrm{m}}\varphi_{\mathrm{m}}}_{T1} + \underbrace{S_{\mathrm{air}}}_{T2} = \sum_{j=0}^{n} \underbrace{Q_j}_{T3} \cdot \underbrace{S_j}_{T4} \quad j = 0, 1, 2, \ldots n \tag{11.2.4}$$

where $T1$ and $T2$ denote the terms of input from rivers and atmospheric deposition, respectively; $T3$ (Q_j) and $T4$ (S_j) denote the terms of external source intensity and single unit source, respectively; $j = 0$ denotes the atmospheric deposition, and $Q_0 = 1$, $S_0 = S_{air}$. The pollutant response concentration ϕ_j can be calculated using the water quality model with single unit source S_j. Considering the linearity of differential equation of the water quality model under the same hydrodynamic conditions, the relationship between the overall concentration ϕ and the response concentration ϕ_j can be written as

$$\phi = \sum_{j=0}^{n} Q_j \cdot \phi_j \tag{11.2.5}$$

If the response concentration ϕ_j of each pollutant is calculated beforehand, the actual overall concentration can be obtained by linear superposition, by which the layout design and selection can be carried out.

In addition, under various restrictive conditions such as hydrological conditions and socio-economic environment conditions, the allowable amounts of pollutant discharge can be estimated by means of optimization method. It should be noted that the same hydrodynamic conditions and the approximation of first-order pollutant decay are the important premises of the above superposition method.

In the calculation of the pollutant response concentration in the PRE and adjacent coastal areas, the hydrological conditions in 2006 were used. In general, the year of 2006 was a normal flow year, which is representative to some extent. The hydrological conditions of the rivers flowing into the sea were selected as follows. For nutrients, the multiyear average monthly hydrological data in 12 months were used. For oxygen consumption and toxic pollutants, the hydrological data in the month with the lowest water level in dry season were selected to represent monthly runoff discharge with a probability of 90%. Considering the evaluation criteria of water quality, three pollution indicators including dissolved inorganic nitrogen (DIN), active phosphate (PO4-P), and chemical oxygen demand (CODMn) were selected. Among them, DIN is a key pollution indicator affecting the water quality of the seawaters as DIN concentration exceeding the standard is common and serious. Figure 11.17 shows the distributions of DIN response concentration in the PRE with a single source at Hu-men in March, July, October, and December.

11.2.5 Improving Water Quality in Pearl River Estuary

The main purpose of this study was to provide scientific references for developing effective water management measures to improve water quality in the PRE. DIN was chosen as a key indicator due to its high concentration in the PRE.

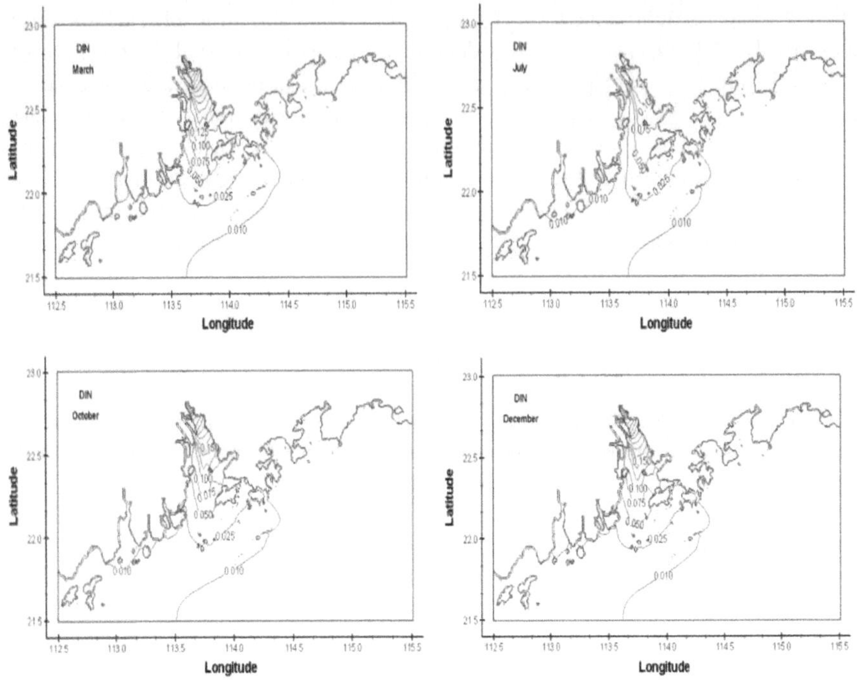

Fig. 11.17 DIN response concentration in the PRE with a single source at Hu-men in March, July, October, and December

The single-source response distributions of all pollution loads were used to determine the environmental capacity and optimal layout of DIN flux with an optimization method, in which the existing socio-economic status was taken into account.

Table 11.4 presents the categories of water quality level (WQL) according to DIN concentration. The existing WQL distributions in the PRE and adjacent coastal areas in the wet and dry seasons are shown in Fig. 11.18. It can be seen that the water quality in the whole PRE and some adjacent areas reaches WQL 5 with the worst polluted water. Thus, DIN reduction measures should be taken.

The modeling results indicated that, in the optimal layout, the total DIN load needs to be reduced by 66% to make the WQL in different functional areas in the PRE meet the requirements of the water quality standard. Figure 11.19 shows the WQL distributions in the wet and dry seasons in this optimal layout with the allowable

Table 11.4 Categories of water quality level (WQL) according to DIN concentration

	WQL 1	WQL 2	WQL 3	WQL 4	WQL 5
DIN concentration (mg/L)	<0.2	>0.2 <0.3	>0.3 <0.4	>0.4 <0.5	>0.5

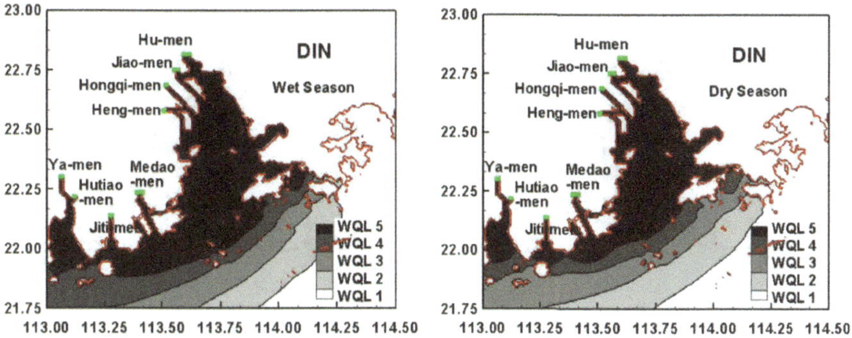

(a) Wet season (b) Dry season

Fig. 11.18 Modeled water quality level distribution of DIN with existing load

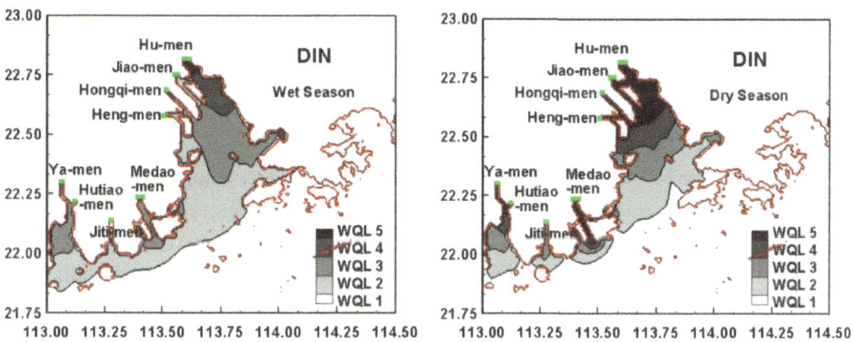

(a) Wet season (b) Dry season

Fig. 11.19 Modeled water quality distribution of DIN with the allowable load

load. The water quality reaches WQL 1 and WQL 2 in most areas of the domain except for the local area in the northern PRE.

However, even assuming that all planned reduction measures for DIN were implemented in the relevant catchments for the next 18 years (until 2030), the maximum DIN load reduction is only about 35% being about half the required reduction. Figure 11.20 shows the WQL distributions in the wet and dry seasons in this scenario with the planned DIN load. The water quality in most areas in the PRE would still remain WQL 5 in this scenario even though the water quality is improved significantly compared with that in the existing condition.

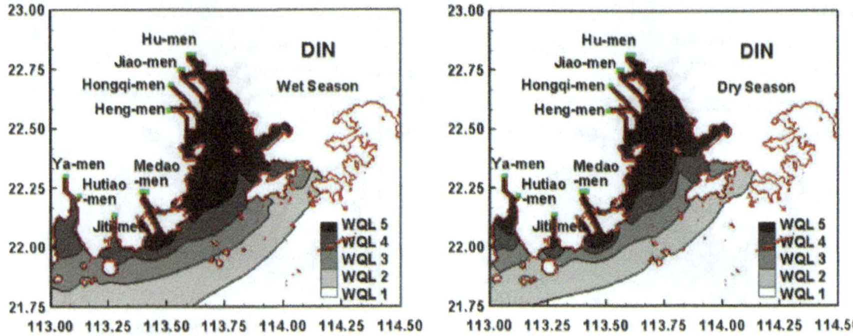

(a) Wet season (b) Dry season

Fig. 11.20 Modeled water quality distribution of DIN with planned load

11.2.6 Conclusions

1. Based on the in situ survey data, a numerical water quality model was used to investigate the distributions and seasonal variations of DIN and COD concentrations in the Pearl River Estuary and adjacent coastal areas. The model results generally agree well with the observed data from the present and previous studies.
2. According to the national seawater quality standard, the water quality in the whole PRE and some adjacent coastal areas reaches WQL 5 in the existing conditions. Model results show that a 66% DIN load reduction is required to enable the WQL in all the functional areas in the PRE to meet the requirements of water quality standard. Taking all the planned measures on DIN load reduction into account in the catchments until 2030, namely a 35% DIN load reduction ratio, the water quality in most areas in the PRE would still be WQL 5. Further, reduction measures are therefore still required.

11.3 Numerical Simulations of Tidal Flow and Sediment Transport for Design of a Deepwater Port in East China Sea

11.3.1 Models for Two-Dimensional Sediment Transport and Quasi-Three-Dimensional Tidal Flow

1. **Two-dimensional model for sediment transport**

Movements of suspended sediments dominate the overall sediment transport in this application case, thus allowing the use of the following transport model with decoupled two-dimensional tidal flow,

$$\frac{\partial(\mathrm{SH})}{\partial t}+\frac{\partial(pS)}{\partial x}+\frac{\partial(qS)}{\partial y}-\frac{\partial}{\partial x}\left(\mathrm{HD}_{xx}\frac{\partial S}{\partial x}+\mathrm{HD}_{xy}\frac{\partial S}{\partial y}\right)$$
$$-\frac{\partial}{\partial y}\left(\mathrm{HD}_{yx}\frac{\partial S}{\partial x}+\mathrm{HD}_{yy}\frac{\partial S}{\partial y}\right)=E \quad (11.3.1)$$

where S is the depth-averaged concentration of suspended sediment; E is a source term due to sediment erosion–deposition; D_{xx}, D_{xy}, D_{yx}, and D_{yy} are the depth-averaged effective diffusion coefficients.

Bed level changes satisfy

$$\frac{\partial z_\mathrm{b}}{\partial t}+\frac{1}{1-\phi}\left(\frac{\partial q_x}{\partial x}+\frac{\partial q_y}{\partial y}\right)=0 \quad (11.3.2)$$

where z_b is the bed level; ϕ is the bed porosity; q_x and q_y are the sediment fluxes in the x- and y-directions.

The determination of source term follows the proposed formula in the Technical Regulation of Modeling for Tidal Current and Sediment on Coast and Estuary (JTJ/T 233-98) [13],

$$E=\alpha\omega(S_*-S) \quad (11.3.3)$$

where α is the coefficient of saturation recovery; ω is the settling velocity of sediment; S_* is the sediment carrying capacity.

As recommended by the Code of Hydrology for Sea Harbor (JTJ 213-98) [14]. The sediment carrying capacity is given by a formula proposed by Liu Jiaju

$$S_*=0.0273\gamma_\mathrm{s}\frac{(|u_\mathrm{c}|+|u_\mathrm{w}|)^2}{\mathrm{gh}} \quad (11.3.4)$$

where γ_s is the bulk density of sediment particles; u_c is the sum of time-averaged tidal current velocity and the time-averaged wind-induced current velocity; u_w is the averaged horizontal wave particle velocity; h is the water depth; g is the gravitational acceleration.

Assuming the sediment gradation under the silt condition, the weighted mean sediment grain size is

$$D_k=\sum_1^n D_n f_n \quad (11.3.5a)$$

the weighted mean settling velocity is

$$\omega_k=\sum_1^n \omega_n f_n \quad (11.3.5b)$$

and the modified concentration of suspended silt sediment is

$$S_k = S_* \left[\sum_1^n F_n^{1/F_n} f_n \right] \tag{11.3.5c}$$

where f_n is the weight percentage of grouped sediment, F_n is the modified coefficient of grain size grouping.

$$F_n = D_m/(D_n + A/D_n)$$

The specified grain size is $D_m = 0.11$ mm and specified area is $A = 0.0024\ \text{mm}^2$.

The governing Eqs. (11.3.1–11.3.4) are numerically solved using the ADI-TOASOD (third-order convection and second-order diffusion) difference schemes, which have second-order accuracy in time and third-order accuracy in space.

With no measurement data available for deposition of suspended sediments, we rely on Liu Jiaju formula to make a prediction,

$$P = \frac{\omega_k S_k T}{\gamma_s}\left\{K_1\left[1-\left(\frac{d_1}{d_2}\right)^3\right]\sin\theta + K_2\left[1-\frac{d_1}{2d_2}\left(1+\frac{d_1}{d_2}\right)\right]\cos\theta\right\} \tag{11.3.6}$$

where d_1 and d_2 are the water depth in meters under the tide action before and after dredging; θ is the angle between the current direction and the shoreline in degrees; K_1 and K_2 are empirical constants with $K_1 = 0.35$ and $K_2 = 0.13$.

2. **Quasi-three-dimensional model for tidal flow**

As the domain of simulation related to the project area is large, a quasi-(layered) three-dimensional mathematical model of tidal flow is used. The governing equations in the form of layered integration (the kth layer is taken as an example) are as follows.

The continuity equation can be written as

$$w_{k-1/2} = -\sum_{k=k}^{K}\left\{\frac{\partial(hu)}{\partial x} + \frac{\partial(hv)}{\partial y}\right\} \tag{11.3.7}$$

Integration along the vertical direction gives

$$\frac{\partial\zeta}{\partial t} + \sum_{k=1}^{K}\left\{\frac{\partial(hu)}{\partial x} + \frac{\partial(hv)}{\partial y}\right\} = 0 \tag{11.3.8}$$

where k is the index for a layer with K being the total number of layers in the vertical direction.

The momentum equation in the x-direction can be written as

$$\left.\frac{\partial q_x}{\partial t}\right|_k + \left[\frac{\partial \bar{u} q_x}{\partial x} + \frac{\partial \bar{v} q_x}{\partial y}\right]_k = f q_y|_k - \Delta Z g \left.\frac{\partial \zeta}{\partial x}\right|_k - \Delta Z \frac{g}{\rho_0} \int_z^{\zeta} \frac{\partial \rho}{\partial x} \mathrm{d}s$$
$$+ \left\{\frac{\partial}{\partial x} \varepsilon_{\mathrm{h}} \Delta Z \left[\frac{\partial \bar{u}}{\partial x} + \frac{\partial \bar{u}}{\partial x}\right] + \frac{\partial}{\partial y} \varepsilon_{\mathrm{h}} \Delta Z \left[\frac{\partial \bar{u}}{\partial y} + \frac{\partial \bar{v}}{\partial x}\right]\right\}_k$$
$$+ (w\bar{u})_{k+1/2} - (w\bar{u})_{k-1/2} + \frac{\tau_{xz}|_{k-1/2} - \tau_{xz}|_{k+1/2}}{\rho_0} \tag{11.3.9}$$

The momentum equation in the y-direction can be written as

$$\left.\frac{\partial q_y}{\partial t}\right|_k + \left[\frac{\partial \bar{u} q_y}{\partial x} + \frac{\partial \bar{v} q_y}{\partial y}\right]_k = - f q_x|_k - \Delta Z g \left.\frac{\partial \zeta}{\partial y}\right|_k - \Delta Z \frac{g}{\rho_0} \int_z^{\zeta} \frac{\partial \rho}{\partial y} \mathrm{d}s$$
$$+ \left\{\frac{\partial}{\partial x} \varepsilon_{\mathrm{h}} \Delta Z \left[\frac{\partial \bar{v}}{\partial x} + \frac{\partial \bar{u}}{\partial y}\right] + \frac{\partial}{\partial y} \varepsilon_{\mathrm{h}} \Delta Z \left[\frac{\partial \bar{v}}{\partial y} + \frac{\partial \bar{v}}{\partial y}\right]\right\}_k$$
$$+ (w\bar{v})_{k+1/2} - (w\bar{v})_{k-1/2} + \frac{\tau_{yz}|_{k-1/2} - \tau_{yz}|_{k+1/2}}{\rho_0} \tag{11.3.10}$$

where h is the thickness (water depth) of each layer; f is the Coriolis parameter; u, v, and w are the layer-average velocity components in the x-, y-, and z-directions, respectively; τ_{xz} and τ_{yz} are the stress components.

The free surface boundary conditions are

$$\tau_{xz}|_{\zeta} = \rho_{\mathrm{a}} C^* W W_x \tag{11.3.11a}$$

$$\tau_{yz}|_{\zeta} = \rho_{\mathrm{a}} C^* W W_y \tag{11.3.11b}$$

where C^* is the friction coefficient; ρ_{a} is the density of air; Wx and Wy are the wind velocities in the x- and y-directions; W is the wind speed.

The bottom boundary condition is

$$\varepsilon_{\mathrm{v}} \frac{\partial u}{\partial z} = \frac{\tau_{\mathrm{b}}}{\rho} = (u_*)^2 \tag{11.3.12}$$

where τ_{b} is the bottom stress; u_* is the bottom friction velocity.

11.3.2 Determinations of Calculation Domain and Boundary Conditions

1. **Site descriptions**

There are many large or small islands near shore in the surrounding area, forming the anchors for the deepwater port to be constructed. In the design phase, the tidal flow field and sediment transport before and after constructions of various layouts were performed using numerical simulations, which provided the initial assessments of the port designs.

The deepwater port is located at the mouth of Hangzhou Bay, not distant away from the Yangtze Estuary with large flood discharge in the north. Figures 11.21 and 11.22 show the overall deepwater port area and a zoom-in project site, respectively. In order to determine the proper domain of simulations, investigations into whether the flood discharge of the Yangtze Estuary affects the project set are deemed to be necessary at first.

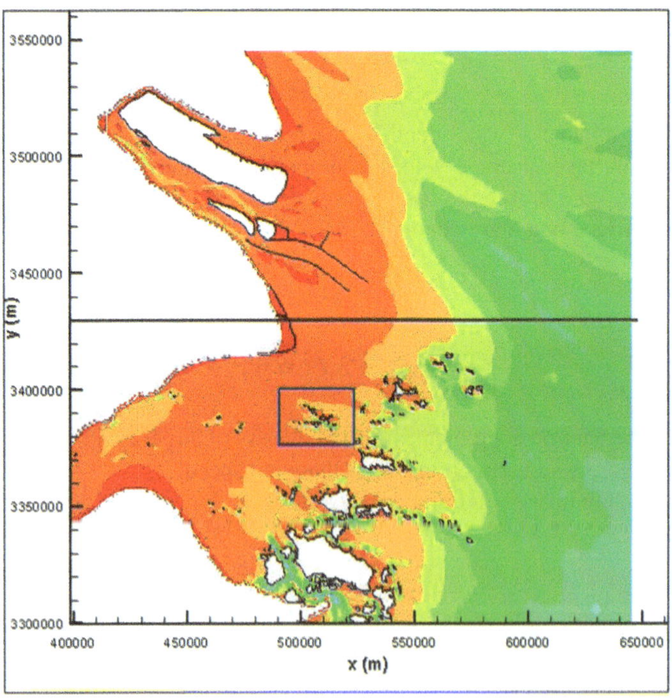

Fig. 11.21 Location of the project area (the rectangular area) at the mouth of Hangzhou Bay near the Yangtze Estuary

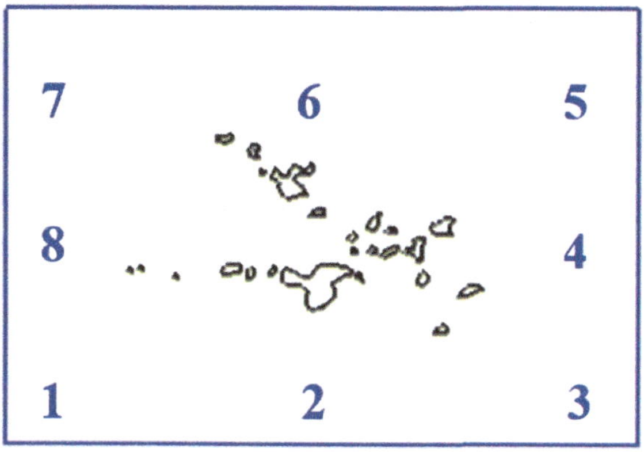

Fig. 11.22 Zoomed project area with control points

2. **Effect of the Yangtze River runoff on the current in the project area**

Based on historical data, the average annual runoff in the Yangtze Estuary is 924 billion cubic meters per year, and the average discharge is 29,300 m^3/s. The observed maximum discharge is 92,600 m^3/s in 1954 and the minimum discharge is 4620 m^3/s in 1979. In order to analyze the effect of the Yangtze River runoff on the current in the project area, a scenario with the maximum effect, i.e., the maximum discharge in neap tide, is selected to compare against a case without discharge in neap tide. Figure 11.23 shows the changes vs time of water level, current speed, and direction at Control Point 5 extracted from the simulation results for the two scenarios selected for comparison.

It is indicated that the calculated results with and without discharge have not demonstrated any appreciable disagreement. Since peak-discharge and no-discharge are used in the calculation, all other scenarios of the Yangtze River runoff will be bounded by these two extreme cases. A conclusion is reached that the effect of the discharge from the Yangtze Estuary on the hydrodynamics characteristics in the project area is small enough to be neglected. Therefore, the discharge of the Yangtze River will not be included in all subsequent numerical simulations.

3. **Model calibration for large region**

Nested models with a large region nesting a small area are combined to analyze the flow field in the project area. A model for the large region with coarse grid serves to provide boundary conditions for a refined tidal flow model with open boundaries in the local project area.

The simulation domain of the large model is shown in Fig. 11.24, in which there are a number of tidal stations. The tidal flow model for the large region can be calibrated and validated using the water-level observations at these stations. Figure 11.25 shows the simulated water levels at two tidal stations from the calibrated model as compared against observations. The numerical results agree with the observations well. The water level, current speed, and direction at the open boundaries of the smaller project site (see Fig. 11.22) are extracted from the numerical results of the large model as

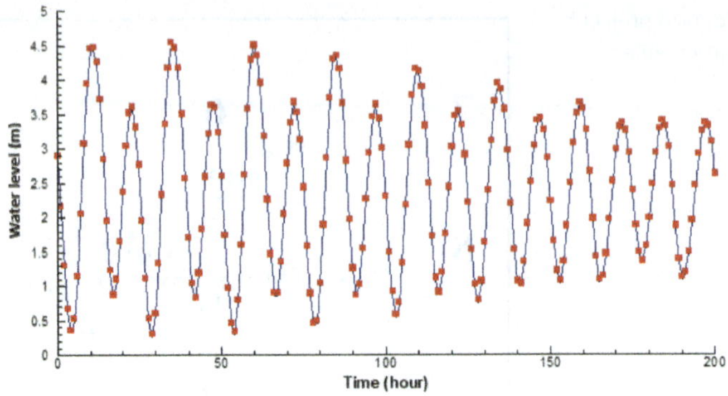

(a) Water level

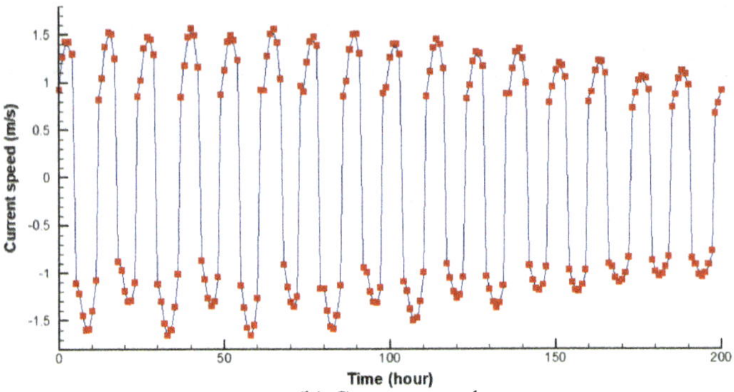

(b) Current speed

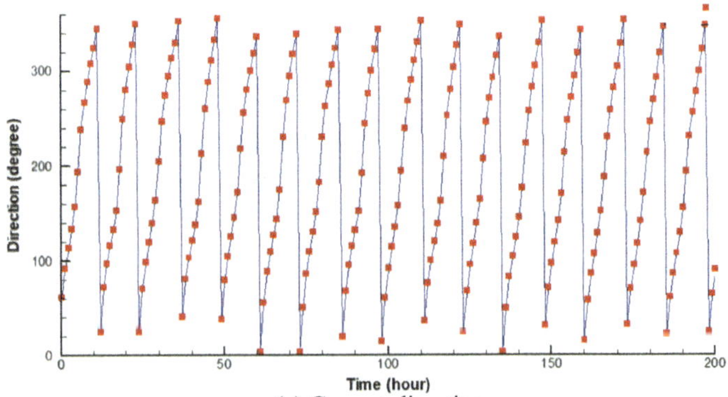

(c) Current direction

Fig. 11.23 Effect of the Yangtze River runoff on hydrodynamic characteristics at Control Point 5 (line: without discharge from the Yangtze River, points: with the discharge from the Yangtze River)

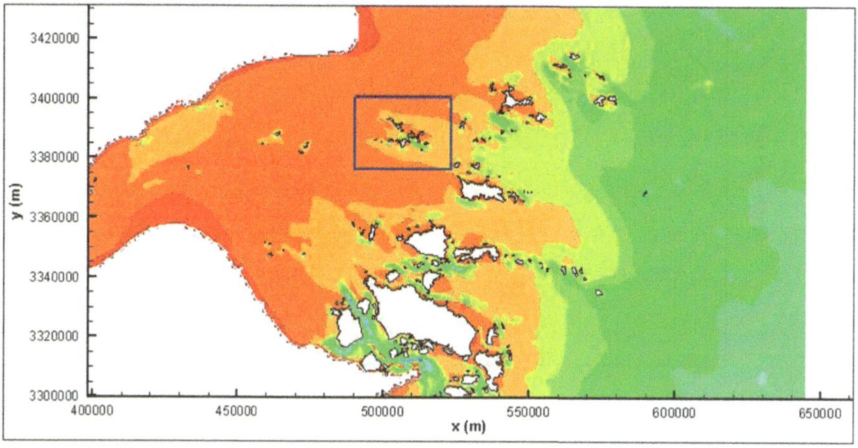

Fig. 11.24 Calculating domain of the large region model

the imposing boundary conditions of the refined tidal flow model in the local project area.

11.3.3 Validation of Tidal Flow Model in the Local Project Area

1. **Validity of the boundary conditions of the local model**

In order to verify the accuracy of the open boundary conditions for the local model obtained by means of interpolating data extracted from the large region model results, the local model results and the large region model results at eight picked points (see Fig. 11.22 for their locations) with the same positions in the two models are compared. If the two model results at these points are in agreement, it is deemed that the numerical results of the large region model and the small local model are consistent. Figure 11.26 shows the comparison of water-level results at Control Point 1 and Point 5 simulated by the region model and the local model. It can be seen that the simulated results from the two models are identical, confirming that the two models are indeed in a desired consistency.

2. **Validation of the local tidal flow model using observations**

There are small amount of water-level data from tidal station observations in the local model area that can be used to further validate the refined local model. Figure 11.27 shows the simulated water-level results against the observations in typical spring tide and middle tide at a station within the area. It is shown that the numerical results are in good agreement with the observation data in tidal shape, tidal range, and phase, which confirms the refined local numerical model.

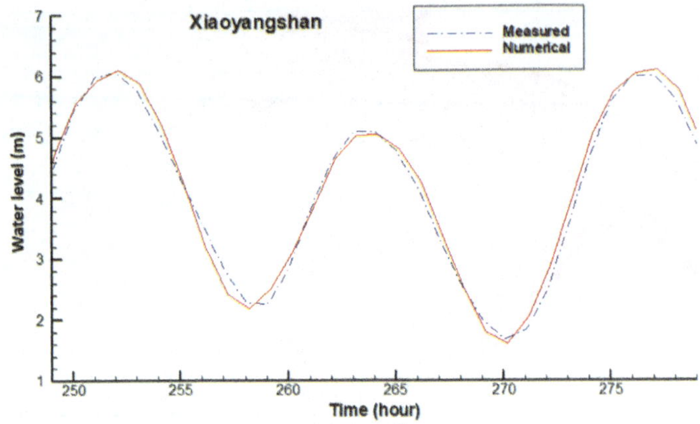

(a) A station

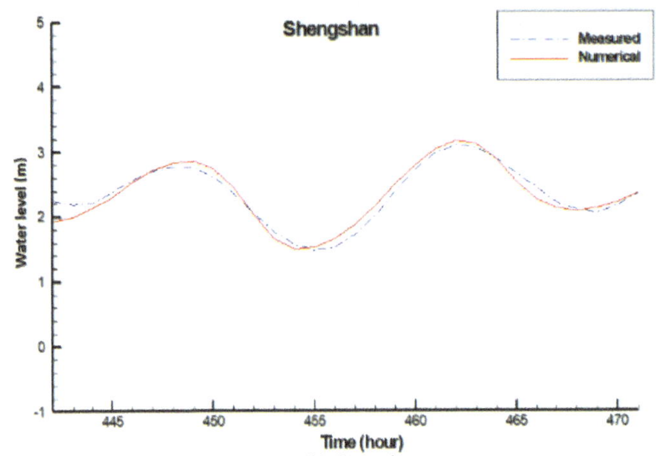

(b) B station

Fig. 11.25 Comparison of numerical and measured water-level results at two stations in the domain of the large region model

3. **Validation on current speed and direction using ADCP survey data**

To aid the design and engineering of the port, extensive data of tidal current and sediment have been collected in the project area, which included some from shipboard ADCP survey. Figure 11.28 shows a sketch of partial ADCP transects. These ADCP survey provided additional data for further validation of the numerical models. Figure 11.29 shows comparison of tidal speed and direction at typical positions on transects of ADCP09 and ADCP12 in spring tide. Figure 11.30 shows the comparison of tidal flux through ADCP09 and ADCP12. It can be seen that the local model reproduces the observed flow field and tidal transport in the local project area.

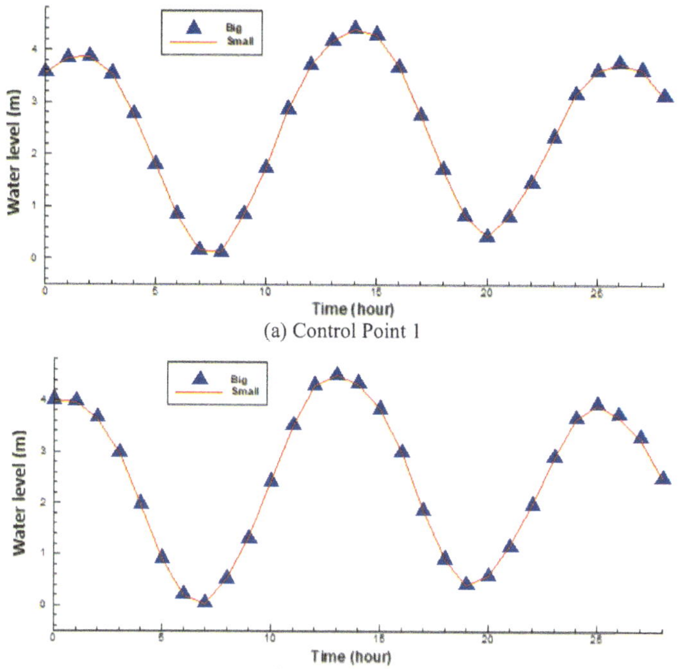

(b) Control Point 5

Fig. 11.26 Comparison of water-level results simulated by the region model (Big) and the local model (Small)

11.3.4 Analyses on Tidal Current in Design Layouts

1. **Design layouts**

Two planned layouts (Layout 1 and Layout 2) are selected to analyze the tidal current and sediment transport using the aforementioned flow and sediment transport models. In Layout 1, several islands are linked up together to form a stretch. Layout 2 is similar to Layout 1 with the exception of a waterway in the north being retained. For the convenience of description, the existing condition before constructions is labeled as Existing layout.

2. **Characteristic analysis of depth-averaged flow fields in design layouts**

Flow fields in the Existing and the two proposed layouts are simulated. Figure 11.31 shows the comparison of local flow fields at maximum flood in spring tide in the design layouts before and after dredging the navigation channel and the port basin. It can be seen that before and after dredging the current direction is generally parallel to the shoreline of the passageway in Layout 1. In Layout 2, due to the offload effects of the waterway, the main current direction is redirected to the midline of the waterway in the area near the west side of the waterway.

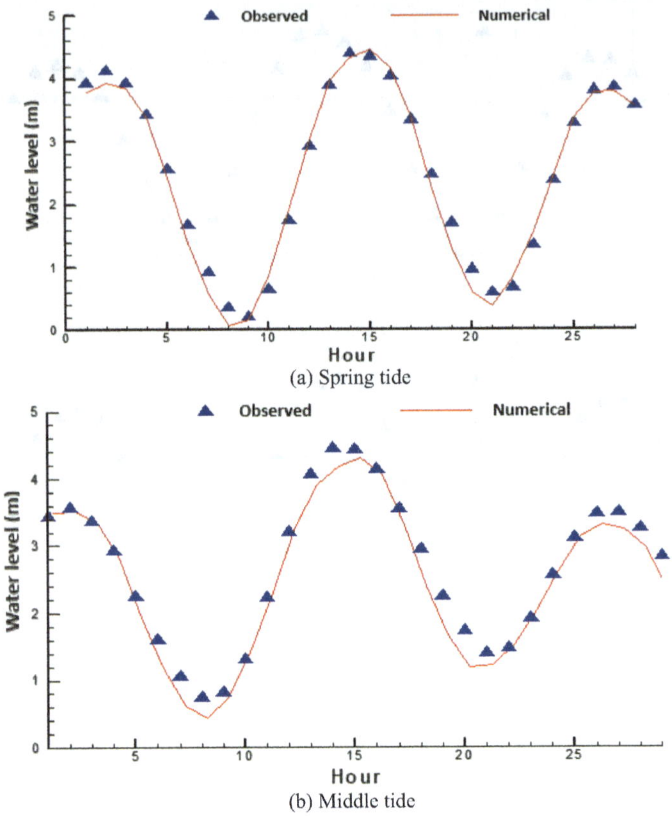

Fig. 11.27 Numerical water-level results by the local model compared with observations

3. **Analysis of 3D flow fields in the project area**

3D flow field in the project area provides critical information for design and construction. Using the validated 3D tidal flow model, the 3D flow fields in spring, middle, and neap tides, before and after dredging the navigation channel and the port basin in each layout, are simulated. Figure 11.32 shows the comparison of current velocity in the Existing and two design layouts after dredging in spring tide, at the surface, intermediate, and bottom layers, respectively. It can be seen that in each layout, the current speed at the surface layer is the largest among all layers, and the speed in an intermediate layer is larger than that in the bottom layer. The current direction basically remains the same in all layers.

4. **Comparisons of tidal transport**

The tidal flux through each ADCP transect can be calculated based on the flow fields in flood and ebb tides. Figure 11.33 shows the tidal flux through transects of ADCP09 and ADCP12 in spring tide. The tidal discharge and the maximum tidal flux in spring, middle, and neap tides in each layout can be further analyzed.

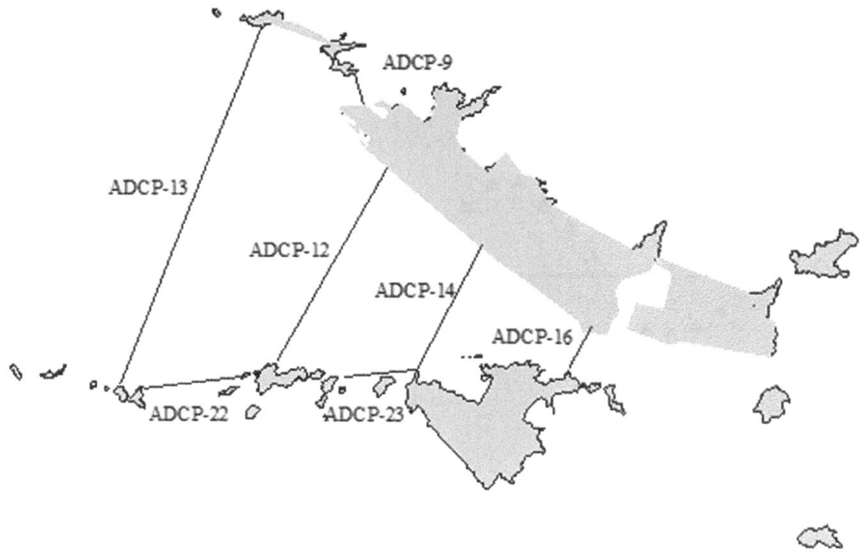

Fig. 11.28 Sketch of ADCP transects in the project area

Tables 11.5 and 11.6 present comparisons of the maximum tidal flux and the tidal discharge through ADCP09 and ADCP12 in flood and ebb tides, respectively. Compared with the results in the other two layouts, the maximum tidal flux and the tidal charge in Layout 2 change slightly in both flood and ebb tides, except that the values through ADCP09 decrease greatly from Existing layout due to the effect of the waterway. By comparing Layout 1 and Layout 2, it is seen that all the discharge amounts in flood and ebb tides in spring, middle, and neap tides in Layout 1 are larger than those in Layout 2. Overall, the discharge in Layout 1 is larger than that in Layout 2.

11.3.5 Analyses of Sediment Transport in Design Layouts

Changes in tidal dynamics introduced by the engineering projects will affect sediment transports and local erosion–deposition balances in the project area and in turn affect the cost of operation and maintenance of the port and safety of navigation. Based on the simulation data for tidal flows, sediment transport and local erosion–deposition in the project area are further analyzed using the sediment transport model.

1. **Validation of the sediment transport model**

Before proceeding to the simulation of sediment transport and local erosion–deposition, it is necessary to calibrate and validate the sediment transport model parameters based on the observed data of suspended sediment concentration (SSC) in the

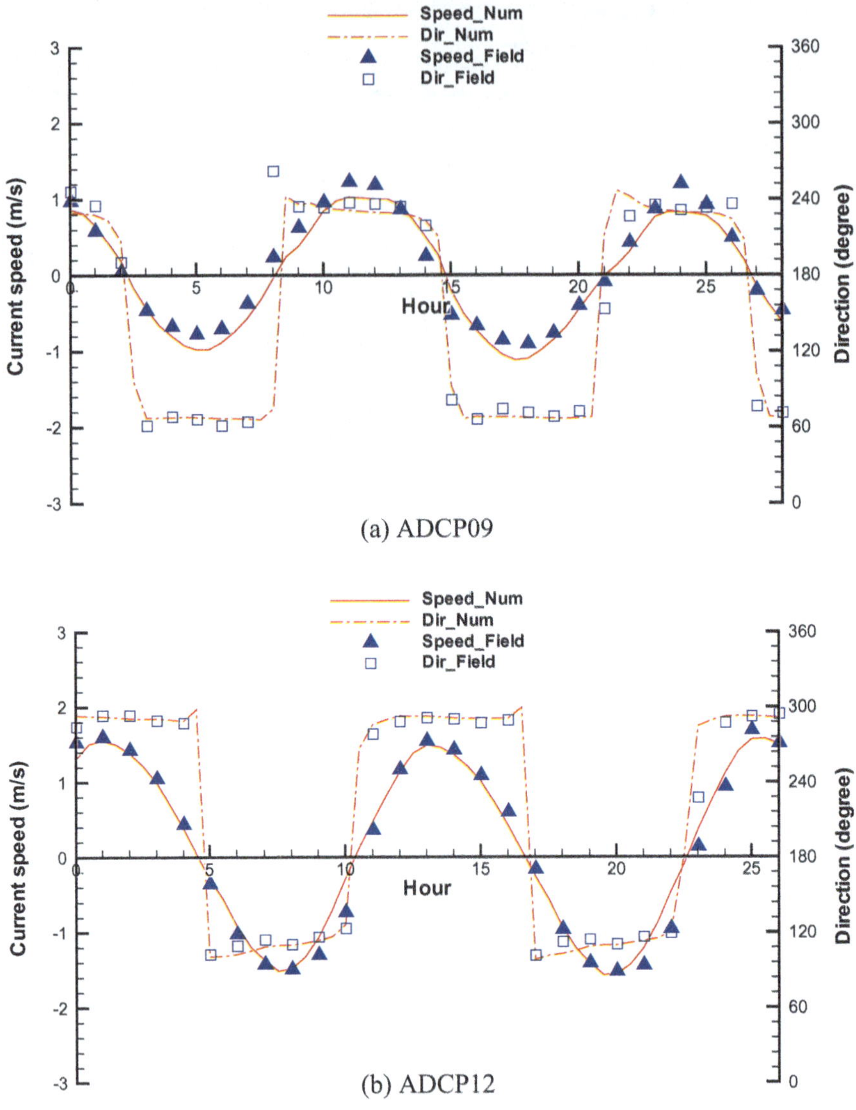

Fig. 11.29 Validation on current speed and direction at certain points on ADCP09 and ADCP12

project area and surrounding waters and to adjust the boundary conditions of the model accordingly. Figure 11.34 shows the simulated SSC results compared with the observations at two stations. It can be seen that the simulated and observed SSC results are in agreement in general.

The observed distribution of seabed erosion–deposition in one year before and after blocking the waterway in the local observation area is shown in Fig. 11.35a. This observation area is relatively remote from the area of large-scale construction.

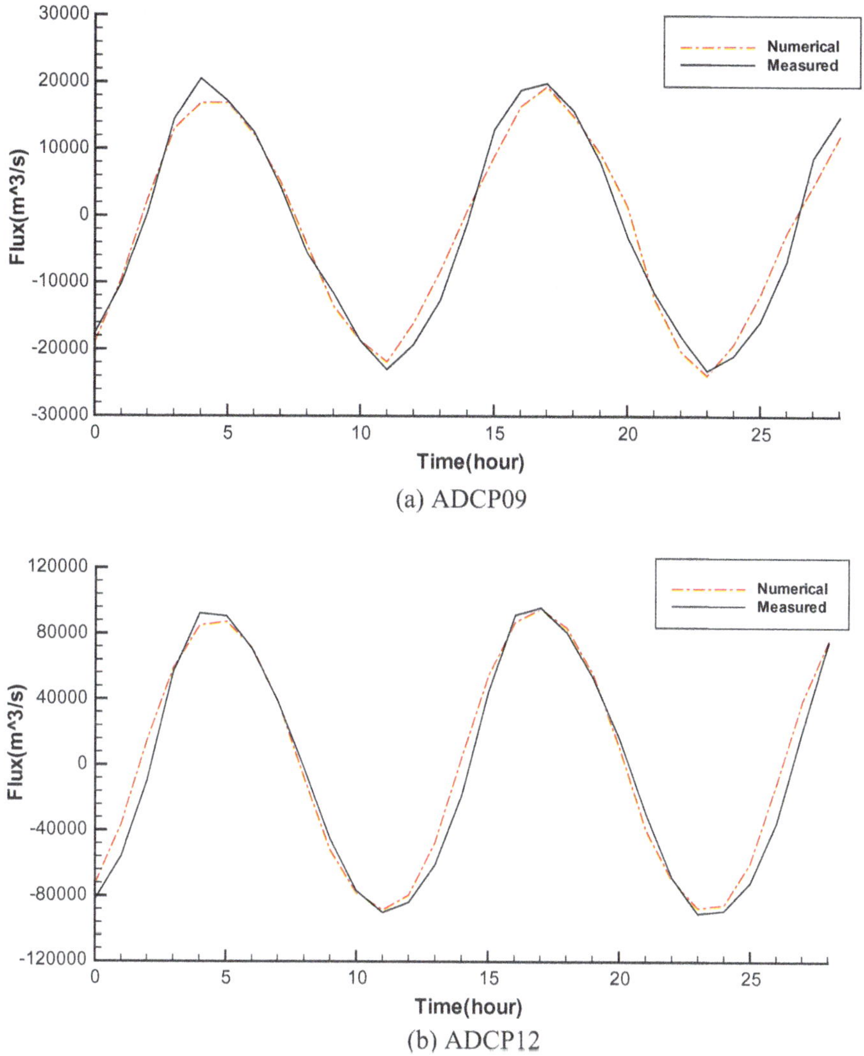

Fig. 11.30 Comparison results of tidal flux through ADCP09 and ADCP12

The seabed erosion–deposition in this area is less affected by engineering changes in the construction area and reflects the natural influence on the morphological changes.

The observed data show that there is no significant deposition in this area before blocking the waterway, but the significant morphological change occurs after its blockage. Therefore, it can be used to verify the sediment transport model for simulating local erosion–deposition.

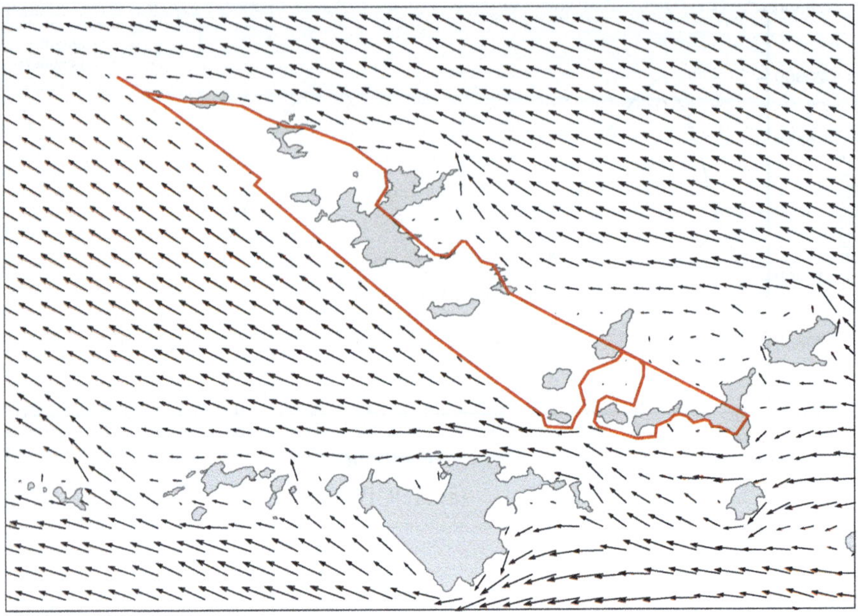

(a) Layout 1

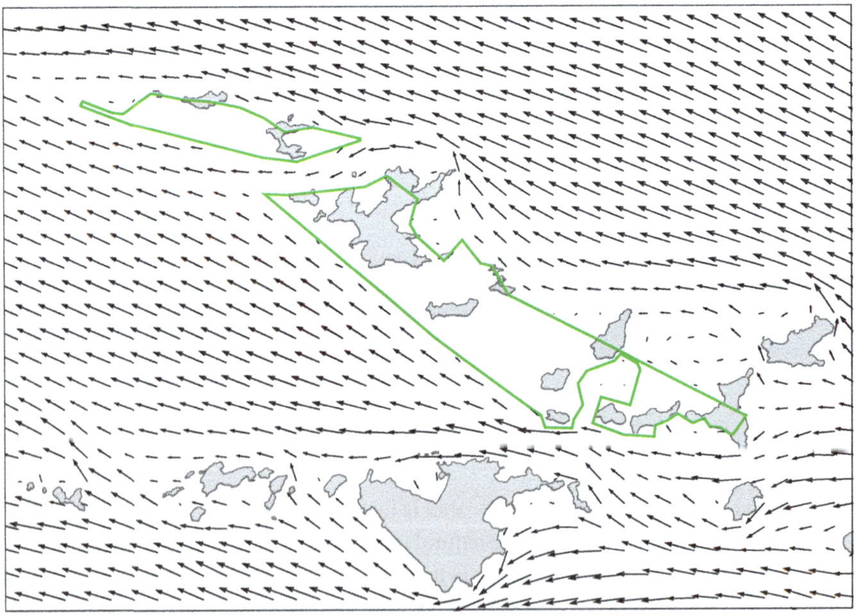

(b) Layout 2

Fig. 11.31 Flow fields at maximum flood in spring tide in the design layouts before (red) and after (black) dredging

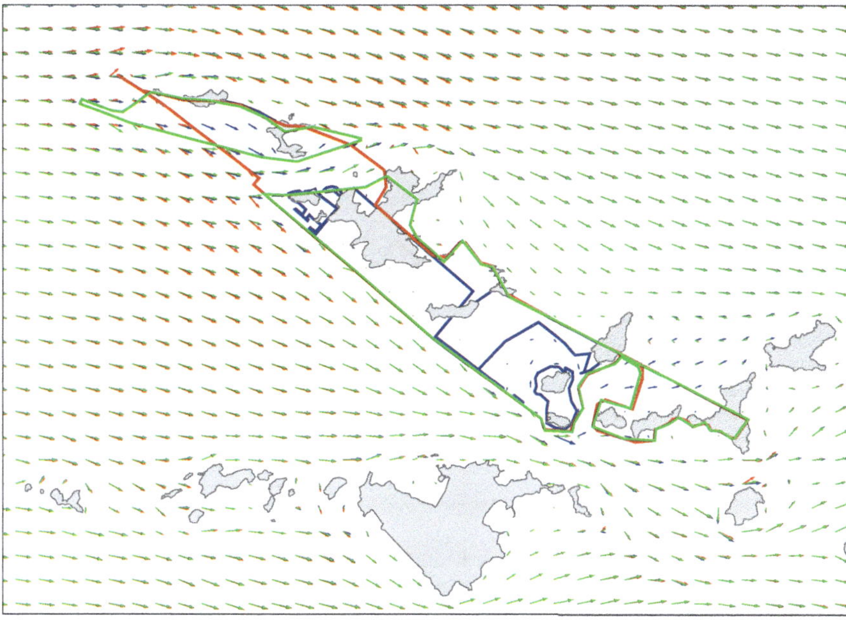

(a) Bottom layer

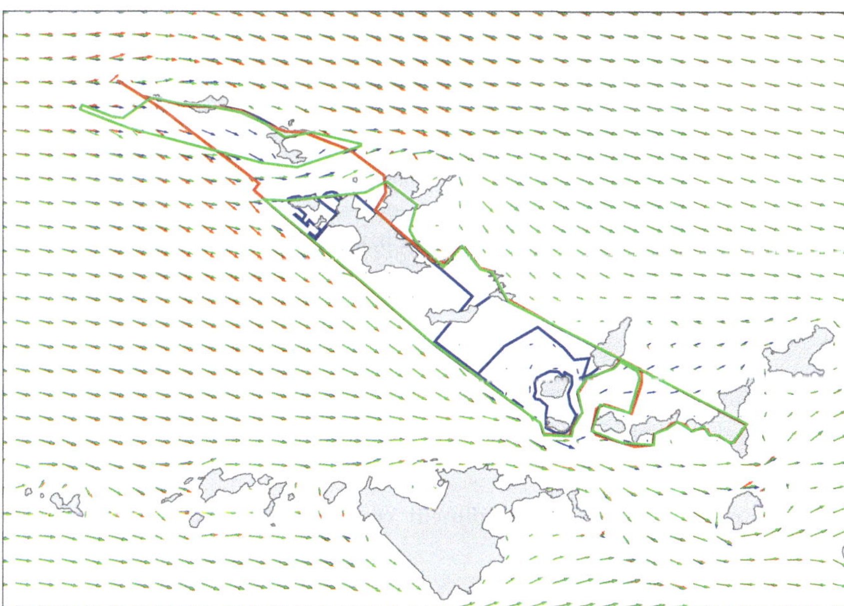

(b) Intermediate layer

Fig. 11.32 Comparison of 3D flow fields in three different layouts obtained by 3D model (at maximum ebb in spring tide; blue: Existing layout, red: Layout 1, green: Layout 2)

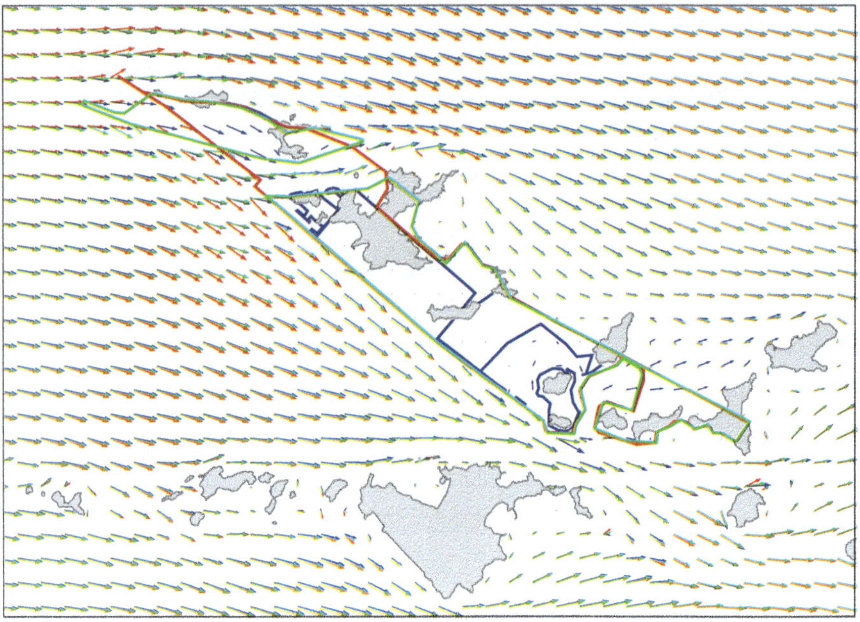

(c) Surface layer

Fig. 11.32 (continued)

As shown in Fig. 11.35a, the change of seabed erosion–deposition in one year can be obtained by subtracting the bathymetric survey data before blocking the waterway from those after blocking within the same time frame. Figure 11.35b shows the simulated distribution of seabed erosion–deposition from the sediment transport simulation model. By comparing Fig. 11.35a, b, it can be seen that the simulated erosion–deposition trends well with the observation in general. The simulated amounts of erosion–deposition and the corresponding impact area are smaller than the actual observations, but the deviations are not large enough to render the simulations useless.

The above validations indicate that the sediment transport model can successfully simulate the suspended sediment concentration and the morphological changes caused by sediment erosion–deposition in the waters near Yangshan under natural conditions.

2. **Comparison of suspended sediment concentration distributions in two layouts**

The distributions of suspended sediment concentration in flood and ebb tides during spring, middle, and neap tides in each layout are analyzed, respectively, using the numerical models. Figures 11.36 and 11.37 show the distributions of suspended sediment concentration at maximum flood and maximum ebb in spring tide in Layout 1 and Layout 2, respectively.

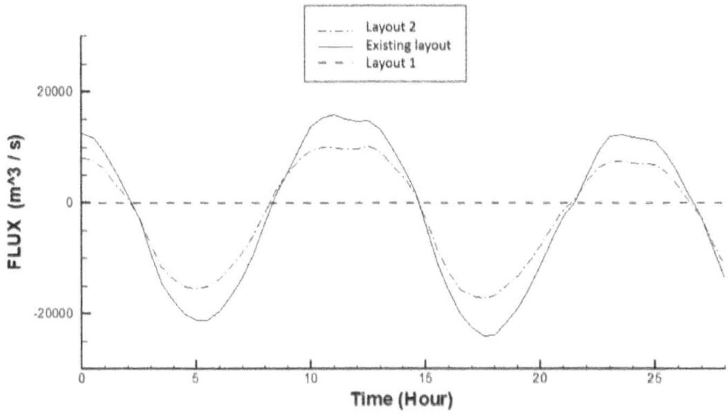

(a) ADCP09

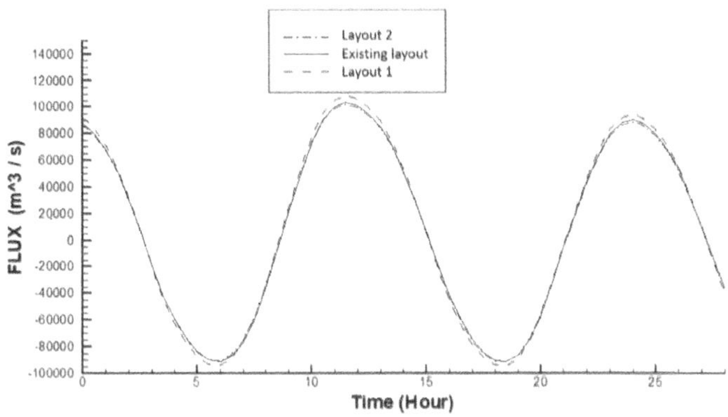

(b) ADCP12

Fig. 11.33 Time series of tidal flux through ADCP09 and ADCP12 in spring tide

3. **Comparison of local seabed erosion–deposition in two layouts**

The local seabed erosion–deposition in each layout is analyzed, respectively, using the numerical model. Figure 11.38 shows the difference of the seabed erosion–deposition profiles between the two layouts. Due to the effect of blocking the waterway, the intensity of sediment deposition along the flow path of the original waterway in the Western port area (at the top of the northwestern side) is increased in Layout 1 compared to Layout 2. However, because the shoreline in Layout 1 is uninterrupted and relatively streamlined as a result of blocking the waterway, the flow tends to go parallel along the shoreline to form a more regularized pattern with less particle decelerations, which is conducive to the transport of suspended sediment. Thus, the intensity of deposition at the wharf apron of the Western port and at the port basin area in the south of the waterway is actually decreased in Layout 1 in comparison with Layout 2. The trend and magnitude of erosion–deposition in other areas are generally the same in the two layouts.

Table 11.5 Comparisons of the maximum tidal flux through ADCP09 and ADCP12 in flood tide and ebb tide after dredging the navigation channel and port basin

ADCP09						
Ratios of the maximum tidal flux between two layouts	Flood tide			Ebb tide		
	Spring tide	Middle tide	Neap tide	Spring tide	Middle tide	Neap tide
Layout 2/existing layout	0.70	0.72	0.80	0.75	0.76	0.78
ADCP12						
Layout 1/existing layout	1.04	1.07	1.12	0.98	1.01	1.04
Layout 2/existing layout	0.98	0.98	1.02	0.95	0.95	0.97
Layout 1/layout 2	1.06	1.09	1.10	1.03	1.07	1.07

Table 11.6 Comparisons of the tidal discharge through ADCP09 and ADCP12 in flood tide and ebb tide after dredging the navigation channel and port basin

ADCP09						
Ratios of the tidal discharge between two layouts	Flood tide			Ebb tide		
	Spring tide	Middle tide	Neap tide	Spring tide	Middle tide	Neap tide
Layout 2/existing layout	0.76	0.78	0.81	0.76	0.74	0.83
ADCP-12						
Layout 1/existing layout	1.06	1.09	1.11	0.99	1.04	1.00
Layout 2/existing layout	0.99	1.00	1.01	0.96	0.97	0.93
Layout 1/layout 2	1.07	1.09	1.10	1.04	1.07	1.08

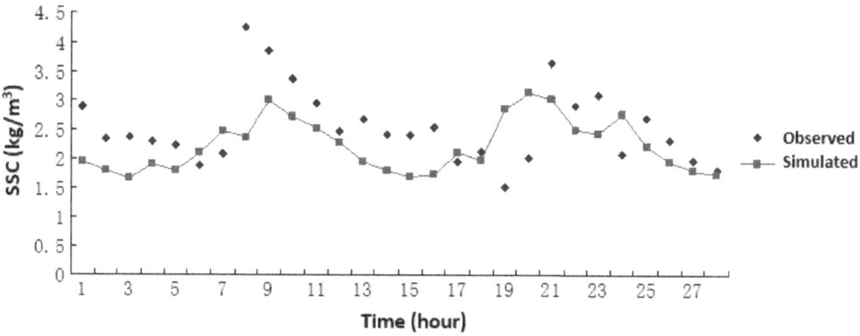

(a)W1 station

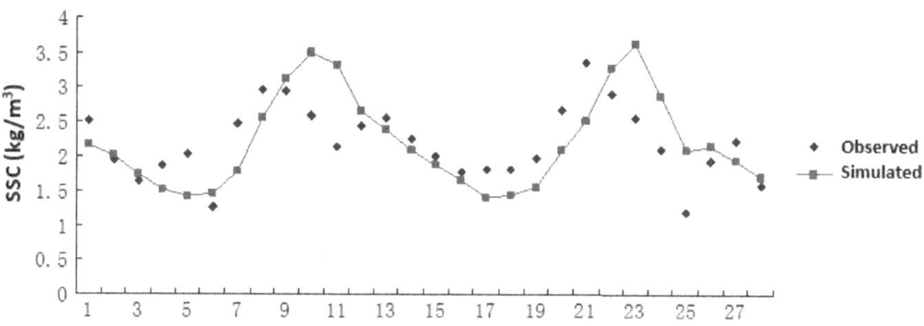

(b)W3 station

Fig. 11.34 Simulated SSC results compared with the observations at two stations

In conclusion, the above analyses of the change of suspended sediment concentration and the prediction of seabed erosion–deposition suggest that Layout 2 is superior to Layout 1 in the Western port area (on the northwestern side of the waterway) along the streamline of the waterway. Layout 1 is superior to Layout 2 in the port basin area located in the south of the waterway, and the difference between the two layouts in other areas is insignificant.

11.4 Comparisons of Numerical Predictions of Shoreline Evolutions Against Satellite Images of Friendship Port (Port of Nouakchott) in Mauritania

11.4.1 Introduction

In the 1970s, the Friendship Port (Port of Nouakchott) in Mauritania assisted by the Chinese government was built. Due to the urgent time of commencement, the

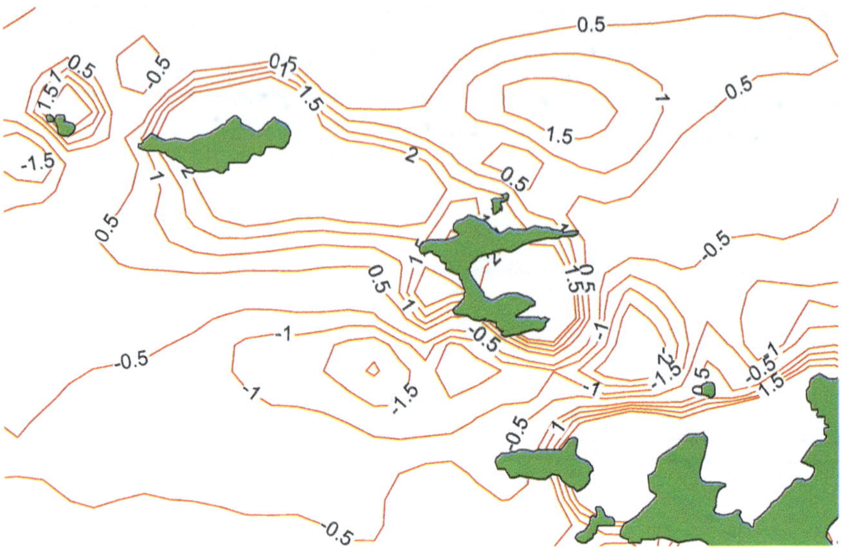

(a) Observed seabed erosion-deposition

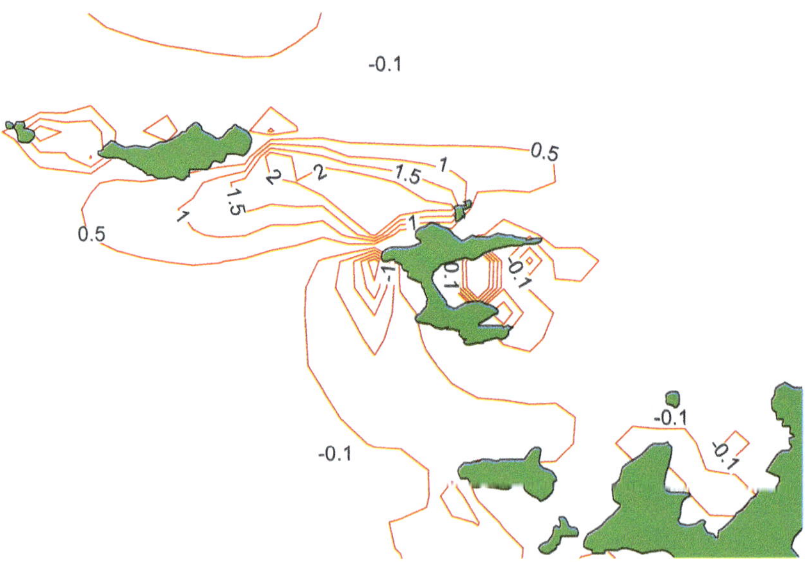

(b) Simulated seabed erosion-deposition

Fig. 11.35 Simulated and measured distributions of seabed erosion–deposition in one year before and after blocking the waterway in the local observation area (solid line: deposition; dotted line: erosion; unit: meter)

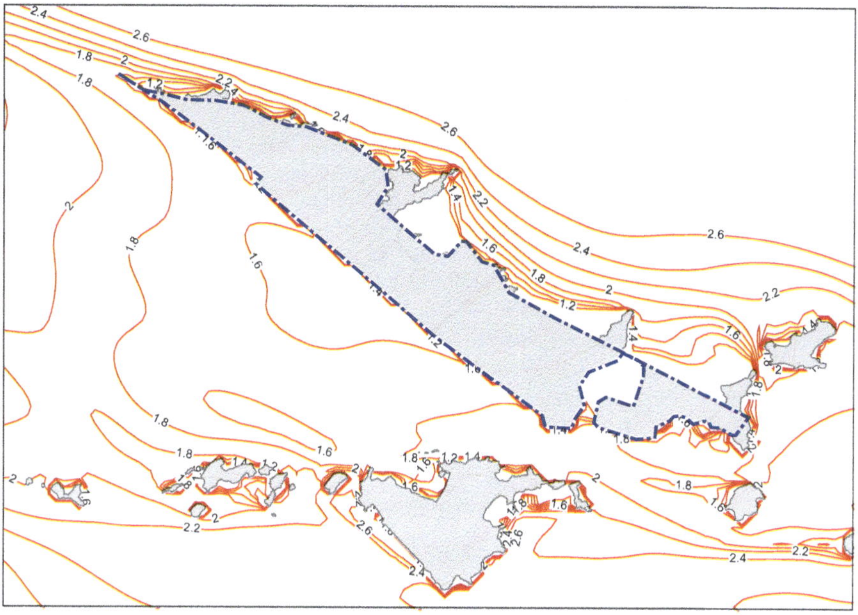

(a) Maximum flood

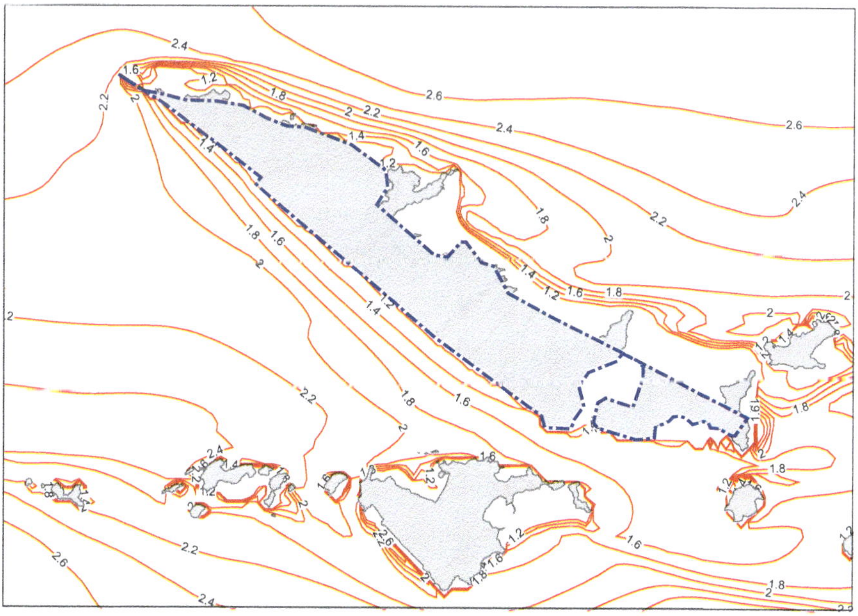

(b) Maximum ebb

Fig. 11.36 Distributions of suspended sediment concentration at maximum flood and maximum ebb in spring tide in Layout 1, unit: kg/m^3

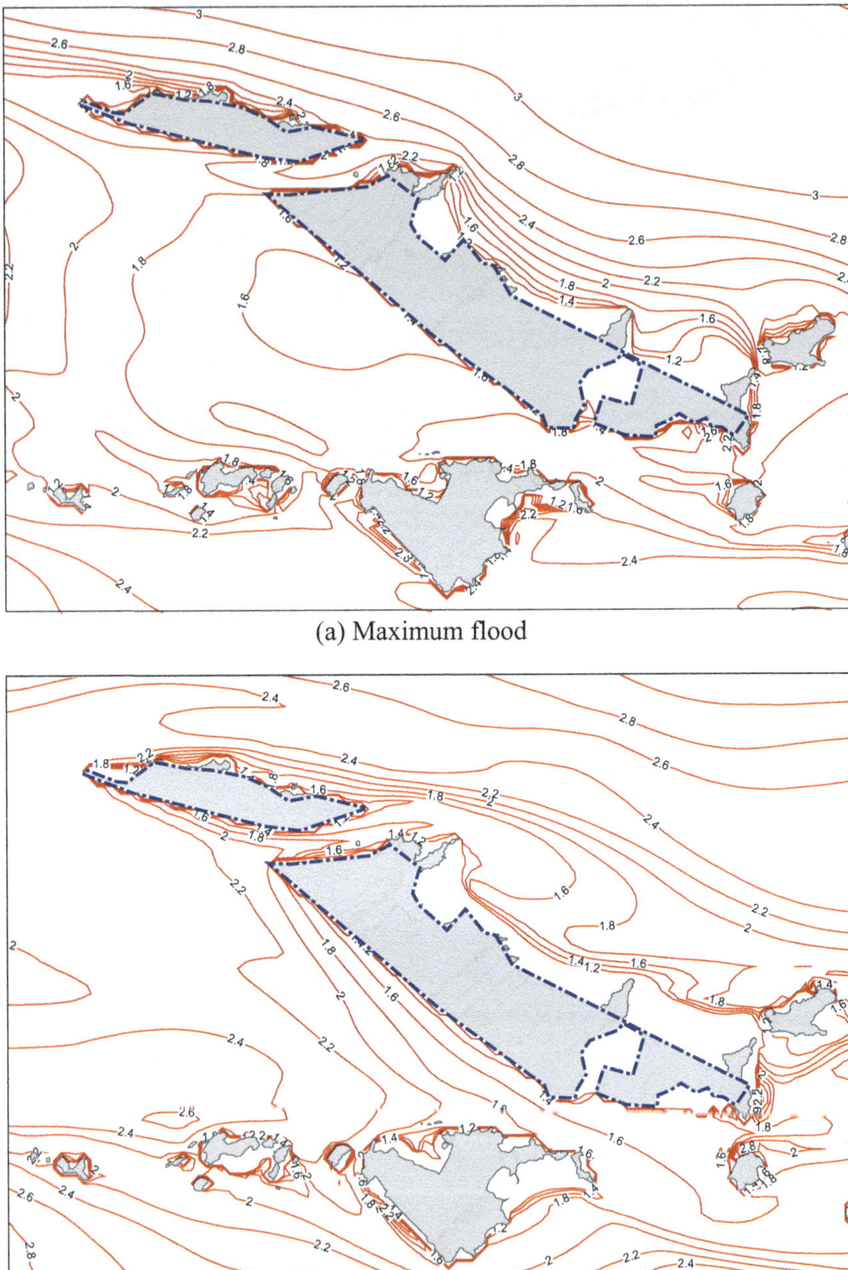

(a) Maximum flood

(b) Maximum ebb

Fig. 11.37 Distributions of suspended sediment concentration at maximum flood and maximum ebb in spring tide in Layout 2, unit: kg/m^3

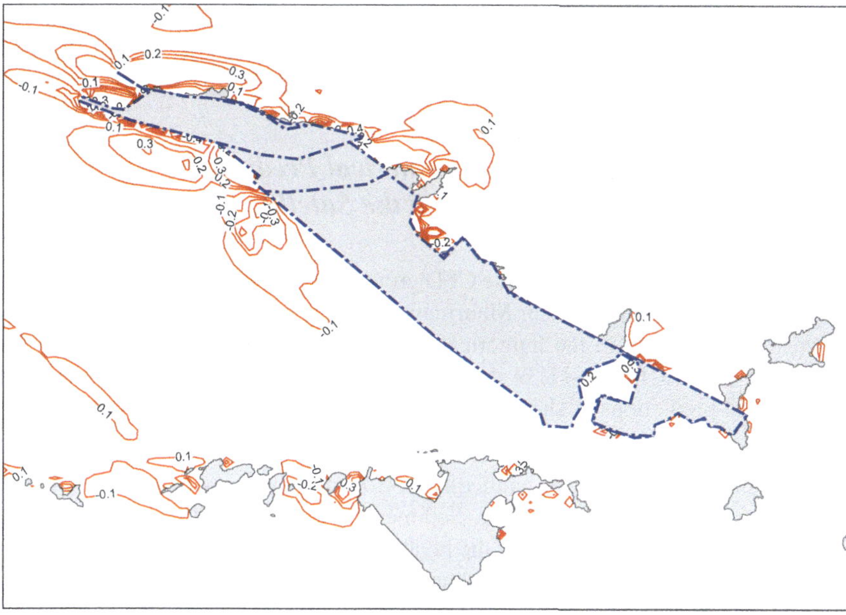

Fig. 11.38 The difference of the seabed erosion–deposition between Layout 1 and Layout 2 (Layout 1 minus Layout 2; solid line: relative deposition; dotted line: relative erosion; unit: m/a)

experiments by physical models on the design layout had not been able to carry out. In the early 1980s, the shoreline erosion along the downstream of the port had started to occur, endangering the safety of the port. In 1986, the Mauritanian government requested the Chinese government to develop a timely solution to safeguard the future operations of the port. This task was assigned to the First Harbor Engineering Investigation and Design Institute (FDINE) of Ministry of Transportations which originally undertook the design task of the Friendship Port.

Due to the urgent nature of the project, Academician Xie Shileng, engineering chief of the FDINE, decisively requested the participation of the Computational Fluid Dynamics (CFD) group in the Department of Mechanics at Tianjin University, which was just established, to study the repair plans using numerical model.

After discussions on the treatment of shoreline erosion between the FDINE and the CFD group at Tianjin University, an idea of building a groin at a certain appropriate position at the downstream of the port was put forward. The purpose was to move the shoreline erosion further away from the port in the downstream so that the port safety was not affected by the shoreline erosion. Various layouts with different locations to build the groin were investigated using numerical model (see Sect. 8.2.2 in Chap. 8 of this book). The numerical predictions of shoreline evolution over a duration of 30 years from 1986 to 2016 with an interval of two years for the proposed layout have been shown in Fig. 8.7b. Three months later, the proposed layout and the numerical predictions of shoreline evolution were submitted to the government of

Mauritania and approved. The engineering plan was eventually conducted on the basis of numerical studies by the CFD group at Tianjin University.

11.4.2 Comparisons Between Numerical Predictions of Shoreline Evolution and the Satellite Images

In 2017, Professor Dekui Yuan in the CFD group occasionally looked at the satellite images of Port of Nouakchott in Mauritania on Google Earth and found that they bear visual similarities to the numerical predictions of shoreline evolution by the CFD group in 1986. Figure 11.39 shows the comparison of the satellite picture and the numerical predictions of shoreline evolution.

It can be seen by analyzing the satellite images that

1. At Mauritanian coast, the wind direction is mainly from north to south for three quarters out of a full year. Sediments carried by the longshore current are deposited in the upstream of the breakwater, which creates a large new beach (around ④ in Fig. 11.39) in more than 30 years.
2. Due to a large amount of sediment deposition in the upstream of the breakwater, the south-going current increases the erosion capability in the downstream of the breakwater. The setting of the groin ① effectively moved the erosion of shoreline to the downstream of it. The shoreline around the port is thus protected.
3. It can be seen that, after the setting of another groin ⑤ built by Mauritania in 2011–2012, the shoreline in the upstream of this groin is not eroded, but the shoreline erosion occurs in the downstream of it.

The above are consistent with the judgment by numerical predictions 30 years ago.

In order to further verify the numerical prediction results, historical satellite images are downloaded from Google Earth and compared with the numerical simulation results. The comparison results are shown in Fig. 11.40. Due to the lack of data before 2004, the sequence of comparisons only goes forward from 2004, ending at 2012. The reason for not going beyond the year 2012 is that another groin located at about 2000 m downstream of the first groin was built in 2011–2012, as ⑤ in Fig. 11.39a, to further protect the downstream shoreline of the port.

It can be seen from the figures that the shoreline evolutions predicted 30 years ago are in agreement with the historical satellite images. The discrepancy between the prediction results and the images in the area close to the port (below the port in the figure) is due to the artificial trimming of the shoreline which is not the result of natural evolution (this can be seen clearly from the satellite images). The predicted shoreline in 2008 seems to be quite different from the satellite image in the lower half of the figure. But if we observe the satellite image carefully, we can find that the satellite image in 2008 is spliced by two pictures which are not exactly aligned. If the error of image splicing is removed, the prediction result in 2008 would be in agreement with the satellite image.

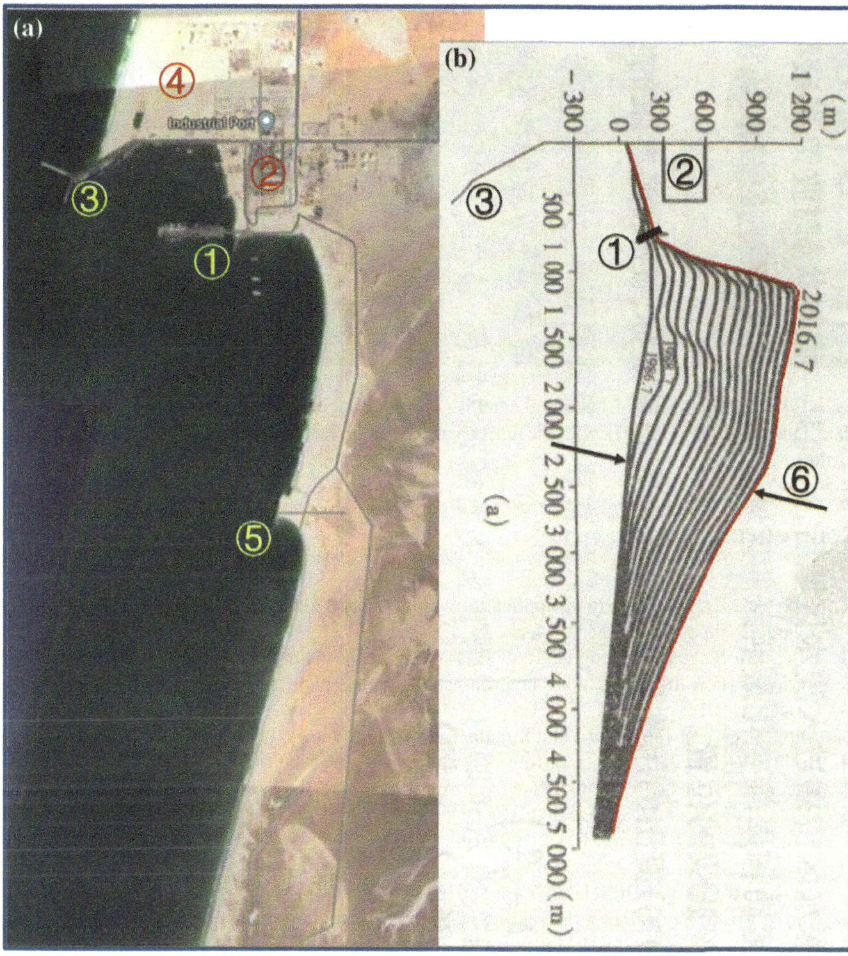

Fig. 11.39 Comparison of the satellite image and numerical prediction of shoreline evolution around Port of Nouakchott in Mauritania. ①: groin built in 1986; ②: port area; ③: breakwater; ④: new beach created by sediment deposition after the construction of breakwater; ⑤: new groin built in 2011–2012; ⑥: predicted shoreline after 30 years from 1986

The above comparison results using the observed data in nearly 30 years show that the mathematical model and numerical calculation method established by the CFD group in 1986 (described in Chap. 8) are proper to simulate the shoreline evolution of such sandy coast.

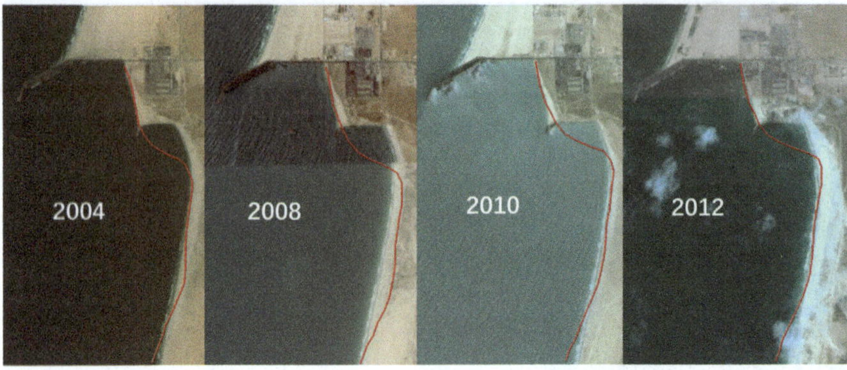

Fig. 11.40 Comparisons of historical satellite images of shoreline around Port of Nouakchott in Mauritania from 2004 to 2012 with the numerical prediction results. Red lines: numerical prediction results

References

1. Li X, Yuan D, Tao J. Study on application of random walk method to calculate water exchange in large-scale bay. Appl Math Mech. 2011;32(5):621–34.
2. Lv X, Liu B, Yuan D, Feng H, Teo F-Y. Random walk method for modeling water exchange: an application to coastal zone environmental management. Journal of Hydro-environment Research. 2016;13:66–75.
3. Admiralty Tidal Table and Tidal Stream Tables. H.o.t. Navy; 1990.
4. Bolin B, Rodhe H. A note on the concepts of age distribution and transit time in natural reservoirs. Tellus. 1973;25(1):58–62.
5. Delhez EJM, Campin J-M, Hirst AC, Deleersnijder E. Toward a general theory of the age in ocean modelling. Ocean Model. 1999;1(1):17–27.
6. Meier HEM. Modeling the pathways and ages of inflowing salt- and freshwater in the baltic sea. Estuar Coast Shelf Sci. 2007;74(4):610–27.
7. Ren Y, Lin B, Sun J, Pan S. Predicting water age distribution in the pearl river estuary using a three-dimensional model. J Mar Syst. 2014;139:276–87.
8. Zhang S, Xi P, Feng S. Numerical simulation of circulation in the Bohai Sea. J Shandong College of Oceanol 1984; 14(2):12–19.
9. Tao J, Sun J, Yuan D, Li C, Nie H. Technical report of the study on 2D numerical modelling on the water quality in the Pearl River Estuary and adjacent water. Tianjin: Tianjin University; 2008.
10. Sun J, Lin B, Jiang G, Li K, Tao J. Modelling study on environmental indicators in an estuary. Water Manage. 2014;167:141–51.
11. HoN (Hydrographer of the Navy). Admiralty Tidal Table and Tidal Stream Tables. London, UK: HoN; 1990.
12. Mao Q, Shi P, Yin K, et al. Tides and tidal currents in the Pearl River Estuary. Cont Shelf Res. 2004;24(16):1797–808.
13. Technical Regulation of Modelling for Tidal Current and Sediment on Coast and Estuary (JTJ/T 233-98). The Ministry of Communications of the People's Republic of China (1998).
14. Code of Hydrology for Sea Harbour (JTJ 213-98). The Ministry of Communications of the People's Republic of China (1998).

The manufacturer's authorised representative in the EU is Springer Nature Customer Service Centre GmbH, Europaplatz 3, 69115 Heidelberg, Germany. If you have any concerns regarding our products, please contact ProductSafety@springernature.com

Printed and bound by CPI Group (UK) Ltd, Croydon, CR0 4YY
30/09/2026
02225292-0011